ISBN 978-1-5277-7711-8
PIBN 10892405

1 MONTH OF
FREE
READING

at

www.ForgottenBooks.com

By purchasing this book you are eligible for one month membership to ForgottenBooks.com, giving you unlimited access to our entire collection of over 1,000,000 titles via our web site and mobile apps.

To claim your free month visit:

www.forgottenbooks.com/free892405

English
Français
Deutsche
Italiano
Español
Português

www.forgottenbooks.com

Mythology Photography **Fiction**
Fishing Christianity **Art** Cooking
Essays Buddhism Freemasonry
Medicine **Biology** Music **Ancient
Egypt** Evolution Carpentry Physics
Dance Geology **Mathematics** Fitness
Shakespeare **Folklore** Yoga Marketing
Confidence Immortality Biographies
Poetry **Psychology** Witchcraft
Electronics Chemistry History **Law**
Accounting **Philosophy** Anthropology
Alchemy Drama Quantum Mechanics
Atheism Sexual Health **Ancient History**
Entrepreneurship Languages Sport
Paleontology Needlework Islam
Metaphysics Investment Archaeology
Parenting Statistics Criminology
Motivational

ORNAMENTAL PLASTICS. These useful and beautiful objects are gifts from the chemist to you.

CHEMISTRY
AND ITS WONDERS

By OSCAR L. BRAUER, *Professor of*
Chemistry and Physics, San Jose State College
San Jose, California

AMERICAN BOOK COMPANY, *New York*
Cincinnati, Chicago, Boston, Atlanta, Dallas, San Francisco

PREFACE

The present generation has witnessed some notable achievements in chemical theory. Discoveries have been made which already exert a profound influence on the thinking of educated men. Of possibly even greater significance is the growing alliance of chemistry with industry. This interdependence of industry and science — the former sponsoring research, while the latter directs production — has far-reaching implications for society as a whole. It is primarily for the purpose of pointing out some of the remarkable ways in which modern chemistry affects life that this book has been written. In its preparation an effort has been made to depart somewhat from the cold formality of a textbook; that is, along with the presentation of facts and theories an effort has been made to give the book a conversational touch, as though one person were speaking to another.

From his twelve years of teaching chemistry in high schools, the author has learned the viewpoints, capabilities, and cultural needs of those students who will take no more chemistry than this one course; from a like period of teaching chemistry in college, he has learned the needs of those students who will take college chemistry.

Those students who will not take another course in the subject — those who will shortly become bankers, merchants, farmers, housewives, and laborers — will find that the carefully explained principles and the colorful descriptions give them an understanding of many hitherto mysterious phenomena. They will also find the specialized topics, dealing with the applications of chemistry to agriculture, cooking, health, metallurgy, warfare, etc., both interesting and valuable.

For the college-preparatory student — he who will become a chemist, a pharmacist, a physician, a dentist, or an engineer — there are such topics as the gas laws, chemistry and electricity, the periodic system, the internal structure of the atom, and chemistry and energy. The molecular and atomic hypotheses, with formulas and equations, are presented early in the text, thus giving the student command of these useful tools for learning almost from the start.

Some parts of chemistry are necessarily more difficult than others. Young students are likely to become discouraged on their first attempts to solve problems, write formulas, balance equations, and picture atomic structures. The author has, therefore, devoted much study to the best methods of introducing these topics. As a result, he has developed methods of presentation which give decidedly superior results.

In recognition of the varying abilities of students, the work has been made very flexible. Study helps include "Additional Exercises for Superior Students" and "References for Supplementary Reading." The supplementary references are all interesting and easy to read, and they cover work beyond the scope of the text. A new feature is offered in the questions on the references, questions which give direction to the student's outside reading.

The illustrations have been selected for the stories they tell. Several of the drawings are really cartoons, and a touch of humor drives home the point in an impressive way.

The biographies, with portraits, of a score of eminent chemists not only familiarize the student with the contributions of these men to the science but also make him aware of the fact that they too had to surmount all the obstacles that confront him today. It is hoped, moreover, that these life stories may inspire the better student to accomplishments really commensurate with his superior talents.

This text meets the requirements of the Committee of Chemical Education of the American Chemical Society, the College Entrance Examination Board, the Syllabus of the New York Board of Regents, the Pennsylvania Course of Study, and other courses of study.

Acknowledgment is gratefully made to Mr. E. M. Cunningham, Head of the Science Department, San Jose High School, who has read the manuscript very carefully and made many valuable suggestions from the wealth of his long experience in teaching beginning chemistry. Dr. E. D. Botts of San Jose State College also read part of the manuscript and made helpful suggestions. To other members of the Department of Natural Science, San Jose State College, the author is indebted as follows: Dr. P. Victor Peterson, who aided him in obtaining many of the illustrations; Miss Gertrude Witherspoon and Miss Lee Sauvé, who helped type the manuscript; and Dr. E. S. Greene, Dr. Albert Schmoldt, and Dr. J. Wilfred Richardson, who helped with the problems and questions. Dr. L. I. Brauer, Mr. E. B. Brauer, and Mr. Ralph Montoya helped with the illustrations. Many commercial firms generously supplied photographs. Credit to them is given in connection with the pictures.

CONTENTS

CONTENTS

CHEMISTRY AND ITS WONDERS

PACKING SYNTHETIC CAMPHOR. Synthetic camphor, which has largely replaced natural camphor, is but one of the thousands of useful products the chemist has given us.

$$\overset{\longleftarrow}{\longrightarrow}$$

INTRODUCTION

On taking up a new subject, a student often asks questions about it such as these:

"What is it about? What is it good for? Should I take it? Is it interesting? Is it hard?"

What is chemistry about? The answer to the first question when applied to chemistry is that it deals with all of the substances that are known on the earth — rocks, water, air, and plant and animal products. It studies the natural processes of change, such as burning, rusting, and decay. It deals with the multitude of changes in living things upon which our health depends. Chemistry teaches us, among other things, how valuable metals are obtained from their ores, how apparently worthless substances are made valuable, and how new medicines are found. The story of chemistry is as full of miracles as is that of Aladdin of the wonderful lamp. In the story of *The Arabian Nights*, when Aladdin wished help, he rubbed his magic lamp, and a jinni appeared to do his bidding. Instead of rubbing a lamp, the modern Aladdin hires a jinni, the chemist, and puts him to work. And the things the chemist can do are no less wonderful than those performed by the mythical jinni.

What is chemistry good for? In answer to the second question, it will only be possible to cite a few examples at this time.

The chemist makes new substances to order. Someone sees the need of a new substance with special characteristics. The chemist sets out to find it or, more frequently, to make it out of other substances, and he usually succeeds. For instance, automobile engineers learned that it would be possible to get considerably more power out of gasoline if the mixture of air and gasoline vapor were compressed more before it was exploded. However, when the mixture was so compressed, the

COOKING FAT. Dr. Wesson's discovery that hydrogen can be made to react with cottonseed oil to produce a palatable, wholesome fat created a new industry and almost drove the animal fats out of use in the kitchen.

engine knocked at the time of explosion. At this point the chemist was asked to find an antiknock substance, a few drops of which dissolved in each gallon of gasoline would stop the knocking, yet permit the vapor to be compressed to the point where additional power could be obtained. Thomas Midgely, Jr., the chemist who had this problem to solve, gave us tetraethyl lead, of which we hear a great deal. Tetraethyl lead is so effective as a knock preventer that it is necessary to put only four ounces of it into fifty gallons of gasoline to eliminate knocking completely at pressures much greater than could be used successfully without it.

Some time ago an electric refrigerating system sprang a leak, and several people were asphyxiated. This accident emphasized the need of a perfectly safe liquid for use in such refrigerators, a liquid that would not be poisonous and would not explode. Careful research by skilled chemists led to the production of a compound of carbon, chlorine, and fluorine that is both nonpoisonous and incombustible, yet which has the right characteristics for use in an electric refrigerator.

Chemistry makes waste products into useful substances. A study of chemistry tells us how waste products from the industries are made into valuable and useful materials. In the early days of the cotton industry the seeds, which represented over half the weight of the freshly picked cotton, were practically a waste product. The first step in using this waste consisted in pressing out the oil. However, the housewife did not like it for cooking, and it was not worth much for other purposes. A chemist by the name of Wesson found that this dark yellow oil could be changed into a snow-white grease by causing it to react with hydrogen. Housewives eagerly accepted the new fat. An industrial waste had become a valuable product.

In the production of cane sugar, the cane fiber from which the sweet sap had been pressed was formerly a useless by-product. It did not even pay to dry it out for fuel. It had to be hauled away and dumped on vacant land to rot. Moreover, it was slow to decay, and remained an unsightly mass for years. In time, a chemist saw in this tough, slow-rotting fiber a possible raw material for the manufacture of wallboard. A successful process was worked out. Today celotex wallboard is sold all over the United States. In this case the raw material

Courtesy of Celotex Co.

BALES OF BAGASSE. Bagasse is the tough fiber of the sugar cane. It used to be a nuisance, but now it is dried, baled, and made into wallboard.

costs really less than nothing, because previously there had been the expense of hauling the fiber away and paying rent on the land where it was dumped. Frequently in times of business depression the utilization of by-products for profit is the only thing that can save a company from financial ruin.

Chemistry has cheapened expensive substances so they are brought within the reach of all. About the time of Napoleon III, aluminum was a rare novelty. A chemist sponsored by the

Courtesy of Celotex Co

THE MANUFACTURE OF WALLBOARD. The chemist has created a new industry from the waste fiber from the sugar cane. Wallboard is built into many modern homes.

Emperor was able to make a little card tray out of this new metal. Napoleon was actually paying for scientific research, but in terms of what it cost, the tray was worth $100,000. Today a similar article could be bought for thirty cents. Chemistry has brought about this remarkable reduction in price.

A century ago only the very wealthy could afford gorgeously colored clothes. One dye in particular was known as royal purple, because only kings could afford robes dyed with it. Today chemists are making thousands of dyes more beautiful and more lasting than royal purple. The shop girl of today has a more varied assortment of colors to choose from than had a queen of the eighteenth century. The dye chemist can now make dyes of every possible color, shade, or tint. In daintiness and beauty his creations rival the flowers.

The chemist has produced new substances. Certain new chemical substances have become the basis of whole industries. Bakelite, the material of which radio panels, dials, and many other

Courtesy of Bakelite Corp

BAKELITE. Thousands of industrial and domestic accessories are molded from bakelite. The new telephone, the twenty-five pound insulator (center), and the automobile distributor head are typical.

electrical parts are made, is a comparatively new chemical substance. Dr. Baekeland made it out of phenol, the active substance in carbolic acid, and another chemical called formaldehyde. The electrical industry leans heavily on bakelite.

The white curd of milk used to be used only for making cheese and for feeding pigs. The chemist learned that it also could be toughened with formaldehyde and made into an imitation ivory, capable of being dyed in a variety of colors. It is used in the manufacture of combs, brushes, and similar articles.

Chemistry has improved many useful things. Millions of dollars' worth of iron and steel products are lost every year from rust. The chemist is rapidly reducing this loss by introducing rustless and corrosion-proof steel. The addition of nickel and chromium to steel has brought about the desired effect. Nearly every product we have is being improved from year to year — better automobile paint, better tires, better foods, better cloth, better cement, better radios, and better fuels. These are just a few products selected at random. A study of chemistry tells us how these things have been improved.

Should I study chemistry? To the student's third question, it may be said that if he wishes to become any one of the following he *must* take chemistry: physician, dentist, druggist, assayer, analyst, mining engineer, lawyer, architect, civil

CHEMISTRY HELPS THE FARMER. This photograph shows how the application of chemical fertilizer may stimulate the growth of clover.

engineer, mechanical engineer, electrical engineer, geologist, science teacher, homemaking teacher, dietitian, nurse, industrial chemist, or chemical research worker. Even if the student does not expect to take up a life work directly connected with chemistry or demanding chemistry as a preparatory subject, his life will be enriched by its study, and his chances of success will be bettered by knowing it.

Moreover, chemistry will help him understand how nature is working all about him. It will help him to be more intelligent in his diet and health habits and consequently will enable him to live more happily and longer.

SHOULD I STUDY CHEMISTRY?

The girl who has studied chemistry will be a better housewife; she can cook more intelligently and get uniformly better results. She will feed her family more nearly correctly. She will better understand many things that are used about the home. The chemically trained farmer will preserve the fertility of his soil better than if he lacked this training. He will be more successful in running his machinery. He will be able to raise better crops. The livestock man similarly trained will grow better animals at a cheaper cost. The dairyman will produce better butter if he knows chemistry. It is impossible to think of a person who will not find a knowledge of chemistry a help to him. This knowledge becomes increasingly valuable from year to year with the growth of chemical science.

Many industries must maintain a corps of chemists to insure products of uniformly high quality. Managers, superintendents, and directors of companies should know enough about chemistry and its possibilities to keep from being outdistanced by their rivals who may put out cheaper or better products.

Before the World War a German and a Frenchman each patented a process for making wood alcohol from coal, hydrogen, and oxygen. The yields were poor, which naturally made the product expensive, so no one paid much attention to what was being attempted. However, the German kept working on his process to cheapen it and increase the yield. A few years after the close of the war, a lot of this synthetic wood alcohol was shipped into the United States from Germany. It was sold for forty-six cents per gallon after paying a duty of ten cents per gallon. The wood distillation industries in this country had previously produced all of the wood alcohol we used by distilling hard wood. This they had to sell for seventy cents per gallon in order to make a profit. The arrival of the cheap German product gave them a distinct shock. If the directors of other large companies fail to keep up to date chemically, they too may wake up some day to the fact that they are being undersold and that their plants are out of date and practically worthless.

The person who understands chemistry will be awake to its

possibilities in every line.   Chemical research can improve any
product, can help any business, and ofttimes is the only thing
that can save an industry.   Bankers and other investors should
know chemistry in order to invest intelligently the funds in
their care.   Many persons have invested in schemes which any
chemist could have told them were impossible.   Others have
failed to support legitimate industries from the same lack of
chemical information.   The study of chemistry is profitable

A HIGH SCHOOL LABORATORY.   Here students acquire first-hand information
concerning the properties and reactions of chemicals.   Much of this training will
prove useful to them throughout the remainder of their lives.

for anyone.   Any person who wishes to be up to date and generally informed must know some chemistry.   Those who do
not understand elementary chemistry are a hundred years behind the times.

Is chemistry interesting?   In answer to the fourth question,
chemistry is interesting.   The laboratory part is fascinating;
working with new and strange substances gives one the thrill of
the discoverer.   To see iron burn in oxygen or phosphorus take
fire in the air, or to prepare substances with beautiful colors

and strange odors is indeed interesting. Would you like to solve nature's riddles? Would you be interested in man's struggle with great problems? All of these and more are outlined in the study of chemistry.

Is chemistry hard? Some parts of chemistry may seem hard at times; other parts will be found very simple. In this respect it is like many other subjects. What difference does it make if parts of it are difficult? All worth-while things take more or less effort. Is it easy to win football games? Was it easy for the pioneers to explore and develop this country? Why should we as students expect to drift through school without ever having to work hard? The chance that some parts of chemistry may require a little thought and study does not worry the ambitious, red-blooded young person. However, to one who is fearful, it can be truthfully said that chemistry is no harder than other high-school subjects. Any student with ordinary intelligence and willingness to work will be able to pass the course. Most of the students who get poor grades in chemistry do so because it seems easy and they think that reading a lesson over once is all that is necessary. The result is that what they have read has not been deeply impressed on the mind; hence, it soon slips away. Would you like to try a simple method of overcoming this difficulty? It requires a little effort, but it is guaranteed. The method is to write out an abstract of each paragraph as it is read — that is, to say the principal thought in one or two sentences. It is impossible to condense a thought unless one understands it. This is the chief value of this scheme. Perhaps the student may be studying at night when his mind is tired. He reads over a paragraph. It does not seem to tell him anything. The mind seems blank. There are only two things to do, either go to bed and finish studying in the morning or else wake up and read that paragraph over until it does say something that can be written down. A fatal mistake that students sometimes make is to let several days go by without thorough study, hoping to make it up later. It is consistent effort that counts, rather than occasional spurts.

REVIEW QUESTIONS

1. Name four sources of chemical substances.

2. Name three processes explained in chemistry.

3. Discuss one illustration of what chemistry has done in the following cases: (a) finding substances to fill definite needs; (b) changing waste products into useful materials; (c) cheapening costly substances; (d) making new substances; (e) improving common articles.

4. Make a list of ten occupations that demand that you know something of chemistry.

5. Tell one way a knowledge of chemistry will benefit: (a) a housewife; (b) a farmer; (c) a banker.

REFERENCE FOR SUPPLEMENTARY READING

1. Clarke, Beverly L. *Marvels of Modern Chemistry*. Harper and Brothers, New York, 1932.
   (1) Relate the story of the importance of chemistry to one firm. (p. 3)
   (2) Tell of the need for chemists. (p. 4)

CHAPTER 2

OXYGEN

The importance of oxygen. In beginning the study of chemistry, our first thought is, Where shall we begin? Of the thousands of substances from which to choose, we shall select oxygen, the substance most essential to our existence. Without oxygen to breathe, a person could scarcely live ten minutes. In fact, he would be most uncomfortable if he held his breath for two minutes. All living plants and animals, except a few one-celled plants called bacteria, are dependent upon oxygen to maintain life. Even a fish living in the water must have oxygen. If the water were boiled until all of the dissolved oxygen were driven out, the fish would smother. When a person drowns, he really dies from lack of oxygen. Even if he had gills similar to those of a fish, he could not get enough oxygen out of water to keep him alive. The fish, being a cold-blooded animal, requires much less oxygen in proportion to its size than the warm-blooded animals.

Oxygen is important for other reasons than that of maintaining nearly all forms of life. Without it there could be no fire, one of man's greatest blessings in that fire is needed to heat our homes, cook our food, and make possible scores of industrial processes, such as making glass, preparing fuel gases, and separating metals from their ores. Oxygen also helps purify the water in streams, thus furnishing us with wholesome drinking water, in spite of occasional pollutions.

How much oxygen is there? Probably the first question that comes to us after we learn the importance of oxygen is, How much is there of it? Is there any danger that the myriads of people and animals that are breathing oxygen twenty-four hours per day can exhaust the supply? In addition to animals there are untold numbers of plants that also use it some of the time, together with many fires both large and small, which

11

remove this essential substance from the atmosphere. However, we need not fear the exhaustion of the supply. Since man first analyzed the air, there does not seem to have been a measurable decrease in the part that is oxygen. When we try to estimate the amount of the substance in the atmospheric envelope around the earth, we see that it is very large. Above each square inch of ground there are about 15 pounds of air containing nearly one-fourth oxygen. Above a city lot 50

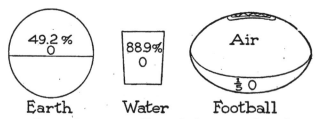

| Earth | Water | Football |

OXYGEN IN COMMON SUBSTANCES. The earth, the water, and the air constitute the bulk of matter as we know it. Oxygen constitutes a large part of these substances.

feet wide and 150 feet deep there would be nearly four million pounds. On the whole earth's surface there must be over 1,500,000,000,000,000 tons. There are several processes that restore oxygen to the air. Considering the enormous amount of this substance in our atmosphere and the fact that more or less of that which is removed is being returned, we need fear no depletion of the supply.

Combined oxygen. The oxygen in the air is called free oxygen, not because we do not have to pay for it, but because it is not chemically united to other substances. Most of the oxygen is in other substances, such as water and rocks. An analysis of all the things we know on the earth's surface and beneath the surface as far down as we have been able to go shows that the earth's crust is about one-half oxygen. Water, another very abundant substance, is eight-ninths oxygen. A marble tombstone is 48 per cent oxygen; sand, over 53 per cent; and the human body about 72 per cent oxygen.

Preparing pure oxygen. The problem of preparing pure oxygen is one of finding a substance which contains an abundance of it and which will at the same time let it go. Many substances contain oxygen but will not part with it under conditions easily produced in the laboratory. The common laboratory method of preparing uncombined oxygen is to heat a mixture of potassium chlorate and manganese dioxide in a test tube provided with a glass delivery tube. The end of this delivery tube reaches into a pan of water below an inverted bottle full of water. As the oxygen enters the bottle, the water is pushed out. The po-

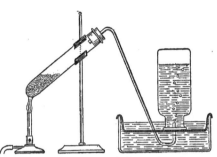

LABORATORY METHOD of Preparing Oxygen. The test tube at the left contains a mixture of potassium chlorate and manganese dioxide. When the tube is heated, oxygen is liberated. The oxygen passes into the bottle on the right, where it replaces the water.

tassium chlorate has been broken down into oxygen and potassium chloride. Whenever a change results in the formation of new substances, it is said to be a chemical reaction.

Chemical reactions and physical changes. Whenever we observe the world about us, we see changes. Plants and animals grow, die, and decay; buildings and fences get old and fall to pieces; knives get dull; milk sours; and in time our hair turns gray. Only part of these changes can be classed as chemical changes. To learn the distinction between the two classes of changes, let us consider a few simple typical illustrations of each. In the preparation of oxygen as described in the preceding paragraph, we had potassium chlorate before the change, and in its stead the two substances oxygen and potassium chloride after the change. The formation of new substances with characteristic properties different from those of the original substance is the test for a chemical reaction. Another typical chemical reaction takes place when iron is heated with sulfur. A new substance, iron sulfide, is formed. Its brown color and its brittleness are very different from the properties of both iron and sulfur; hence we are sure it is a new substance.

As a typical physical change we might break a stick of wood. Although there has been a marked change, the substance is still wood. No new substance has resulted; hence the change is physical. As another illustration let us heat the stove poker in the fireplace. Its temperature is soon changed. If the heating continues it will become red hot, the color having changed also. However, if the heating is not too severe, the substance is still the same iron it was before; hence the change is just a physical change and not a chemical reaction. Suppose again a bottle is broken to pieces, the pieces are still glass. No new substance has been formed. It is just another physical change. Let us sum up these facts into a definition of chemical reaction and physical change. *When a change results in new substances, it is called a chemical reaction. When no new substance results from the change, it is a physical change.*

In chemistry we are primarily interested in chemical reactions, in the formation of new substances.

Other methods of preparing oxygen. One method of preparing oxygen is of historical interest because it was the method that resulted in its discovery. Of course, oxygen has been used ever since the first man drew breath. However, we do not think of its having been discovered until someone prepared it in the pure form and learned how it behaves when freed from other substances. In 1774 Joseph Priestley, an English scientist, was amusing himself by focusing the sun's rays with a burning glass on various substances. One of the substances he tried was the red mercuric oxide. Immediately the oxide lost its red color and left in its place a drop of mercury. This aroused Priestley's curiosity. Replacing the oxide and poking at it with a pencil, he was surprised to see the wood take fire and burn energetically. This and further study led to the realization that he had found a new substance. This Lavoisier named *oxygen*. Because of the high cost of mercuric oxide, this method is of little importance.

Another method of preparing pure oxygen, which is of

Joseph Priestley

1733–1804

As a student, Joseph Priestley excelled in both ancient and modern languages, including Chaldee, Hebrew, Syriac, Greek, Latin, French, and German. His interests were varied, and his study time was divided between languages, mathematics, religion, and scientific research.

After spending three years at an academy, he began preaching at the age of twenty-two. Although quite unorthodox in his religious belief, and although often in literary controversy with high church officials, he was able to keep in the ministry from 1755 until 1772, when he became librarian and literary companion of Lord Skelburne at a salary of £250 (about $1250) per year.

In his new position Priestley was able to devote considerable time to research. In 1774 — having recently acquired a lens, which interested him greatly — he focused the sunlight upon some red mercuric oxide which happened to be on his desk. Seeing it darken under the heat of the ray of light, he touched it with the end of a pencil. Imagine his surprise when the pencil caught on fire. He had accidentally discovered oxygen. Priestley himself did not fully realize the significance of his discovery but still clung to the old phlogiston theory. However, one of his successors, a Frenchman named Lavoisier, used this discovery to overthrow forever the old theory of combustion and put chemistry on a solid foundation.

An unpopular religious publication by Priestley in 1777 resulted in his dismissal by Lord Skelburne. The following year he became the minister in a dissenting chapel at Birmingham. Because he took a stand on the issues at stake in the French Revolution, which was very unpopular in England, in 1791 a mob attacked his chapel and home and burned all his books, manuscripts, and scientific instruments. This ill treatment caused him to move to America.

practical importance in cases where small amounts of oxygen are needed, is to allow sodium peroxide to react with water. Physicians used to use an oxygen generator for administering oxygen to very weak patients. This generator produced the oxygen by allowing powdered sodium peroxide to drop into water. Here again the usefulness of the process is limited by the cost of sodium peroxide.

The oxygen of commerce is obtained by liquefying air and allowing it to evaporate. The nitrogen and rare gases pass off first, leaving the oxygen. The process will be explained in more detail in a subsequent chapter.

QUESTIONS OF FACT

Which of the first four statements are true and which false?

1. A fish, being a cold-blooded animal, does not need oxygen.
2. Free oxygen means uncombined oxygen and it occurs in the air.
3. So many animals and so many fires are using oxygen that the supply is likely to be exhausted in one thousand years.
4. The oxygen of commerce is obtained from potassium chlorate.
5. Name three substances that are rich in combined oxygen.
6. What two substances were used in the laboratory preparation of oxygen?
7. From which substance did the oxygen come?
8. How was oxygen discovered by Priestley?
9. How close was Priestley's discovery of oxygen to the time of the American Revolution?
10. Why did Priestley come to America?
11. Explain the scientific terms: (a) free oxygen; (b) chemical reaction; (c) physical change.

QUESTIONS OF UNDERSTANDING

Choose the response which most suitably completes the first two statements.

1. When fish are taken out into the air they soon die because :
They get too much oxygen.
They are scared to death.
Their gills are not capable of absorbing oxygen from the air.
The temperature is too high.

2. When we consider the facts that in the laboratory oxygen is collected over water and that fish need oxygen, we conclude that oxygen is:

Insoluble.
Slightly soluble.
Very soluble.

3. Why is air continually bubbled through aquariums where fish and other animal life are kept?

4. In preparing oxygen, which is more important, that the substance contain a high percentage of oxygen or that it readily give off part or all of its oxygen?

5. Tell which of the following are physical changes and which chemical reactions:

Rusting of iron.	Freezing of ice cream.	Washing clothes.
Sharpening a knife.	Chopping wood.	Explosion of powder.
Lighting a match.	Burning wood.	Setting a wagon tire.
Boiling meat.	Making lemonade.	Decay of an apple.
Souring of milk.	Fading of color.	

6. In what sense is oxygen the most important substance for life?

ADDITIONAL EXERCISES FOR SUPERIOR STUDENTS

1. If a person has a lung capacity of 250 cubic inches and breathes 20 times per minute, how many hours would the air in a room 12 ft. × 12 ft. × 8 ft. last him if none of the air was breathed over?

2. At 15° C. .034 gallon of oxygen can be dissolved in 1 gallon of water. How many gallons of oxygen are dissolved in a fish tank of 100 gallons if the water is fully aerated?

REFERENCES FOR SUPPLEMENTARY READING

1. Holmyard, Eric John. *Makers of Chemistry.* Oxford University Press, New York, 1931.
   Fairly complete and very interesting.
2. Kendall, James. *At Home among the Atoms.* Bell and Sons, London, 1932. Chap. VII, "Oxygen, the Working Girl."
   Gives brief biographies of Priestley and Lavoisier.
3. Jaffe, Bernard. *Crucibles.* Simon and Schuster, New York, 1930. Chap. IV, pp. 51–72, "Priestley."
   (1) Write a biography of Priestley.
   (2) Name the incidents not related in the text.

4. Martin, Geoffrey. *Triumphs and Wonders of Modern Chemistry.* D. Van
   Nostrand Co., New York, 1922. Chap. VIII, pp. 171–194, "Oxygen,
   the Life-Supporting Element."
   (1) In what sense is the substance in our bodies changing? (p. 172)
   (2) How is oxygen carried by the blood through our bodies? (pp. 172–
   173)
   (3) Account for the fact that a strip of iron when once ignited burns in
   liquid air. (p. 176)
   (4) Discuss the importance of oxygen to life. (p. 175)
   (5) Name four substances which give off oxygen when heated. (p. 185)
   (6) Describe the largest generation of oxygen on record. (pp. 186–187)
   (7) Describe the methods of concentrating the oxygen of the air. (p. 188)

Courtesy of U. S. Army Air Corps

CARRYING OXYGEN INTO THE AIR. These aviators are preparing to ascend to
a high altitude where the air is much less dense than at sea level. Hence they
are taking along metal cylinders filled with the oxygen they will need to breathe.

Uses of oxygen. Oxygen has many uses. Most of them can
be supplied directly by the oxygen of the air. However, there
are a few uses for the pure element. Men engaged in mine
and fire rescue work and aviators flying at very high altitudes
find it necessary to carry oxygen for breathing.

One form of breathing apparatus consists of a mouthpiece
connected by a series of tubes with the breathing bag, the

cylinder of compressed oxygen gas, and the regenerating can containing pieces of potassium hydroxide to absorb the water vapor and carbon dioxide exhaled from the lungs. The supply of oxygen needed by the wearer can be regulated by a valve on the cylinder; the nitrogen originally inhaled and in the apparatus at first is breathed over and over. The cylinder contains oxygen sufficient for two hours. Provided with such an arrangement, a man can safely enter places where the air contains smoke or poisonous gas and make repairs, extinguish fires, or rescue persons who have been overcome. Extensive use is made of this sort of safety device in mine disasters, largely through the efforts of the United States Bureau of Mines.

Courtesy of U S Bureau of Mines

GIBBS BREATHING APPARATUS. This breathing apparatus was designed by the Bureau of Mines for use in mine rescue work. Fire fighters also use this apparatus for entering buildings filled with smoke. The oxygen tank is strapped on the back.

Physicians sometimes administer pure oxygen to patients who are too weak to be sustained by the oxygen of the air. Also it has to be given along with some anesthetics, especially nitrous oxide. The chief industrial use of pure oxygen is in oxyacetylene welding. Due to the rapid burning of acetylene with oxygen, a very high temperature is obtained. The oxyacetylene blow torch readily cuts through the toughest metal,

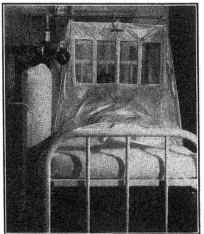

ADMINISTERING OXYGEN. · When the patient is very weak, the physician often administers pure oxygen to hasten oxidation processes in his body. The picture shows an oxygen tent.

Photo by F. H. King

OXYACETYLENE WELDING. This very hot flame can be used either to cut up large pieces of metal or to fuse broken parts together.

which makes it the best method of removing old bridges and other steel structures. Oxygen has recently been used in the making of quartz glassware from quartz, a substance very difficult to melt. Ordinary fuels burned in air do not produce enough heat to fuse it. However, by burning acetylene in pure oxygen, it is possible to melt quartz and make a window glass which will admit the ultraviolet light of the sun.

A catalyst. In the laboratory method of preparing oxygen, you may have wondered why we added the black manganese dioxide and did not mention it in discussing the chemical reaction involved. Its purpose was to speed up the reaction. If we had left it out, we could have melted the glass test tube without getting much oxygen. A substance that helps a chemical reaction without itself being used up is called a *catalyst* or a *catalyzing agent*. It has been said that "you cannot eat your cake and have

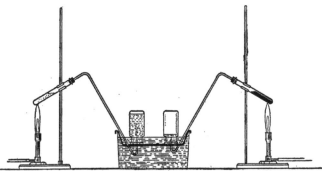

EFFECT OF A CATALYST. On the left potassium chlorate alone is being heated. At the temperature of the Bunsen flame practically no oxygen is liberated. On the right black manganese dioxide has been mixed with the potassium chlorate. An abundance of oxygen is given off in the presence of this catalyst under the same temperature conditions as before.

it too," but one can use a catalyst and still have it for future use. The catalysts are wonderful tools in the hands of the chemist. The German who invented the process for making synthetic wood alcohol succeeded because he found the right catalyst. The catalyst he used was also the secret of Dr. Wesson's success in making a clean white cooking fat out of cotton seed oil. Today when a chemist tries to bring about a new chemical reaction or improve an old one, he usually begins by seeking a catalyst.

Burning and oxidation. When we experiment with pure oxygen, the thing that impresses us most is how much more vigorously things burn in the pure oxygen than in the air. For instance, a stick of wood with only a small glowing speck still on it immediately bursts into flame when thrust

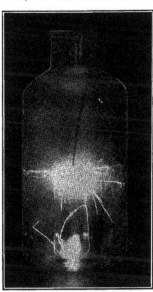

BURNING IRON IN PURE OXYGEN. Iron burns vigorously in pure oxygen and about as rapidly as pitch burns in air.

into a bottle of oxygen. Left in the air, it goes out in a few seconds. Even iron burns readily in pure oxygen. Scientists up until about the first of the nineteenth century did not understand burning. As they glanced at the wood in the fireplace, the dancing flames seemed to be something escaping from the wood. This something they called *phlogiston.* This explanation or theory that burning consists of the escape of phlogiston served all right until the discovery of a new set of facts which could not be harmonized with it. The stubborn fact that overthrew the theory is that when metals, such as iron and magnesium, burn, the resulting ash weighs more than the original metals. Common sense tells us that something cannot have escaped and still leave a residue weighing more than the unburned substance. This called for a new theory or explanation.

EXPERIMENT

To show that oxygen is removed from the air when iron rusts.

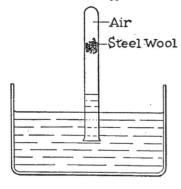

Insert a wad of moist steel wool into a 100 ml. eudiometer tube or long test tube and invert this over a dish of water and mark the surface of the water in the tube. After 24 hours the iron will be seen to be covered with rust, and the water will have risen in the tube. The rusting iron removes oxygen from the air inside the tube. The atmosphere then pushes water up into the tube to replace the oxygen used up.

A Frenchman by the name of Lavoisier studied the process and arrived at a new explanation, which is still satisfactory. Lavoisier found (1) that a substance burns better in pure oxygen than in the air but the final product or products are the

Antoine Laurent Lavoisier

1743–1794

Antoine Lavoisier, a young French lawyer, became interested in chemistry through attending a series of lectures. His interest became so great that he deserted law and plunged into chemical research. To obtain a living he accepted a job as tax collector, a position greatly hated by the common people, for most of the tax collectors were grafters.

About 1770 Lavoisier began a series of researches on combustion, which were destined to overthrow the theory of phlogiston. In these researches he established several facts. (1) When such substances as phosphorus and tin burn, the resultant ash weighs more than the original substance. (2) A given volume of air will burn only a definite weight of phosphorus. (3) When phosphorus burns, it removes part of the air. From these and similar researches, he came to the conclusion that burning is not the escape of phlogiston, but the union with oxygen. In fact, he concluded that there was no such substance as phlogiston. Up to this time the products of combustion were thought to be the simple substances. Lavoisier concluded that tin is an element but that the result of its combustion is a complex substance — just the reverse of the accepted theory.

Shortly after the complete victory of Lavoisier, the French Revolution broke in its full fury. Although Lavoisier had never grafted as a tax collector but had used every economy to make the tax burden lighter, the popular hatred was so great against the tax collectors that he was rounded up with the others and condemned to be executed.

One of Lavoisier's friends pleaded with the revolutionists for his life on the grounds that he was a great scientist. "The Republic has no use for scientists," was the cynical reply; and one of the world's greatest scientists was put to death.

same, and (2) that burning in a closed jar stops when the oxygen is all gone. Where there is no oxygen, there is no burning. The new explanation is that *burning is a chemical reaction between the fuel substance and the oxygen.* This explanation harmonizes all of the facts concerning burning. The sub-

stance burns better in pure oxygen than in air because in the latter the oxygen is diluted by nitrogen to one-fifth its full strength. Reactions in air are slower than those in pure oxygen for another reason. In air, part of the heat is wasted in heating the nitrogen instead of helping the burning. Since the difference between the two processes is only one of speed, it is to be expected that the final products of burning will be the same in each instance.

Wide World Photo

BURNING OIL WELL. Burning is the uniting of the elements of a substance with oxygen with the evolution of light and heat. The final products of burning are usually the oxides of the elements composing the substance.

The second fact discovered by Lavoisier, that a substance only burns a short time in a closed jar and then stops, can also be explained by the new theory. The closed jar contains only a limited amount of air (and less oxygen); hence as soon as this oxygen is used up, the chemical reaction cannot continue.

That a burned metal weighs more than it did before burning naturally follows as a consequence of this theory. If burning is the addition of oxygen to the metal by chemical reaction, the metal plus the oxygen necessarily weighs more than the metal alone.

The law of conservation of mass or matter. The success of Lavoisier in determining the nature of burning caused scientists to pay more attention to the weights involved in chemical reactions. As a result of carefully weighing the substances

BURNING ADDS WEIGHT TO IRON. The figure at the left shows a crucible containing powdered iron counterbalanced with weights. The center figure shows this iron being burned. The figure at the right indicates that the resulting iron compound weighs more than the original iron.

entering into chemical reactions and the products resulting from them, chemists have come to a very fundamental conclusion that has been named the *law of conservation of mass*.

The word *law* as used in science has a very different meaning from the common use of the word in reference to enactments of Congress or a state legislature. Law in the civil code means something that people must obey or risk fine or imprisonment.

LAW OF CONSERVATION OF MATTER. In every chemical reaction the weight of the products of reaction equal the weight of the reacting materials. The illustration shows that 2 lbs. of sulfur react with 3.5 lbs. of iron to form 5.5 lbs. of iron sulfide.

A scientific law is a general statement that summarizes or describes a large number of facts. The law of conservation of mass states the facts about the weights involved in all of the hundreds of thousands of chemical reactions already studied and undoubtedly all of those yet to be studied. The law may be stated thus: *The mass of all the substances that result from a chemical reaction is the same as the mass of those substances that enter into the reaction.* The term *mass* may be new to many beginning-chemistry students. If so, we may substitute the word *weight* for *mass.* So long as we remain on the surface of the earth, *weight* is a fairly satisfactory substitute for *mass.* Strictly, however, weight is the earth's pull on a given mass, while mass itself represents the unvarying quantity of *matter*, which is the general term for all substances.

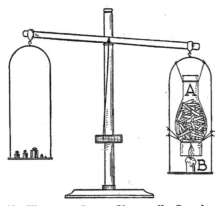

No Weight is Lost. If a candle B and a lamp chimney containing sticks of potassium hydroxide A are counterbalanced with weights and the candle lighted, that side of the balance containing the candle — instead of becoming lighter as one might expect — actually gains in weight, due to the potassium hydroxide catching the products of combustion, including oxygen removed from the air.

The expression *law of conservation of mass* means really the law of constancy of mass. Another way of expressing it would be *law of indestructibility of matter*, which would imply also the uncreatability of matter. In other words, we can neither create nor destroy matter.

Sometimes it looks as if matter has been destroyed. For instance, as a candle burns, it seems to disappear. Gasoline burning in a dish will leave the dish empty. Smokeless powder seems to vanish when it explodes. Careful experiments, however, will prove that in all these cases nothing has been destroyed. The reason the substances seem to vanish is that the products of the reaction are invisible gases, which escape into the air unnoticed. The method of making such a proof in the case of a burning candle is shown in the illustration.

Sticks of potassium hydroxide catch and hold the invisible gaseous products of the reaction. The side of the balance containing the vanishing candle is getting heavier. The increase in weight is the weight of the oxygen that has taken part in burning. In the case of the laboratory preparation of oxygen, the weight of the potassium chlorate equals the weight of the potassium chloride plus the weight of the oxygen.

EXPERIMENT

To show there is no change of weight during a chemical reaction. Put some silver nitrate into a very small test tube. Let this test tube down into a flask of salt water. Counterbalance the flask and its contents on the scales. Tip the flask so the silver nitrate solution spills

Silver Nitrate Solution

Salt Solution

into the salt water. That chemical reaction takes place is shown by the formation of a white precipitate. The scales remain balanced after the reaction takes place.

QUESTIONS OF FACT

Which of the first six statements are true and which are false?

1. If the supply of oxygen is shut off from a fire, the burning ceases.
2. A catalyst is a substance that is used up in a chemical reaction.
3. Burning, or combustion, is the reaction with oxygen.
4. The law of conservation of mass was passed by Congress.
5. When gasoline burns, mass is destroyed.
6. When iron burns, it gains in weight.

7. Describe the Gibbs breathing apparatus.

8. Discuss the use of oxygen in connection with (a) sick people, (b) oxyacetylene welding, (c) wrecking buildings, (d) quartz glass.

9. What was the catalyst in the laboratory preparation of oxygen?

10. Name two cases where a chemist succeeded in a noteworthy process by finding the right catalyst.

11. What was the phlogiston theory of burning?

12. What facts threw this theory into discard?

13. Apply the law of conservation of matter to the reaction in the laboratory method of preparing oxygen.

14. How do we know the substance of a candle is not destroyed when it is burned?

15. If a fruit jar is turned over a burning candle, the candle soon goes out. Explain.

16. Why do substances burn faster in pure oxygen than in the air?

17. State all of Lavoisier's discoveries that would not harmonize with the phlogiston theory.

18. Why was Lavoisier put to death?

19. What answer did the French Revolutionists make to the plea for Lavoisier's life on the grounds that he was a scientist?

QUESTIONS OF UNDERSTANDING

In the first two exercises choose the response which most adequately completes the thought in the statement.

1. We paint our houses and machinery:
   (a) To cause spontaneous combustion.
   (b) Only to make them more beautiful.
   (c) To make them more beautiful and to prevent oxidation.

2. Industrial chemists make exhaustive searches for suitable catalysts:
   (a) To increase the yields of the desired product.
   (b) To prevent physical changes.
   (c) To prevent other reactions from taking place.

3. Suppose that there were two unlabeled bottles on the table, one containing oxygen and one containing air; how could we test which is oxygen and which is air?

4. If a lighted candle were lowered into a jar of pure oxygen, compare its burning with that in a jar of air : (a) speed of burning; (b) quantity of candle burned before going out.

5. Suggest a reason for mixing oxygen with gaseous anesthetics.

6. Why is there no ash when gasoline burns?

7. Explain this fact : A magazine is quite difficult to burn unless its leaves are torn out and thrown into the fire individually.

ADDITIONAL EXERCISE FOR SUPERIOR STUDENTS

1. Industrial furnaces requiring very hot fires have tall chimneys to their coal burners. The object is to get a strong draft. Explain why a strong draft is desired.

REFERENCES FOR SUPPLEMENTARY READING

1. Holmyard, Eric John. *Makers of Chemistry*. Oxford University Press, New York, 1931. pp. 197–213, "Antoine Laurent Lavoisier."
2. Jaffe, Bernard. *Crucibles*. Simon and Schuster, New York, 1930. Chap. VI, pp. 93–113, "Lavoisier."
3. Harrow, Benjamin. *The Making of Chemistry*. The John Day Co., New York, 1930. Chap. VI, pp. 34–36, "The Phlogiston Theory"; Chap. VII, pp. 37–42, "Priestley and Oxygen"; Chap. VIII, pp. 43–52, "Lavoisier and the Introduction of Quantitative Ideas."
    (1) Write the life of Lavoisier a little more fully than given in the text.
    (2) Name the incidents related in the supplementary reading that are not mentioned in the text.
4. Foster, William. *The Romance of Chemistry*. D. Appleton-Century Co., New York, 1936. Chap. V, "Oxygen, Fire, and Flame."
    (1) What was oxygen called by the early chemists?
    (2) Why do you think they chose this name?
    (3) How does oxygen help in sewage disposal? (p. 74)
    (4) What is the source of body heat?
    (5) What did the ancients consider a flame to be?
5. Martin, Geoffrey. *Triumphs and Wonders of Modern Chemistry*. D. Van Nostrand Co., New York, 1922. Chap. VIII, pp. 171–194, "Oxygen, the Life-Supporting Element."
    (1) How would coal burn in pure oxygen? (p. 177)
    (2) Relate the story of one disaster resulting from lack of oxygen. (pp. 179–183)
    (3) Relate the effects of absence of oxygen and its restoration. (p. 184)
    (4) Relate the sensations of the first person to breathe pure oxygen. (p. 185)

6. Holmes, Harry N. *Out of the Test Tube.* Ray Long and Richard R.
   Smith, Inc., New York, 1934. Chap. III, pp. 29-40, "With Fire Man
   Rose above the Beasts."
   (1) What did the ancients consider the flame of a fire to be?
   (2) When Lavoisier discovered that metals gain in weight when burning,
       what inconsistency was there in the suggestion of the phlogistonists
       that phlogiston had negative weight? (p. 30)
   (3) Describe Lavoisier's apparatus which proved that burning was the
       chemical reaction with oxygen. (p. 31)
   (4) Tell how Lavoisier's experiment proved the point.
   (5) Who first recognized that Priestley's "good air" from mercuric oxide
       is the same substance as the oxygen of the air? (p. 32)

Photo by Dreyfous

EXTINGUISHING FIRE WITH WATER. Water stops fire by cooling the burning
material to below its kindling temperature.

Kindling temperature. If we should put a piece of wood in
the oven and gradually heat it while we watched a very sensitive thermometer, we would find that at a definite temperature
it would burst into flame. This temperature is called the
kindling temperature of the wood. Some substances ignite
more readily than others because they have lower kindling
temperatures.

Putting out fires. The preceding paragraph suggests one
method of putting out fires, namely, that of cooling the burning

material below the kindling point. As surely as the stick in the oven bursts suddenly into flame as the kindling temperature is crossed, so will the burning stop when the temperature is again lowered below this point. The fireman puts out the fire on a burning building by cooling it below the kindling temperature with a stream of cold water.

EXPERIMENT

To show that oxygen is needed for burning. Invert a fruit jar over a burning candle. In a short time the candle goes out.

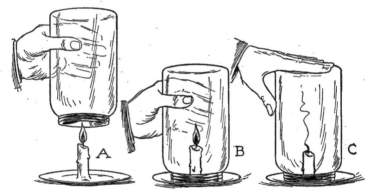

BURNING REQUIRES OXYGEN. As soon as the oxygen of the air in the jar is used up, burning ceases.

One might think that there is something peculiar about water that stops the burning. This is disproved by the fact that a fire can be started with water.

"Start a fire with water! That is absurd," one may say. On the contrary, it is quite possible. One of the members of an expedition of the National Geographic Society to the Valley of Ten Thousand Smokes in Alaska stuck his walking stick into a vent from which steam was escaping. When he pulled the stick out of the hole, it began to burn. The steam was hotter than the kindling temperature of wood; hence the

EXTINGUISHING FIRE WITH A BLANKET. The blanket puts out fire by cutting off the oxygen supply.

stick caught on fire. Since steam is only hot water-vapor, we can rightly say the fire was started with water. However, one must not get the impression that all steam will ignite paper or wood. The steam escaping from the teakettle, although hot enough to burn us severely, will not even char paper. The steam from the volcano was superheated to several hundred degrees.

The second method of putting out a fire depends upon shutting off the supply of oxygen. The fire in a stove may be ever so hot and burning ever so vigorously; but if the stove is closed up air tight, the burning will stop as soon as the oxygen already in the stove is used up.

EXPERIMENT

To make a working model of a fire extinguisher. Seal the glass part of a medicine dropper into the top of a quart mason jar with sealing wax as shown in the figure. With a pint of water in the jar, stir in about ten teaspoonfuls of baking soda. Now hang a test tube of dilute hydrochloric acid into the jar as shown in the drawing. To use this fire extinguisher, invert and point towards the fire.

Dilute Hydrochloric Acid

Baking Soda Solution

Fire extinguishers. Most fire extinguishers employ both methods to some extent, but they depend mostly upon the method of shutting off the oxygen supply. One common extinguisher uses a heavy, incombustible liquid called carbon tetrachloride. When this liquid is sprayed over a flame, it forms about the burning substances a low cloud of incombustible gas, which crowds away the lighter oxygen and so puts out the fire.

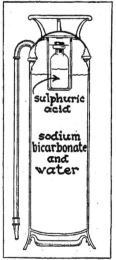

The ordinary hand fire extinguisher shown in the figure is filled with a solution of baking soda. Above the soda solution is a bottle of sulfuric acid with a loose lead stopper. In case of fire, the extinguisher is inverted. Immediately the lead stopper drops out, allowing the acid to run into the soda solution. A chemical reaction between the acid and soda liberates a large volume of carbon dioxide gas. This gas exerts a great pressure, which forces the water and carbon dioxide out through the hose. The water cools the burning substances below the kindling point, while the incombustible carbon dioxide hovers low about the fire to shut out the oxygen of the air. The carbon dioxide cannot burn, since the carbon has already united with all the oxygen it is capable of holding. Thus both methods of fire extinction are used together. The chemical engine used by the fire department uses the same method of fighting fire as the hand extinguisher.

CHEMICAL FIRE EXTINGUISHER. To use, invert the extinguisher. The stopper drops out of the sulfuric-acid bottle at the top, allowing the acid to mix with the soda solution and generate carbon dioxide. The large volume of this gas creates a pressure in the tank, which forces out the solution and the carbon dioxide gas. The water cools the burning material, and the heavy incombustible gas excludes the oxygen.

Fighting oil fires. Burning oil offers a problem of special difficulty. Water cannot be sprayed on it, since the oil will pop and sputter out into the air, where it will burn more vigorously than ever. Furthermore, if large amounts of water are run into the burning oil, the water being heavier than oil will settle to the bottom of the tank, thus causing the oil to overflow and spread the fire. To extinguish an oil fire, it is advisable to cover the surface of the oil with a material called *foamite*. Foamite is a mixture of chemicals which give off a great deal of carbon dioxide when heated and also form an incombustible foam. This foam and the carbon dioxide crowd back the air and stop the burning.

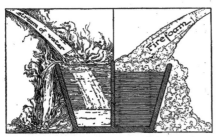

PUTTING OUT OIL FIRES. The left half of the illustration shows the result of trying to extinguish an oil fire with water. The heavier water settles to the bottom of the tank, causing the oil to overflow and burn more vigorously than before, as it is in contact with more air. The right half of the illustration shows the effect of spraying carbon-dioxide foam on the fire. The light foam shuts off the oxygen and so stops the burning.

How burning starts. All fires must be started in some way. When once started, burning is progressive. The match, which starts the burning, heats the paper above its kindling temperature. The burning paper then heats the shavings above the kindling point. The heat from the burning shavings raises part of the wood above its kindling temperature; and, in like manner, the burning wood ignites the coal. Fire travels from one part of a burning structure to another by heating each new part above its kindling point. The match itself is tipped with a phosphorus compound which has a very low kindling point. When the match is struck, the heat of friction causes this compound to burst into flame.

Spontaneous combustion. Sometimes oily rags catch fire without anyone having applied a flame to them. This is spoken of as *spontaneous combustion*. Painters using linseed oil must be very careful not to roll oily rags together and drop them where they might start a fire. Many fires have been started by spontaneous combustion in this way. The reason the oily rags take fire is this: The oil oxidizes as in burning but not quite so rapidly. This slower oxidation also gives off heat. If the rags are rolled up so that the heat cannot escape into the air, the heat accumulates until finally the kindling temperature is reached, the rags burst into flame, and the fire is started. Hay that is put into the barn while damp often takes fire by spontaneous combustion. The dampness aids oxidation, which produces heat, which in turn produces more oxidation, which produces more heat, until the kindling temperature is reached and the hay is afire. Yellow phosphorus oxidizes so readily and its kindling temperature is so low that it must be kept under water or it will take fire spontaneously.

Slow oxidation. A stick of wood may burn in a few minutes to carbon dioxide and hydrogen oxide, giving off a great deal of heat. A similar stick out in the woods may decay, or slowly oxidize, over a period of forty years or more. The final products will be the same in either case. The heat also will total the same. However, being scattered over forty years, it is never noticeable at any time. Thus oxidation may be slow or rapid. Moreover, the oxygen may attach itself to the other substance permanently, or it may tear part of the substance off and depart with it. Oxidation rapid enough to produce noticeable heat and light is called burning or combustion.

SPONTANEOUS COMBUSTION. The drawing shows the effect of pouring a solution of yellow phosphorus in carbon disulfide over a piece of paper. The carbon disulfide readily evaporates, leaving a thin layer of phosphorus on the paper. The phosphorus oxidizes so rapidly that it is soon heated to the kindling temperature, and bursts into flames.

Slow oxidation is both a hindrance and a help to man. It rusts his farm implements and rots down his buildings. On the other hand, the slow oxidation of food in our bodies furnishes energy and keeps us warm. The oxygen we breathe attaches itself to the red corpuscles of the blood, which carry it to the different parts of the body. Slow oxidation renders another great service in improving soil fertility and removing fire hazards by the decay of fallen trees and other vegetable

and animal residues. Millions of tons of wood and leaves in our forests are oxidized each year.

The products of oxidation. It has been pointed out that burning, combustion, or oxidation includes a large number of chemical reactions in which oxygen unites with the substance being oxidized. In those cases in which the substance is simple, the oxygen remains attached to it, and the new substance is usually called an oxide. For instance, when sulfur burns, a sulfur oxide results; when carbon burns, a carbon oxide results; and when iron burns, an iron oxide results. If the burning substance is complex as in the case of the candle, more than one oxide usually results.

SLOW OXIDATION. The rusting of iron (upper) and the decay of wood (lower) are both examples of slow oxidation.

QUESTIONS OF FACT

1. Define "kindling temperature."
2. What two processes will stop burning?
3. Describe the hand fire extinguisher.
4. Explain starting a fire from the striking of the match until the coal is burning.
5. Name two ways slow oxidation aids man and two ways it hinders.
6. If oxygen and any combustible substance are mixed, what will happen if the temperature is raised above the kindling temperature of the substance?

7. In the preceding question would it make any difference how the temperature was raised?

8. What are the products of combustion called?

9. Explain the scientific terms: (a) burning; (b) spontaneous combustion; (c) kindling temperature.

QUESTIONS OF UNDERSTANDING

In the first exercise choose the most suitable ending for the statement.

1. The conditions necessary to produce spontaneous combustion are:

(a) A non-oxidizing oil in rags.

(b) A rapidly oxidizing substance insulated so the heat will not escape.

(c) Damp hay spread out in the field.

(d) Linseed oil painted on the wall.

2. Explain why it is true that blowing a candle puts it out, while blowing a small fire makes it burn more brightly.

3. A ship carrying coal is found afire at sea. Might the fire have been caused by spontaneous combustion?

4. Some mineral ores containing sulfur oxidize readily. Account for frequent fires in such mines.

5. Discuss the sensible thing to do if your clothing were afire.

6. Explain why the oxidizable oil in paint never catches fire on the exposed surface.

7. Which is the more inclusive term, oxidation or combustion?

8. Are combustion and burning practically synonymous?

ADDITIONAL EXERCISES FOR SUPERIOR STUDENTS

1. Serious explosions are caused by gas leaks. Show the relation of combustion to the explosion.

2. Explain the running of an automobile engine in terms of oxidation.

3. A defective window pane focuses the sunlight on the carpet and sets it afire. Is this a case of spontaneous combustion?

4. If 265 pounds of oxygen are required to burn 100 pounds of coal, how many cubic feet of air will be needed? (1 cu. ft. of air contains approximately .017 lb. of oxygen.)

REFERENCES FOR SUPPLEMENTARY READING

1. Foster, William. *The Romance of Chemistry*. D. Appleton-Century Co.,
   Inc., New York, 1936. Chap. V, pp. 58–78, "Oxygen, Fire, and
   Flame."
   (1) Name three metals that do not combine with oxygen.
   (2) What substance did the ancients call "copper calyx"?
   (3) Name two illustrations of spontaneous combustion not mentioned in
       the text. (p. 73)
   (4) Name six ways nature has started fires. (p. 72)
   (5) What is the kindling temperature of white phosphorus?
2. Holmes, Harry N. *Out of the Test Tube*. Ray Long and Richard R.
   Smith, New York, 1934. Chap. III, pp. 29–40.
   (1) Compare the human body with a furnace. (p. 33)
   (2) Name ten possible explosive dusts.
3. Weed, Henry T., and Rexford, Frank A. *Useful Science, Book I*. The
   John C. Winston Co., Philadelphia, 1931. Chap. VIII, "Air in the
   World"; Chap. IX, "What Is Air"; Chap. X, "Oxygen, Oxides,
   Oxidation"; Chap. XI, "Oxygen at Work"; Chap. XII, "Air Is
   Necessary." Written for junior high school.

SUMMARY

Oxygen is the substance most essential to living things.
Fortunately there is an enormous amount of free oxygen in
the air, an amount which shows no signs of decreasing. Still
greater quantities are combined in the waters of .the oceans
and the rocks which compose the earth's crust.

Pure oxygen can be prepared in the laboratory by heating
potassium chlorate with manganese dioxide.

Oxygen may also be prepared by: (1) heating mercuric
oxide, the method by which Priestley accidentally discovered
the substance; (2) mixing sodium peroxide and water; (3) liquefying air and allowing the nitrogen to evaporate off.

Pure oxygen is employed in the breathing apparatus for mine
rescue work, is administered by physicians to very weak
patients, and supports combustion in the oxyacetylene blow
torch.

All changes in nature can be divided into two classes: chemical reactions, in which new substances are formed; and physical changes, from which no new substances result.

A *catalyst* is a substance that aids a chemical reaction without being used up itself.

The products of combustion weigh more than the fuel supplied, the addition of oxygen from the air being the explanation.

A fundamental fact of science is that the total amount of matter involved in a chemical reaction is neither increased nor decreased.

A substance when heated in the air does not burst into flame until it has reached a characteristic temperature known as the *kindling temperature.*

Two principles are used in putting out fire: (1) cooling the burning substance below its kindling temperature; and (2) shutting off the supply of oxygen.

One type of hand fire extinguisher uses the incombustible liquid, carbon tetrachloride, to form a cloud and hold the oxygen away until burning ceases. Another type of extinguisher, which is similar to the chemical engines used by fire departments, uses a chemical reaction to produce incombustible carbon dioxide, which together with the water both cools the burning material and holds the oxygen away. Burning oil in tanks should be treated with the bubbly mixture from a foam extinguisher.

Starting fires is a progressive process. The more readily combustible kindling catches first and heats the adjoining fuel to its kindling temperature.

Spontaneous combustion results from the accumulation of heat in easily oxidized substances.

Slow oxidation is both a help and a hindrance; it supplies our body with energy, but it rusts our machinery.

⇄

CHAPTER 3

WATER

Importance of water. Next to oxygen, the substance most essential to our existence is water. The traveler lost on the desert soon begins to suffer greatly for want of water, and even dies before the need for other food is very insistent. The

Courtesy of American Museum of Natural History,

AN AFRICAN WATER HOLE.

human body is about 70 per cent water. Nearly all of the body processes require water for their accomplishment. The blood, which carries in building materials and removes waste products, is largely water.

Not all animals require water with the same frequency. On a hot day a working man may require a drink every hour, while the horse he drives may get along nicely with a drink three times per day. A snake, a lizard, or a camel may go for days without water.

Plants show more variety than animals in regard to their needs for water. Some plants, such as sea weeds, water cress, and water lilies, must live in water always. Others, such as willows and alders, must have their roots dipping into the underground water. In the desert we find the cactus, which may receive water only a few times each year. Sometimes on a hot day, vines, such as squash, may wilt because they cannot get water from the soil as fast as it evaporates from their leaves. If the wilting plant is irrigated, it will stiffen up in a few minutes, which shows that the plant literally drinks in the water. A plant can get some moisture from soil that is anything but bone dry, but often this will not be sufficient. A large tree may evaporate 300 pounds of water per day from its leaves. Desert plants can get water only occasionally when there happens to be a shower. They are equipped to take in and store enough to last for months. Whatever the habits or capabilities of plants and animals, the great outstanding fact remains that every living thing must have more or less water.

Most plants in arid regions where water is scarce have structures that conserve their moisture. Usually they have a very limited surface area exposed to the air. Furthermore, they are usually covered with a wax which prevents evaporation.

The importance of water is not confined to drinking, irrigation, and transportation; but, as a chemical substance, it is involved in more reactions and processes than any other substance. Cooling the radiator of the automobile, fighting fire, and washing of all kinds are processes already familiar to the student, which require water.

Occurrence of water. Water occurs almost everywhere on the surface of the earth. To the person who has sailed across the ocean, it seems to be the most abundant substance, which it probably is. The ocean is deeper in places than the mountains are high. On land, aside from the great polar ice caps and the lakes and streams, there is a vast underground supply that can be tapped almost anywhere by boring wells. All animal and plant products contain some water. The following table shows its percentage in a few common foods.

WATER IN COMMON FOODS

FOOD	WATER (Per Cent)
Steak	60.7
Salmon	57.9
Cheese	33.0
Butter	13.0
Milk	87.0
Oatmeal	7.2
Peas	78.1
Squash	86.5
Tomatoes	96.0
Cucumbers	96.0
Corn (green)	81.3

PERCENTAGE OF WATER IN FOODS. The shaded part of each bar shows the percentage of water in the accompanying item.

Water is a constituent of many substances that are apparently dry; for instance, dry soil. The particles of dry soil are covered with a film of moisture. Bacteria and certain disease germs live in this film. Although it is so thin that it is invisible, this film of *hygroscopic* moisture, as it is called, is thick enough for a bacterium to swim in. Almost everything contains some hygroscopic moisture on its surface. Moisture may also be absorbed by porous substances which appear dry; for instance, wood.

EXPERIMENT

Heat a pyrex test tube half full of apparently dry soil and note the collection of moisture in the upper part of the tube. This is hygroscopic moisture.

Another form of water that frequently escapes notice is that known as water of crystallization. Many crystals require water as part of their make-up. If the water is driven out with heat, the crystal crumbles into a powder. Ordinary alum crystals are glassy in appearance. If heated, these crystals lose a great deal of water and crumble into a white powder, often called burnt alum. This water of crystallization is quite a feature in the cost of many industrial chemicals. The laundryman uses tons of washing soda. If he buys it in the form of crystals, he buys 63 per cent water, while if he buys it in the white powdered form, it contains very little moisture. Any person buying large quantities of chemicals must know how much water of crystallization they contain if he is to buy intelligently. Sometimes a person wants to

WATER OF CRYSTALLIZATION. When crystalline copper sulfate is heated, it gives off a great deal of water and changes to a white powder.

mix up a formula, such as a photographic developer. The directions call for a definite weight of a certain chemical that may be obtained in either the crystalline form or the *anhydrous* (without water) form. If the formula calls for the anhydrous material and he uses the crystalline variety, his solution will be too weak, since a large part of what he weighs will be water. The druggist, also, must know whether a prescription calls for the crystalline or anhydrous form of the ingredients if he is to compound it correctly. Whole industries are built around this water of crystallization, or *water of hydration* as it is sometimes called when the substance does not appear distinctly in the form of crystals. Portland cement, which goes into our sidewalks and concrete buildings, is made by driving off with intense heat the water of hydration from a mixture of rocklike materials. When the dry powder comes in contact with water again, the anhydrous powder changes again into rock by taking up water of hydration.

"SWEATING." Moisture from the surrounding warm air condenses on the cool surface of the pitcher.

Invisible water vapor. A large quantity of water is about us as invisible water vapor. If a glass pitcher is filled with ice water on a warm day, it will soon be covered with moisture. The cold water has condensed the moisture out of the air. The refrigerating pipes in the butcher shop are usually covered with ice, which has first appeared as dew from the invisible moisture of the air, and then been frozen into ice. Invisible moisture is the source of much of our inland fogs, dew, frost, and rain. In semiarid regions there is very little invisible moisture in the air. This makes it easy for the body to evaporate water from the skin and keep it at 98.6° F., the normal temperature, even when the outside air is as warm as 110° F. On the other hand, humid air is so filled with invisible moisture that it is very difficult for more to enter by evaporation; hence the natural cooling processes of the body do not work, and the person suffers greatly at relatively low temperatures. .

The invisible moisture in the air influences many industrial processes. The rate of all drying processes is largely determined by the humidity of the air. Figs dry on the trees. In humid climates it is impossible to produce dried figs satisfactorily, since they mold instead of drying. In curing lumber in very dry climates it will often dry out so fast that it warps. The condition of many chemicals is affected by the humidity of the air. If the air is low in invisible moisture, many crystalline substances will lose their water of crystallization and crumble into a powder. These substances are said to *effloresce* and are called efflorescent substances. On the other hand, some substances absorb moisture from the air. A piece of sodium hydroxide will take enough water out of moist air to dissolve in and make a liquid. Such substances are said to be *hygroscopic*. In damp weather, table salt often becomes damp and refuses to pour from the shaker. This is because it contains an impurity that is hygroscopic. Salt itself is not hygroscopic, and if it has been freed from all hygroscopic substances, it will pour in all kinds of weather.

A dealer in chemicals must know the condition of hydration of his materials, that is, whether they have lost or gained moisture from the air. Suppose a dealer has purchased a lot of a chemical that is one-half water of crystallization. Suppose that while he has the chemical in his warehouse it effloresces, thereby losing half its weight. Imagine his perplexity when he finds out that he has only half as much to sell as he thought he had. However, if he is chemist enough to understand what has happened, he will realize that he will have to double his price, since for all purposes for which this chemical is used, it now has twice as much capacity for reaction, pound for pound, as it did have.

QUESTIONS OF FACT

1. What percentage of the human body is water?
2. (a) Name a plant that needs to stand in water.
   (b) Name a plant whose roots must reach water.
   (c) Name a plant that gets water only at rare intervals.
3. State the importance of water to chemical and industrial processes.
4. Name two foods containing over 90 per cent water.
5. What is hygroscopic moisture?
6. Explain the term water of crystallization.
7. What is meant by the anhydrous form of a chemical?
8. How is the invisible water vapor related to our comfort?
9. What is an efflorescent substance?
10. What will a hygroscopic substance do when left in the air?
11. Explain the new scientific terms: (a) water of hydration; (b) humidity.

QUESTIONS OF UNDERSTANDING

1. Can there be any life without water?

2. Calculate the weight of water in 100 lbs. of tomatoes? (See table, p. 42.)

3. Chemically, what are frost, dew, fog, and snow?

4. People wearing glasses sometimes have them fog when they enter the kitchen. Explain.

5. One writer said water is the most ubiquitous substance on earth. Look up the meaning of ubiquitous in the dictionary and say whether you agree with him or not.

6. What places out in nature can you think of where there is no water in any form?

REFERENCES FOR SUPPLEMENTARY READING

1. Holmes, Harry N. *Out of the Test Tube*. Ray Long and Richard R. Smith, New York, 1934. Chap. VII, pp. 79–89, "The Elixir of Life."
   (1) What justification is there for calling water the "elixir of life"?
   (2) Name five things water has done for us.
   (3) What peculiarity of water makes ice float?
   (4) How many gallons of water are used daily by a city of 100,000?
2. Clarke, Beverly L. *Marvels of Modern Chemistry*. Harper and Brothers, New York, 1932. Chap. XII, pp. 168–177, "Water."
   (1) What is nature's greatest solvent? (p. 168)
   (2) What has been said to be twelve pounds of ashes and eight buckets of water? (p. 168)
3. Martin, Geoffrey. *Triumphs and Wonders of Modern Chemistry*. D. Van Nostrand Company, New York, 1922. Chap. V, pp. 88–112, "Water."
   (1) What is the average ocean depth? the greatest depth? (p. 89)
   (2) Discuss the occurrence of water on Venus and Mars. (p. 93)
   (3) How much water does the average man throw out daily? (p. 97)
   (4) What per cents of land plants are water? of water plants? (p. 97)
   (5) How much water does an oak tree give out in 1000 years? (p. 101)
   (6) What would the earth be like without water? (pp. 101–103)
   (7) Describe some snow crystals. (pp. 108–109)
   (8) Under what pressure would water boil at red heat? (p. 110)
4. Weed and Rexford. *Useful Science Book I*. Chap. III, "The Water of the Earth"; Chap. IV, "Water Changes the Surface of the Earth."
5. Darrow, Floyd L. *The Boy's Own Book of Science*. The Macmillan Co., New York, 1935. Chap. 11, "The Examination of Water." Partially experimental.

Water an equalizer of our climate. Aside from being absolutely necessary as a food for every living organism, water is the great equalizer of climate. Without it there would be wide extremes of hot and cold with the changing seasons. Such extremes are common in inland regions, where the winters are cold and the summers hot. Places near the ocean usually have mild climates.

EFFECT OF SUNLIGHT ON LAND AND WATER. The heat necessary to raise the temperature of ocean water 1 degree would raise the temperature of an equal weight of land 5 degrees.

Water has a very high specific heat. One property of water that enables it to equalize climatic conditions is its high *specific heat*. By specific heat we mean the quantity of heat necessary to raise the temperature of 1 gram of a substance 1° C. (See Appendix for the meanings of gram and degree centigrade.) The table (page 48) compares water with other common substances. The *calorie*, the unit used in this discussion, is the quantity of heat necessary to raise the temperature of 1 gram of water 1 degree centigrade.

EXPERIMENT

Weigh out equal quantities of lead shot and tap water. Set the water aside while you heat the shot in another vessel of water until the water boils. This will raise the temperature of the shot to 100° C. (212° F.). Now take the temperature of the tap water. Quickly pour *all* of the boiling water off the shot and drop the shot into the tap water. Stir well and again take the temperature of the tap water, which is also the temperature of the immersed shot.

How many degrees has the temperature of the tap water risen?

How many degrees has the temperature of the shot fallen?

Divide the former figure by the latter, and you should have approximately the specific heat of lead (0.03).

(Caution: Do not put an ordinary household thermometer into boiling water, or you may burst it.)

SPECIFIC HEATS

SUBSTANCE	SPECIFIC HEAT	SUBSTANCE	SPECIFIC HEAT
Water	1.00	Wood	.42
Chalk	.21	Lead	.03
Charcoal	.16	Mercury	.028
Clay	.22	Iron	.11
Granite	.19	Copper	.09
Quartz	.188	Air	.24

A study of the table shows that the heat that will raise the temperature of water 1° C. will raise the common rock materials of the earth's surface, such as chalk, clay, granite, and quartz, four or five degrees. If the sun shines on land and water in summer, the temperature of the land will rise nearly five degrees while that of the water rises only one degree. This explains why a coast region is cooler in summer than the interior of a continent. On the other hand, during the long winter nights heat is being lost from the earth's surface, and the temperature of the land will drop almost five times as rapidly as that of the water; hence in winter interior towns are much colder than coast towns.

Water has a very high heat of fusion. Another property of water that enables it to modify a climate is its abnormally large *heat of fusion*. By heat of fusion is meant the heat necessary to melt one gram of a solid or, what amounts to the same thing, the heat that must be removed from one gram of a liquid at its freezing point in order to solidify it. The accompanying table gives the heats of fusion of some common substances.

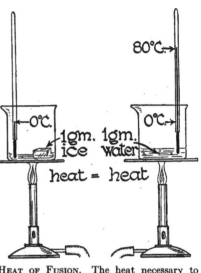

HEATS OF FUSION

SUBSTANCE	HEAT OF FUSION
Copper	42
Iron	48.0
Ice	79.2
Lead	5.8
Mercury	2.8
Nickel	73.9
Phenol	24.9
Paraffin	35.1
Sulfur	9.4

HEAT OF FUSION. The heat necessary to melt 1 g. of ice is equal to that which would heat the resulting ice water to 80° C.

EXPERIMENT

Heat of fusion of ice. Heat a hundred grams of water to 90° C. in a beaker, which is then placed on some asbestos or wood. Drop in a 100 g. piece of ice, which has been wiped dry with a rag, and measure the final temperature. If the experiment is carefully done, the final temperature is 5° C. In cooling, the water loses 100 × 85 = 8500 cal. This heat melted the ice and raised the resultant ice water from 0° C. to 5° C. This last step requires 100 × 5 = 500 cal., which leaves 8000 cal. to melt 100 g. of ice or 80 calories to melt one gram.

If we should freeze the bulb of a thermometer into a block of ice, heat the ice slowly in a dishpan on the stove, and place the bulb of another thermometer into the water from the melting ice, both thermometers would read 0° C. until the ice was all melted. Thus we see that many calories of heat have been added to the mixture without any rise of temperature. The heat has merely melted the ice, hence is called *heat of fusion.*

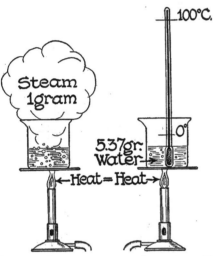

HEAT OF VAPORIZATION. The heat that will vaporize 1 g. of water is equal to that which would raise the temperature of 5.4 g. from 0° to 100° C.

This experiment shows us how melting ice in the refrigerator or on the mountain top can take up heat and keep us cool. If, however, we should desire to refreeze the above water, we would have to remove the same amount of heat. In other words, whenever water freezes, the heat of fusion is given to something else. To freeze a gram of water requires the removal of 79.2 calories of heat, and to melt a gram of ice requires the restoring of an equal amount. The freezing of water tends to keep our climate from getting too cold. It can be noticed that when it begins to snow at night, it turns warmer. The reason is that for every gram of water that changes to snow, 79.2 calories of heat are liberated. In cold weather open pans of water are sometimes placed in cellars to keep potatoes from freezing. The exposed water in the pans freezes before that in the protected cells of the potato. As soon as it begins to freeze, the heat liberated by the freezing water slows down the rate of cooling so it never gets cold enough to freeze the potatoes.

Heat of vaporization of water. The greatest factor of all in the equalization of climates by water is its large *heat of vapori-*

zation. The heat of vaporization is the number of calories necessary to change 1 gram of a substance from the liquid to the gaseous state. In the case of water the value is around 540 calories; that is, if one gram of water is evaporated, 540 calories will be taken up. On a day when the temperature of the air is 110° F., the body is still able to hold its temperature at 98.6° F. by rapidly evaporating water from the skin. Sprinkling the porch and yards cools the immediate vicinity by evaporation. The heat used up in evaporating water is put away so it no longer is felt. The thermometer does not recognize it. When there is a great deal of moisture in the air, a killing frost is not likely. Killing frosts come with dry air. Why?

EXPERIMENT

To determine the heat of vaporization of water. The heat of vaporization of water is the number of calories necessary to vaporize 1 g. of water or, what amounts to the same thing, the number of calories given off when 1 g. of steam condenses. Into 200 g. of water at room temperature in a glass beaker set on a piece of wood, pass steam from a steam generating flask until the temperature reaches 80° C. Disconnect the steam and weigh to get the weight of steam condensed.

In this experiment the condensing steam gives its heat to the water, thus raising its temperature. In other words:

Calories gone into the water = calories given up by the steam. Or 200 × degrees rise in temperature of the water = grams of steam × (H + degrees fall in temperature of water condensed from steam), where H represents the number of calories per gram given up by condensing steam.

Suppose that the room temperature were 20° C., and that 21.4 g. of steam were condensed,

then
$$200 \ (80 - 20) = 21.4 \ (H + 100 - 80)$$
$$200 \times 60 = 21.4 \ (H + 20)$$
$$12000 = 21.4 \ (H + 20)$$
$$\frac{12000}{21.4} = H + 20$$
$$560 = H + 20$$
$$H = 540 \text{ cal. per g.}$$

HEATS OF VAPORIZATION
(Calories per Gram)

SUBSTANCE	HEAT OF VAPORIZATION	SUBSTANCE	HEAT OF VAPORIZATION
Air	51	Nitrogen	47.6
Ammonia	327	Water at boiling	539.6
Carbon dioxide	87	Water at rm. temp.	
Helium	6	(20° C.)	584.9
Hydrogen	108	Alcohol (grain)	204
Mercury	65	Ether	83.9

QUESTIONS OF FACT

1. What substance in the table on page 48 has the lowest specific heat?

2. What substance in the table has the specific heat nearest to that of water?

3. If chalk, granite, and quartz are representative rocks of the earth's crust, what must be the approximate specific heat of the earth's rock crust?

4. How do the specific heats of metals compare with the specific heats of other substances?

5. Define heat of fusion.

6. What substance in the table, p. 49, has the highest heat of fusion?

7. What substance in the table has the lowest heat of fusion?

8. What substance has a heat of fusion nearest to that of water?

QUESTIONS OF UNDERSTANDING

1. Which heats the faster, water or wood, when receiving the same quantity of heat? (See table, p. 48.)

2. What substance listed in the table will rise in temperature the fastest, assuming that equal weights of the different substances receive heat at the same rate.

3. If the heat from the sun falls alike on a lake and a granite boulder, through about how many degrees would you expect the surface of the granite to rise for each degree of rise in the temperature of the water?

4. If a hammer with a wooden handle is lying in the sun, and the handle warms up two degrees, about how much would you expect the metal head to warm up?

5. Explain how the high specific heat of water makes the coast climate more mild than that of the interior regions.

6. How does the heat of fusion of ice keep potatoes in the cellar from freezing?

7. The heat that would melt 1 g. of ice would melt how many grams of sulfur?

8. If 79.2 cal. of heat are added to 1 g. of ice, it becomes 1 g. of water? Is this heat gone forever?

9. How could the heat of ex. 8 be obtained again for other use?

10. What temperature is maintained by the human body even when the summer air reaches 110° F.?

11. If the body temperature should rise to 101° F., what would the doctor say we had?

12. What substance in the table, p. 52, has a heat of vaporization nearest that of water?

13. Which requires the more heat, to evaporate 1 g. of water at boiling or at room temperature?

14. What substance in the table has the least heat of vaporization?

15. Explain why killing frosts are not likely if the atmosphere is very moist.

PROBLEMS

1. How many calories are needed to heat 100 g. of water 1° C.?

2. How many calories are needed to heat 1 g. of water 10° C.?

3. How many calories of heat are needed to heat 20 g. of water 5° C.?

4. How many calories are needed to heat 30 g. of water from 30° C. to 32° C.?

5. Calculate the number of calories necessary to melt 2 g. of ice. (See p. 49.)

6. How many calories would be required to raise 10 g. of lead 100° C.? (See table, p. 48.)

7. How many calories must be taken from 10 g. of steam in order to make it condense?

8. Change a reading of 100° F. to the centigrade reading. (See Appendix.)

9. Change the reading + 10° C. to the Fahrenheit reading.

ADDITIONAL EXERCISES FOR SUPERIOR STUDENTS

1. How many degrees centigrade could the heat necessary to melt 2 g. of ice heat 10 g. of water?

2. If 100 g. of lead at 500° C. are dropped into 100 g. of water at 20° C., what will be the final temperature?

3. How many grams of steam would have to be condensed in order to heat 500 g. of water 80° C.?

4. Explain how the human body maintains its normal temperature of 37° C. when the prevailing temperature is 40° C.

5. Change a reading of − 10° C. to the Fahrenheit reading. (See Appendix.)

6. Change the desirable room temperature 68° F. to the centigrade reading.

Composition of matter. In order to understand the next part of the study on water, we need to digress a little and consider different kinds of matter in the world around us. One of the first things we notice is that some substances, such as sugar, salt, pure water, cooking fat, copper, soap, and glass, seem to be homogeneous throughout; that is, each part is exactly like every other part. Other substances, including dirt, soup, stew, wood, many rocks, and concrete, are spotted or streaked, one part being different from some other part. These are undoubtedly mixtures of substances which are quite dissimilar. Some substances that appear to the eye to be alike in all parts can be shown to be mixtures. Fresh milk, for instance, appears to be homogeneous, yet we know that cream separates out from the rest of the milk. Clear ocean water looks uniform throughout, yet when we taste it we know that there are other substances in it besides water.

When we separate out a single substance from a mixture, we say we have purified it. Nature gives us mixtures as a rule. Occasionally a naturally occurring substance may be said to be pure, but such occurrences are rare. Out of the mixtures supplied by nature, man has been able to separate hundreds of thousands of pure substances. One person could not study all of them in a whole lifetime. However, we will select a few of the more important ones for our study.

Crystallization. If a concentrated solution is allowed to stand quietly and slowly evaporate, the dissolved substance

separates out in definitely shaped *crystals*. While crystals of different substances are of many different shapes, they may all be regarded as modifications of six general forms. Crystals frequently crowd one another so that few of them are perfect; but when they are perfect, the crystals of a substance usually all have the same shape. Thus the perfect crystals of salt are cubes. Occasionally a substance, such as sulfur, exists in two forms, each of which has a differently shaped crystal. Ice seems to have at least five different crystalline forms. However, of the thousands of crystalline substances only a few have more than one crystalline form.

CRYSTALS. Lead nitrate (top), copper sulfate (middle), and several minerals (bottom). Most pure chemicals which are solids at ordinary temperatures separate from solution in definite geometric crystals. All the crystals of any one substance are much alike, but the crystals of different substances are usually very unlike.

EXPERIMENT

Crystallization. Set very concentrated solutions of the following aside in shallow dishes to crystallize for examination of the crystals : (*a*) copper sulfate ; (*b*) potassium dichromate ; (*c*) hypo ; (*d*) chrome alum.

Each kind of crystal has its individual shape and in the following cases different colors : copper sulfate, blue ; potassium dichromate, orange ; hypo, white ; and chrome alum, purple.

Purification of substances. Liquids can be separated from insoluble impurities by filtering ; from soluble ones by evaporating off the liquid and condensing the vapor. The latter

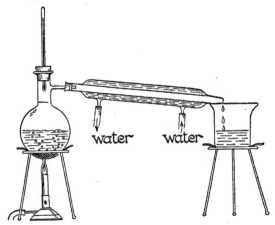

water   water

PURIFICATION OF WATER BY DISTILLATION. The water to be purified is heated in the flask. The resulting vapor passes through the condenser, where it is cooled by water running through the jacket and changed back to liquid. All mineral impurities remain in the flask.

process is called distillation. A modification of the distillation process is also made use of in separating the various constituents in a mixture of liquids. Two liquids with different boiling points can usually be almost completely separated by *fractional distillation.* When they are heated, the boiling point gradually rises. The liquid with the lower boiling point tends to come off first, the other liquid evaporating more freely at the higher temperatures. If the portions that come off at different short intervals of temperature are kept separate and distilled over several times, the two original constituents of the mixture become more and more nearly separated. To purify a soluble solid, it is necessary to crystallize it from the concentrated solution. The crystals are very pure, since all the impurities remain in the liquid part.

Solutions. One class of mixtures is of special importance, the solutions. If some salt is put into a glass of water and stirred, the salt will distribute itself uniformly throughout the water. This is called a solution. Solutions resemble compounds in appearance, since they appear the same in all parts. A solution is clear so one can see through it, but it may have color. When we come to separate the constituents, we shall

A Saturated Solution. At a definite temperature a given amount of water can dissolve just so much of a solid. Such a solution is called a saturated solution. If a saturated solution cools or if some of the water evaporates, crystals of the solid separate out.

see that solutions are not compounds. In many cases a liquid, such as ether or chloroform, can remove a substance from a water solution, showing that the substance is not held chemically.

EXPERIMENT

Solvent and solute. Take a beaker of water, the *solvent*, and add a crystal of potassium permanganate, the *solute*. Shake until the permanganate dissolves to form a purple *solution*.

Also, the liquid in most solutions can be evaporated off from the solid without involving either more or less heat than that needed to evaporate that much liquid, showing the absence of a chemical reaction, which always either absorbs or evolves heat. There are solutions of gases in liquids, as ammonia water; liquids in liquids, as tetraethyl lead in gasoline; and solids in solids, as the alloys of metals. In any solution of one liquid in another the ingredient that is present in the larger amount is called the *solvent;* and the one in the smaller amount, the *solute.* Substances like iron show little tendency to distribute themselves in other substances like water and are said to be *insoluble.* At any temperature a given amount of most solvents will hold only a certain amount of solute. A solvent that has all of the solute it can hold when in contact with a crystal of the solute is said to be *saturated.* Sometimes

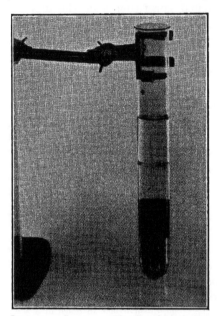

IMMISCIBLE LIQUIDS. Liquids which do not dissolve in each other are said to be immiscible. When shaken together, such liquids separate into two distinct layers, with the lighter liquid on top. The illustration shows four immiscible liquids. From bottom to top they are: mercury, aniline, water, and ether.

by carefully cooling a solution from a higher temperature and in the absence of a crystal of a solid solute, a solution can be made to hold a little more of the solute than it could at the same temperature in contact with a crystal. Such a solution is *supersaturated.* If a supersaturated solution is shaken or a crystal of solute is dropped into it, the excess will immediately separate out, leaving an ordinary saturated solution.

Supersaturation. Saturate 100 cc. of water at about 50° C. with crystallized sodium sulfate and filter out any undissolved salt. Put some cotton in the mouth of the flask and allow it to cool down to room temperature. Drop into this solution one small crystal of $Na_2SO_4 \cdot 10\ H_2O$ and observe the sudden formation of crystals.

Liquids like ether and water that dissolve each other only to a limited extent are said to be *partially miscible*. Others like water and alcohol that dissolve each other in unlimited amounts and never become saturated are said to be *miscible in all proportions*. All gases are miscible in one another in all proportions.

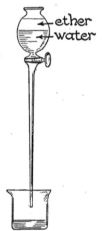

ether
water

EXPERIMENT

Partially miscible liquids. If half its volume of ether is added to water in a separatory funnel and shaken, a surface soon forms between the liquids. The water is now drained off through the bottom of the separatory funnel. It has a strong ether odor, which shows some ether has dissolved in it. Add some white copper sulfate to the ether in the separatory funnel. It turns blue, which is a test for water, showing that some water has dissolved in the ether.

Natural waters. After rain has fallen, it comes into contact with the ground, running off in streams or sinking into the soil, in the latter case usually to appear at some other point in a spring. All soils contain substances that dissolve to a greater or less extent in water; hence natural waters always contain an appreciable quantity of foreign material. Waters flowing through land containing limestone or gypsum dissolve small quantities of these substances and are said to be hard. Mineral springs contain dissolved substances with an odor or a taste. Some mineral waters are said to be beneficial to health, but their effect is probably due to the water itself rather than to the material which it contains, for most of us drink too little water. Disease germs are another impurity that may get into water. They are dangerous to man, and are most difficult to detect. The water may taste ever so good and appear perfectly clear yet be unfit for drinking because of disease germs. The only safe plan is never to drink without boiling water that by any chance might be contaminated.

Purification of water. Pure water — water containing no other substances — is absolutely necessary for some purposes. The garageman puts only distilled water into your battery, as certain impurities in naturally occurring water would soon ruin the battery. Water with certain mineral constituents forms boiler scale, soon making the boiler unfit for service. Mineral constituents are wasteful of soap in laundries; hence the water often has to be given a partial purification.

To purify mineralized water completely, it has to be distilled. This process is so expensive that it cannot be used where large quantities of water are needed. Therefore it is not feasible for man to purify the mineral-laden ocean water for industrial purposes. Nature, however, is continually evaporating water from the ocean, water which will later condense and fall as rain and snow. Distillation will not completely separate water from such liquids as alcohol. Since these do not naturally occur in waters, their removal is not a public problem. Many large cities have difficulty in getting drinking water, which they need in millions of gallons daily. The supply must be free enough from minerals so that it will taste good. Suspended matter and disease germs must also be removed. Their removal is often accomplished by sand and gravel filters. The filters screen out the suspended mud, and amoebas destroy the germs. Amoebas are small animals that seem to establish themselves in the sand filter and eat the bacteria as they go by. At any rate, after water of high bacterial count has been passed through a sand filter, the bacteria are mostly gone.

When the problem of purification is only one of removing the suspended matter from muddy water, it is often done by precipitating small quantities of a tasteless, jellylike substance in the water. This jellylike material settles to the bottom of the reservoir, carrying the mud with it and thus clearing the water. When the water is clear but has a high bacterial count, it can be purified by chlorine, which kills bacteria in dilutions that are safe to man and have no taste.

Courtesy of Cleveland Water Works

FILTER GALLERY.   Large cities spend enormous sums in providing wholesome drinking water for their inhabitants.

Water in relation to health.   Water is related to our state of health in more ways than most of us realize. Many people employed in offices or at other work not requiring vigorous exercise do not drink enough water to keep their bodily processes functioning as they should. This is especially true in the winter when the water is cold and not particularly inviting. The first result of skimping the body for water is constipation. This gives rise to numerous other ills. No one knows how much suffering and ill health has its beginning in not drinking enough water. This lack of desire for enough water seems to be hereditary and may run in families. Because of its high heat-holding capacity, water is being made the basis of a system of treatments known as hydrotherapy. One form of treatment is as follows. Hot packs stimulate the flow of blood to the affected part. Short cold applications between hot packs increase the reaction. A final cold friction rub closes the pores and leaves the part all aglow.

EIGHT GLASSES A DAY. Each person should drink two quarts (eight glasses) of water daily. The water taken in soup, milk, and other beverages may be counted as part of the requirement.

Water in chemical reaction. Water is involved in as many known chemical reactions as any other single substance. It reacts vigorously with several metals even when cold, and with several others on heating. It is a necessary ingredient of many crystals, which crumble into powder when the water escapes. It reacts with fats in the formation of soap. In making taffy from cane sugar, water is the necessary reacting substance. It is true that in the making of soap, alkali is used to cause the water to react, but largely as a catalyst. In taffy making, also, a catalyst is necessary — namely, vinegar. Nevertheless, without water neither reaction will take place. These two reactions are representative of hundreds of other similar ones. The hardening of Portland cement and plaster of Paris also are caused by reactions with water.

In hundreds of other chemical reactions, water is one of the products of reaction. For instance, freshly plastered walls dry more slowly than they ought because of the formation of water in the reactions of hardening. In the making of plaster of Paris, water is given off. In the formation of many synthetic flavors, dyes, perfumes, medicines, explosives, and industrial chemicals, water is one of the products. If a person thoroughly understood all of the chemical reactions in which water is involved even as just one of the reacting substances or a product of reaction, he would perhaps be the wisest chemist who ever lived. He would be sought by industrialists the world over and offered rich rewards for his advice.

QUESTIONS OF FACT

1. How could one prepare a supersaturated solution?
2. Ether and water are partially miscible; what does this mean?
3. Describe distillation.
4. What can a sand filter remove from water?
5. What does the precipitation method of water purification remove?
6. How is water often related to crystals?
7. Discuss water as a product of chemical reactions.
8. Name three substances which will react with water.
9. Explain these terms: (a) crystals; (b) fractional distillation; (c) solution; (d) solvent; (e) solute; (f) insoluble; (g) partially miscible; (h) miscible in all proportions.

QUESTIONS OF UNDERSTANDING

1. Classify each of the following as to whether it is a mixture or a single substance: milk, ice, soil, cream, chalk, distilled water, alkali water, white sugar, brown sugar, cake, rubber, cotton, ashes, lemonade, and soup.
2. Name the solvent and solute in sugar sirup.
3. Could one ounce of water dissolve a barrel of alcohol?
4. How could one prepare a perfect crystal of sugar?
5. If there were sand in the sugar, how could the sugar be purified?
6. How could one distinguish between a saturated, an unsaturated, and a supersaturated solution? (Hint: By introducing a crystal of the solute and watching for changes in size.)
7. Invent a health slogan pertaining to drinking water.

REFERENCES FOR SUPPLEMENTARY READING

1. Holmes, Harry N. *Out of the Test Tube.* Ray Long and Richard R. Smith, New York, 1934. Chap. VII, pp. 79–89, "The Elixir of Life."
   (1) Discuss water as a chemical. (pp. 82–83)
   (2) How is milk powder produced?
2. Sadtler, Samuel Schmucker. *Chemistry of Familiar Things.* J. B. Lippincott Co., Philadelphia, 1924. Chap. VII, pp. 82–97, "Water."

(1) Make a list of substances water will dissolve. (p. 86)
(2) Name six liquids that do not dissolve in water.
(3) What is meant by total solids in natural waters?
(4) Name seven things that make water unfit to drink. (p. 87)
(5) Discuss bacteria in relation to drinking water. (pp. 89–90)
(6) What is nature's distilled water? (p. 95)
(7) What is the highest mineral content water may have and still be used for domestic purposes? (p. 96)

SUMMARY

Water forms about 70 per cent of the human body. Most of the processes within the body take place by the aid of water. Some animals live in water, while others may get a drink only occasionally. Some plants also must be in the water, others must have their roots in water, while desert plants get water at infrequent intervals. Desert plants are able to store water from the occasional showers for times of drought.

In addition to its occurrence in oceans, lakes, and streams and underground, water is found in plant and animal tissues, on the surface of soil particles as hygroscopic moisture, within crystals as water of hydration, and in the air as invisible moisture.

Because of its high specific heat, high heat of fusion, and high heat of vaporization, water is very efficient in equalizing the climate.

All matter exists either as pure substances or as mixtures, usually as the latter. A pure substance may be distinguished from a mixture by examining it for homogeneity or by applying a test for the suspected impurity.

When a substance dissolves in water (or other liquid), the mixture is called a solution. The water is called the solvent; and the solid, the solute. Crystals form when the solutions of some substances evaporate. Such substances can be purified by crystallization; others require distillation.

When a substance refuses to distribute itself in the solvent, it is said to be insoluble. Liquids which dissolve each other in any amounts are said to be miscible in all proportions. If they dissolve to a limited extent, they are said to be partially miscible.

MUNICIPAL AERATION PLANT

Mineral waters have little medical value, but they are a great nuisance to industry.

Large cities filter their drinking water through sand filters. Filtration removes suspended matter and reduces the bacterial count.

Pure drinking water is of very great importance to health. Each adult should drink about two quarts of water daily.

CHAPTER 4

HYDROGEN AND ITS OXYGEN COMPOUNDS

Hydrogen from water. In the preceding chapter pure water was mentioned as a single substance, not a mixture. If a little sulfuric acid is added to some water and a direct current of electricity is passed through the solution, the water is decomposed into oxygen and hydrogen. This shows us that water consists of oxygen and hydrogen. We can arrive at the same conclusion by burning hydrogen in oxygen and cooling the resulting water vapor. Thus we see oxygen and hydrogen can unite to form water. The uniting of substances to form a new substance we call *synthesis;* and the reverse process, *decomposition.* In this instance both operations teach the same fact, namely, that water is composed of oxygen and hydrogen in chemical union.

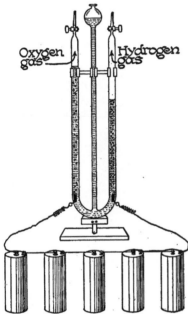

Oxygen gas

Hydrogen gas

ELECTROLYSIS OF WATER. The apparatus is filled with water containing a little sulfuric acid, the latter being necessary to help the electric current through the water. The passage of the current through the solution slowly breaks down the water into oxygen and hydrogen gases. Oxygen forms in one branch of the apparatus and hydrogen in the other. There is twice as much of the hydrogen as of the oxygen.

Element and compound. In the previous paragraph we concluded that water is a more complex substance than either hydrogen or oxygen. Most of the hundreds of thousands of pure substances can be decomposed into simpler substances. The ways that this can be done are various. Heat does it in some cases; electricity in other cases. Some substances are easily decomposed, while others break down only after most strenuous treatment. Ninety-two substances have resisted all ordinary chemical methods of decomposition into simpler substances. These are called *elements*. The world and everything in it is built of substances which are themselves combinations of these elements. Oxygen and hy-

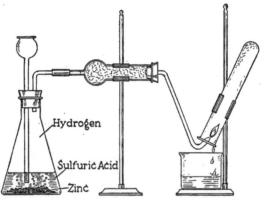

SYNTHESIS OF WATER. As the hydrogen burns at the end of the tube, it unites with the oxygen of the air to form water, which condenses in the test tube. The horizontal tube contains calcium chloride, which prevents any water being carried over from the generating flask.

drogen are in this list of elements. Other common elements are sulfur, gold, silver, copper, aluminum, and tin. In the back of the book will be found a list of the names of the elements. These elements range in abundance all the way from oxygen, which forms half of the accessible portions of the earth, to illinium, which has been found only in minute traces. The most abundant elements in the earth's crust are listed below.

COMPOSITION OF THE EARTH'S CRUST

ELEMENT	PER CENT	ELEMENT	PER CENT
Oxygen	46.71	Sodium	2.75
Silicon	27.69	Potassium	2.58
Aluminum	8.07	Magnesium	2.08
Iron	5.05	All others	1.42
Calcium	3.65		

Substances that have been found to be composed of two or more elements chemically united are called *compounds*. There are literally hundreds of thousands of known compounds. Water, sugar, quartz, salt, chalk, and quinine are just a few familiar ones.

Law of constant composition. A pure chemical compound always has the same composition. The common products of commerce do not have constant composition. Meat does not always carry the same proportion of fat, nor wheat the same

SOME COMMON ELEMENTS. Many common substances are elements — that is, substances which cannot be separated into simpler substances by known chemical reactions. The elements shown in the illustration are: A, sulfur; B, copper; C, mercury; D, aluminum; E, tin; F, lead; G, gold; H, zinc; I, silver; and J, iron.

proportion of bran. Peaches do not always have the same percentage of pits. All these products are variable in composition because they are mixtures and not single pure substances. Pure sugar, however, is a single substance and always contains the same elements in the same proportions — carbon, oxygen, and hydrogen. A great deal of misleading advertising is given over the radio. We are told to use a certain brand of aspirin. Pure aspirin is identically the same substance regardless of who made it. When 18 grams of water are decomposed, there result 16 grams of oxygen and 2 grams of hydrogen. It makes no difference where the water comes from so long as it has been purified and separated from all other substances. Water from rain, from springs, from meat, from fruit, or from any other source yields the same amounts of oxygen and hydrogen respectively from a given weight of the compound. This fact illustrates a general law of chemistry called the *law of constant composition*, or sometimes the *law of definite proportions*. A statement of this law is: *Every pure chemical compound always contains the same percentage of its constituent elements.*

Symbols. In the working of problems, writing of chemical reactions, and tabulation of data, it is not convenient to write out the names of the elements. For this reason chemists have adopted symbols to represent the elements. In a number of cases the symbol is merely the (capitalized) first letter of the name of the element, as A for argon, B for boron, C for carbon, etc. The symbol for tungsten (W) comes from wolfram, a less common name for this element. Where two or more elements have the same initial letter, symbols for the additional elements are formed by adding a (small) letter to the initial; thus, Al is aluminum, As is arsenic, and Ba is barium. Finally, there are ten elements — most of them known since ancient or medieval times — whose symbols are derived from their Latin names. They are given in the accompanying table.

LATIN NAMES OF ELEMENTS

MODERN NAME	LATIN NAME	SYMBOL	MODERN NAME	LATIN NAME	SYMBOL
Antimony	Stibium	Sb	Mercury	Hydrargyrum	Hg
Copper	Cuprum	Cu	Potassium	Kalium	K
Gold	Aurum	Au	Silver	Argentum	Ag
Iron	Ferrum	Fe	Sodium	Natrium	Na
Lead	Plumbum	Pb	Tin	Stannum	Sn

The occurrence of hydrogen. As the free or uncombined element, hydrogen is found only in traces in the atmosphere and in small quantities in a few underground regions. However, in chemical combination, hydrogen is widely distributed

Henry
Cavendish

1731–1810

Henry Cavendish was a wealthy Englishman of a very bashful and retiring nature, who lived contemporaneously with Priestley and Lavoisier. Not having any desire for fame, Cavendish followed his great interest in experimental science for its own sake. In 1766, however, his friends persuaded him to publish his discoveries in *Philosophical Transactions*, the publication of the Royal Society of England.

In these articles he announced the preparation of *inflammable air*, as he called the substance we now call hydrogen, which name comes from another discovery by Cavendish that hydrogen burns to form water. Hydrogen means water former. The method of Cavendish for preparing hydrogen is the same one as we now use to prepare it in the laboratory; namely, by the action of sulfuric or hydrochloric acid on zinc, iron, or tin. Cavendish, however, made two serious mistakes in his interpretation of the experiment. His first mistake was to think that the hydrogen came out of the metal. This was a reasonable conclusion to come to since the hydrogen is first observed on the surface of the metal. We now know that the metal is an element; hence no ordinary chemical reaction can get another element out of it. The second mistake made by Cavendish was in thinking that hydrogen is practically pure phlogiston. At this same time Lavoisier was proving that there is no such substance as phlogiston.

In order to find out whether or not the composition of the air is constant, Cavendish made a large number of analyses. In one of them he learned that after all of the water vapor, carbon dioxide, and nitrogen are removed from a sample of air, there remains a small residue of gas which will not react with anything. Years later this inactive part was discovered to be mostly argon, which, as Cavendish discovered, will not react.

in the things about us.   Aside from being one-ninth of the
water, it forms a high percentage of all oil products from gasoline to asphalt or paraffin.   It is in all wood and wood products,
such as paper and pasteboard; it is in all kinds of cloth; it
is in leather and leather products; and it is in almost all kinds
of food, including fruits, vegetables, meats, cereals, sugars,
jams, jellies, milk, and bread.   Besides these it occurs in chewing gum, most medicines, most cos-
metics, celluloid, ivory, cellophane,
rubber, all plants, all animals, and
practically all fuels.

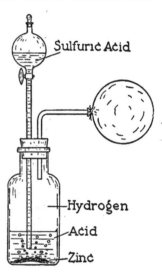

EXPERIMENT

*To demonstrate the lightness of hydro-
gen by inflating a balloon with it.*   Fill
a toy balloon by the method illustrated
in the drawing.   Tie a string to it and
note that it tries to rise to the roof.
The reason for this is that, being lighter
than air, the balloon is buoyed up
as is an empty barrel pushed under
water.

EXPERIMENT

To make soap bubbles that explode when ignited.   To prepare soap
solution, cut into thin shavings 100 g. of pure alkali-free soap and
put into 1 l. of water in a bottle.   Shake until a saturated solution
of soap is obtained.   After it has stood for some time, pour off some
of the soap solution from the sediment and add half its volume of
glycerol.

Fill the large bottle as shown in the drawing two-sevenths full of
hydrogen by displacement of water.   Lift the bottle until the water
runs out and it fills with air.   While the bottle is still inverted, insert
the stopper equipped with inlet tube connected to the funnel for

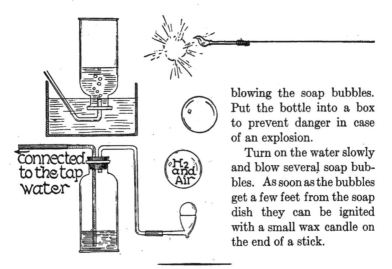

blowing the soap bubbles. Put the bottle into a box to prevent danger in case of an explosion.

Turn on the water slowly and blow several soap bubbles. As soon as the bubbles get a few feet from the soap dish they can be ignited with a small wax candle on the end of a stick.

Other sources of hydrogen. Only a few of the compounds in which it occurs are useful as sources of hydrogen. Water has been mentioned as being one-ninth hydrogen. Electricity can free the hydrogen from chemical combination with oxygen. The high cost of electricity limits the usefulness of this method for many commercial uses of this element. Very active metals, such as sodium, when placed in contact with water will liberate hydrogen rapidly. Here again the usefulness of the reaction is limited by the high cost of sodium. The most practical method of getting hydrogen from water consists of spraying steam on red-hot carbon in the form of coke. The carbon takes the

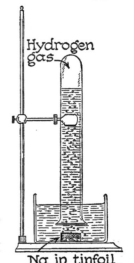

Hydrogen gas

Na in tinfoil

PREPARING HYDROGEN BY THE ACTION OF SODIUM ON WATER. A piece of sodium is wrapped in some tinfoil to prevent too rapid reaction. A few small holes are punched into the foil, so that the water can slowly enter and the hydrogen get out. If the reaction is not thus controlled, the hydrogen forms so rapidly that it will blow the water out of the container.

oxygen away from the hydrogen to form carbon monoxide. The hydrogen usually has to be separated from the carbon monoxide with which it is mixed before it can be used for anything except for burning.

When hydrogen is needed in the laboratory, we get it from the reactions between certain metals and acids. The metals crowd the hydrogen out of the acids and take its place in these compounds. Zinc or iron serve very well for this purpose. This method of preparation is not cheap enough for extensive commercial use. Even if we chose the cheapest metal, iron, and

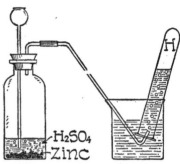

LABORATORY PREPARATION OF HYDRO-GEN. For use in the laboratory, hydrogen is usually prepared by the action of acid on zinc.

the cheapest acid, sulfuric, this process cannot compete with the "water-gas method," as the steam-coke method is usually called, where large quantities of hydrogen are needed in industry.

QUESTIONS OF FACT

1. Name the five most abundant elements in the crust of the earth in the order of their abundance.

2. How can hydrogen be obtained from water by: (a) electricity; (b) sodium; (c) carbon?

3. Name two other methods of obtaining hydrogen.

Which of the next two statements is true and which false?

4. Water is separated by electrolysis into hydrogen and oxygen.

5. The sources of industrial hydrogen are hydrogen peroxide and zinc.

6. Name an element discovered by Henry Cavendish. What was his method?

7. How did Cavendish prove the composition of water?

8. Name two beliefs of Cavendish that are now known to be incorrect.

9. State and explain the law of constant composition.

10. Why is not the method of obtaining hydrogen by zinc and acid suitable as a commercial source of hydrogen?

11. Define the following scientific terms, which have already been introduced in this chapter: (a) synthesis; (b) decomposition; (c) element; (d) compound; (e) symbol; (f) water gas.

12. What is another name for the law of constant composition?

13. Why are some chemical symbols so unlike the present names of the elements?

14. What are the extreme limits of the occurrence of the elements?

15. Name ten substances that contain hydrogen in chemical combination.

QUESTIONS OF UNDERSTANDING

1. Make a list of all the substances you can find around your home that are uncombined elements.

2. Name some substances you have seen which you think are compounds. Give your reasons for so thinking.

3. Turn to the list of elements in the back of the text and make a list of all of them that you have ever heard of. Bring the list to class.

4. Charcoal is largely the element carbon. Sugar is a compound of carbon. In this case which costs the more, the element or the compound?

5. In a copper mine there occurs the compound copper sulfide, which has practically no industrial uses. Copper, however, has many uses. In this case which is worth the more, the element or the compound?

PROJECT

On a large sheet of paper draw a large circle to represent the earth with a radius of 15.9 cm. The circumference of this circle will measure approximately 100 cm., which may be taken to represent 100 per cent. Using the data on page 67 of the text, take a small string and measure off arcs on the circle in centimeters the same in number as the per cents of the elements. From the ends of the arcs draw lines to the center of the circle so each element will be represented by an area looking like a slice of pie. Label each slice with the name of the element. Begin with the smallest, "The rest of the elements," whose arc will be only 1.42 cm. and whose slice will be very small. Join to this one for magnesium with an arc of 2.08 cm. and so on through and including silicon. What is left of the circle can be labeled oxygen.

REFERENCES FOR SUPPLEMENTARY READING

Biographies of Henry Cavendish.
1. Holmyard, Eric John. *Makers of Chemistry.* Oxford Press, 1931. Pp. 177–186.
2. Jaffe, Bernard. *Crucibles.* Simon and Schuster, New York, 1930. Chap. V, pp. 73–92.
3. Leonard, Jonathan Norton. *Crusaders of Chemistry.* Doubleday, Doran, and Co., Inc., Garden City, N. Y., 1930. Pp. 241–261. Very complete and interesting.
   (1) Write a biography of Cavendish, telling more than is told in the text.
4. Foster, William. *The Romance of Chemistry.* D. Appleton-Century Co., Inc., New York, 1936. Chap. VII, "Hydrogen and Water."
   (1) What is the meaning of the root words from which the name hydrogen was derived? (p. 94)
   (2) Who discovered hydrogen? When? (p. 94)
   (3) Where does hydrogen exist besides on the earth?
5. Kendall, James. *At Home among the Atoms.* The Century Company, N. Y., 1929. Chap. IX, pp. 131–140, "Hydrogen, 'Nize Baby.'"
   (1) Explain why Cavendish is said to have discovered hydrogen, whereas Paracelsus had prepared it previously. (p. 132)
6. Martin, Geoffrey. *Triumphs and Wonders of Modern Chemistry.* D. Van Nostrand Co., N. Y., 1932. Chap. VI, pp. 113–131, "The Element, Hydrogen."
   (1) Relate the discovery of inflammable air. (p. 113)
   (2) What was inflammable air? (p. 113)
   (3) Describe hydrogen flames on the sun. (p. 119)

EXPERIMENT

Other methods of preparing hydrogen. (a) *From sodium metal.* Cut a small piece of sodium about 5 mm. on each side. Roll this in a piece of lead foil from a tea wrapper. Press the foil tightly against the sodium and fold over one end, leaving the other end open. Drop into a dish of water and catch the hydrogen by displacement of water as usual.

(b) *From aluminum and sodium hydroxide.* Into a 250 cc. Erlenmeyer flask fitted with a funnel tube and outlet tube put 50 cc. of dilute sodium hydroxide and a few pieces of aluminum. Heat until the reaction starts, and then stop heating. Collect as usual, or a toy balloon may be attached and filled as shown on page 71.

Properties of hydrogen. Like oxygen, hydrogen is a colorless, tasteless, odorless gas. Also, like oxygen, it is only slightly soluble in water. It is the lightest substance known. When mixed with oxygen, hydrogen is very explosive. In the laboratory, students should be careful not to ignite the escaping hydrogen before all the air is driven out of the generating flask,

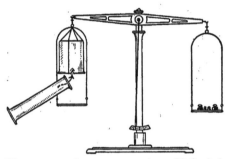

HYDROGEN IS LIGHTER THAN AIR. If the balances are balanced when the beaker on the left is filled with air, the left side becomes lighter when hydrogen replaces the air. (Hydrogen being lighter than air, the beaker must be inverted and the hydrogen poured upward.)

or an explosion may result. Explosions of this kind have been known to blow the flask to small pieces and endanger the worker and nearby students from flying glass and acid. The reaction between oxygen and hydrogen, although of explosive violence when once ignited, does not seem to begin until ignited with a flame. Other chemical reactions of hydrogen will be discussed under subsequent topics.

Uses of hydrogen. Because of its lightness hydrogen has long been used to inflate all kinds of balloons, from children's toys to great Zeppelins. The large dirigibles of England and France are still filled with hydrogen. Although hydrogen has greater lifting power than any other gas, it is very combustible. Because of its combustible nature, several of the large balloons have caught fire and burned. Some were set on fire by lightning; others were lost because of leaking gas, which was ignited by sparks from the engine. The danger of fire limits the use of the hydrogen-filled dirigible for war purposes. The United States Government is recovering the inactive gas helium from natural gas for its dirigible balloons. Helium is twice as heavy as hydrogen and has only about 92 per cent of the lifting power of that element. However, helium will not burn, and all danger of explosion is gone. The United States Army has chosen the safety of helium in preference to the higher lifting power of hydrogen. The Euro-pean nations, being without helium, have no choice in the matter.

Hydrogen as a fuel. Hydrogen is sometimes used as a fuel to get a very high temperature. If a stream of oxygen is led into a stream of burning hydrogen, there is produced one of the hottest flames that is obtainable by combustion. The reason hydrogen gives such a hot flame is that it gives more heat units per gram than any other fuel.

The device used for burning hydrogen in oxygen is called the oxyhydrogen blowpipe. The oxyhydrogen blowpipe is used to melt platinum, which is difficult to melt by other methods. It is also used in the manufacture of artificial gems from difficultly fusible materials. It will cut through steel faster than a hack saw.

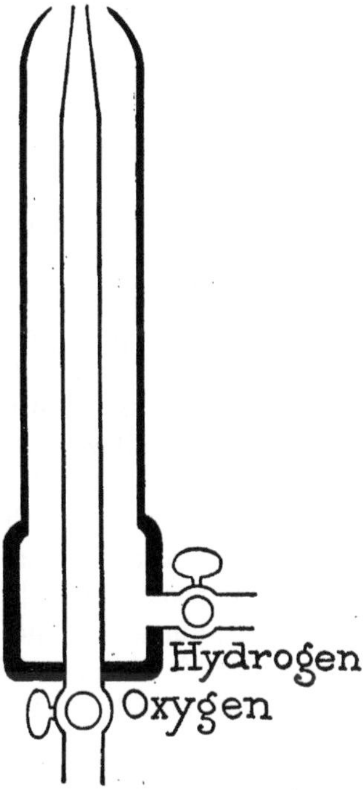

Hydrogen is used as a fuel, moreover, in conjunction with carbon monoxide, in *water gas*, so

OXYHYDROGEN BLOWPIPE. Hydrogen comes through the tube at the right, and oxygen enters at the bottom. The oxygen comes out as a stream inside the cone of hydrogen. The hydrogen burns both inside and outside, the rapid combustion producing an intense heat.

called because it is made by spraying steam upon white-hot coke (carbon). The following reaction takes place :

carbon + steam = carbon monoxide + hydrogen.

Both of the products are combustible and are excellent fuels. As yet, we cannot produce hydrogen cheaply enough to use it alone for ordinary heating purposes.

Hydrogen has one property that makes it rather poorly suited for industrial fuel purposes. It will leak out through small openings more rapidly than any other. gas. It escapes from containers that safely hold other fuel gases. The loss of the substance itself is not the worst thing about this tendency to escape. Mixtures of hydrogen and air are very explosive, so that the use of hydrogen is often accompanied with danger.

Hydrogen used to make new cooking fats. Hydrogen finds an industrial use in the hydrogenation of oils. As was mentioned in Chapter 1, cottonseed oil is produced in large quantities. Formerly there was little market for it. The housewife was encouraged to use it for cooking purposes, but she found it too oily in taste and texture. She liked the semisolid fats like lard and tallow much better. Dr. Wesson found that if this oil is heated with hydrogen in the presence of certain catalysts, it unites with the hydrogen and changes into a clean, snow-white fat a little harder than lard but softer than tallow. This new cooking fat shows less tendency to become rancid than the animal fats and has won almost complete favor with the housewife for all cooking requirements.

Hydrogen used in making ammonia. Each year sees hydrogen more in demand by industry. In the last score of years all of the leading nations have developed plants for making nitrogen compounds out of the nitrogen of the air. This is called *fixation of nitrogen*. All explosives are nitrogen compounds, and nearly all fertilizers contain them. Hence they are very essential to a nation's progress in peace and war. If it had not been for their ability to fix atmospheric nitrogen, the Central Powers of Europe could not have held out six months in the World War. The most successful process for fixing nitrogen is the Haber process. The *Haber process* uses hydrogen to unite with the nitrogen as the equation shows.

$$\text{nitrogen} + \text{hydrogen} \longrightarrow \text{ammonia}.$$

The success of this process also was due to finding the right catalyst. The reaction was known for years before it was made a commercial success. The catalyst turned out to be a mixture which is principally finely divided iron and aluminum.

Sources of cheap hydrogen. The big problem today in connection with this process is that of getting the hydrogen at a price cheap enough to make a cheap product. One of the cheapest sources of hydrogen in the United States is natural gas. Natural gas is mostly methane, a compound of hydrogen and carbon. Methane can be made to break up and release its hydrogen under certain conditions of temperature and pressure. Over a trillion cubic feet of natural gas are obtained

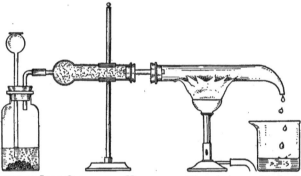

REDUCTION OF IRON OXIDE WITH HYDROGEN. There is iron oxide in the tube above the flame. Hydrogen from the generator (left) is passed through the drying tube into the heated tube where it reduces the iron oxide, leaving iron in the tube and carrying away the oxygen as water vapor, which partially condenses into the beaker (right).

in the United States yearly. Some of this is used in nitrogen fixation. When the ammonia is once obtained by the fixation process, other nitrogen compounds are readily made from it.

A German chemist named Bergius has worked out a process whereby hydrogen and low-grade soft coal can be heated together in the presence of catalysts to form oillike products. This oil mixture can be used in place of gasoline. A few years ago automobile engineers were worried lest our gasoline supply should give out. Today they are not so worried; when production from our oil wells falls off, the Bergius process can be used to make up the deficiency. The synthetic gasoline will be nearly as cheap as the natural product.

QUESTIONS OF FACT

1. Discuss hydrogen as a fuel.
2. Give two uses of hydrogen in addition to its use as a fuel.
3. Name two commercial processes of getting hydrogen from its compounds.

Are the following statements true or false? Judge each one individually.

4. Water gas used for fuel purposes is steam.
5. The cheapness of hydrogen is the principal reason it is used for inflating balloons.
6. Its explosive nature and its tendency to escape through the bag make hydrogen unsuited for balloons.
7. The United States Navy uses helium in its balloons in preference to hydrogen because it is not explosive.
8. Two uses of the oxyhydrogen blowpipe are to cut through steel and to melt quartz. ·
9. Vegetable shortening is made by the hydrogenation of cooking fats.
10. Nitrogen fixation is causing nitrogen to react with other elements.
11. The Bergius process is the fixation of nitrogen by causing it to unite with hydrogen.
12. The Haber process is the formation of fuel oils by treating coal with hydrogen.

QUESTIONS OF UNDERSTANDING

1. Upon what does the value of the oxyhydrogen blowpipe depend?
2. Why does a nation want plants for nitrogen fixation?
3. What does the Bergius process mean to the future of the automobile industry?
4. Name three industrial processes which use hydrogen and which require catalysts for their success.
5. What does the fact that we collect hydrogen over water tell us of its solubility in water?
6. What is the advantage in being able to collect a gas over water?
7. What kind of gases cannot be collected over water?
8. List five physical properties of hydrogen.
9. Why does pure hydrogen burn quietly in air while a mixture of air and hydrogen explodes?

REFERENCES FOR SUPPLEMENTARY READING

1. Foster, William. *The Romance of Chemistry.* D. Appleton-Century Co., Inc., New York, 1936. Chap. VII, "Hydrogen and Water."
   (1) At what temperature does a liquid boil? (p. 97)
   (2) Describe what happens when hydrogen gas blows upon finely divided platinum exposed to the air. (p. 97)
2. Holmes, Harry N. *Out of the Test Tube.* Ray Long and Richard R. Smith, New York, 1934. Chap. VI, pp. 66–78, "The Lightest Substance Known."
   (1) Tell about a foolish reply of a college student and the professor's humorous comment. (p. 71)
   (2) How is most industrial hydrogen obtained in the United States? (p. 71)
   (3) Tell of the first balloon flights. (pp. 73–74)
   (4) Tell of the relation of hydrogen to fertilizers. (p. 74)
   (5) The cartoon on page 76 reads: "Seven million hogs resent hydrogenation." Explain why. (p. 76)
   (6) How may hydrogen help the oil industry? (p. 77)
3. Kendall, James. *At Home among the Atoms.* The Century Company, N. Y., 1929. Chap. IX, pp. 131–140, "Hydrogen, 'Nize Baby.'"
   (1) Tell how Priestley used to amuse his friends after he learned of Cavendish's discovery. (p. 132)
   (2) How close did Priestley come to being the discoverer of the fact that water is formed when hydrogen burns? (p. 132)
   (3) When Cavendish did make this discovery, in what predicament did he find himself, in trying to explain his discovery in terms of the phlogiston theory? (p. 133)
   (4) What name means "water producer"? (p. 134)
   (5) What is the "water gas" reaction? (p. 138)
4. Martin, Geoffrey. *Triumphs and Wonders of Modern Chemistry.* D. Van Nostrand Company, N. Y., 1922. Chap. VI, pp. 113–131, "The Element, Hydrogen."
   (1) Explain the cause of a serious explosion. (p. 117)
   (2) What could the burning of one pound of hydrogen do? (p. 125)
   (3) How would a cork act in liquid hydrogen? (p. 126)

Reduction. When the blacksmith heats a piece of iron too hot, the outside iron burns and forms black brittle scales. In our study of oxygen we learned that this is iron oxide. If we heat this iron oxide in a tube and at the same time pass hydrogen gas over it, the hydrogen takes the oxygen away and sets the iron free. This removal of oxygen from a compound is called *reduction.* Reduction is the opposite of oxidation, which, as will be recalled, is the addition of oxygen. Anything which removes oxygen from its compounds is called a *reducing agent*. Hydrogen is often used as a reducing agent in the laboratory.

At the instant hydrogen is released from a compound, it is more active as a reducing agent than it is a little later. During this time when the hydrogen is abnormally active, it is said to be in the *nascent* condition and is called *nascent hydrogen*.

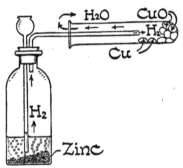

The explanation of this increased activity is that the hydrogen momentarily exists as separate particles before it has time to join with other particles to form hydrogen gas.

In the formation of hydrogen gas a reaction takes place; hence further chemical reaction is impossible until the hydrogen is again separated into unattached particles. In practice the substance to be reduced is mixed with some acid and zinc. The zinc and acid react to free the hydrogen, which reduces the other substance.

REDUCTION OF COPPER OXIDE BY HYDROGEN. When the hydrogen comes in contact with heated copper oxide, it removes the oxygen, leaving metallic copper.

Hydrogen-oxygen compounds. There are two compounds that are composed of hydrogen and oxygen only — water and hydrogen peroxide. Water is by far the more common of the two; but as it was studied in some detail in the previous chapter, we will now turn our attention to hydrogen peroxide.

Hydrogen peroxide. Hydrogen peroxide is a substance richer in oxygen than water. The former is 94.1 per cent oxygen as compared with 88.9 per cent for the latter. This extra oxygen seems to be loosely held in the peroxide. In the presence of oxidizable substances or even on standing, the extra oxygen is released and only water remains. In other words, hydrogen peroxide has decomposed into oxygen plus water. As this behavior would suggest, this substance is not very stable or permanent. To reduce this tendency to decompose, the manufacturers add a little sulfuric acid to the peroxide. In spite of this, however, after the bottle has stood some time, the cork may pop out because of the pressure of the oxygen that has been released in the bottle. The end of the cork in a bottle of peroxide is always bleached white and often partially decomposed by the oxidizing action of the oxygen, which is slowly released from the hydrogen peroxide.

EXPERIMENT

Preparation and properties of hydrogen peroxide. In a beaker stir a few grams of barium peroxide with some crushed ice. Add enough dilute sulfuric acid to cover the ice. Filter off the hydrogen peroxide solution.

(a) To a very dilute solution of potassium dichromate containing a few drops of dilute sulfuric acid, add 1 cc. of hydrogen peroxide solution. The intense blue of perchromic acid is the test for hydrogen peroxide.

(b) To a small quantity of freshly prepared lead sulfide suspended in water add a few cubic centimeters of hydrogen peroxide solution. The black lead sulfide is oxidized to white lead sulfate.

As in the case of nascent hydrogen, freshly released or nascent oxygen is more active than the ordinary oxygen gas. Nascent oxygen is one of the most effective bleaches we have.

Uses of peroxide. The principal uses of hydrogen peroxide are as a bleach and as a germ killer. Hydrogen peroxide is often used for bleaching the hair. Many of the extreme blondes are "peroxide blondes." The germicidal action of this substance is also due to the nascent oxygen. Although the hair usually survives bleaching, this process is fatal to many disease bacteria. When used in a mouth wash, peroxide is characterized by foaming, the foam consisting of bubbles of oxygen.

QUESTIONS OF FACT

1. The following scientific terms were introduced in this division of the chapter. Define each: (a) reduction; (b) nascent.

2. Compare oxidation and reduction.

3. Compare hydrogen peroxide and water: (a) as to stability; (b) as oxidizing agents.

4. Why is a little sulfuric acid added to the commercial peroxide?

5. Commercial hydrogen peroxide is a solution. What is the solvent?

6. What is one of the most effective bleaching agents we have?

7. Name the substances in your home that contain hydrogen.

8. Give two uses of hydrogen peroxide.

QUESTIONS OF UNDERSTANDING

1. How does nascent hydrogen differ in its properties from gaseous hydrogen?

2. Account for this difference.

3. Considering the fact that three-fourths of the surface of the earth is covered with water which is five miles deep in spots, and water is one-ninth hydrogen, how can hydrogen be listed with 83 other elements which make up less than 2 per cent of the earth's crust?

4. Does the fact that a catalyst is not consumed when it causes reactions to take place prove that it has not been in the reaction and then emerged again whole?

ADDITIONAL EXERCISES FOR SUPERIOR STUDENTS

1. Prepare a report on catalysts.

Howe, H. E. *Chemistry in Industry*, Vol. 2. The Chemical Foundation, Inc., New York, 1926. Chap. I, "Catalysis — A New Factor in Industry" by Elwood Hendrick.

2. Calculate the weight of hydrogen in the water that will be stored behind Boulder Dam if the water will weigh 4,000,000,000 tons. (Water is one-ninth hydrogen.)

3. The United States produces over one trillion cubic feet of natural gas yearly. Each cubic foot of natural gas could produce two cubic feet of hydrogen gas. What volume of hydrogen might be produced yearly?

4. Assume that each cubic foot of hydrogen of problem 3 would weigh .0055 lb. Calculate the total weight.

SUMMARY

Electricity decomposes water into two gaseous substances, oxygen and hydrogen. The oxygen and hydrogen can be made to unite again to form water.   This process is called synthesis. It is the opposite of decomposition.

An element is a substance that cannot be further simplified by chemical reaction.   When two or more elements unite chemically, the resulting substance is called a compound. Every compound always has the same composition, regardless of how it is formed.   This fact, that compounds always have the same composition, is called the *law of definite proportions.*

To simplify the writing of the names of the elements, symbols are adopted of one or two letters of the English or Latin name.

Hydrogen may also be obtained by the action of sodium or another active metal on water.   Steam acting on hot iron also produces hydrogen.

Hydrogen is the lightest substance known.   It is explosive with oxygen.   Its principal uses are to inflate balloons, as fuel in the oxyhydrogen blowpipe, to make cooking fats, and to synthesize ammonia from nitrogen.

Reduction is defined as the removal of oxygen from a substance.   Freshly released hydrogen, being more active than it is later, is called nascent hydrogen.

Hydrogen peroxide is a rather unstable compound of hydrogen and oxygen only.

It is used to bleach hair and to kill germs.

←
→

CHAPTER 5

MOLECULES AND ATOMS

Molecules. When one tablespoonful of water is changed into steam, more than twenty-five quarts of water vapor are formed. We ask ourselves how it is possible for one tablespoonful of liquid to expand to twenty-five quarts of vapor. We know that if steam is kept in a smaller volume than that just mentioned, its tendency to expand exerts a large pressure, enough to drive the piston of a steam engine. All other gases possess very large volumes compared to their volumes in the liquid state.

There is a theory of gases, which explains how a gas can occupy so large a volume compared to that of the liquid from which it was produced. In terms of this theory every gas consists of a large number of moving particles, usually at great distances apart as compared to their diameters. The pressure of the gas against the walls of the containing vessel is produced by the collisions of these particles with the walls of the vessel. An automobile is held up by a very large number of air particles in the tire continually striking the fabric and keeping it pushed apart. These particles are called molecules. Molecules are too small to be seen, even under the most powerful microscope. This explains why air and water vapor are invisible. In a liquid, where the molecules are quite close together, we do not see molecules as individuals, but we do see large collections of them. In a tablespoonful of water there are about 500 sextillion molecules, a number far beyond our comprehension. A consideration of this number gives us a better idea of how a gas can exert a pressure. The magnitude of the pressure can be explained in terms of the large number of impacts in a very short time. Molecules explain other interesting facts. For instance, heat is partly the motion of these molecules. When something burns us, the injury is caused by the molecules striking our skin like a multitude of small bullets.

From the fact that gases, such as steam, can be changed into liquids, such as water, we conclude that liquids, too, are composed of molecules. Liquids can be changed into solids, as when water freezes to ice. It is natural, therefore, to conclude that solids also are composed of molecules. The idea that a dense solid such as glass should be composed of a large number of separate, individual particles seems a little surprising. Yet there is evidence that glass really is porous and not continuous as it appears. If a sealed, empty glass bottle is lowered to the bottom of the ocean, it will be found to contain a little water when it is brought to the surface. The water could only have come through invisible pores in the glass — that is, between the molecules.

In terms of the molecular theory, a molecule of water is a small piece of water, really the smallest piece of water capable of existing. The situation is the same with other substances. A molecule of chalk is the smallest possible piece of chalk, a molecule of salt is the smallest possible piece of salt, and a molecule of sugar is the smallest possible piece of sugar. All molecules of a pure substance are thought to be alike, but different from the molecules of all other substances. Although there are scores of facts that prove the existence of molecules, scientists still call the idea a theory, since the molecules cannot be seen as individuals.

The atomic theory. We learned that electricity decomposes water into hydrogen and oxygen. If water is made up of molecules which are all alike, each molecule must have some hydrogen and some oxygen. The pieces of oxygen and hydrogen that compose the molecule must undoubtedly be smaller than the molecule. These small pieces of the elements that go to make up the molecules are called the atoms of the elements.

About 1804 John Dalton announced his atomic theory. According to this theory an atom is defined as: *The smallest particle of an element that can unite with particles of other elements to form molecules of a compound.* The Dalton atomic theory contains several points:

(1) Every element is ultimately composed of very small particles called atoms.

(2) The atoms of the same element are practically all alike.

(3) The atoms of one element are different from the atoms of every other element.

(4) The atoms of different elements combine chemically to form the molecules of the compounds.

(5) The atoms have definite, fixed, and unchangeable weights.

(6) The atoms are indestructible. (As will be shown in Chapter 19, this idea has recently been modified.)

For about seventy years after the atomic theory was stated, it was thought that every atom of any element had exactly the same weight as every other atom of this element. The most recent research shows, however, that there are atoms of slightly different weights of the elements. For instance, chlorine has some of its atoms weighing 35 atomic units and some 37 atomic units. Bromine is composed of atoms some of which weigh 79 atomic units and some 81. Tin has atoms weighing 112, 114, 115, 116, 117, 118, 119, 120, 121, 122, and 124 atomic units. However, these atoms are nearly always found in the same proportions, so we can think of their having only one atomic weight — that is, the average of the mixture. In the case of chlorine it is 35.46 atomic units. Bromine and tin have 79.92 and 118.7, respectively, as average atomic weights.

Atoms of the same element with different atomic weights are called *isotopes*. Usually one of the isotopes of an element is much more abundant than the others. As a consequence of this fact the average atomic weight is nearer that of one of the isotopes than that of any of the others. For instance, hydrogen with isotopes of atomic weight 1, 2, and 3 has so many more atoms of atomic weight 1 than of the other two that the average atomic weight is 1.008.

All atoms and molecules are too small to be seen with the unaided eye. It would probably take billions of them to make a pile large enough to become visible even under the microscope, with the possible exception of the largest molecules, which may be visible in the ultramicroscope. Even then we are not sure but that we are looking at a cluster of molecules instead of a single one.

Although the individual atoms are too small to be seen (and there is reason to believe that we never will be able to see them), they are just as real to the scientist as a tree or an elephant. There are thousands of scientific facts that can be explained only in terms of atoms. Over one hundred years of intensive testing by the world's best chemists has established the atomic theory as a reality. There seems little reason for calling it a theory now, since it has been so firmly established; but since the atoms are too small to be seen, most scientists still speak of it as the *atomic hypothesis* or theory. Perhaps this is being overly cautious, and we should speak of it as a fact or law.

Molecules of an element. The student is already familiar with the expressions "atoms of an element" and "molecules of a compound," but he may be a bit confused by the term "molecules of an element." Just as a molecule of water, H_2O, is composed of two atoms of hydrogen and one atom of oxygen, so is a molecule of hydrogen, H_2, composed of two atoms of hydrogen joined together. Similarly, a molecule of oxygen, O_2, is made up of two atoms of that element. All elementary gases have two atoms to the molecule except a few inactive ones. We do not know how many atoms are in the molecules of solid elementary substances so we assume one atom in each molecule.

QUESTIONS OF FACT

1. What are the particles of a gas called?
2. Give the reason why liquids and solids are also thought of as being composed of molecules.
3. Define molecule and atom.
4. Why do we speak of the atomic theory, if so many things show the reality of atoms.
5. State the six points of the atomic theory.
6. Which points have been modified since Dalton's time.
7. What constitutes a molecule of hydrogen?

John Dalton

1766–1844

Following close on the heels of Priestley, Lavoisier, and Cavendish came a humble Quaker schoolmaster, who turned out to be one of the outstanding scientists of all time, John Dalton. Born into a poor family, which made it necessary for the children to earn something as soon as possible, John Dalton at the age of twelve with his older brother opened a school, which announced that English, Latin, Greek, French, writing, arithmetic, bookkeeping, and mathematics would be taught. Because of his great enthusiasm and gentle nature, John was liked better by the pupils than his older brother, Jonathan.

In 1793 young Dalton obtained the position as tutor in mathematics and natural philosophy at the Manchester Academy. Natural philosophy was a subject more like physics than any other one subject taught today, but it overlapped on to what we now call chemistry.

After six years at the academy, Dalton resigned to become a private tutor, which would allow more time for his researches. His simple, unassuming life did not require much. Because his life was so simple and his dwelling so inconspicuous, his fellow townsmen were hardly aware of his existence. When a great French scientist who had heard of the new theories of Dalton came to see him, the Frenchman had difficulty in learning where Dalton lived.

Dalton's interest in the study of gases was aroused by his hobby of making observations on the weather. The atomic theory, which was Dalton's greatest achievement for science, resulted from his study of gases. Dalton's atomic theory, like all great scientific conceptions, developed from a few hazy suggestions into clear-cut ideas. The theory itself consists of these ideas (see page 88).

By 1822, scientists the world over had accepted Dalton's theory. Honors and awards now began to be showered upon the modest schoolmaster.

8. What is the molecular structure of most elementary gases?

9. Explain the term isotopes.

10. What are the isotopes of hydrogen?

11. The experimental atomic weight of chlorine is 35.45. Which isotope occurs in the larger number?

QUESTIONS OF UNDERSTANDING

1. Explain the difference between a molecule and an atom. Illustrate in the case of water.

2. How does a gas exert a pressure?

3. Has anyone ever seen an atom?

4. Has anyone ever been able to isolate a single atom?

5. What reason is there for thinking a molecule of water is heavier than an atom of oxygen?

ADDITIONAL EXERCISES FOR SUPERIOR STUDENTS

1. Write a three-page essay on John Dalton.

2. The number of molecules in 18 ml. of water is often written as 6.06×10^{23}. Write this out in the usual way.

3. Using the figures in the previous question, calculate the number of molecules in 1 cc. (ml.) of water.

4. Assuming that there are 25 drops of water in 1 cc., calculate the results of question 3 in terms of the number of molecules in a drop of water.

5. Assuming that on a clear night 6000 stars are visible to the unaided eye, calculate how many times this number expresses the molecules in a drop of water.

REFERENCES FOR SUPPLEMENTARY READING

1. Holmyard, Eric John. *Makers of Chemistry.* Oxford University Press, New York, 1931. Pp. 221–240, "John Dalton."

2. Harrow, Benjamin. *The Making of Chemistry.* The John Day Company, New York, 1930. Chap. IX, pp. 53–59, "Dalton and the Atomic Theory."

3. Jaffe, Bernard. *Crucibles.* Simon and Schuster, New York, 1930. Chap. VII, pp. 114–135, "A Quaker Builds the Smallest of Worlds." (Dalton)

4. Fisk, Dorothy M. *Modern Alchemy.* D. Appleton-Century Co., New York, 1936. Chap. I, pp. 24–39, "The Discovery of the Atom."

(1) What two theories as to the nature of matter are possible? (p. 24)

(2) What was the one confusing point in Dalton's discussion of atoms? (p. 30, bottom)

(3) Who cleared up this confusing point? (p. 30, bottom)

(4) Describe Dalton's conception of the atom. (p. 31)

(5) Give experimental evidences of atoms. (pp. 35–39)

5. Kendall, James. *At Home among the Atoms.* The Century Co., New York, 1929. Chap. VI, pp. 75–90, "The Mighty Atoms"; Chap. X, p. 141 to the middle of page 147, "Valencia."

(1) State the atomic hypothesis of Democritus. (pp. 76–77)

(2) Was this hypothesis the outcome of experimentation?

(3) State the opposite hypothesis of other Greek philosophers. (p. 77)

(4) State President John Adams's understanding of atoms and molecules. (pp. 77–78)

(5) Name two assumptions regarding atoms that Dalton added to those of Democritus. (p. 78)

6. Langdon-Davies, John. *Inside the Atom.* Harper and Brothers, New York, 1933. Pp. 2–56, A picturesque story of molecules and atoms, written from the point of view of the beginner.

(1) Express one great question that confronts us. (pp. 4–6)

(2) What is the true answer to this great question? (pp. 6–9)

(3) Compare the arrangement of nature's bricks to that of man's bricks. (pp. 13–14)

(4) How often do the dancing atoms of a gas collide? (p. 20)

(5) Under what condition do the molecules of substance dance? (pp. 22–25)

(6) Why does ice melt? (pp. 25–27)

(7) How can we make molecules go faster? (pp. 27–29)

(8) Tell about the dancing building bricks in an automobile tire. (pp. 41–44)

(9) How can we make the molecules do useful work? (pp. 47–48)

Restatement of laws.   We are now in a position to restate the law of definite proportions in terms of atoms and molecules. The reason a pure substance always has the same composition is that all of its molecules are made of the same number and kinds of atoms. The law of conservation of mass can be stated thus: The total number of all kinds of atoms in the substances that react are the same as the total number in the products of the reaction; that is, atoms are neither created nor destroyed.

Formulas.   A formula expresses the composition of a molecule of a substance in chemical symbols. It gives the number

John Jacob Berzelius

1779–1848

At the university Berzelius became interested in experimental chemistry. Although he wanted more time for experiments, the instructor allowed him to work only during the regular periods with the class. One day, being very anxious to try some experiments, he slipped into the laboratory when no one was there. The instructor came in unawares and watched him for a while without making himself known. Soon, however, he spoke to the startled student, who expected to be expelled. However, the instructor only pretended to scold him, and gave him permission to work in the laboratory overtime. In these experiments Berzelius discovered a new gas, nitrous oxide.

Next, while repeating Galvani's experiment of making frogs' legs twitch by touching them with metals, he discovered many of the facts of electrolysis. The scientific world, however, failed to notice his discoveries until Davy repeated them and used the ideas to discover sodium. No notice was taken of Berzelius' discovery of nitrous oxide either until Davy discovered its peculiar intoxicating properties and named it laughing gas. In 1802 he published some chemical researches which both made him famous and created for him a position as assistant professor of chemistry and pharmacy in the medical school at Stockholm. Five years later he became a full professor. In 1818 he was made a nobleman by the king, and in 1835 a baron.

The greatest contribution to science by Berzelius was the introduction of modern chemical symbols and formulas. These made Dalton's atoms more real and useful than they had been previously.

Berzelius determined a large number of atomic weights with remarkable accuracy. Before his time, many of the atomic weights were very uncertain, due both to inaccurate methods of analysis and to lack of knowledge of the number of atoms in the molecule.

of each kind of atom. When there is more than one atom of a kind in the molecule, a number is placed to the right and below the symbol of that element. As we have already seen, H_2O, the formula for water, means that the molecule of water consists of two atoms of hydrogen combined with one atom of oxygen. Likewise, $KClO_3$, the formula for potassium chlorate, tells us that the molecule of potassium chlorate contains one atom of potassium, one atom of chlorine, and three atoms of oxygen chemically united. The formula for hydrogen peroxide, H_2O_2, means that each molecule consists of two atoms of hydrogen and two atoms of oxygen in chemical union. Zinc sulfate has the formula $ZnSO_4$, meaning one atom of zinc, one atom of sulfur, and four atoms of oxygen are in each molecule.

Valence. Because hydrogen is the lightest element known, we often compare the other elements with it. One atom of hydrogen combines with one atom of chlorine. In comparing the capacity of atoms for reacting with one another, we begin by saying the holding power of hydrogen for other atoms is one. This holding power of an atom for other atoms is called its valence. The valence of hydrogen then is one. The chlorine atom and every other atom that unites with a hydrogen atom, in a one-to-one ratio, also has a valence of one and is said to be *monovalent*. The calcium atom unites with two chlorine atoms, hence its valence is two. It and all other atoms that unite with two chlorine atoms are called *divalent*, meaning they have a valence of two. Aluminum chloride has the formula $AlCl_3$, hence the valence of aluminum is three. The chloride of carbon is CCl_4, which shows the valence of carbon is four. For some cause, not known, no common element has a valence above seven. However, there are two quite rare elements that have valences of eight. For theoretical reasons scientists think there are no valences higher than eight.

Positive and negative parts of compounds. As a general rule, dissimilar elements react readily and form stable compounds, while similar elements either do not react or, if they do, form only unstable compounds. This has led chemists to divide the elements into two groups, one group spoken of as the positive and the other the negative elements. This classi-fication is not absolute; that is, an element may be negative when compared with a more positive element, yet positive when compared with a more negative one. This means that

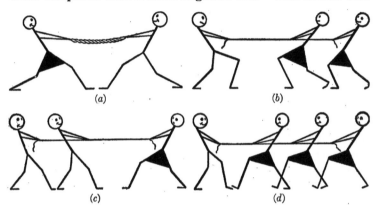

VALENCE. (*a*) This illustration shows a tug of war between a man and woman of equal strength or holding power. A compound, such as hydrogen chloride (HCl), in which one positive atom, hydrogen, holds one negative atom, chlorine, represents a situation similar to that shown in the cartoon. Many elements com-bine with each other in a one-to-one ratio. All those elements that are equiv-alent in holding power to hydrogen are said to be *monovalent* — that is, having a valence of one. (*b*) This cartoon shows a man strong enough to hold his own against two women in a tug of war. The atom of a positive element, such as calcium, unites with two negative chlorine atoms or with two atoms of any other monovalent element. Any positive element that unites with two monovalent negative elements is said to have a positive valence of two and to be *divalent*. (*c*) Here we have a strong woman able to hold two men. This is supposed to illustrate such compounds as water (H_2O), in which one negative atom is united with two positive atoms, each with a valence of one. Oxygen and all negative elements with equal holding power are said to be divalent. Thus we see there are both divalent positive elements and divalent negative elements. (*d*) Here we have a man with strength equal to that of three women. This man represents one of those positive elements with a valence of three. Aluminum is such an element and forms such compounds as aluminum chloride ($AlCl_3$). Atoms with a valence of three are said to be *trivalent*.

we can speak of negative valences and positive valences. Each element with positive valences is often labeled with a $+$ over it for each positive valence, as $\overset{+}{N}a$, $\overset{++}{C}a$, and $\overset{+++}{A}l$. Like-wise the negative elements are often marked with a negative sign for valences, as $\overset{-}{C}l$, $\overset{=}{O}$, and $\overset{=}{S}$. This knowledge of the number and kind of valences is necessary when writing the formulas of the different compounds. As will be shown in a future chapter, the positive and negative charges have more meaning than we have stated as yet. For the present we can think of the numbers of plus and minus charges as giving the valence.

Elements having more than one valence. Some elements have different valences in their compounds. For instance, mercury has a valence of one in part of its compounds and two in the rest. Copper also has a valence of both one and two. Iron has a valence of two in some of its compounds and three in the rest. Tin has a valence of two in some of its compounds and four in the rest. In all of these cases of two valences for the same element, the lower valence form is given the ending *ous* and the higher valence form the ending *ic*.

COMMON ELEMENTS WITH TWO VALENCES

VALENCE	ELEMENT	SYMBOL	COMPOUNDS
1	mercury	$\overset{+}{Hg}$	mercurous
2	mercury	$\overset{++}{Hg}$	mercuric
1	copper	$\overset{+}{Cu}$	cuprous
2	copper	$\overset{++}{Cu}$	cupric
2	iron	$\overset{++}{Fe}$	ferrous
3	iron	$\overset{+++}{Fe}$	ferric
2	tin	$\overset{++}{Sn}$	stannous
4	tin	$\overset{++++}{Sn}$	stannic
3	phosphorus	$\overset{+++}{P}$	phosphorous
5	phosphorus	$\overset{+++++}{P}$	phosphoric

Radicals. In many chemical compounds, a group of elements may take the place of a single element. There are hundreds of compounds with the same group of elements occurring in them. In chemical reactions these groups may move from one compound to another without the individual elements separating from one another. These groups are called *radicals*. We might illustrate the manner in which a group of several individuals may take the place of one by using horses. Although one horse might serve to pull a cart, a team of two are needed to draw a farm wagon and a team of four to haul a large load. Such a team might be unhitched as a group and hitched to another wagon. It would correspond to the radical. Some common radicals are OH, the hydroxyl radical; NH_4, the ammonium radical; NO_3, the nitrate radical; and SO_4, the sulfate radical.

Common atoms and radicals. The following is a list of common elements and radicals that the student will need to know before being able to write formulas for the common compounds.

Valences of Common Elements and Radicals

Atoms and radicals whose valence is one.

$\overset{+}{H}$	hydrogen	$\overset{+}{Ag}$	silver
$\overset{+}{Na}$	sodium	$\overset{-}{Cl}$	chloride
$\overset{+}{K}$	potassium	$\overset{-}{Br}$	bromide
$\overset{+}{Hg}$	mercury (mercurous)	$\overset{-}{I}$	iodide
$\overset{+}{Cu}$	copper (cuprous)	$\overset{-}{OH}$	hydroxide
$\overset{+}{NH_4}$	ammonium	$\overset{-}{NO_3}$	nitrate

Atoms and radicals with a valence of two.

$\overset{++}{Hg}$	mercuric	$\overset{++}{Sr}$	strontium
$\overset{++}{Cu}$	copper (cupric)	$\overset{++}{Pb}$	lead
$\overset{++}{Zn}$	zinc	$\overset{=}{S}$	sulfide
$\overset{++}{Ca}$	calcium	$\overset{=}{SO_3}$	sulfite
$\overset{++}{Mg}$	magnesium	$\overset{=}{SO_4}$	sulfate
$\overset{++}{Fe}$	iron (ferrous)	$\overset{=}{O}$	oxide
$\overset{++}{Sn}$	tin (stannous)	$\overset{=}{CO_3}$	carbonate
$\overset{++}{Ba}$	barium		

Other atoms and radicals.

$\overset{+++}{Fe}$	iron (ferric)	$\overset{++++}{Sn}$	tin (stannic)
$\overset{+++}{Al}$	aluminum	$\overset{++++}{C}$	carbon
$\overset{+++}{P}$	phosphorus (phosphorous)	$\overset{++++}{Si}$	silicon
$\overset{=}{PO_4}$	phosphate	$\overset{+++++}{P}$	phosphorus (phosphoric)

Naming substances whose formulas are given. The name of each substance is usually composed of the name of the positive part followed by that of the negative part. The greatest problem in naming is to tell whether it is the *ous* compound or an *ic* compound in those cases where it may be either. This can be told from the valence of the negative part with which it is paired. The student should study the following table.

NAMES OF COMMON COMPOUNDS

FORMULA	NAME	FORMULA	NAME
$HgCl$	mercurous chloride	$Ca(NO_3)_2$	calcium nitrate
$HgCl_2$	mercuric chloride	$Fe(NO_3)_2$	ferrous nitrate
$CuCl$	cuprous chloride	$Fe(NO_3)_3$	ferric nitrate
$CuCl_2$	cupric chloride	Na_2SO_4	sodium sulfate
$SnCl_2$	stannous chloride	$(NH_4)_2SO_4$	ammonium sulfate
$SnCl_4$	stannic chloride	$AlPO_4$	aluminum phosphate
PBr_3	phosphorous bromide	$Ba(OH)_2$	barium hydroxide
PBr_5	phosphoric bromide		

QUESTIONS OF FACT

1. If we need to show three oxygen atoms in a formula, where do we put the 3?

2. How could one show four sulfur atoms in a formula?

3. Define valence.

4. When an element has two different valences, how is this fact recognized in naming its compounds?

5. What chemist first used modern chemical symbols?

6. Name five elements which have two valences, and give the valences in each case.

7. In the customary naming of chemical compounds, which is named first, the negative or the positive part?

8. Define monovalent and divalent elements.

9. Define the term radical.

10. What is the highest negative valence shown on page 97?

11. What is the highest positive valence shown on page 97?

QUESTIONS OF UNDERSTANDING

1. Explain the difference in meaning between: (*a*) O_2 and $2\,O$; (*b*) O_3 and $3\,O$.

2. Show how the law of constant composition is explained by the atomic theory.

3. State the law of conservation of matter in terms of Dalton's atomic theory.

4. How many atoms of each kind are in the formula $Na_2B_4O_7$?

5. Explain when the name of a substance is given the ending *ous* and when *ic*.

6. Pick out the radicals listed on page 97 and give the valence of each.

7. In the list what is the only positive radical?

8. What is the highest valence any element ever has?

9. What is the valence of P in the formula PCl_5?

10. What is the valence of radium in radium oxide, whose formula is RaO?

11. What is the valence of sulfur in SO_3?

12. What is the valence of osmium in OsO_4?

13. Explain how the chemical symbols introduced by Berzelius were superior to those used up to his time.

REFERENCES FOR SUPPLEMENTARY READING

1. Holmyard, Eric John. *Makers of Chemistry*. The Clarendon Press, Oxford, 1931. Pp. 221–240, "John Dalton"; pp. 240–248, "Johann J. Berzelius."

2. Jaffe, Bernard. *Crucibles*. Simon and Schuster, New York, 1930. Chap. VII, pp. 114–135, "Dalton (A Quaker Builds the Smallest of Worlds)"; Chap. VIII, pp. 136–156, "Berzelius (A Swede Tears Up a Picture Book)." Life and work of Berzelius.
   (1) Write a more extensive biography of Berzelius than is given in the text.

3. Kendall, James. *At Home among the Atoms*. The Century Co., New York, 1929. Chap. VI, pp. 75–90, "The Mighty Atoms"; Chap. X, pp. 141–147, "Valencia."
   (1) State the situation which demands the idea of valence. (p. 142)
   (2) Give one way of representing valences in a compound. (p. 143)
   (3) Describe the cartoons representing valences. (pp. 144–145)
   (4) What kind of element or radical forms the positive part of a compound? (p. 146)
   (5) What kind of element or radical forms the negative part of a compound? (pp. 146–147)
   (6) How does the atomic hypothesis account for two oxides of carbon? three oxides of iron? (pp. 79–80)

(7) What does Kendall refer to as the telephone number of an element? (p. 80)

(8) What mistake did Dalton make which made it hard for contemporary chemists to accept his theory? (p. 81)

(9) How did poor experimental technique hinder the adoption of the atomic theory? (pp. 82–83)

(10) If we cannot isolate and weigh a single atom, how can we compare atomic weights? (p. 83)

Writing formulas when the name is given. The problem in writing formulas is largely one of choosing the right number of each part to make the positive valences the same number as the negative. We might state as the fundamental rule of writing formulas:

There must be the same number of (+) valences on the positive parts of the compound as there are (−) valences on the negative parts.

When the negative element or radical has the same number of minus charges as the positive element or radical has plus charges, the problem is easy, as: $\overset{+}{N}\overset{-}{a}Cl$; $\overset{++}{C}\overset{=}{a}O$; $\overset{+++}{Al}\overset{=}{P}O_4$; $\overset{++}{B}\overset{=}{a}S$. Only one of each is taken and no subscript is needed.

When the positive and negative valences are not the same, the problem is to choose the right number of atoms to make the charges balance, such as $\overset{++}{C}\overset{-}{a}Cl_2$, $\overset{+++}{Al}(\overset{-}{N}O_3)_3$, $\overset{++++}{S}\overset{-}{n}I_4$, $\overset{+++}{Al}_2\overset{=}{O}_3$, $\overset{++++}{S}n_3(\overset{\equiv}{P}O_4)_4$. This is not a very difficult problem in arithmetic. However, it often takes considerable practice before a student is sure of himself.

A look at the above formulas will show the customary place for putting the figure that tells how many atoms are used. A small figure below and a little to the right means that the atom before the figure is to be taken that number of times. If several atoms of a radical are needed, a parenthesis is placed around the radical and the figure placed outside the parenthesis.

A second rule of formula writing is:

Take as many of the positive atoms or radicals as there are negative charges on each negative part and as many of the negative atoms or radicals as there are positive charges on each positive part.

Suppose we want to write the formula for ferric sulfide. As a start we put down one of each part, $\overset{+++}{Fe}\overset{=}{S}$. Next we take 2 of the ferric part and 3 of the sulfide as $\overset{+++}{Fe_2}\overset{=}{S_3}$.

As another illustration let us write the formula for stannic phosphate. The stannic part is $\overset{++++}{Sn}$, while the phosphate part is $\overset{\equiv}{PO_4}$. Following the above rule we write $\overset{++++}{Sn_3}(\overset{\equiv}{PO_4})_4$. This makes the formula have 12 of each kind of charge, the necessary condition for a correctly written formula.

It will be seen that when it is necessary to use more than one of a radical, a parenthesis is placed around the radical and the number placed to the right and below as in the case of a single element.

In a few formulas the second rule results in subscripts which are both divisible by two. In these cases two is divided out and the subscripts are reduced to lowest terms, unless molecular weight determinations demand a double formula.

Suppose we wish to write the formula for stannic oxide. By the second rule we get $\overset{++++}{Sn_2}\overset{\equiv}{O_4}$. Dividing by 2 gives SnO_2 as the correct formula.

Chemical equations. The chemist uses a shorthand method of stating the facts of chemical reactions. In the case of the electrolysis of water the equation, as it is called, would be :

$$2\,H_2O \longrightarrow 2\,H_2 + O_2$$

This equation tells us a great deal very briefly. In short, it says water has been changed into hydrogen and oxygen. More than this, this equation, which may be considered as a sample of reactions involving molecules, tells us that two molecules of water decompose into two molecules of hydrogen gas and one molecule of oxygen. The equation for the preparation of oxygen from potassium chlorate is :

$$2\,KClO_3 \longrightarrow 2\,KCl + 3\,O_2$$

Here we find that one molecule of potassium chlorate separates into two molecules of potassium chloride and three molecules of oxygen.

Writing chemical equations. Any chemical reaction can be written in the form of an equation if we know the correct formulas for all the substances that react and all the products of the reaction. The constituents of the reaction are learned only by experiment. Sometimes we can guess how a given pair of substances will react by analogy. If we have learned from experiment how two similar substances react, we are often reasonably safe in assuming that the two given substances will react in a similar way. However, experiment is the ultimate court of appeal when in doubt about the products of a reaction.

Balancing a chemical equation. Let us examine the reaction for the formation of hydrogen by the action of sulfuric acid on zinc. The correct formulas are:

$$Zn + H_2SO_4 \longrightarrow H_2 + ZnSO_4$$

When we check up on the numbers of each kind of atom, we find that there are the same number of atoms of each kind on each side of the equation, which means that it is complete or balanced. Not all equations are made so easily. Let us try another. Potassium reacts with water to form potassium hydroxide and hydrogen. Stating this in symbols, we have:

$$K + H_2O \longrightarrow KOH + H_2$$

When we compare the two sides of the equation, we see that there are three hydrogen atoms on the right side and only two on the left side. This is impossible, since we have not created any hydrogen. What really happened was that we used more molecules of water. It is always possible to juggle the number of atoms and molecules until the two sides balance, or have the same number of all of the atoms. This is called *balancing the equation*. In the equation above, we see that water contains hydrogen atoms in pairs; if we take 1 molecule, there will be 2 hydrogen atoms; if 2 molecules, there will be 4, etc. This means that we must do something to make an even number of hydrogen atoms on the right side. Suppose we take 2 KOH, then 4 hydrogen atoms are needed on the right side, 2 in the hydrogen gas and 2 in the 2 KOH. This calls for 2 H_2O to furnish the 4 hydrogen atoms, and 2 potassium atoms for the KOH. We now have:

$$2 \, K + 2 \, H_2O \longrightarrow 2 \, KOH + H_2$$

This is completely balanced.

We could summarize equation writing in two rules :

(1) Write the correct formulas for the substances entering into reaction and those coming out of the reaction, on the two sides of the arrow.

(2) Balance the equation.

There is one thing to be guarded against. *Never change a formula to make the atoms balance.* A formula is the unvarying symbolic expression for a definite compound.

If a formula is written incorrectly, the equation usually cannot be balanced. If a given equation cannot be balanced, it would be well to check up on the formulas to see if they are correct.

In balancing chemical reactions, it is necessary to understand thoroughly how the figures in the formulas and preceding the formulas apply to the different atoms. A number of formulas are analyzed for the purpose of clarifying these points. Thus $Ca(OH)_2$ indicates 1 atom of calcium, 2 atoms of oxygen, and 2 atoms of hydrogen; $Fe_2(SO_4)_3$, 2 atoms of iron, 3 atoms of sulfur, and 3×4 (or 12) atoms of oxygen; and $Al(NO_3)_3$, 1 atom of aluminum, 3 atoms of nitrogen, and 3×3 (or 9) atoms of oxygen. On the other hand, $2 \, H_2O$ means 2 molecules of water; $3 \, HCl$, 3 molecules of hydrogen chloride; and $4 \, H_2SO_4$, 4 molecules of sulfuric acid. The figure preceding a formula multiplies every atom in it; thus $4 \, H_2SO_4$ indicates 4×2 (or 8) hydrogen atoms, 4 sulfur atoms, and 4×4 (or 16) oxygen atoms.

QUESTIONS OF FACT AND RULE

1. State the fundamental rule of formula writing.

2. State the second rule of formula writing.

3. Name three things that must be considered in order to write a chemical equation.

4. State the three things referred to in question 3 in the form of the two rules for writing a chemical equation.

5. What must never be done in balancing a chemical equation?

QUESTIONS OF UNDERSTANDING

1. What is the valence of arsenic as shown by the formula As_2O_3? by the formula As_2O_5?

2. When are there no subscripts written in a formula?

3. In writing a chemical equation, how do you know the products of the reaction?

4. If an equation will not balance, what should be checked?

5. Which rule in formula writing applies only to a limited number of formulas?

6. Which of the following formulas are incorrectly written?

(a) $\overset{++\,-}{Ca}\,Cl$ (b) $\overset{+++}{Al_2}\,\overset{=}{O_3}$ (c) $\overset{++++}{Sn_2}\,\overset{=}{O_4}$ (d) $\overset{+++}{Al}\,\overset{-}{Cl_2}$

7. Given the reaction :

$$8\,HNO_3 + 3\,Cu \longrightarrow 4\,H_2O + 2\,NO + 3\,Cu(NO_3)_2$$

(a) Nitric acid is the first substance, how many molecules of it react?

(b) How many atoms of copper react?

(c) How many atoms of oxygen are on the left side of the reaction?

(d) How many molecules of copper nitrate are formed?

(e) How many atoms of nitrogen are in one molecule of copper nitrate?

(f) How many atoms of oxygen are in three molecules of copper nitrate?

(g) How many atoms of hydrogen are in the four molecules of water?

8. Which of the following equations are balanced and which not?

(a) $Ca + HI \longrightarrow CaI_2 + H_2$

(b) $Zn + 2\,HBr \longrightarrow ZnBr_2 + H_2$

(c) $Mg + O_2 \longrightarrow MgO$

(d) $2\,Ca + O_2 \longrightarrow 2\,CaO$

SUMMARY

All matter, whether solid, liquid, or gaseous, is thought to be made up of separate individual particles called molecules. In compounds these molecules are made up of smaller particles of the constituent elements, which we call atoms. The pressure of a gas is produced by the continual bombardment of the walls of the containing vessel by the molecules.

Dalton's atomic theory embodied six points :

(1) All elements are composed of atoms.

(2) All the atoms of an element are alike.

(3) Different elements have different kinds of atoms.

(4) Atoms combine to form molecules.

(5) The atoms have fixed and unchangeable weights.

(6) The atoms are indestructible.

In recent years the atomic theory has been changed in two points. An element may have atoms of slightly different masses, and some atoms may be changed into others by being struck by other high-speed atoms or similar missiles. The atoms of an element of different atomic weights are called *isotopes*.

The laws of *indestructibility of matter* and of *constant composition* are more understandable when expressed in terms of atoms.

The capacity of an atom for holding other atoms in chemical union is called its valence. Valence is expressed in terms of the holding power of hydrogen.

Compounds are composed of a positive part and a negative part. A group of elements taking the place of a single element in a compound is called a radical.

When an element has more than one valence, its name is given the ending *ous* in compounds where it has the smaller valence and *ic* in compounds where it has the larger one.

In naming a chemical substance, the positive part is mentioned first.

Three rules of writing formulas. (1) There must always be as many (+) valences as there are (−) valences.

(2) Take as many of the positive atoms as there are negative charges on the negative atom and as many negative atoms as there are charges on the positive part.

(3) If from rule (2) there results an even number of each kind of atom, the subscripts are divided by two.

Two steps in writing a chemical equation. (1) Write the correct formulas for the substances entering into reaction and those coming out of the reaction. (2) Balance the equation.

←
→

CHAPTER 6

THE ATMOSPHERE

Introduction. We live at the bottom of an ocean of air, whose weight presses down upon us with a force of approximately fifteen pounds per square inch. Our bodies are designed in such a manner that we are not aware of this pressure. In fact, we experience inconvenience when it is reduced. People climbing high mountains, where the atmosphere is less dense, often are troubled with nose bleeding and dizziness. Men ascending to great heights in balloons become unconscious unless they take along apparatus for breathing oxygen.

The air is a mixture. The air is a mixture of several gases. There are various ways of proving that the constituents are not chemically united.

(1) The composition varies. The air in a stuffy room is considerably different in composition from that out in the open. This could not be true if the elements were chemically united. Chemical compounds have unchanging composition, a fact we expressed as the law of constant composition.

Air will dissolve to some extent in water, from which it can be expelled by boiling. When this recovered air is analyzed, it is found that there is a different ratio between the nitrogen and the oxygen, its two most abundant constituents, than there was before dissolving. If the oxygen and nitrogen were tied together by chemical union, they would have the same ratio after solution that they had before.

(2) There is no heat change when air is prepared out of its constituents. Whenever two substances react chemically, heat is either absorbed or given out. The constituents of air can be mixed in the proper proportions, without any heat change.

The conclusion we draw from these facts is that the constituents of air are only mixed together.

MOUNTAIN AIR IS PURE AND INVIGORATING.

Composition of air. The constituents of the atmosphere can be divided into three groups: (1) those that are always present in practically unvarying proportions; (2) those that are always present but whose concentrations vary between wide limits; and (3) those that are present only occasionally, such as the sulfur dioxide found in the vicinity of volcanoes and smelters and the ozone noticed after an electrical storm.

CONSTITUENTS OF THE AIR

CONSTANT CONSTITUENTS	BY VOLUME DRY AIR (PER CENT)	VARIABLE CONSTITUENTS	BY VOLUME (PER CENT)
Nitrogen	78	Water	1.4
Oxygen	21	Carbon dioxide . .	00.03
Argon	00.94	Ozone	traces
Helium	00.0005		
Neon	00.0012		
Krypton	00.00005		
Xenon	00.0000006		

Nitrogen dilutes the oxygen. Nitrogen, the most abundant constituent of air, is useful in that it dilutes the oxygen so it does not react too rapidly. If the oxygen were not diluted with an inactive substance like nitrogen, practically everything combustible would burn up. When a fire once got started, nothing could put it out. The fire department would be helpless, for a fire would spread so rapidly and burn so intensely that water would never cool it below the kindling temperature of burning wood. In a previous chapter we learned that burning is uniting with oxygen. When oxygen molecules from the air crowd around to react with a burning substance, each oxygen molecule is accompanied by four inactive molecules of nitrogen. Thus the nitrogen crowds the oxygen away and absorbs part of the heat, both of which actions slow down the rate of burning.

All living things must have nitrogen. Nitrogen is also valuable for itself. Every plant and animal has to have nitrogen in its living cells. We can say, then, for nitrogen, as we did for oxygen, that it is absolutely necessary for all life. A queer situation, however, exists in regard to nitrogen. Although every living animal and plant must have nitrogen, not a single animal and only few kinds of plants can use the nitrogen of the air. Animals live in an ocean of air, mostly nitrogen; tons and tons of it are everywhere, yet if an animal had to depend upon the air for its nitrogen, it would soon perish. Animals are helpless except as they can eat the nitrogen-containing substances in plants.

Ordinary plants, too, would die for the lack of nitrogen if they had to get their supply directly from the air. Nearly all plants have their nitrogen prepared in a special form. It has to be united with oxygen and some metal to form a nitrate before they can use it. A few plants, such as the potato and corn, can use the nitrogen in ammonium compounds, but the vast majority can use it only in the form of nitrates, such as potassium nitrate (KNO_3) and sodium nitrate ($NaNO_3$).

Making the nitrogen of the air available to plants. The process of causing the nitrogen of the air to unite with other elements is called the "fixation of nitrogen." Nitrogen is by nature an inactive substance, which makes its fixation a difficult problem. During the present century, chemists have found several ways of artificially fixing nitrogen.

Nature has been carrying on the fixation of nitrogen ever since there were plants. Man, however, has only discovered her process within the last few decades. The flash of lightning causes some oxygen and nitrogen of the air to unite. Chemists wondered if it were possible that all the nitrogen now found in plants and animals was fixed by the lightning. All measurements, however, showed this to be impossible. The real secret of nature's process was later discovered in the activities of certain bacteria, little one-celled plants. One gram of soil often contains millions of bacteria. A limited number of these are nitrogen-fixing organisms, which live in the soil and fix the atmospheric nitrogen, using it for their life processes and when they die leaving it in the soil, where it is absorbed by plants.

Another kind of nitrogen-fixing bacterium lives in nodules or sores in the roots of the legumes, a family of plants including beans, peas, clover, vetch, and alfalfa. The bacteria in the roots of a legume are really a benefit to the plant. To illustrate this fact, suppose we prepare two cans of soil containing all of the plant foods except nitrates. We plant a clover plant in each can. In one can we inoculate the roots with bacteria, but not in the other can. The plant in the first can grows

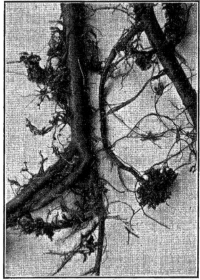

nicely, but the one in the second does not grow and finally dies. The bacteria evidently supply the missing nitrate to the plant, which would otherwise starve for nitrogen. Careful analysis of the soil from time to time shows that the legumes restore nitrogen to it.

Farmers employ a *rotation of crops* to keep up the nitrogen fertility of their land. For example, a field may be planted with a legume crop every third year. The theory back of the plan is that the legume will restore the nitrogen removed by the other crops during the other two years. Even though most of the legume

Courtesy of Western Soil Bacteria Co.

ALFALFA ROOT WITH NODULES. Inside these nodules are colonies of bacteria, which supply nitrogen compounds for the plant and take sugars in return. Alfalfa and the legumes do much·better when their roots are covered with colonies of nitrogen-fixing bacteria than when they are not.

crop may be removed and marketed, part of the leaves and roots remain to decay and form nitrates.

Orange trees require a great deal of nitrogen. The growers have found that one of the cheapest ways of getting this nitrogen is to grow a crop of vetch between the trees each summer and then to plow it under, where bacteria decompose it and change it into nitrogen compounds that are usable by the orange trees.

The presence of bacteria in the roots of a plant is profitable to both plant and bacteria. The plant is able to prepare plenty of sugar by means of the green chlorophyll in its leaves. The bacteria cannot do this; hence they get their sugar out of the plant sap and give nitrate in exchange. The bacteria can fix extra nitrogen, so the two forms of life can live together better than either could live alone. The nitrogen-fixing bacteria which live in the roots of legumes are the most effective and hence the most important of all nitrogen-fixing bacteria.

QUESTIONS OF FACT

1. Name the three most abundant regular constituents of the air in the order of their amounts.

2. Which one constituent is practically four-fifths of the total volume of the air?

3. Name one important substance in the air that varies widely in amount.

4. Name two accidental constituents sometimes found in the air.

5. Can animals get the nitrogen they need for cell building from the air?

6. Where do the animals get their nitrogenous substances?

7. In what form must most plants get their nitrogen?

8. What agency does nature use for restoring nitrogen to worn-out soils?

9. Are nodules on the roots of clover plants a help or a hindrance?

10. Why does it not injure the leguminous plants to have sugars taken from them by the bacteria?

11. Are there any living things that do not have nitrogenous substances in their cells?

QUESTIONS OF UNDERSTANDING

1. Give two proofs that air is a mixture and not a compound.

2. Name two ways the nitrogen of the air retards burning.

3. Explain how plants get their nitrogen.

4. How do we know legumes restore nitrogen to the soil?

5. What does the farmer mean by the rotation of crops?

6. Of what value is the rotation of crops?

7. Why does an orchardist sometimes grow a crop and then plow it under?

8. Virgin soil is soil upon which man has never grown crops but upon which grasses, herbs, bushes, and trees have grown since time immemorial. How can it still be very fertile?

9. Where would you expect to find air much richer than the average in carbon dioxide?

10. If air is much above average in percentage of carbon dioxide, what is it likely to be in oxygen?

REFERENCE FOR SUPPLEMENTARY READING

1. Martin, Geoffrey. *Triumphs and Wonders of Modern Chemistry.* D. Van Nostrand Company, New York, 1922. Chap. VII, pp. 132–170, "The Air." Discussion of air, liquid air, and the atmosphere.
   (1) What does a cubic yard of air weigh? (p. 134)
   (2) What is the weight of air in an average-sized lecture hall? (p. 134)
   (3) What would the earth be like if there were no air? (p. 135)
   (4) Discuss the composition of air as illustrated on page 140.
   (5) How does water vapor compare with dry air in weight? (p. 141)
   (6) How would it be here upon earth if water vapor were heavier than dry air? (p. 142)
   (7) What weight of moisture is breathed out each day by all the people of earth? (p. 143)

The nitrogen cycle. Bacteria fix the nitrogen of the air. Plants use the resulting nitrates and build them up into proteins, very complex compounds containing nitrogen. When the plants and animals die, the bacteria of decay work the proteins down into simpler ammonium compounds, such as $(NH_4)_2SO_4$. Another set of bacteria, called nitrifying bacteria, change these ammonium compounds into nitrates. Living plants now use this nitrate over again, and the nitrogen begins a new cycle. Thus, the same nitrogen may appear in generation after generation of plants and animals. Sometimes in the process of decay this nitrogen is sidetracked into the air as nitrogen gas, but the bacteria are fixing more to take its place and to keep up the cycle.

Occurrence of nitrogen. Most of the nitrogen on the earth occurs in the air. The supply is practically inexhaustible, as a little calculation will show. There is also a vast amount of nitrogen stored in the bodies of living and dead plants and animals. The crust of the earth, except for the presence of plant and animal matter, contains little nitrogen. However, in Chile, in what seems to be a dried-up lagoon of the sea, there is a large deposit of sodium nitrate ($NaNO_3$), which is called "Chile nitrate" or sometimes "Chile saltpeter." Before the success of artificial nitrogen fixation, Chile supplied the whole world with nitrate for fertilizers and explosives.

THE NITROGEN CYCLE. Nitrogen from the air is fixed by the bacteria in the nodules (1) on the roots of leguminous plants. The sap carries the nitrates to the leaves (2) for use in building the cells of the plant. In autumn the leaves drop to the ground (3) where the bacteria of decay change the nitrogen compounds of the plant back to nitrates. The growing grass (4, 5) of the next season uses these nitrates in its cells. An animal (6) eats the grass and converts the nitrogen compounds to animal proteins. When the animal dies (7), the bacteria of decay change most of this protein back to nitrates (8) to be used again as plant food.
The remaining nitrogen is returned to the air, thus completing the cycle.

Preparation of pure nitrogen. The nitrogen for nitrogen fixation is obtained from the air on a commercial scale by removing the oxygen, either by passing the air over hot copper, or liquefying it and evaporating the nitrogen. In both cases the nitrogen, although pure enough for most practical uses, contains several other substances in small amounts. Pure nitrogen, however, can be prepared by heating ammonium nitrite.

$$NH_4NO_2 \longrightarrow N_2 + 2\,H_2O$$

In practice, the ammonium nitrite is prepared from more stable compounds at the same time that it is decomposed.

$$NH_4Cl + NaNO_2 \longrightarrow NH_4NO_2 + NaCl$$
$$NH_4NO_2 \longrightarrow N_2 + 2\,H_2O$$

Courtesy of Nitrate Corp. of Chile

THE CHILE SALTPETER INDUSTRY. This view shows part of one plant where sodium nitrate is prepared for shipment to all parts of the world.

Chemical reactions of nitrogen. Nitrogen is a very inert substance; it does not take part in many reactions. Nearly every one of its reactions is the basis for a method of nitrogen fixation. An exception is its reaction with magnesium metal. When heated with magnesium, nitrogen combines with it. Since metallic magnesium is costly, the only value of this reaction is to remove nitrogen from certain gaseous mixtures in the laboratory. An interesting reaction of nitrogen is the one which takes place when nitrogen is fixed by the lightning.

$$N_2 + O_2 \longrightarrow 2\,NO$$

This has been made into a commercial process, called the Birkeland-Eyde process. Where electric power is very cheap, an electric spark is passed through a mixture of nitrogen and oxygen, thus effecting a union of the gases. The longer the spark, the greater the fixation. By means of a magnetic field, the spark may be drawn out into the shape of a horseshoe, making it much longer than it would otherwise be. When the oxide is once formed, it will dissolve in various chemicals to make a large number of useful compounds. This process has

Courtesy of General Electric

ARGON-FILLED GLOBES. These lamps range from 150 to 1000 watts. Inert gas in the globe keeps the metal filament from evaporating and darkening the glass.

recently been discontinued, since, as a commercial proposition, it cannot compete with the cheaper Haber process.

The Haber process was described when we studied hydrogen. It is the most successful process today. Plants are in operation in all the leading countries of the world. These plants are so successful that Chile now finds it difficult to market her nitrate.

Oxygen. Oxygen, the second constituent of the air in abundance, has been described in detail in the second chapter, so at this time we will pass it by, with merely a mention.

The inert gases. Lord Rayleigh once found that the nitrogen prepared from ammonium nitrite, as described previously in this chapter, has a different density from the nitrogen obtained from the air by the removal of the other known constituents — oxygen, water vapor, and carbon dioxide. This difference in density suggested that possibly there are other gases with the nitrogen which are so inactive as to escape notice up to that time. His friend, Sir William Ramsay, began a careful study of the air and discovered that there are several inactive gases in it besides nitrogen. These inert gases have been named argon, helium, neon, krypton, and xenon.

Argon, helium, neon, krypton, and xenon are alike in their chemical properties. None of them has ever been found to

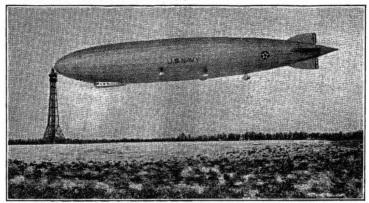

HELIUM-INFLATED DIRIGIBLE.

take part in a chemical reaction. This fact has given them several uses. Krypton and xenon are so rare that they have not as yet been put to any use.

Argon. Argon is used industrially for filling incandescent light bulbs. It is forced into the bulb under pressure. Its inertness prevents reaction with the filament, while its presence keeps down evaporation from the filament. Thus, the use of argon preserves the brilliancy of the light and adds to the length of its life.

Helium. Helium got its name from the sun (Greek *helios*) because it was found to exist in that body before its presence was discovered on the earth. The explanation of how it was possible to discover a substance on the sun is a long story.   In the first place, the only information we can get about the sun comes in the sunlight.   To say that each element in the sun sends us its signature in the sunlight sounds fanciful, yet practically this is what happens.

The spectroscope and spectra. In order to understand how an element in the sun may be identified from an examination of the sunlight, it is necessary to refer briefly to the spectroscope. A common form of this instrument has as its principal part a triangular piece of glass. When the light from an incandescent solid is passed through a slit, a narrow beam results. If such a beam is then caused to pass through the prism of a spectroscope, it is spread out into a wide band. Even though the slit be not over one sixty-fourth of an inch wide, a good spectroscope may widen it until the light becomes a band three feet across. Examination of

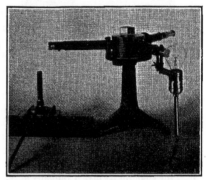

SPECTROSCOPE. This form of the spectroscope consists of three tubes leading to a chamber containing a triangular prism. One tube containing a very narrow slit is pointed toward the light source whose spectrum is being studied. The tube with the light globe at the end is used to carry light to illuminate a scale alongside the spectrum. The third tube contains a telescope for examining the spectrum. The prism makes the spectrum visible by bending the different colored rays so that they fall at different positions along the scale.

this wide band of light shows that the white light has been separated into all of the colors of the rainbow — red, orange, yellow, green, blue, and violet — each color blending into the colors next to it. The light from every solid body heated to incandescence can be made to give a similar spectrum, which is called a *continuous spectrum*. (See plate facing p. 120.)

If instead of an incandescent solid an incandescent gas is used as the source of light, the spectroscope shows not a continuous spectrum but only a few colored lines. Each element will give

Irving Langmuir

1881–

While he was still a youth, Irving Langmuir had his interest in chemistry aroused by his brother, an industrial chemist, who taught him qualitative analysis.

After Langmuir had graduated from Columbia University, he spent three years under Professor Nernst in Germany, where he got his Doctor's degree. In order to devote himself to research, on his return to the United States he accepted an instructorship at Stevens Institute of Technology.

In the summer of 1909 Langmuir was invited to do research for the General Electric Company at Schenectady, New York. Here he became interested in the difficulty the company was having due to the brittleness of tungsten wire filaments for light globes. A little experimenting convinced Langmuir that the trouble was caused by water, which was condensed on the inner glass surface of the globe, reacting with the heated filament. The outcome of this study was the discovery that gas-filled globes are about twice as efficient as evacuated ones.

Now being offered a chance to devote his entire time to research, Langmuir decided to quit teaching. His next discovery was the atomic-hydrogen blow torch, which produces the hottest flame available for welding.

Then his research led to improving the filaments of radio tubes. In the field of radio communication alone, he took out sixty patents.

Langmuir also opened up a new field of chemistry now known as "surface chemistry," which has to do with the behavior of soap films, oils, and similar substances. This study has resulted in a better understanding of many common phenomena, such as catalyses, emulsions, cleansing, and adsorption of gases on surfaces.

In 1932 Langmuir received the Nobel Prize in chemistry.

a different set of lines. Sodium vapor gives a pair of lines in the yellow part of the spectrum. These lines are the signature of sodium for the reason that no other element can duplicate them. Potassium has a signature of five lines, two in the red, two in the green, and one in the blue. If we should put a scale alongside the spectrum so as to locate different positions, we would find again that no other element can produce the potassium lines. Thus it is with every element. Each one gives a combination of colored lines in the spectrum that cannot be imitated by other elements; that is, each element has a signature that cannot be forged. These bright lines, so characteristic of the different elements, are spoken of as *bright-line spectra*.

The elements do not always use color in writing their signatures. They have another way of writing with black lines. Suppose we pass light having a continuous spectrum — that is, light from an incandescent solid — through sodium vapor at a temperature too low to give out light. When the light coming through this vapor is examined, it is found to be still practically continuous except that the sodium line is missing; that is, there is now a dark line instead. If potassium vapor, also below the temperature of incandescence, is used in place of the sodium vapor, the five potassium lines are darkened. Similar experiments prove that every element in the vapor state either emits characteristic lines or absorbs the same lines. It emits these lines when heated enough to give out light, and absorbs them when at a lower temperature.

Careful examination of the solar spectrum shows that it is not really continuous but that it is crossed by a large number of fine dark lines. It is believed that these Fraunhofer lines, as they are called, are caused by vapors in the relatively cool outer envelope of the sun absorbing their characteristic lines from the continuous spectrum emitted by the dense interior. In 1869 scientists tried to account for all the Fraunhofer lines on this theory. However, there were a few lines in the sun's spectrum that could not then be identified with any known element. It was then assumed that there was an element in the sun not known on the earth. Later a new element was discovered on the earth, which was shown by its spectrum to be identical with the helium of the sun.

Helium is the second lightest of all substances, hydrogen being the lightest. Since helium is also incombustible, it is highly suitable for inflating balloons. We have already seen in Chapter 4 how in this country it is being recovered from natural gas for this purpose. The United States is able to produce all of the helium it needs at a reasonable cost.

NEON SIGNS. Without neon it would require dangerously high voltages to cause the electrical discharge to pass through the long tubes of electric signs.

Neon. Neon has peculiar electrical properties. It carries the electric discharge more readily than any other substance. This property is turned to commercial use in the manufacture of neon signs. An electric discharge through neon and other substances in long tubes makes a most brilliant display. Without neon, a dangerously high voltage would have to be used to drive the electrical discharge through the tubes, making such electrical signs impracticable.

SODIUM SPECTRUM. The bright yellow line gives the sodium flame its characteristic color.

MERCURY SPECTRUM. The bright line pattern constitutes the "signature" of the element.

CALCIUM SPECTRUM. Another bright line spectrum of a rather more complicated pattern.

SPECTRUM OF AN INCANDESCENT SOLID. The continuous spectrum, in which the colors blend together.

SOLAR SPECTRUM. An absorption spectrum, whose dark lines correspond to the bright lines of hot gases.

QUESTIONS OF FACT

1. Trace the progress of nitrogen through the nitrogen cycle.
2. What do the nitrifying bacteria do?
3. What changes are brought about by the bacteria of decay in relation to nitrogenous substances?
4. Where does most of the nitrogen occur?
5. Where is a large nitrate deposit in the earth?
6. What chemical substance is called "Chile nitrate"?
7. How can chemically pure nitrogen be formed?
8. Give the formula for ammonium nitrite.
9. In what respects does the formula for ammonium nitrite differ from that of ammonium nitrate?
10. What can you say of the stability of ammonium nitrite towards heat?
11. Define nitrogen fixation.
12. Name two reactions of nitrogen which are the basis of nitrogen fixation processes.
13. Why is not the reaction between nitrogen and magnesium the basis of a commercial nitrogen fixation process?
14. Would you characterize nitrogen as chemically active or inactive?
15. In trying to prepare pure nitrogen from the air what three substances did Lord Rayleigh make provision for removing?
16. What did the difference in density between the nitrogen obtained from chemical reaction and that from the air suggest to Sir William Ramsay?
17. Name five gases discovered in explaining the discrepancies between the two densities of nitrogen.
18. In what connection is argon used, and what purpose does it serve in this use?
19. Where was helium first discovered?
20. By what means was helium discovered on the sun?
21. Under what condition will an element give out its characteristic bright lines?
22. What are the Fraunhofer lines in the sun's spectrum?
23. What do the Fraunhofer lines teach us?
24. How does helium compare with other substances in lightness?
25. How does it compare with hydrogen in combustibility?
26. What use has resulted because of these properties?
27. What is the commercial use of neon?

QUESTIONS OF UNDERSTANDING

1. Assuming the air to be 75.5 per cent nitrogen and the air over 1 square inch to weigh 15 pounds, calculate the weight of nitrogen above 1 square yard of ground.

(144 sq. in. = 1 sq. ft.; 9 sq. ft. = 1 sq. yd.)

2. Explain how helium was discovered on the sun.

3. Discuss the advantages and disadvantages of helium over hydrogen for inflating balloons.

4. Under what conditions will an element write its signature in dark lines?

5. What must be done to the light in order to read this signature?

6. If potassium chloride were thrown into a Bunsen flame while you were observing it with a spectroscope, which kind of potassium lines would you see, bright lines or dark lines?

7. What property of neon makes it useful in electric signs?

ADDITIONAL EXERCISES FOR SUPERIOR STUDENTS

1. Why are traces of nitrogen compounds found in the atmosphere after thunderstorms?

2. State why carbon would not be as satisfactory as phosphorus in removing the oxygen from the air to determine the nitrogen.

3. Tell what the following substances remove from the air when exposed to it: phosphorus, sodium hydroxide sticks, hot magnesium metal, and lime water.

4. Review the illustration in Chapter 2, page 26, with the explanation as to how it is shown that the products formed when a candle burns weigh more than the original candle. Now plan a method of removing both carbon dioxide and water vapor from the air in a container.

5. Describe chemical tests which could identify a bottle of each of the following gases: carbon dioxide, oxygen, nitrogen, and hydrogen.

6. What is meant by air conditioning?

References for Supplementary Reading

1. Clarke, Beverly L. *Marvels of Modern Chemistry*. Harper and Brothers, New York, 1932. Chap. XI, pp. 159–167, "Nitrogen."
   (1) Who discovered an element on the earth, that was previously discovered on the sun? (pp. 159–160)

(2) Who discovered neon, krypton, and xenon? (p. 160)

(3) Why did the early chemists pay little attention to nitrogen? (p. 160)

(4) What is the status of nitrogen today? (p. 160)

(5) Name the important substances which contain nitrogen. (pp. 160–161)

(6) What is the chemical name of "sal ammoniac"? (p. 163)

(7) Under what name did the ancients know nitric acid? (p. 163)

2. Holmes, Harry N. *Out of the Test Tube.* Ray Long and Richard R. Smith, New York, 1934. Chap. XIII, pp. 159–167, "The Romance of the Lazy Elements.'

(1) Tell how Cavendish showed that there is an inactive residue of gases in the air. (p. 159)

(2) Tell how Lord Rayleigh rediscovered this fact. (p. 160)

(3) How did Ramsay separate argon?

(4) What was the first clue to helium on the earth? (p. 164)

(5) What was the first price of helium? (p. 164)

(6) Describe the peculiar natural gas at Dexter, Kansas. (p. 165)

(7) Describe the growth of the helium industry. (pp. 165–166)

3. Weeks, Mary Elvira. *Discovery of the Elements.* Journal of Chemical Education, Easton, Pa., 1935. Pp. 178–291, "Discovery of the Rare Gases."

(1) Who first suggested that there was a gas more inactive than nitrogen in the air? (p. 278)

(2) When was this suggestion made? (p. 278)

(3) Relate Lord Rayleigh's discovery relating to nitrogen. (p. 279)

(4) State the four possible explanations of the reason why nitrogen from the air is heavier than nitrogen from chemical reaction. (p. 279)

(5) How did Ramsay learn foreign languages as a boy? (p. 280)

(6) Relate the discovery of argon. (p. 282)

(7) How was it proved that argon was not an allotropic form of nitrogen? (p. 282)

(8) Relate the story of discovery of helium on the sun and on earth. (pp. 284–286)

(9) Relate the story of the discovery of krypton, neon, and xenon.

4. Travers, Morris Wm. *The Discovery of the Rare Gases.* Edward Arnold & Co., London, 1928.

(1) Compare the weights of 1800 cc. of nitrogen obtained from chemical reaction with that obtained from the air. (pp. 5–6)

(2) Who suggested that the study of this discrepancy might result in the discovery of a new element? (p. 7, last paragraph, and p. 8, first paragraph)

(3) Describe Ramsay's apparatus. (pp. 8–9 and pp. 12–13)

(4) Give the density of the new gas. (Density of nitrogen is 14.) (p. 15, bottom)

(5) Describe the attempts of Ramsay to make argon react. (p. 31, bottom)

(6) Describe the spectrum of argon. (p. 32)

(7) Describe the discovery of helium. (pp. 56–57)

(8) Who predicted a new gas? (p. 78, bottom par.)

(9) Describe the discovery of krypton. (pp. 89–91)

(10) Relate the discovery of neon. (pp. 94–97)

(11) How long did it take Ramsay and his co-workers to solve the problem of the rare gases of the atmosphere? (p. 103)

(12) Write a summary of the account of the discovery of xenon. (pp. 105–106)

(13) What volumes of krypton and xenon were obtained by Ramsay and his helpers? (p. 109)

Humidity.   Visible drops of moisture as clouds and fog often are seen floating in the atmosphere.   However, there is always more or less invisible water vapor there also.   This is spoken of as the humidity of the air.   Humidity varies a great deal from climate to climate and season to season.   The holding capacity of the air for moisture is dependent on the temperature.   Hot air can hold much more moisture than cold air.   In cold climates you often hear the expression, "It is too cold to snow."   This may be literally true.   The air may be so cold that it will not hold enough water to produce a snowstorm. All our rain and snow are dependent on the humidity of the air. Bodily comfort, too, depends on humidity; high humidity means slow evaporation and slow body cooling, while low humidity means a dry, harsh air.

Temperature scales.   Throughout this book, temperatures are frequently expressed in degrees centigrade.   The centigrade thermometer is discussed and compared with our common household thermometer in the Appendix.   The centigrade scale is somewhat more convenient than the ordinary, or Fahrenheit, thermometer, for which reason it is used almost entirely by scientists for expressing temperatures.   (Compare the two thermometers.)

Carbon dioxide.   Although the average occurrence of carbon dioxide in the atmosphere is only 3 parts in 10,000, it is of fundamental importance.   It is the source of all the carbon in the tissues of plants and animals. The green chlorophyll of the plants enables them to use the carbon dioxide to prepare sugars and other organic substances. Many processes, such as burning, the breathing of animals, decay, and fermentation, are adding carbon dioxide to the air, while the plants are removing it. The two kinds of processes are about in equilibrium, so in general the percentage of carbon dioxide remains approximately the same.

Under side of leaves breathes CO_2

$12\,CO_2 + 11\,H_2O = sugar + 12\,O_2$

HOW PLANTS USE CARBON DIOXIDE. All green plants are able to absorb carbon dioxide from the air. The carbon dioxide is taken in on the under side of the leaves through openings called stomata. Within the leaves, by means of energy supplied by the sun, the carbon dioxide and water are combined chemically into sugars.

EXPERIMENT

The test for carbon dioxide in the air. With a bicycle pump force some air into some freshly filtered lime water. A milkiness shows the presence of carbon dioxide. The milkiness is due to the formation of white calcium carbonate, often called precipitated chalk.

Ozone. Sometimes after a heavy electrical storm a peculiar odor is noticed in the atmosphere. This odor is especially noticeable around any electrical machine when a spark is jumping through the air. The substance responsible for this odor is called ozone. Ozone is another form of oxygen, since it can change entirely into oxygen. Its formula is O_3, while that of oxygen is O_2. As it is an oxidizing agent, ozone is an excellent germ killer. It is used sometimes to purify drinking water, and its presence in the air is thought to keep it pure. For commercial use ozone is prepared by passing oxygen through a tube across which there is an electrical discharge.

Ozone emerges from the other end of the tube. Ozone and oxygen gas are said to be *allotropic forms* of the element oxygen.

EXPERIMENT

Testing for ozone. In the Priestley experiment for oxygen a little ozone is formed at the same time. Heat some mercuric oxide in a pyrex test tube and test the vapor for ozone with a strip of red litmus paper which has been dipped in potassium iodide solution. Ozone makes the paper turn blue, due to the formation of potassium hydroxide by a reaction with the potassium iodide.

Ozone from an ozone generator will turn starch iodide test paper blue. It also decolorizes by oxidation paper that has been blackened with lead sulfide by dipping into lead acetate solution and exposing to hydrogen sulfide fumes.

Liquid air. Under suitable conditions air can be changed to a liquid looking somewhat like water but possessing some very unusual properties due to the fact that its temperature is from $-183°$ C. to $-195°$ C. If a teakettle containing liquid air is set on a block of ice, the liquid air will boil and fume like the water in a teakettle on a red-hot stove. The block of ice is as hot compared with liquid air as the stove is compared with the water. If liquid air is poured over a piece of meat, the meat is frozen as hard as a chunk of ice. If

MUCH COLDER THAN ICE. Liquid air is so much colder than ice that it actually boils when heated by contact with ice.

the frozen meat is now struck with the hammer, it breaks into pieces. If a bouquet of flowers is put into liquid air, the flowers will crush in the hands like dry eggshells. A hollow rubber ball, when dipped into liquid air, will crumble to pieces. Liquid air will freeze mercury until it is hard enough to use as a hammer for driving nails. Liquid air can be kept only in a double-walled container with a vacuum between the walls. Even then the lid must not be clamped on or the container will explode.

Liquid air has three important commercial uses. It is used in producing extremely low temperatures; it is the commercial source of oxygen (see page 16); and it also serves as a source of nitrogen. Oxygen and nitrogen are only mixed in liquid air. Nitrogen boils or vaporizes at $-195°$ C., while oxygen will not vaporize to any great extent until its boiling point of $-183°$ C. is reached. This means that the nitrogen may be separated from the oxygen so each can be used separately.

EXPERIMENT

Testing liquid air. In the large cities and around the universities it is often possible to buy liquid air at approximately $1 per quart. This can be used in very spectacular experiments:

(a) Geraniums dipped in liquid air crush like eggshells.

(b) Pour over beefsteak. When hit with a hammer, the frozen beefsteak flies to pieces.

(c) Cheap hollow rubber balls, when frozen, pop when thrown against the wall.

(d) Into a wooden form put enough mercury to make a hammer. Stick a handle into the mercury and freeze with liquid air. Drive a nail into a board with the mercury hammer.

(e) Set a teakettle on a block of ice. Pour liquid air into it and watch it fume.

(f) Into a test tube pour alcohol and let down a string for a wick. Freeze with liquid air and ignite as a candle.

(g) Fit a small piece of iron pipe with a metal valve on one end. With the valve open pour in a few cc. of liquid air. Quickly fit a cork into the open pipe. On closing the valve it pops out the cork with a sharp report.

(h) With some liquid oxygen from the residue of liquid air ignite a small piece of cotton and drop in. It will burn violently. Charcoal also may be burned.

How liquid air is made. By this time the student probably wonders how it is possible to cool air down to −183° C. so it will liquefy. This is quite a problem. Before we can understand the process, we must learn something about heat and cold. According to modern theory, heat, in a gas, is the helter-skelter motion of the molecules, together with the vibrations within the molecules. The effects of heat are produced by the impact of the vibrating molecules. Suppose we compress a large volume of air at room temperature. After the molecules are all crowded into a small space, the concentration of the

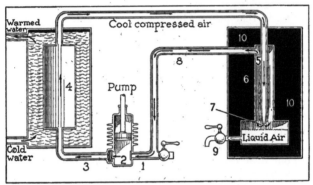

MANUFACTURE OF LIQUID AIR. Air is allowed to enter the apparatus through valve (1), which is then closed. The air is compressed by the pump (2) and passes to chamber (3), where the heat of compression is removed by cool water. The cooled air is allowed to expand through tiny opening (7). Part of it becomes liquid, while the rest repeats the cycle.

agitation will be much more than it was when distributed through the larger volume. In other words, compressing a gas heats it. Experiment tells us this. Perhaps you have pumped up an automobile tire or bicycle tire and noticed that the pump got quite warm. Most of this heat came from the compression and not from the friction. The soft leather valve in the pump creates very little friction. Let us suppose that the compressed air heats up to 150° C. Now this heated air can be cooled with ice water. It now has lost a large part of its heat. Suppose we release the pressure and let it expand back to the original volume. The molecules have lost most of their agitation, or heat, by being cooled, so now when with their lessened amount of agitation they are scattered again, the intensity of agitation, or temperature, will be much less than it was originally. Let us suppose this reduced temperature to be −150° C. If some more air is heated by compression and now cooled to −150° C. with this cool air before it escapes, it must drop still lower in temperature. By using this colder air for cooling, next time the temperature drops some more. When the liquefying point of oxygen is reached, some of the air begins to liquefy. From then on, part of the air liquefies, and the rest escapes to keep up the cooling. Drops of liquid fall from the escaping nozzle into a double-walled *Dewar* flask.

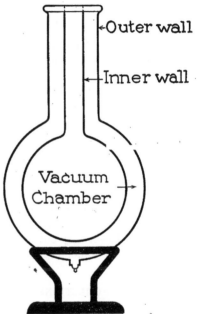

The Dewar flask. The Dewar flask, like the thermos bottle, consists of a double-walled flask with a vacuum between the walls. Heat, being the agitation of molecules, cannot pass

THE DEWAR FLASK. This flask is similar to the common thermos bottle. It is used for holding liquid air and similar substances. It has double walls with a vacuum between. The walls are silver plated inside and out. The vacuum keeps out heat energy, and the plating keeps out light energy.

through a vacuum where there are no molecules to agitate. Radiant energy, which is similar to light, can pass through a vacuum and then be changed into heat. In order to keep radiant energy out of the flask, it is silvered on the outside. In spite of these precautions, some heat gradually works its way into the flask around and through the glass. Such a flask must not be closed with a tight-fitting stopper, or the evaporating liquid will form enough gas to cause an explosion. A pad of wool felt is usually placed over it instead. If pressure forms, the gas can escape under the felt pad.

QUESTIONS OF FACT

1. Why does high humidity in summer cause discomfort?
2. Why did scientists invent the centigrade thermometer? (See Appendix)
3. What does the freezing point of water read on the centigrade thermometer?
4. What does boiling water register on the centigrade thermometer?
5. Name the processes which add carbon dioxide to the air.
   Name a process that removes carbon dioxide from the air.
   Of what importance is carbon dioxide to life on earth?
   How does ozone get into the air?
   Name two differences between ozone and oxygen.
10. Describe the Dewar flask.
11. How does it keep out radiant energy such as light?
12. Give the difference between the formulas of oxygen and ozone.

QUESTIONS OF UNDERSTANDING

1. Explain how liquid air is made.
2. Give some of the properties of liquid air.
3. Give three uses of liquid air.
4. How is the temperature of a gas affected by compressing it?
5. If air expands, what will the temperature do?
6. Why would it not be safe to clamp a tight-fitting lid on a vessel containing liquid air?
7. Explain how it can be "too cold to snow."
8. Might there be more moisture in the air on a hot day than on a cold one?

REFERENCES FOR SUPPLEMENTARY READING

1. Holmes, Harry N. *Out of the Test Tube.* Ray Long and Richard R. Smith, New York, 1934. Chap. XIII, pp. 159–167, "The Romance of the Lazy Elements."

(1) Who first liquefied helium and at what temperature? (p. 162) (Absolute zero is − 273.13° C.)

(2) How near to absolute zero does helium freeze? (p. 163)

2. Martin, Geoffrey. *Triumphs and Wonders of Modern Chemistry.* D. Van Nostrand Company, New York, 1922. Chap. VII, pp. 132–170

  (1) Discuss the occurrence of ozone. (p. 143)

  (2) What evidence have we that the upper air contains more hydrogen and more helium than at the earth's surface? (p. 143)

  (3) What is cosmic dust? (p. 144)

  (4) How much cosmic dust reaches our atmosphere yearly? (p. 145)

  (5) What is the relation of dust to rain? (p. 145)

  (6) Discuss the relative abundance of the gases in the atmosphere. (p. 146)

  (7) How many molecules are in 1 cc. of air? (p. 147)

  (8) Describe an explosive using liquid air. (p. 154)

  (9) In what sense is this explosive safer than dynamite? (p. 154)

  (10) Discuss burning substances in liquid air. (pp. 155–156)

  (11) Discuss the possibility of bacteria going from one planet to another. (pp. 156–158)

3. Sadtler, Samuel Schmucker. *Chemistry of Familiar Things.* J. B. Lippincott Company, Philadelphia, 1924. Chap. VI, pp. 65–81, "Air, Oxidation, and Ventilation."

  (1) What is meant by the dew point of the air? (p. 69)

  (2) What is hail? (p. 69)

  (3) Compare the weights of water in 1 cubic yard of saturated air at 14° F. with the same at 86° F. (p. 70)

  (4) What is the upper limit of carbon dioxide in air for satisfactory breathing? (p. 76)

SUMMARY

That the air in which we live is a mixture of substances is shown by the fact that its composition varies from time to time, that the ratio of its constituents is different after dissolving in water, and that it can be duplicated by mixing its constituents without the formation of heat.

Nitrogen is the most abundant constituent of the atmosphere. It serves the double purpose of diluting the oxygen — thereby preventing many serious fires — and of serving as a source for the nitrogen required by plants and animals.

The nitrogen-fixing bacteria make the nitrogen of the air into nitrates which can be absorbed by plants. The most important of these bacteria live in the roots of legumes. Legumes are therefore grown in regular rotation with other crops to keep up the nitrogen fertility of the soil.

Animals feed upon plants and get their nitrogen from them. The waste products of animals — including their dead bodies — decay and return nitrates to the earth. This cycle of changes is known as the nitrogen cycle.

Most of the nitrogen of the world occurs in the air. The rest is in the nitrogen cycle and a deposit of sodium nitrate in Chile.

The nitrogen of commerce comes from the air. Pure nitrogen, however, is obtained by heating ammonium nitrite. Although nitrogen is a rather inactive substance, it will react with a few substances, including magnesium, oxygen, and hydrogen. These reactions are mostly the bases of the methods of artificial nitrogen fixation.

Other substances present in the air in small amounts are argon, neon, helium, krypton, and xenon. Argon is used for filling electric bulbs. Neon makes possible our brilliantly colored electric signs. Helium was originally discovered on the sun by means of the spectroscope. It is now recovered in considerable quantities from the natural gas of certain American wells. Helium from this source is used for inflating dirigibles.

Still other constituents of the air are carbon dioxide (the source of the carbon in the plants), moisture, and ozone, a form of oxygen. Liquid air has remarkable properties because of its low temperature. Liquid air is made by compressing air, removing the heat generated by compression, and allowing the air to cool further by expansion.

CHAPTER 7

⇄

THE COMPOUNDS OF NITROGEN

Introduction. In the last chapter it was pointed out that nitrogen is an inactive element. Only at high temperatures will it combine with a few of the active elements to form compounds. When once combined with other elements, however, nitrogen no longer shows a hesitancy for reaction. As a result of the great reactivity of its compounds, nitrogen is a constituent of thousands of substances, including most of the explosives, such as T.N.T. and nitroglycerin; many of the strong poisons, such as nicotine, morphine, and prussic acid; many of our most useful drugs, such as quinine, strychnine, and atropin; and protein, the part of food that makes possible growth and the renewal of body cells. In this beginning course of chemistry we will study only a few of these many important compounds.

Ammonia. In the Haber process for the fixation of nitrogen, mentioned in the hydrogen chapter, the product is ammonia, a compound of nitrogen and hydrogen only, with one atom of the former united to three atoms of the latter in each molecule. The balanced equation for this reaction is:

$$N_2 + 3 H_2 \longrightarrow 2 NH_3$$

From the various nitrogen fixation plants in the United States hundreds of thousands of tons of liquid ammonia and ammonium compounds are made and sold each year.

Besides the plants where nitrogenous substances are the chief products, there are chemical processes from which ammonia is obtained as a by-product. In the manufacture of fuel gases from coal there is always produced a small percentage of ammonia, which in the course of a year amounts to thousands of tons. Coal, formed from the wood of prehistoric forests, contains a little nitrogen. This often forms ammonia in the chemical utilization of the coal. Frequently the ammonia from a gas plant is made to react with sulfuric acid to form ammonium sulfate, which, being a solid, is easier to handle than the gaseous ammonia.

$$2 NH_3 + H_2SO_4 \longrightarrow (NH_4)_2SO_4$$

The manufacture of coke, another product formed by heating coal, yields many times as much ammonia as the gas industry.

Formerly coke was made in "beehive" ovens, so called from their shape. No provision was made in these ovens for saving the ammonia, which was burned with the rest of the combustible gases. In time it was realized that the ammonia was too valuable to waste in this fashion. A new type of coke oven called a "by-product oven" was invented to replace the wasteful beehive ovens. Today by-product ammonia pays a large part of the cost of producing coke. Although the ammonia is recovered as ammonium sulfate, it can easily be released again as ammonia, according to the following reaction:

$$(NH_4)_2 SO_4 + 2 NaOH \longrightarrow 2 NH_3 + Na_2SO_4 + 2 H_2O$$

LABORATORY METHOD OF PREPARING AMMONIA. A solution of ammonium chloride and sodium hydroxide liberates ammonia. Heating hastens the process. Since ammonia is lighter than air, the collecting bottle must be inverted.

For the preparation of ammonia in the laboratory, it is only necessary to heat any ammonium salt with sodium hydroxide solution and catch the ammonia gas which is given off as the heating progresses.

Properties of ammonia. Ammonia was first discovered by its characteristic odor at the heathen temple of Jupiter Ammon, where a large amount of organic refuse was buried. The hot sun caused a peculiar odor to arise, which was named ammonia

from the last name of the temple.   Odors, as a rule, are hard to
describe and perhaps harder to recognize from the description.
One's nose can learn more from one whiff of a gaseous substance than from a page of description, so we'll let the reader
learn the odor of ammonia in the laboratory.

Ammonia gas is very soluble in water.   Its molecules crowd
in among the water molecules, and part of them react with the
water to form the very soluble ammonium hydroxide :

$$NH_3 + H_2O \longrightarrow NH_4OH$$

The total volume of ammonia that will dissolve in one volume
of water is remarkable, being 1100 times that of the water at

Courtesy of Ingersoll Rand

LARGE SCALE MANUFACTURE OF AMMONIA. This photograph shows some of
the machinery used in the industrial production of liquid ammonia.

zero degrees centigrade. Heating a solution of ammonium
hydroxide reverses the reaction shown above for the formation
of the ammonium hydroxide. Not only does heat decompose
the ammonium hydroxide, but it also expels all of the dissolved
ammonia from the water, so that after the solution is boiled
for a short time only water remains. Because of its instability,
ammonium hydroxide never exists out of solution.

Besides reacting with water, ammonia reacts with chlorine and loses its nitrogen, with magnesium and loses all its hydrogen, with sodium and loses part of its hydrogen, and with acids without losing anything. These reactions are all given in equation form below :

$$2\,NH_3 + 3\,Cl_2 \longrightarrow 6\,HCl + N_2$$
hydrochloric acid

$$2\,NH_3 + 3\,Mg \longrightarrow Mg_3N_2 + 3\,H_2$$
magnesium nitride

$$2\,NH_3 + 2\,Na \longrightarrow 2\,NaNH_2 + H_2$$
sodamid

$$NH_3 + HCl \longrightarrow NH_4Cl$$
ammonium chloride

The last reaction of this series is representative of a large number of reactions for the formation of ammonium salts. Ammonia gas can react with any acid to form the corresponding salt.

Ammonia in refrigeration. In practically every butcher shop there is an ammonia refrigeration system. Sometimes frost-covered pipes can be seen in the show cases. These are the cooling pipes of the ammonia refrigerating system. The ice plant also uses ammonia in its cooling system.

The principles underlying ammonia cooling systems are practically the same as those employed in the manufacture of liquid air. However, ammonia liquefies at about $-33.4°$ C., instead of about $-190°$ C. for air. Liquid ammonia results from the compression and the cooling. When it evaporates and expands into coils that run through tanks of brine, the temperature of the brine drops below the freezing point of water. Large square-cornered cans of water are let down into the brine tanks. The water in these cans freezes into the large cakes of ice we see the ice man hauling around. The efficiency of ammonia refrigeration is increased by a further heat change. When any liquid evaporates, it uses up a great deal of heat and produces cooling. We have already seen how the evaporation of water cools the human body. In liquefying the ammonia, heat had to be removed from it. When the ammonia evaporates, it must get an equal amount of heat back or else drop to a much lower temperature. In the ice machine it drops to a low temperature and absorbs heat from the water being frozen. Ammonia itself is usually sold as liquid ammonia in strong iron tanks.

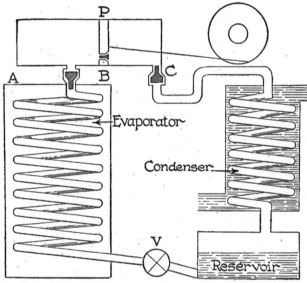

AMMONIA REFRIGERATION. The ammonia is compressed by the pump P and forced into the condenser at the right, where it is liquefied by cooling. The liquid ammonia drops into the reservoir, where it remains liquid under high pressure. For refrigeration, the ammonia passes through a small valve V into the evaporator at the left. The evaporation of the ammonia requires a great deal of heat, which is taken from the brine surrounding the evaporator pipes. When the brine has been cooled below the freezing point of water, large rectangular cans are filled with water and placed in the brine to freeze.

Other uses of ammonia. Besides its use in household laundering and in chemical laboratories, large quantities of ammonia are used in the production of baking soda, washing soda, nitric acid, and ammonium compounds, which in turn find uses in chemical laboratories, in industries, and especially in agriculture.

Bases. Ammonium hydroxide is typical of a whole class of substances called *bases*. For the present, we will define bases in terms of their characteristic properties:

(1) Water solutions of bases turn pink litmus paper blue. (Litmus paper is filter paper that has been dipped into a vegetable dye called litmus.)

(2) Water solutions of bases change colorless phenolphthalein red. (Phenolphthalein is a complex organic compound.)

(3) Bases taste bitter.

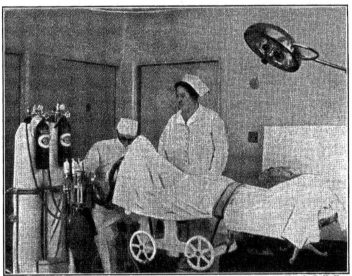

Photo by F. H. King

ADMINISTRATION OF LAUGHING GAS. Nitrous oxide was the first general anesthetic. It is still used for extracting teeth and in other light operations.

Ammonium vs. ammonia. We are now in a position to notice the difference between *ammonia* and *ammonium*. Ammonia is a gas, a chemical substance whose formula is NH_3. Ammonium, on the other hand, is the name of a radical (NH_4), which is only part of the formula for a compound. This radical must always be joined to something else. It cannot exist alone. There are a large number of ammonium compounds, such as ammonium nitrate (NH_4NO_3), ammonium hydroxide

(NH₄OH), and ammonium sulfate $(NH_4)_2SO_4$. Ammonia (NH_3) does not turn pink litmus paper blue when perfectly dry, while ammonium hydroxide, which can only exist in solution, does change the color of this vegetable dye.

Oxides of nitrogen. Nitrogen unites with oxygen to form several different oxides, in which the nitrogen has different valences. When ammonium nitrate is heated, there results a gas, *nitrous oxide*, which is sometimes called "laughing gas" because of its peculiar effect when breathed. If given in large amounts, it produces unconsciousness. It was the first anesthetic used in surgery, and is still used to some extent for light operations.

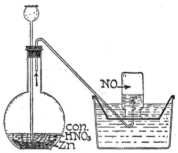

$$NH_4NO_3 \longrightarrow N_2O + 2 H_2O$$

The reaction for producing nitrous oxide is very similar to the reaction for producing pure nitrogen, described on page 114.

PREPARATION OF NITRIC OXIDE. When any metal dissolves in nitric acid, nitric oxide results.

When the ammonium nitrite was heated, the hydrogen went with the oxygen to form water, which left the nitrogen by itself. In the present reaction ammonium nitrate has one more oxygen atom than the ammonium nitrite, so when the hydrogen has taken all of the oxygen it can hold in chemical union, there still remains one oxygen atom attached to the nitrogen.

$$NH_4NO_2 \longrightarrow N_2 + 2 H_2O$$
$$NH_4NO_3 \longrightarrow N_2O + 2 H_2O$$

When copper dissolves in nitric acid, away from the air, there results a colorless gas, called nitric oxide, NO.

$$3 Cu + 8 HNO_3 \longrightarrow 3 Cu(NO_3)_2 + 4 H_2O + 2 NO$$

When any metal dissolves in concentrated nitric acid, nitric oxide is formed. It will be noticed that this reaction is different from the reactions of other acids on metals. In the chapter on hydrogen we noted the reaction between zinc and sulfuric acid, which was one way of obtaining hydrogen in the labora-tory. Nitric acid, especially when concentrated, acts differently from the dilute sulfuric acid.

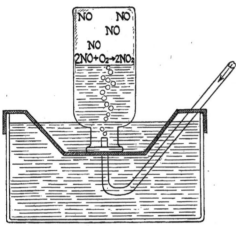

Nitric oxide read-ily unites with the oxygen of the air to form a reddish-yellow gas, called nitrogen dioxide, which has a pecul-iar odor.

$$2\,NO + O_2 \longrightarrow 2\,NO_2$$

This oxide is useful as an oxidizing agent. The yellow

NITROGEN DIOXIDE FROM NITRIC OXIDE. If oxygen (or air) is brought into contact with the colorless nitric oxide, the reddish-brown fumes of nitrogen dioxide appear.

color in bottles of concentrated nitric acid is due to nitrogen dioxide.

QUESTIONS OF FACT

In the first two questions choose the answer most suitable for com-pleting the statement.

1. Ammonium sulfate is used : (a) as a medicine; (b) as a source of ammonia; (c) for neutralizing acids; (d) for preparing nitrous oxide.

2. Two sources of ammonia are (a) the Haber process; (b) oxida-tion of metals; (c) by-product coke ovens; (d) combustion of gas; (e) dissolving copper in nitric acid.

3. Name two explosives which are nitrogen compounds.

4. Name three very poisonous nitrogen compounds.

5. What other nitrogen compounds are useful drugs?

6. What class of foods contain nitrogen?

7. Name two industries that may produce ammonia as a by-product.

8. Where was ammonia first noticed?

9. Does ammonia ever react in such a way that its hydrogen atoms are replaced?

10. In reaction does ammonia ever add itself bodily to the other substance?

11. How many volumes of ammonia will dissolve in one volume of water at zero?

12. In what sort of containers is ammonia usually sold?

13. Give five uses of ammonia.

14. Name a base that contains nitrogen.

15. Which oxide of nitrogen is an anesthetic?

16. Which oxide of nitrogen forms when metals dissolve in concentrated nitric acid?

17. Which oxide of nitrogen has a dark orange color?

QUESTIONS OF UNDERSTANDING

1. How could one get ammonia gas from NH_4OH solution?

2. What advantage does NH_4OH have over $NaOH$ for neutralizing acid on clothes? (Does not dissolve wool; consider the volatility.)

3. If there were four unlabeled bottles containing oxygen, nitrogen, nitrous oxide, and nitric oxide, how could they be identified by chemical or other tests? (Assume enough of the gases to test the physiological effect when breathed.)

4. Why is ammonia sometimes called the volatile alkali? (See the dictionary for the meaning of alkali.)

5. If two dishes, one containing ammonium hydroxide solution and the other containing salt water, were left exposed to the air until they were evaporated to dryness, what would be left in each of the dishes?

6. If some ammonium hydroxide solution were boiled slowly in an open dish for twenty minutes, and then cooled, what would it taste like?

7. Explain the difference between ammonia and ammonium.

8. Name three separate substances that are in ammonium hydroxide solution.

9. Can ammonia be collected over water as was oxygen, hydrogen, and nitrogen? Give reasons for your answer.

10. Account for the odor of ammonia around heaps of garbage.

11. White fumes form when a stopper from a bottle of HCl is held near the mouth of an open ammonium hydroxide bottle. What is the white substance?

12. Explain the working of an ammonia ice machine.

13. What is meant by a by-product coke oven?

14. How can ammonia be obtained from ammonium sulfate?

15. When ammonia dissolves in water, what two kinds of solubility are there?

16. Which oxide of nitrogen oxidizes spontaneously when exposed to the air?

17. What is the formula for ammonium nitrate?

18. Compare the numbers of oxygen atoms in ammonium nitrate and ammonium nitrite.

ADDITIONAL EXERCISES FOR SUPERIOR STUDENTS

1. Devise a method for distinguishing between NaCl and NH₄Cl.

2. Devise an experiment to show ammonia is lighter than air. (See Chapter 4.)

3. Mention one thing that is against ammonia for use in ice machines. (Consider the odor.)

4. If concentrated ammonium hydroxide is equivalent to 35 per cent ammonia, and each cubic centimeter weighs 0.942 gram, how many grams of ammonia are in 1000 cc. of solution?

5. In ammonia refrigeration, explain how the temperature is reduced below that of the cooling agency, water.

6. Write a detailed report on one of the following references:

   (a) Larson, "Synthetic Ammonia by Catalysis," *Journal of Chemical Education*, Vol. III, March 1926, pp. 284–290.

   (b) Mittasch and Frankenberger, "The Historical Development of Ammonia Synthesis," *Journal of Chemical Education*, Vol. VI, December 1929, pp. 2097–2103.

   (c) Turpentine, "Synthetic Ammonia in the Fertilizer Industry," *Journal of Chemical Education*, Vol. VI, May 1929, pp. 894–898.

REFERENCES FOR SUPPLEMENTARY READING

1. Martin, Geoffrey. *Triumphs and Wonders of Modern Chemistry*. D. Van Nostrand Company, New York, 1922. Chap. IX, pp. 195–211.

   (1) Why is nitrogen found largely free in the atmosphere? (p. 195)

   (2) Comment on the statement that nitrogen is dead. (p. 198)

   (3) What can overcome the deadness of nitrogen gas? (p. 198)

   (4) What are nitrides? (p. 198 bottom)

   (5) Why must new nitrogen compounds be prepared? (pp. 199–200)

(6) Name one process that makes soil barren. (p. 201)
(7) Why are synthetic fertilizers needed more now than in centuries past? (pp. 201–203)
(8) How much nitrates are supplied to the earth by lightning? (p. 204)
(9) Describe man's attempt to imitate the nitrogen fixation of the lightning. (pp. 204–206)
(10) Give the chemistry of the cyanamide method of nitrogen fixation. (pp. 207–208)
(11) About what fraction of the atmospheric nitrogen is fixed annually by bacteria? (p. 209)

2. Slosson, Edwin E. *Creative Chemistry*. The Century Co., New York, 1919. Chaps. II and III, pp. 14–44, "Nitrogen Preserver and Destroyer of Life"; "Feeding the Soil."

(1) Summarize modern wars in terms of a chemical reaction. (p. 14)
(2) With what invention did nitrogen compounds become a factor in war? (p. 14)
(3) In what sense might nitrogen be called an active element? (p. 15)
(4) What group or radical is active in many explosives? (p. 17)
(5) How did the Germans supply themselves with nitrates during the World War? (pp. 29–32)
(6) Tell of the efforts of the United States to get a supply of nitrogen compounds. (pp. 33–36)
(7) What effect did the World War have upon the fertility of the soil? (p. 37)
(8) How much nitrogen does a ton of wheat remove from the soil? (p. 38)
(9) Describe the work of bacteria on plant residues. (p. 39)

Nitrates. The nitrates were mentioned in connection with plant fertility. They are also useful in many manufacturing processes, such as the manufacture of explosives and rayon. Potassium nitrate, KNO_3, is sometimes called saltpeter. It is used in gunpowder, and in curing corned beef. It is also one of the most valuable constituents of fertilizers, since it contains two of the three most important plant foods, potassium and nitrogen.

Source of nitrates. Today all industrial nitrates come either from the Chile nitrate beds or from the substances prepared by artificial nitrogen fixation. Nitrates were not always easy to obtain. It is said that Napoleon levied a tax on the French people in potassium nitrate. He needed it above all else to manufacture gunpowder for his armies.

"How could the people get it?" you ask. "Wasn't this an unreasonable demand?"

In the light of the science known then, the requirement was reasonable enough. It was known that if animal and vegetable waste products were buried in the soil, nitrates could be leached out of this soil with water after the substances had completely decayed. When the water was then evaporated, the nitrates were left. People did not know what made the nitrates at that time. Since then, chemists have learned that the decay bacteria, followed by the nitrifying bacteria, made the nitrates out of the nitrogenous substances in the animal and plant substances.

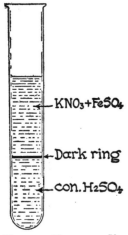
KNO₃+FeSO₄

Dark ring

con. H₂SO₄

TEST FOR NITRATES. If a nitrate is dissolved in dilute ferrous sulfate and concentrated sulfuric acid let down under the other solution, a dark ring will form where the two liquids meet in the test tube.

Nitric acid. One nitrate is of special importance. This is hydrogen nitrate, HNO_3, or nitric acid as it is usually called because its reactions are acidic. Hydrogen nitrate is a member of a large class of substances known as acids. All acids have certain properties in common. These are:

(1) Their water solutions taste sour if not too concentrated.

(2) Their water solutions change blue litmus to pink.

(3) Their water solutions change pink phenolphthalein to colorless.

When we look at the formulas of several of the common acids, HNO_3, HCl, HBr, H_2SO_4, and H_2S, we see that any common properties must be caused by the hydrogen part, since it is the only common constituent of them all.

Methods of preparing nitric acid. Commercial nitric acid is made from the fixed nitrogen of the fixation processes. It can also be prepared from Chile nitrate by heating with concentrated sulfuric acid.

$$NaNO_3 + H_2SO_4 \longrightarrow NaHSO_4 + HNO_3$$

Under the conditions of this reaction, the mixture of the nitrate and the sulfuric acid is boiled vigorously. The nitric acid becomes a gas and distills over into the receiver, leaving the other constituents of the reaction behind. This reaction is typical of a method by which any volatile acid, or acid that can become a vapor, can be made :

Take the salt that has the other part of the acid desired, and heat it with concentrated sulfuric acid. If HCl is wanted, take some chloride, such as NaCl, KCl, CaCl₂, BaCl₂. If HBr is desired, it will be necessary to take some bromide, such as KBr, NaBr, MgBr₂, or CaBr₂.

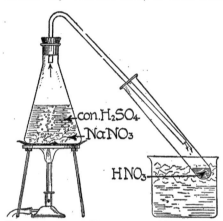

PREPARATION OF NITRIC ACID. Nitric acid is prepared from its salts the same as any other volatile acid — namely, by heating the salt with concentrated sulfuric acid and distilling off the volatile acid.

Nitrate properties of HNO₃. Nitric acid possesses acid properties, which are due to the H part, and nitrate properties, due to the NO_3 part. As a nitrate, it is a very strong oxidizing agent. It shows its acid properties as the outstanding properties in dilute solutions, while in the concentrated acid the nitrate properties far overshadow the acid properties. Concentrated nitric acid oxidizes all organic materials, such as wood, cloth, skin, or feathers. It will dissolve certain metals, such as silver, which will not dissolve in other acids.

EXPERIMENT

Oxidizing sawdust with nitric acid. Heat some sawdust in a small evaporating dish until it is just about to char. Pour 2 or 3 cc. of fuming nitric acid upon the charring sawdust, which undergoes vigorous combustion.

Uses of nitric acid. Large quantities of nitric acid are used in the manufacture of the following materials :

(1) Explosives, such as nitroglycerin, guncotton, and trinitro-toluene.

(2) Dyes.

(3) Rayon.

(4) Lacquers for automobiles.

Neutralization of an acid and a base. If we should take some acid solution and put a piece of blue litmus paper into it, the paper would turn pink as we should expect ; and the acid

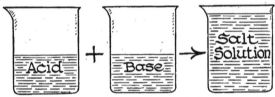

NEUTRALIZATION OF ACID AND BASE. If equivalent amounts of hydrochloric acid and sodium hydroxide are poured together, all the acid properties and all the basic properties disappear.

would taste sour. Another piece of litmus in some base solution would remain blue if it were blue and would turn blue if it were originally pink. This solution would taste bitter. If, however, the base is now slowly poured into the acid and the solution tested with blue litmus from time to time, a condition is finally reached where both colors of litmus paper will remain in solution side by side without change. If we taste the solution now, we find it neither sour nor bitter. We say the acid has neutralized the base and the base has neutralized the acid. We call the process *neutralization*. In neutralization the base has destroyed the acid and the acid has destroyed the base. The substance that results from neutralization we call a *salt*. A typical reaction would be :

$$NaOH + HNO_3 \longrightarrow NaNO_3 + H_2O$$

It is possible to write other neutralization reactions by analogy — that is, by their similarity to the one illustrated by the equation. In every case, the hydrogen of the acid unites with the hydroxyl of the base to form water. This leaves the positive part of the base to unite with the negative part of the acid to form the salt.

Neutralization is an important process. If the garageman spills acid from the battery on his clothes, he must get the acid neutralized quickly or it will make holes in the cloth. He could do this with ammonia solution or baking soda or washing soda. Lye is a very strong base, which will destroy wool cloth and dissolve the skin until it becomes raw and sore. If lye is spilled on the clothes or skin, it must be neutralized soon. Vinegar or lemon juice is suitable for this purpose.

Salts. One of the products of every neutralization reaction is called a salt. Table salt is the salt from the neutralization of hydrochloric acid and sodium hydroxide. There is a different salt from the neutralization of each pair of bases and acids. Potassium bromide is a salt used as a medicine; silver nitrate is a salt used by doctors to cauterize sores. The salts include a large list of useful substances.

QUESTIONS OF FACT

In the first two questions choose the answer most suitable for completing the statement.

1. Three natural processes which add nitrates to the soil are: (a) the decay of rocks; (b) the solvent action of water; (c) lightning; (d) the soil bacteria; (e) the decay of plants.

2. The gas given off when copper dissolves in hot concentrated nitric acid is: (a) nitrous oxide; (b) oxygen; (c) nitric oxide; (d) ozone; (e) nitrogen; (f) hydrogen.

3. Name an acid which contains nitrogen.

4. Tell how the farmers in Napoleon's time got nitrates with which to pay taxes.

5. Name three properties common to all acids.

6. Name the uses of nitric acid.

7. Give the name the meat packer uses for KNO_3.

8. Why is KNO_3 one of the most valuable constituents of fertilizers?

9. In neutralization what two kinds of solutions are mixed?

10. Give two properties of the acid which enter into neutralization. Also two properties of the base.

11. Do any of these properties remain in the mixed solution after neutralization?

12. If after neutralization the solution is evaporated to dryness, what do we call the substance that remains?

13. Define a salt in terms of the neutralization of an acid with a base.

QUESTIONS OF UNDERSTANDING

1. In the preparation of HNO_3 from $NaNO_3$, where does the hydrogen of the nitric acid come from?

2. What kind of acids are successfully formed by this method?

3. State the general reaction in the form of a rule.

4. Explain why pink litmus paper turns blue when ammonia is loose in the room.

5. When KOH and HNO_3 neutralize each other, what becomes of the H of the acid? of the OH of the potassium hydroxide? of the K of the base? of the nitrate, NO_3?

6. Under what conditions do acid properties prevail over nitrate properties, and under what conditions do nitrate properties prevail over acid properties in nitric acid?

7. Explain why soils become deficient in nitrates.

8. Copper is dissolved in concentrated nitric acid. What gas escapes from the reaction? What salt of copper remains in the solution? What becomes of the hydrogen of the HNO_3?

ADDITIONAL EXERCISES FOR SUPERIOR STUDENTS

1. Why does concentrated nitric acid turn yellow when exposed to sunlight? (Consider the possible oxides of nitrogen.)

2. Why cannot nitric acid be used to prepare hydrogen by the usual method?

3. If concentrated nitric acid is 68 per cent HNO_3 and each cubic centimeter weighs 1.4048 grams, what weight of the acid is in 500 cc. of solution?

SUMMARY

Nitrogen, an inactive element, when once in a compound, becomes active and is a constituent of explosives, poisons, and medicines.

Ammonia, one of the common important nitrogen compounds, is obtained from the Haber nitrogen fixation process and as a by-product in making coke and fuel gas from coal. Ammonia is a gas of a strong characteristic odor. It will dissolve 1100 volumes in 1 volume of water at zero degrees centigrade. Ammonia may react with chlorine to lose its nitrogen, with metals to replace its hydrogen, and with acids with neither loss nor replacement. The principal uses of ammonia are in refrigeration, in the manufacture of the sodas, in the manufacture of nitric acid, and in making its compounds.

When ammonia dissolves in water, it forms ammonium hydroxide, a base which turns litmus blue and turns phenolphthalein pink.

The name *ammonia* applies to the colorless gas, while the term *ammonium* refers to the radical NH_4, the positive part of ammonium compounds.

With oxygen, nitrogen forms nitrous oxide, N_2O, commonly called laughing gas, nitric oxide, NO, and nitrogen dioxide, NO_2. Nitrous oxide is obtained by heating ammonium nitrate, nitric oxide by dissolving a metal in concentrated nitric acid, and nitrogen dioxide by the oxidation of nitric oxide by the oxygen in the air.

Of the nitrates, potassium nitrate is important for use in black gunpowder and as a plant food. Nitrates can be obtained either from deposits in Chile or by nitrogen fixation.

Nitric acid, like other acids, has three characteristically acid properties: sour taste, ability to turn blue litmus to pink, and ability to change phenolphthalein from pink to colorless. Nitric acid is obtained by heating any nitrate with concentrated sulfuric acid. A similar method can be used to prepare any acid that may be volatilized — that is, heating a suitable salt with concentrated sulfuric acid. In addition to its acid properties, which are due to the hydrogen part of the substance, nitric acid has other properties due to the nitrate part. The principal property of the nitrate part is a strong tendency to oxidize substances.

An acid and a base may neutralize each other, causing the properties of both to disappear. Salts are the products of neutralization.

CHAPTER 8

THE LAWS OF GASES

Introduction. Many of the important laws of science have been learned from gases. The changes that take place when gases are acted upon by heat or pressure are quite large. These large changes can be measured with considerable accuracy, and the results are highly significant.

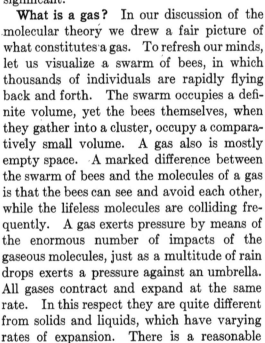

GASES EXPAND WITH HEAT. Ordinarily when a gas is heated, both pressure and volume increase. The illustration shows how increased pressure has forced out the cork and allowed the gas to expand.

What is a gas? In our discussion of the molecular theory we drew a fair picture of what constitutes a gas. To refresh our minds, let us visualize a swarm of bees, in which thousands of individuals are rapidly flying back and forth. The swarm occupies a definite volume, yet the bees themselves, when they gather into a cluster, occupy a comparatively small volume. A gas also is mostly empty space. A marked difference between the swarm of bees and the molecules of a gas is that the bees can see and avoid each other, while the lifeless molecules are colliding frequently. A gas exerts pressure by means of the enormous number of impacts of the gaseous molecules, just as a multitude of rain drops exerts a pressure against an umbrella. All gases contract and expand at the same rate. In this respect they are quite different from solids and liquids, which have varying rates of expansion. There is a reasonable explanation of the equal rate of expansion and contraction of all gases. Although the molecules of one gas are undoubtedly larger or smaller than those of another, the sizes of the molecules are all so small compared with the vast amount of empty space between them that a little difference in size makes practically no difference as to the space dominated by the rapidly moving particles. Small molecules can dominate the same volume as large ones, since they have greater speeds and make up in speed what they lack in weight and size.

A gas is not completely described until we know its volume, pressure, and temperature. By pressure we mean the force exerted against one unit of area on the walls of the containing vessel. If either volume, pressure, or temperature changes, at least one of the other factors will have to change also. It is often necessary to figure how much one of these factors will change when one of the others changes.

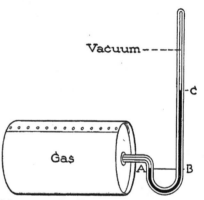

Pressure units. Pressure is frequently expressed in different kinds of units —atmospheres, pounds per square inch, and centimeters (or millimeters) of

MEASURING GAS PRESSURE IN CENTIMETERS OF MERCURY. The pressure of the gas at *A* just balances the weight of the mercury column above *B*. The height of the column above *BC* is given in centimeters.

mercury. Although these units seem to be unrelated to one another, they are simply different ways of saying the same thing. Of the three units, pounds per square inch is the easiest to understand. It is simply the number of pounds with which a gas presses against each square inch of the confining walls, or (what amounts to the same thing) the force in pounds that one square inch of the wall exerts against the gas.

At sea level the atmosphere exerts a pressure of approximately 14.7 lbs. per square inch. This pressure is sometimes taken as the unit and is called an atmosphere. Other pressures can then be expressed in atmospheres, two atmospheres being twice 14.7 pounds per square inch and three atmospheres being three times this value.

Robert Boyle

1626–1691

A century before the discovery of oxygen by Priestley and the overthrow of the phlogiston theory, there lived a rich Irishman named Robert Boyle, who may be said to be the father of modern chemistry.

For two thousand years previous to his time the ideas of Aristotle held sway. Aristotle believed that everything was composed of the four elements — fire, air, earth, and water. The alchemists, who had been diligently searching for ways to transmute the baser metals into gold, with the iatrochemists, who were searching for cure-alls in medicine, had accumulated many facts hard to explain in terms of the old theory of Aristotle. Boldly Boyle began to attack the old theory with logical arguments and forceful ridicule.

Boyle was first to prove the relation between the pressure and volume of a gas that is still called Boyle's law. This law states that the volume of a gas varies inversely as the pressure upon it. Boyle also gave the world the first modern conception of an element. There is no reason, said he, why there should not be many elements. In fact, every substance that cannot be decomposed should be considered as an element. Soon after the publication of a book by Boyle, the elements of Aristotle were dropped and rarely ever mentioned again.

Boyle's idea that an element is a single simple homogeneous substance that does not yield to decomposition was a distinct step in the direction of clearing up the confusion caused by the old understanding of the nature of things. When we think of water as a decomposable compound, air as a complex mixture which varies within limits, fire primarily heat energy and not substance at all, and earth as an extremely complex mixture of substances, varying from place to place, we see why Aristotle's conception of elements did not lead to progress.

For much scientific work pressure is expressed in terms of the length of the column of mercury which a gas can support, measured either in centimeters or in millimeters. At sea level the atmosphere can support a column of mercury 76 cm. or 760 mm. high in a barometer. This is also a method of defining an atmosphere.

Relation between volume and pressure. In order to find out how volume and pressure are related, we will have to keep the temperature constant. If the temperature does not change, when the pressure is increased, the volume gets smaller. This is nothing but common sense. Anyone would know that if he had a bubble of air squeezed between the palms of his hands, the harder he squeezed, the smaller it would become. On the other hand, if the pressure is lessened the elasticity of the gas would make it swell out. Likewise, if one saw an inflated inner tube begin to swell, he would know that the rubber was not able to sustain the air pressure. A scientific statement of these facts is known as *Boyle's law,* after Robert Boyle, one of the early scientists. *The volume of a gas varies inversely as the pressure, if the temperature does not change.*

EXPERIMENT

To demonstrate Boyle's law. The volume V of a gas is measured in centimeters in the eudiometer tube A, which is kept at constant temperature by the circulation of water through the jacket D. The pressure P of the gas (in centimeters of mercury) can be read from the barometer plus the difference in height between M and N. If N is the higher, this column of mercury is added to the atmospheric pressure; if N is the lower, it is subtracted.

For three or four different pressures read the volumes and calculate the products, PV. These should come out practically the same in each case.

Boyle's law is most useful in the form of an equation, using symbols for pressure and volume. Expressed as an equation, Boyle's law is :

$$P_1V_1 = P_2V_2$$

It will be noted that those symbols with like subscripts are grouped together. This equation is true only when the temperature is kept constant. P_1 represents the pressure under which the given gas had a volume V_1, and P_2 the pressure under which the same gas had a volume of V_2. The pressure may be measured in any pressure units just so P_1 and P_2 are measured in the same unit. The volumes V_1 and V_2 may be in any volume units, but must be measured in the same unit. The above equation is useful in working many problems when three out of the four quantities are given.

I think I hear some timid student cry: "Algebra, I never could get algebra." I say to you if you are the timid student, "Don't be afraid of this algebra. It is so simple a child could use it." Let us try a problem and see.

An automobile tire is pumped up to a total pressure of 45 lbs. per sq. in. (P_1) and its volume is 4 cu. ft. (V_1). If it is punctured, and the air leaks out, what volume (V_2) will the air now have under the outside pressure of 15 lbs. per sq. in., which is the P_2 of the formula?

$$P_1 \times V_1 = P_2 \times V_2 \quad \text{(temperature remain-}$$
$$4 \times 45 = 15 \times V_2 \quad \text{ing constant)}$$
$$V_2 = \frac{45 \times 4}{15} = 12 \text{ cu. ft.}$$

REVIEW QUESTIONS

1. What is a gas?

2. State Boyle's law both in words and in symbols.

3. Express a pressure of one atmosphere in (a) lbs. per sq. in.; (b) cm. of mercury; (c) mm. of mercury.

4. At the same price per cubic foot, is gas more expensive on a mountain top or in a valley?

5. Is it possible for the pressure and temperature to increase in such a way that the volume would not change?

PROBLEMS

1. If 100 ml. of air is measured under 1 atmosphere of pressure, what volume would it have if the pressure were changed to 5 atmospheres, the temperature remaining constant?

2. Two hundred ml. of hydrogen under 76 cm. of mercury pressure are changed to 80 cm. of mercury pressure; what must the new volume be? The temperature does not change.

3. An unknown volume of oxygen under 15 pounds per square inch has its pressure changed to 45 pounds per square inch. Its volume is now 1000 cu. ft. What was its original volume? Temperature did not change.

4. A volume of 600 cubic inches under an unknown pressure, P_1, has its pressure changed to 100 cm. of mercury. Its new volume is 400 cu. in. What was P_1 if the temperature did not change?

5. A balloon measures 5,000,000 cu. ft. when filled with air at a pressure of 20 lbs. per sq. in. If it should be opened, what volume would the air have after it had escaped into the outside where the pressure is 15 lbs. per sq. in. The temperature does not change.

REFERENCE FOR SUPPLEMENTARY READING

1. Holmyard, Eric John. *Makers of Chemistry*. The Clarenden Press, Oxford, 1931. Pp. 132-143, "Robert Boyle."
   (1) Write the biography of Robert Boyle in a little more detail than is given in the text.

How pressure is related to temperature. Temperature is principally a measure of molecular and atomic agitation. The pressure of a gas is due to molecular agitation, hence it should be related to the temperature. The temperature at which all gaseous molecules would quit moving and lie helplessly together as dead as bricks is called the *absolute zero*. At absolute zero a gas would have zero pressure. Absolute zero is $- 273.1°$ C. Whenever we use temperature in dealing with gases, we must use absolute temperatures. The law which expresses how temperature and pressure are related is named Charles's law after its discoverer. Stated in symbols, Charles's law is:

$$\frac{P_1}{T_1} = \frac{P_2}{T_2} \quad \text{(volume remaining constant)}$$

It will be noted that symbols with like subscripts are again on the same side of the equality sign.

EXPERIMENT

Verifying Charles's law (pressure–absolute-temperature form). This experiment uses the same apparatus shown in connection with the experiment on Boyle's law.

Stated in words, this form of Charles's law says that if the volume of a gas is kept constant, the pressure varies directly as the absolute temperature. In its simplest algebraic form this law is $\frac{P}{T} =$ a constant.

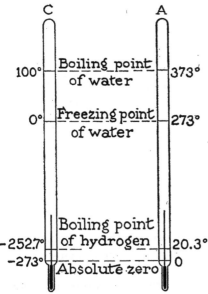

(a) With tap water in the outer tube, and the mercury adjusted so the gas occupies about two-thirds of the volume of the tube, record the centigrade temperature and the pressure on the gas in cm. of mercury. Mark the level of the mercury column with gummed paper on opposite sides of the tube, so when sighting through horizontally the surface is in the same line.

ABSOLUTE TEMPERATURE SCALE. The zero on the absolute scale is − 273.1° C. Absolute degrees are the same size as centigrade degrees. Zero degrees centigrade is 273.1° on the absolute scale.

(b) Run in warm water so the temperature is near 50° C. Increase the pressure on the gas until the volume is brought back to where it was in (a). Record the temperature and pressure.

(c) Repeat (b) with a temperature around 70° C. Add 273 to the centigrade readings to get T, the absolute temperature. Calculate the values $\frac{P}{T}$ in all three cases and compare.

Problem. Suppose the volume of an automobile tire does not change. It is pumped up to 45 lbs. per sq. in. (P_1) in the morning when the temperature is 10° C.; that is, $273 + 10 = 283°$ A. (T_1). In the afternoon when friction heats it up to 40° C. — that is, $273 + 40 = 313°$ A. (T_2) — what will be its pressure (P_2)?

$$\frac{P_1}{T_1} = \frac{P_2}{T_2} \qquad \frac{45}{283} = \frac{P_2}{313}$$

Stated in words, Charles's law says that *if the volume is kept constant, the pressure varies directly as the absolute temperature.* In other words, if the volume is kept constant and the absolute temperature doubled, the pressure is doubled also. If the absolute temperature becomes three times what it was originally, the pressure of necessity becomes three times as great as it was. If the absolute temperature, on the other hand, reduces to two-thirds of what it was, the pressure also reduces to two-thirds of what it was before. All of these facts and many similar ones are concisely stated in the preceding algebraic equation.

Multiply both sides of the equation by 313.

$$\frac{45}{283} \times 313 = \frac{P_2}{313} \times 313$$

The 313 cancels out on the side of P_2 because it is in both numerator and denominator of a fraction.

$$\therefore P_2 = 49.77 \text{ lbs. per sq. in.}$$

EXERCISES

1. What is heat?

2. State in both words and symbols the form of Charles's law that gives the relation between pressures and temperatures.

PROBLEMS

1. Assuming that an automobile tire does not change volume when the pressure of the contained air is increased, what pressure would a tire that had been pumped up to 35 lbs. per sq. in. when the temperature was 280° A. have when the temperature rose to 300° A.?

2. Suppose the first temperature of the tire in problem 1 had been 2° C. and the second had been 17° C., what would the final

Jacques Alexander Caesar Charles

1746–1823

Charles was a French mathematician and physicist. After a few years as clerk in the ministry of finance, becoming tired of routine tasks, which offer no challenge to the mind, he turned to scientific pursuits. He became a very skillful experimenter, demonstrating the law now known as Charles's law. This law states that at constant volume the pressure of a gas varies directly as the absolute temperature, and at constant pressure the volume of a gas varies directly as the absolute temperature. Charles was the first person to use hydrogen gas for inflating balloons. This may sound like a trivial innovation, but actually it was the beginning of a study of the atmosphere which has brought us much valuable information used in weather prediction and aviation.

In 1785, having achieved general recognition for his scientific achievements, he was elected to the French Academy of Sciences, the highest recognition for scientific achievement attainable in France. Later Charles became Professor of Physics at the Conservatoire dès Arts et Métiers.

The work of Charles is remarkable when we consider the number of different fields in which he made outstanding contributions. As a mathematician he contributed many original articles. However, unlike many mathematicians his activities were not solely academic. His introduction of hydrogen as a filler for balloons instead of hot air, which had been used up to this time, paved the way for practical air travel. In fact this method of inflating balloons is so practical that it is still being practiced over one hundred and fifty years later. The discovery of the two forms of Charles's law is the result of skillful and painstaking experimental effort.

pressure have been?   (Remember temperatures must be changed to absolute readings.)

3. If the same tire as used in problem 1 had a first pressure of 40 lbs. per sq. in., and a final pressure of 35 lbs. per sq. in., and a first temperature of 37° C., what was the final temperature on the centigrade scale?

Volume and temperature relations when pressure is constant. When pressure is kept constant, we have another form of Charles's law.

$$\frac{V_1}{T_1} = \frac{V_2}{T_2}$$

In words, this can be stated as follows: *When pressure is constant, the volume of a gas varies directly as the absolute temperature.*

EXPERIMENT

The verification of Charles's law.   (The volume–absolute-temperature form.)

This form of Charles's law is: At constant pressure, the volume of a gas varies directly as the absolute temperature.   In its algebraic form the law now is: $\frac{V}{T} =$ a constant

(a) Using the same apparatus as used in the preceding Charles's law experiment, and jacketing with tap water, adjust the pressure so the volume of the gas occupies about half of the tube.   Record temperature, volume, and pressure.

(b) Filling the water jacket with warm water, and adjusting the pressure so it is the same as in (a), record the temperature and volume.

(c) If time permits, repeat (b) at a high temperature.   Calculate the values for $\frac{V}{T}$ and compare them.

QUESTIONS

1. Would you prefer to have your gas meter in a warm or a cool place?

2. State the other form of Charles's law in words and also in symbols.

3. Can centigrade temperatures be substituted for T_1 and T_2?

PROBLEMS

1. The gas supply of a city is kept under constant pressure, the weight of the lid or covering. A certain tank has 1,000,000 cu. ft. of gas at a temperature of 0° C. What will the volume be if the temperature suddenly changes to 37° C.?

2. If at another time the tank in problem 1 has 900,000 cu. ft. of gas when the temperature is − 3° C., what volume will it have if the temperature changes to 27° C.?

When all three factors change (algebraic method). In reality, all of the factors — pressure, volume, and temperature — may change. This situation is called the *general gas law* and is usually expressed in the form of the following equation:

$$\frac{P_1V_1}{T_1} = \frac{P_2V_2}{T_2}$$

Problem. If a gas has a volume of 4 cu. ft. under a pressure of 20 lbs. per sq. in. and a temperature of 260° A., what volume will it have when the pressure changes to 40 lbs. per sq. in. and the temperature is 300° A.?

Substitute these values in the formula above. Cancel common factors out of the numerator and denominator of each side. Solve as indicated in the example.

$$\frac{\overset{1}{\cancel{20}} \times 4}{\underset{13}{\cancel{260}}} = \frac{\overset{2}{\cancel{40}} \times V_2}{\underset{15}{\cancel{300}}}$$

$$V_2 = \frac{\overset{}{\cancel{4}}}{13} \times \frac{15}{\underset{}{\cancel{2}}} = 2.31 \text{ cu. ft.}$$

Sometimes the temperature is asked for in a problem. This puts the unknown in the denominator. Some students have difficulty in solving in this case.

Problem. Given 100 cc. of air under pressure of 40 lbs. per sq. in. and a temperature of 300° A. The volume changes to 150 cc. and the pressure to 30 lbs. per sq. in.; what is the final temperature?

$$\frac{P_1V_1}{T_1} = \frac{P_2V_2}{T_2} \qquad \frac{40 \times 100}{300} = \frac{30 \times 150}{T_2}$$

Multiply both sides by T_2, which removes it from the denominator of one fraction and puts it into the numerator of the other.

$$T_2 \times \frac{40 \times \cancel{100}}{\underset{3}{\cancel{300}}} = \frac{30 \times 150}{\cancel{T_2}} \times \cancel{T_2}$$

Multiply both sides by 3 and divide by 40 to make T_2 be alone.

$$T_2 = \frac{\overset{3}{\cancel{30}} \times \overset{75}{\cancel{150}} \times 3}{\underset{2}{\cancel{40}}} = \frac{675}{2} = 337.5° \text{ A.}$$

Arithmetic method. It is possible to rearrange this general equation into various forms. When it is once changed, the actual working of problems becomes simply arithmetic.

$$(a) \quad V_2 = \frac{P_1}{P_2} \times \frac{T_2}{T_1} \times V_1$$

When a volume is the unknown, this form of equation is easy to use. It will be noticed that the given volume (V_1) must be multiplied by one fraction made of pressures and one fraction made of absolute temperatures. Common sense will tell how to set up these fractions. If the pressure has increased, make the fraction so the volume will be made smaller; if it has decreased, the fraction must make it larger. If the absolute temperature has increased, the numerator of the temperature fraction must be the larger so the volume will increase. If the gas has cooled down, the fraction must make the volume smaller.

$$(b) \quad P_2 = \frac{V_1}{V_2} \times \frac{T_2}{T_1} \times P_1$$

If the pressure is the unknown, form (b) is the best. Common sense will handle the temperature fraction as before. The volume fraction also can be set up by common sense. If the volume has increased, the pressure must have decreased, hence the fraction must make P_1 smaller.

$$(c) \quad T_2 = \frac{P_2}{P_1} \times \frac{V_2}{V_1} \times T_1$$

When absolute temperature is the unknown, form (c) is best. Common sense can again help us with the fractions. If pressure has increased, the fraction must make the temperature increase. Likewise the volume increases when the temperature increases and decreases when the temperature decreases.

Units. Pressure can be measured in any units so long as the same unit is used for both pressures. Some common pressure units are (a) pounds per sq. in., (b) grams per sq. centimeter, (c) centimeters or millimeters of mercury that the gas can balance in a tube, (d) atmospheres. Volume can be measured in cubic feet, cu. in., cu. centimeters, or liters. *Temperatures can only be used as absolute temperatures, found by adding algebraically 273 to the centigrade reading.*

Suppose the temperature is found to read 10° C. What is the absolute reading? Ans., $T = 10 + 273 = 283$. As another illustration, change 100° C. to absolute: $T = 100 + 273 = 373°$ A.

The most difficult situation to be found in changing centigrade readings to absolute is when the centigrade readings are negative, when adding algebraically is really subtracting. To change $- 5°$ C. to absolute: $- 5 + 273 = 268°$ A.

To change $- 10°$ C. to absolute: $- 10 + 273 = 263°$ A.

EXERCISES

1. What is meant by the absolute zero?
2. How far is the absolute zero below centigrade zero?
3. Room temperature is often 20° C. What is it on the absolute scale?
4. On a very cold night the temperature was $- 8°$ C. What is this on the absolute scale?

PROBLEMS

1. A certain mass of oxygen has a volume of 100 cc. when the pressure is 76 cm. of mercury and the temperature is 7° C. What will be the volume when the pressure is 80 cm. of mercury, and the temperature 100° C.?

2. In dealing with densities of gases, it is desirable to consider them all under the same conditions of temperature and pressure. The temperature and pressure chosen for these *standard conditions* are *zero degrees centigrade* and *76 cm. mercury pressure.*

A certain quantity of hydrogen measures 90 cc. under 74 cm. mercury and a temperature of 10° C. Find the volume under standard conditions.

3. Nine hundred liters of air under standard conditions have the pressure raised to 800 mm. of mercury and the temperature to 27° C. Find the new volume.

4. V_1 = 800 cu. ft.; P_1 = 20 lbs. per sq. in.; t_1 = 17° C.; find the pressure when V_2 = 780 cc. and t_2 is 27° C.

5. If V_1 = 1200 ml.; P_1 = 790 mm. mercury; and t_1 = 10° C.; what must the temperature change to in order that the volume may be 800 cc. when the pressure is 780 mm. mercury?

ADDITIONAL EXERCISES FOR SUPERIOR STUDENTS

1. The body temperature is 98.6° F. What is it on the absolute scale? (See the Appendix.)

2. In the North a thermometer registers − 30° F. Calculate this temperature to the absolute reading.

SUMMARY

A gas is composed of a large number of very small molecules moving at high velocities. It exerts pressure by the continuous impacts of the molecules and occupies a volume many times the absolute volumes of all of its molecules together.

Boyle's law, the first important gas law, states that at constant temperature, the pressure of a gas varies inversely as its volume.

The relation between pressure and temperature, and volume and temperature are both known as Charles's law. Temperature, when used in the gas laws, must be measured from the absolute zero, which is approximately 273° below the centigrade zero. The two forms of Charles's law are: (1) If the volume is kept constant, the pressure varies directly as the absolute temperature. (2) If the pressure is kept constant, the volume varies directly as the absolute temperature.

Expressed as equations, these gas laws are :

Boyle's law $P_1V_1 = P_2V_2$ (temperature constant)

Charles's law $\dfrac{P_1}{T_1} = \dfrac{P_2}{T_2}$ (volume constant)

Charles's law $\dfrac{V_1}{T_1} = \dfrac{V_2}{T_2}$ (pressure constant)

When all three factors change, the general gas law equation is used.

$$\frac{P_1V_1}{T_1} = \frac{P_2V_2}{T_2}$$

In any of these equations, all of the symbols must be represented by known numbers except one, which can be calculated.

Volumes and pressures may be measured in any suitable units, provided both pressures and both volumes are measured in the same units.   The temperatures must always be absolute temperatures.

CHAPTER 9

MOLECULAR AND ATOMIC WEIGHTS

Standard conditions. In problem 2, page 163, it was stated that scientists have adopted a standard pressure and temperature for considering volumes of gases. These *standard conditions* are a temperature of 0° C. and a pressure equal to that exerted by a column of mercury 76 cm. high. Although the gases are seldom measured under exactly standard conditions, it is easy to compute their volumes at standard conditions. Even if a substance changes to a liquid before it can be cooled to standard conditions, its volume can be measured at the higher temperature and calculated for standard conditions.

The problem of weighing a molecule. The problem of weighing a molecule is quite difficult. All ordinary molecules are too small to be seen with the most powerful microscope. Even if one could be seen and placed upon the pan of the most delicate balance in the world, its weight would not have the slightest effect upon it. The chemist, however, has

STANDARD CONDITIONS. In order to compare volumes of gases, it is necessary to reduce them to uniform conditions of temperature and pressure. The conditions agreed upon are a temperature of 0° C, and a pressure of 76 cm. of mercury.

invented a method of getting the molecular weight. His methods are more clever than those of the most skilful detective.

The units of molecular and atomic weights. It is not convenient to express molecular and atomic weights in grams or

Joseph Louis Gay-Lussac

1778–1850

Gay-Lussac one day noticed a seventeen-year-old girl clerk in a store reading a book. On asking her what she was reading, he was told that it was a chemistry. Her interest in chemistry impressed him so much that he cultivated her acquaintance and in time married her.

In the same year as his marriage the young chemist, who was now professor of chemistry at the Jardin des Plantes, announced his *law of the combination of gases by volume*, now called Gay-Lussac's law. He had found experimentally that one volume of oxygen combines with exactly two volumes of hydrogen to form water, that one volume of oxygen unites with exactly one volume of nitrogen to form nitric oxide, that one volume of nitrogen reacts with exactly two volumes of oxygen to form nitrogen dioxide, that two volumes of carbon monoxide unite with exactly one volume of oxygen to form two volumes of carbon dioxide, etc.

The simplicity of these ratios suggested that there must be some deep significance to them. Dalton was asked to explain these facts by his new atomic theory. However, he could not do this because he failed to distinguish between atoms and molecules. According to Dalton's idea, one atom of oxygen gas unites with one atom of nitrogen gas to make one atom of nitric oxide, which does not explain the fact discovered by Gay-Lussac, that one volume of oxygen unites with one volume of nitrogen to make *two* volumes of nitric oxide. Had he realized that each molecule of nitrogen and each molecule of oxygen contain two atoms, he would have been able to account for the two volumes of nitric oxide as found by Gay-Lussac.

It remained for the Italian physicist, Avogadro, to make a clear-cut distinction between atoms and molecules and thereby clear up the general confusion and explain Gay-Lussac's law.

any of the usual weight units. For instance, a molecule of water would weigh about .000,000,000,000,000,000,000,03 g. Such numbers as this are too awkward to use, so chemists have adopted a much smaller unit, one-sixteenth the weight of the oxygen atom.   This unit makes the weight of the oxygen atom 16 and that of the hydrogen atom 1.0081. In general, *the atomic weight of any element is the number of times that its atom is as heavy as this unit.* Likewise, *the molecular weight of any compound is the number of times that its molecule is as heavy as this unit.* There were two reasons for choosing this particular unit : (1) It is smaller than the weight of any atom (the hydrogen atom being the lightest), and (2) it makes more of the atomic weights come out whole numbers than would any other unit of reasonable size.

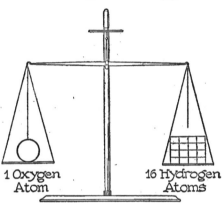

UNITS OF ATOMIC AND MOLECULAR WEIGHT. The accepted unit of atomic and molecular weights is one-sixteenth the weight of the oxygen atom — approximately equal to the weight of the hydrogen atom.

Gay-Lussac's law of combining volumes. We cannot understand how molecular and atomic weights are obtained until we study two more important gas laws. The first of these was discovered by the French scientist Gay-Lussac. The volumes of gases that take part in chemical reactions illustrate Gay-Lussac's law.

1. When hydrogen burns, the reaction is :

$$2\,H_2 + O_2 \longrightarrow 2\,H_2O$$
$$\text{2 liters} + \text{1 liter} \longrightarrow \text{2 liters}$$

that is, 2 liters of hydrogen react with 1 liter of oxygen to produce 2 liters of water vapor.

2 Qts. Hydrogen   +   1 Qt. Oxygen → 2 Qts. Hydrogen Oxide
(Water vapor)

1 Qt. Natural Gas + 2 Qts. Oxygen → 2 Qts. Water + 1 Qt. Carbon Dioxide

1 Qt. Nitrogen + 3 Qts. Hydrogen   ⟶   2 Qts. Ammonia

1 Qt. Nitrogen   + 1 Qt. Oxygen   2 Qts. Nitric Oxide

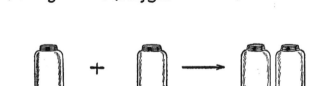

1 Qt. Hydrogen + 1 Qt. Chlorine → 2 Qts. Hydrogen Chloride

GAY-LUSSAC'S LAW OF COMBINING VOLUMES. When gases react, the ratio between any of the volumes involved can be expressed in simple whole numbers. Note that no fractional parts of a quart occur in any of the reactions.

2. Methane, a constituent of natural gas, burns:

$$CH_4 + 2\,O_2 \longrightarrow CO_2 + 2\,H_2O$$
1 liter + 2 liters $\longrightarrow$ 1 liter + 2 liters

3. Nitrogen unites with hydrogen in the Haber process of nitrogen fixation:

$$N_2 + 3\,H_2 \longrightarrow 2\,NH_3$$
1 liter + 3 liters $\longrightarrow$ 2 liters

4. Nitrogen combines with oxygen in another fixation process:

$$N_2 + O_2 \longrightarrow 2\,NO$$
1 liter + 1 liter $\longrightarrow$ 2 liters

5· Hydrogen combines with chlorine:·

$$H_2 + Cl_2 \longrightarrow 2\,HCl$$
1 liter + 1 liter $\longrightarrow$ 2 liters

Further volume relations. When we examine these reactions, we wonder why 1 liter of one gas should react with exactly 1 liter, or 2 liters, or 3 liters of the other gas, never with less than one liter, never with any fractional amount between 1 and 2 or between 2 and 3 liters. There must be some fundamental reason for this exact 1 : 1, 1 : 2, or 1 : 3 relation. Such an exact relationship could not be accidental, as it is so uncommon. For instance, all cats eat meat, but different cats eat different volumes of meat. There is no exact ratio between cats and meat. Any sponge soaks up water. However, different sponges soak up different volumes of water. There is no constant relation between sponges and the water they can hold. We could examine hundreds of processes without finding the participating constituents in any simple 1 : 1, 1 : 2, or 1 : 3 relation.

If we compare the products of the gaseous reactions just given, we find that the volumes of the products also bear simple whole-number relationships to the reacting volumes. These are given in detail below:

(1) hydrogen : oxygen = 2 : 1
hydrogen : water vapor = 2 : 2 or 1 : 1
water vapor : oxygen = 2 : 1

(2) methane : oxygen = 1 : 2
    carbon dioxide : water vapor = 1 : 2
    carbon dioxide : methane = 1 : 1
    water vapor : methane = 2 : 1
    oxygen : water vapor = 2 : 2 or 1 : 1
(3) nitrogen : hydrogen = 1 : 3
    nitrogen : ammonia = 1 : 2
    hydrogen : ammonia = 3 : 2
(4) nitrogen : oxygen = 1 : 1
    nitric oxide : nitrogen = 2 : 1
    nitric oxide : oxygen = 2 : 1
(5) hydrogen : chlorine = 1 : 1
    hydrogen chloride : hydrogen = 2 : 1
    hydrogen chloride : chlorine = 2 : 1

Gay-Lussac summarized these facts into one statement or law :

Whenever gaseous substances react, the ratios between the volumes of the reacting constituents, between the volumes of the products of the reaction, and between the volumes of any constituent and any product can be expressed by small whole numbers.

Although Gay-Lussac discovered this law, he did not know the meaning of it.

Avogadro's hypothesis. The second important gas law was discovered by an Italian scientist named Avogadro. He was the first person to explain what Guy-Lussac's law meant. Although Avogadro's explanation was perhaps little more than a good guess as far as he was concerned, it has stood the test of a hundred years of research. Today scientists are reasonably sure that Avogadro's explanation is approximately correct. This explanation is now known as Avogadro's hypothesis. It is called a hypothesis because it is not susceptible to direct proof. The statement of the hypothesis is as follows :

Equal volumes of all gases and vapors, measured under the same conditions of temperature and pressure, contain the same number of molecules.

This means that it makes no difference whether the molecules are light molecules or heavy molecules, large molecules or small molecules, reactive molecules or inert molecules. It is only the number that counts. It seems strange that the number of molecules in the gas required to fill a football should remain the same whether the molecules are large or small. The reason for this is that the space in the football is mostly empty. If the molecules were packed in close together, they would make only a tiny bit of solid. If the space occupied by the gas is mostly empty, it makes little difference whether the molecules are large or small, since in any case the volume

AVOGADRO'S HYPOTHESIS. The inverted tumblers are supposed to represent equal volumes. The five flies in one glass stand for five molecules of a gas with very small molecules. The other glasses each contain five larger molecules of gas, although no two of the gases have the same size molecules. Regardless of their weight or size, there are always the same number of molecules in equal volumes of gases.

of the molecules is small compared with that of the gas. For all practical purposes large molecules act the same as small ones.

Explanation of Gay-Lussac's law. Let us examine the first gaseous reaction, given to illustrate this law. According to Avogadro's hypothesis, a liter of any gas contains the same number of molecules as long as it is measured under standard conditions. In our illustration we will use the formulas for the different substances, since there isn't room to write the names out in full. Each little square under the equation represents 1 liter of volume. The n represents the number

Amadeo
Avogadro
1776–1856

. After Avogadro had read of
the contradiction between Dalton's atomic theory and Gay-
Lussac's law, he suggested that
perhaps the common gases are
not made up of units of single
atoms. He also became convinced of another fact, which
Gay-Lussac vaguely suspected;
namely, that *in equal volumes of all gases under the same temperature
and pressure, there are the same number of molecules.*

In 1811 Avogadro published his new discoveries. However,
this publication was not noticed by the scientific world until after
his death.

Although his hypothesis was exactly what was needed to clear
the confused thinking of the times, scientists did not make use of it.
By 1860 the confusion in scientific thought was almost unbearable.
Experiments with different compounds seemed to indicate that an
element might have more than one atomic weight. To try to help
the situation, an international congress of scientists was called to
meet at Karlsruhe, Germany. Italy's representative was Cannizzaro, a former pupil of Avogadro's. When his time came to speak,
he got up and presented Avogadro's hypothesis with zeal and enthusiasm. His speech was not fully comprehended and received little
applause. However, he passed around reprints of his article, "Outline of a Course in the Philosophy of Chemistry," based on the
theory of Avogadro.

On his way home from the convention, a young chemist named
Lothar Meyer read and reread Cannizzaro's reprint. "It was as
though the scales fell from my eyes," he wrote. "Doubt vanished
and was replaced by a feeling of peaceful clarity." This was the
beginning of the turning of the tide. In a few years order had
emerged out of chaos, and Avogadro had become one of the im-
mortals of chemistry.

of molecules in the liter and the formulas the kinds of molecules.

$$2 H_2 \; + \; O_2 \; \longrightarrow \; 2 H_2O \qquad \text{(hydrogen burns)}$$

$$\boxed{\begin{array}{c} n \\ H_2 \end{array}}\boxed{\begin{array}{c} n \\ H_2 \end{array}} + \boxed{\begin{array}{c} n \\ O_2 \end{array}} \longrightarrow \boxed{\begin{array}{c} n \\ H_2O \end{array}}\boxed{\begin{array}{c} n \\ H_2O \end{array}}$$

If our formulas are correct, Avogadro's hypothesis explains the above reaction nicely. Let us consider oxygen first. There are n molecules, each containing 2 oxygen atoms. Since each molecule of water vapor contains only one oxygen atom, there would be oxygen atoms enough to make $2 n$ molecules of the vapor, which would have a volume of two liters. Since each molecule of water vapor requires 2 hydrogen atoms, $2 n$ molecules would require $2 n \times 2 = 4 n$ hydrogen atoms, which is exactly what the two liters of hydrogen would afford. This shows us that Gay-Lussac's law has a fundamental reason underlying it.

In order to make this fact more emphatic, let us explain the other gaseous reactions in the same way. Equation (4), page 169, is simple, so we will consider it next.

$$N_2 \; + \; O_2 \; \longrightarrow \; 2 NO \qquad \text{(fixation of nitrogen by lightning)}$$

$$\boxed{\begin{array}{c} n \\ N_2 \end{array}} + \boxed{\begin{array}{c} n \\ O_2 \end{array}} \longrightarrow \boxed{\begin{array}{c} n \\ NO \end{array}} + \boxed{\begin{array}{c} n \\ NO \end{array}}$$

If Avogadro's hypothesis is true, there will be n molecules again in each liter of substance. The n molecules of oxygen, each containing two atoms, would contain $2 n$ atoms. The nitric oxide molecule needs only one oxygen atom, hence there must be $2 n$ of them, which would make two liters. This would call for one liter of nitrogen, which was used. Here again the hypothesis shows us why there had to be a 1 : 1 relation between the volumes of nitrogen and oxygen gases and a 1 : 2 one between either oxygen or nitrogen and nitric oxide, the final product of the reaction.

In equation (5) exactly the same reasoning applies as in reaction (4).

$$\boxed{\begin{array}{c} n \\ H_2 \end{array}} + \boxed{\begin{array}{c} n \\ Cl_2 \end{array}} \longrightarrow \boxed{\begin{array}{c} n \\ HCl \end{array}} + \boxed{\begin{array}{c} n \\ HCl \end{array}}$$

On the assumption of 2 hydrogen atoms per molecule, there are $2n$ hydrogen atoms, just enough to make $2n$ molecules of HCl, each of which contains only one atom of hydrogen. Since each molecule of HCl has only one chlorine atom, the two liters containing $2n$ molecules would require 1 liter of chlorine also containing n molecules or $2n$ atoms of this element.

Let us look at reaction (3).

$$\boxed{\genfrac{}{}{0pt}{}{n}{N_2}} + \boxed{\genfrac{}{}{0pt}{}{n}{H_2}} + \boxed{\genfrac{}{}{0pt}{}{n}{H_2}} + \boxed{\genfrac{}{}{0pt}{}{n}{H_2}} \longrightarrow \boxed{\genfrac{}{}{0pt}{}{n}{NH_3}} + \boxed{\genfrac{}{}{0pt}{}{n}{NH_3}}$$

If we assume with Avogadro that there are equal numbers of molecules per liter of all gases, that the molecules of the elementary gases each have two atoms, and that the experimental simplest formula for ammonia (NH_3) is the correct one, we understand why Gay-Lussac's law is true. The one liter of nitrogen gas containing $2n$ atoms is enough to form 2 liters or $2n$ molecules of ammonia containing one atom of nitrogen per molecule. Since each ammonia molecule contains 3 atoms of hydrogen, the $2n$ molecules would require $3 \times 2n = 6n$ atoms. With two hydrogen atoms per molecule and n molecules per liter, there would be $2n$ atoms of hydrogen per liter. The total volume of hydrogen would be $6n \div 2n = 3$ liters.

Reaction (2) is explainable in a similar way.

$$\boxed{\genfrac{}{}{0pt}{}{n}{CH_4}} + \boxed{\genfrac{}{}{0pt}{}{n}{O_2}} + \boxed{\genfrac{}{}{0pt}{}{n}{O_2}} \longrightarrow \boxed{\genfrac{}{}{0pt}{}{n}{CO_2}} + \boxed{\genfrac{}{}{0pt}{}{n}{H_2O}} + \boxed{\genfrac{}{}{0pt}{}{n}{H_2O}}$$

Here again, if the formulas for the gaseous compounds are as noted, that is, the simplest obtained by quantitative analysis, the volumes are accounted for. In the liter of methane there will be n atoms of carbon. This will permit also n molecules of CO_2, hence also 1 liter. Since 2 hydrogen atoms go into each molecule of water, the $4n$ hydrogen atoms in the liter of methane will make $2n$ molecules of water vapor, or two liters. The right side of the equation calls for $4n$ oxygen atoms, which can be obtained by 2 liters of oxygen.

WEIGHT RATIOS IN ELEMENTARY GASES

ELEMENT	WT. OF 1 L. (n MOLECULES) OF ELEMENT	COMPOUND ANALYZED	WT. OF ELEMENT IN 1 L. OF COMPOUND	RATIO OF WEIGHTS
Nitrogen . .	1.25 g.	Ammonia	.625 g.	$\frac{1.25}{.625} = \frac{2}{1}$
Hydrogen . .	.09 g.	Hydrogen chloride	.045 g.	$\frac{.09}{.045} = \frac{2}{1}$
Chlorine . .	3.12 g.	Hydrogen chloride	1.56 g.	$\frac{3.12}{1.56} = \frac{2}{1}$

We have seen that Gay-Lussac's law of combining volumes as applied to these reactions is easily explained by Avogadro's hypothesis and the formulas for the gases. These simplest empirical formulas for the gaseous compounds can be obtained from the quantitative analysis of the substances. The assumed molecules of two atoms of N_2, O_2, H_2, Cl_2 are based upon experiment also. If we analyze 1 liter of water vapor, we get a certain weight of oxygen. One liter of oxygen gas weighs exactly twice as much. Therefore, on the basis of Avogadro's hypothesis of equal numbers of molecules in the equal volumes, there must be 2 atoms in each molecule of oxygen. The same reasoning can be applied to each of the other elementary gases, as the experimental results in the accompanying table show.

GRAM-MOLECULAR VOLUME. This volume unit (22. 4 l.) was chosen so that there would always be as many grams of the substance as there are units in its molecular weight.

We have assumed that there is only 1 atom of oxygen in the molecule of water vapor, 1 atom each of hydrogen and chlorine in hydrogen chloride, and 1 atom of nitrogen in ammonia. Since everything works out so nicely, it is reasonable to believe that this assumption is correct. Many experimental facts are in harmony with it and none contradict it.

A new volume unit. The liter is not the most convenient unit for measuring volumes, especially for problems in weight. Let us choose a gas such as hydrogen chloride, which contains as little hydrogen as is ever found in a liter of any hydrogen compound. Now let us choose a volume large enough to contain 1 gram of hydrogen. There is .045 gram of hydrogen in 1 liter of HCl, therefore it would take $\frac{1}{.045}$ = 22.4 liters to contain 1 gram. This new volume is called the *gram-molecular volume*. We will not say now why it has been given this peculiar name, but just think of it as a new unit of volume, one that enables us to do wonderful things. This new volume unit is very convenient. With it the weights of gases taking part in reactions are whole grams rather than fractions of a gram.

REVIEW QUESTIONS

1. What are standard conditions of temperature and pressure?
2. What is meant by the atomic weight of oxygen?
3. What is meant by the molecular weight of a substance?
4. State Gay-Lussac's law of combining volumes.
5. State Avogadro's hypothesis. Why is it called a hypothesis instead of a law?
6. How was 22.4 liters determined as a suitable volume?
7. Why is 22.4 liters a more convenient volume than the liter for measuring gases?
8. What name is given to 22.4 liters?
9. Why are atoms not weighed in grams?
10. Why does a given number of small molecules in a gas occupy as much space as the same number of large molecules in another gas under the same temperature and pressure?
11. If n molecules of water are electrolyzed according to the equation $2 H_2O \longrightarrow 2 H_2 + O_2$, how many molecules of oxygen and how many molecules of hydrogen are formed?
12. If lightning fixes n molecules of nitrogen according to the equation $N_2 + O_2 \longrightarrow 2 NO$, how many molecules of nitric oxide result? How many molecules of oxygen would be needed?

13. If the reaction for the Haber process of nitrogen fixation is $N_2 + 3 H_2 \longrightarrow 2 NH_3$, how many molecules of ammonia result from the fixation of n molecules of nitrogen? How many molecules of hydrogen are needed also? What is the ratio between the number of hydrogen molecules and the nitrogen molecules? What is the ratio between the number of ammonia molecules and the number of hydrogen molecules?

14. Methane burns as shown by the reaction:

$$CH_4 + 2 O_2 \longrightarrow CO_2 + 2 H_2O$$

If n molecules of CH_4 were taken, give the number of molecules of each of the other substances involved. Express ratios between the participants of this reaction taken in pairs.

Determining molecular weights. We have defined the molecular weight of a substance as *the number of times its molecule is as heavy as one-sixteenth the oxygen atom.* Let us think of 22.4 liters of HCl. Let N represent the number of molecules in the 22.4 liters. Since one liter of no compound of hydrogen contains less hydrogen than one liter of hydrogen chloride, it is natural that scientists have assumed that there is only 1 hydrogen atom in each molecule.

The weight of the 22.4 liters of any gaseous substance will give us the molecular weight in grams, since, by Avogadro's hypothesis, 22.4 liters will always have N molecules of the gaseous substance.

$$\frac{\text{wt. of 22.4 liters}}{1 \text{ g.}} = \frac{N \times \text{wt. 1 molecule}}{N \times \text{wt. at. unit}}$$

Since the N's cancel out of the numerator and denominator, and the weight of the atomic unit is 1,

$$\frac{\text{wt. of 22.4 liters}}{1} = \frac{\text{wt. 1 molecule}}{1} \text{ (numerically)}$$

and the weight of 22.4 liters in grams equals the molecular weight numerically.

We see now why the volume of 22.4 liters was called the *gram-molecular volume.* There is one gram for each atomic unit in the molecular weight. As an illustration of what the gram-molecular volume means, water vapor has a molecular

Theodore W. Richards

1868–1928

Most of Richards' early training was under the direction of his mother. She being a good teacher and he an apt pupil, he progressed faster than the average child of his age. At the age of thirteen he was qualified to enter Haverford College. Being rather young for a college student, he studied under his mother for another year and then entered as a sophomore. In 1885 he graduated with honors.

Richards was greatly interested in chemistry as a boy, performing many experiments at home. In Haverford he received a thorough training in chemistry. The next fall Richards continued his education at Harvard College Working under Professor Cooke, he became interested in revising the atomic weights of the elements. In this work he showed an exceptional aptitude for taking extreme pains to eliminate every possible source of error. His determination of the relative atomic weights of oxygen and hydrogen has been one of the most accurate determinations of this ratio.

After getting his Doctor's degree in 1888, Richards studied in Europe under Janasch, Victor Meyer, and Hempel. Back at Harvard in 1891 he began teaching at that college, in which work he continued the rest of his life. Still interested in atomic weights, Richards measured the atomic weights of barium, strontium, zinc, and (with the help of his students) twenty other elements. Up to the present time there have been no determinations of the atomic weights more accurate than he made. Many of the methods of procedure invented by him are still the best that are known.

Taking a year's absence from teaching, Richards studied physical chemistry under Ostwald and Nernst in Germany in 1895. After returning home, he did a great deal of work in heat chemistry. In this work, again, he devised some of the most accurate methods known.

weight of 18. The gram-molecular volume of water vapor weighs 18 grams. The molecule of ammonia weighs 17 times one atomic unit. The weight of 22.4 liters of ammonia is 17 g. An equal volume of carbon dioxide weighs 44 g., which tells us the molecule of carbon dioxide weighs 44 times the atomic unit. To summarize the method we would say:

To find the molecular weight of an unknown substance, weigh 22.4 liters of it under standard conditions of temperature and pressure. The resulting weight in grams is numerically equal to the molecular weight.

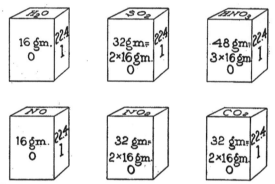

ATOMIC WEIGHT OF OXYGEN. In 22.4 l. of oxygen will be found 16 g. of oxygen, 2 × 16 g., 3 × 16 g., or some other exact multiple of 16 g. It is believed that where there are 16 g. of oxygen in 22.4 l. there is 1 atom of oxygen in each molecule of the compound; where there are 2 × 16 g. of oxygen, there are 2 atoms of oxygen in the molecule, and so on. The atomic weight of oxygen is taken as 16.

Practically, 22.4 liters is too large a volume to weigh on the balances. However, if a smaller amount is weighed, it is easy to calculate the weight of 22.4 liters. Sometimes a substance will not vaporize without decomposition; another method must then be used. The special problems involved in getting molecular weights for nonvolatile substances will be left for study in advanced chemistry.

Measuring atomic weights. The measuring of the atomic weight of an unknown element is very difficult in practice. The theory, however, is not difficult to understand. In the discussion following, we will talk about breaking compounds into their elements and weighing the quantities of the individual elements in them without telling how it can be done.   The method of finding an atomic weight is to *take 22.4 liters of each of the gaseous compounds of that element and tear them to pieces and weigh the portion of element alone.   The smallest weight found in any of the compounds is the atomic weight.*   This assumes that in the compound or compounds having this smallest weight there is only one atom of that element in the molecule.   If several compounds are known, the chances of

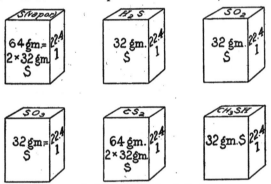

ATOMIC WEIGHT OF SULFUR.   Since in 22.4 l. of all the gaseous compounds of sulfur there are 32 g. of sulfur or some exact multiple of 32 g., 32 is considered the atomic weight of sulfur.

this assumption being wrong are very small, especially if the next larger amount is exactly twice as much and the next in size, three times, etc.

To illustrate the method, suppose a number of carbon compounds are tested.   The 22.4 liters will yield 12 grams of carbon from carbon dioxide, methane, formaldehyde, carbon monoxide, and methyl chloride.   From ethane, acetaldehyde, methyl formate, and acetic acid will be obtained 24 grams of carbon, which is 2 × 12 g.   From propane, ethyl formate, and methyl acetate are obtained 3 × 12 g. in each case.   Thousands of other carbon compounds are known which yield various multiples of 12 grams of carbon, but always an integral multiple of 12 grams.

The conclusion is that the atomic weight of carbon is 12.

Similarly, 22.4 liters of any common sulfur compound will yield 32 grams of sulfur, 64 grams, or some other exact multiple of 32. Hence we conclude the atomic weight of sulfur is 32.

If we examine the bromine compounds, 22.4 liters will yield 80 grams of bromine, 160 grams of bromine, or some other exact multiple of 80. Hence we conclude that 80 is the atomic weight of bromine.

Finding formulas. The volume 22.4 liters was chosen so that a substance which contains only one atom of hydrogen in the molecule would contain 1 gram of hydrogen. For every additional hydrogen atom in the molecule there will be another gram of hydrogen. Likewise every other atom occurring in the molecule will have as many grams as there are units in the atomic weight. When there are 2 or more atoms per molecule, there will be 2 or more times as many grams as there are units in the atomic weight.

What the elements are, is learned by analysis. To find the number of atoms of each of the elements, take the weight of that element in 22.4 liters and divide it by the atomic weight of the element. For instance, a substance is found on analysis to contain H, N, and O. From 22.4 liters of the substance there is obtained 1 g. of hydrogen, 14 g. of nitrogen, and 48 g. of oxygen. Dividing 1 by the numerical value of the atomic weight of hydrogen, 1, we get 1 H in the formula. Dividing 14 by the atomic weight of nitrogen (14), shows that there will be 1 N in the formula also. Dividing 48 by 16, we get 3 oxygen atoms, and the complete formula is HNO_3.

Another substance analyzes to give 24 grams of carbon and 6 grams of hydrogen in the gram-molecular volume. Applying this process in the case of carbon, $24 \div 12 = 2$, and in the case of hydrogen, $6 \div 1 = 6$. The formula, then, is C_2H_6.

Finding valences. Finding the formulas also measures the valences of the elements. HCl shows that chlorine has the same valence as hydrogen, 1. H_2O shows that the valence of oxygen is 2. H_2S shows that the valence of sulfur is also 2. $AlCl_3$ shows that the valence of aluminum is 3. Thus we can continue on until we have found the valences of all the elements.

Conclusion. In conclusion, we note that the determination of molecular weights is easy in practice, while atomic weights are very difficult to determine completely. The underlying theory is not very difficult in either case, and with careful study the student ought to be able to follow it. When we realize that we are able to weigh atoms and molecules so small that they are invisible under the most powerful microscope, so light that they would not make the slightest impression on the most delicate balance in the world, and in such rapid motion that we could not possibly keep track of them, we feel that we really have accomplished a great work. We should be willing to study a little theory, if it can enable us to do such wonderful things.

REVIEW QUESTIONS

· 1. Why does the weight in grams of 22.4 liters of a substance equal the molecular weight, numerically?

2. State the method of finding the atomic weight of an element.

3. How is a formula determined?

PROBLEMS

1. From the data given, compute the molecular weights of the following compounds : (a) carbon dioxide, 22.4 l. weigh 44 g. ; (b) nitric acid, 22.4 l. weigh 63 g. ; (c) ammonia, 22.4 l. weigh 17 g. ; (d) nitrogen, 22.4 l. weigh 28 g. ; (e) water, 22.4 l. weigh 18 g. ; (f) hydrogen sulfide, 22.4 l. weigh 34 g.

2. Calculate the atomic weight of carbon from the following weights of carbon in 22.4 l. of each substance. Carbon dioxide, 12 g. ; ethane, 24 g. ; carbon monoxide, 12 g. ; wood alcohol, 12 g. ; grain alcohol, 24 g. ; propyl alcohol, 36 g. ; ethyl formate, 36 g. ; ethyl acetate, 48 g.

3. Calculate the atomic weight of oxygen from the weights of oxygen obtained from 22.4 liters of each of the following : carbon dioxide, 32 grams ; carbon monoxide, 16 grams ; water, 16 grams ; hydrogen peroxide, 32 grams ; sulfur trioxide, 48 grams ; methyl oxalate, 64 grams.

4. The gaseous compounds of fluorine contain the following weights of fluorine in 22.4 liters of each of several of the fluorine compounds : 38 g. ; 19 g. ; 38 g. ; 19 g. ; 57 g. ; 19 g. ; 76 g. What is the atomic weight of fluorine?

5. Calculate the formulas of carbon dioxide from the following data: 22.4 liters contains 12 grams of carbon and 32 grams of oxygen; the atomic weights are: $C = 12$; $O = 16$.

6. Analysis of a gram-molecular weight of a substance gave 24 grams carbon, 6 grams hydrogen, 16 grams oxygen. Calculate the formula if the atomic weights are $C = 12$, $H = 1$, and $O = 16$.

7. Calculate the formula of the substance the gram-molecular volume of which yields the following: 24 grams carbon, 6 grams hydrogen, and 32 grams sulfur. $C = 12$, $H = 1$, $S = 32$.

ADDITIONAL EXERCISES FOR SUPERIOR STUDENTS

1. How can the volume of a gas be obtained for standard conditions if it condenses to a liquid before it reaches zero degrees centigrade?

2. How could it be proved that hydrogen gas is H_2?

3. Acetylene gas has the formula C_2H_2. Give an experimental reason why it is not written CH.

4. Sugar decomposes without vaporizing. Will the method described in this chapter serve to determine its molecular weight?

5. Would this method be practical for a piece of quartz?

REFERENCES FOR SUPPLEMENTARY READING

1. Jaffe, Bernard. *Crucibles.* Simon and Schuster, New York, 1930. Chap. IX, pp. 157–176, "Amadeo Avogadro."
2. Holmyard, Eric John. *Makers of Chemistry.* The Clarendon Press, Oxford, 1931. Pp. 248–257, "Avogadro."
   (1) Write a 300-word essay on the life of Avogadro.

SUMMARY

Molecules and atoms are weighed in terms of an arbitrary unit (one-sixteenth the weight of the oxygen atom), since the common units of weight are a billion trillion times too large for convenience.

In order to understand how molecules and atoms are weighed, it is first necessary to consider two important gas laws. The first of these is Gay-Lussac's law, which says that whenever gaseous substances react, the ratios between the volumes of the reacting constituents, between products of the reaction, and between any constituent and any product can be expressed by small whole numbers. (It is understood that when volumes are discussed, they are assumed to be measured at 0° C. and 760 mm. of mercury pressure.)

Avogadro's hypothesis is necessary to explain the law of Gay-Lussac. It says: Equal volumes of all gases and vapors measured under the same conditions of temperature and pressure contain the same number of molecules. These two principles enable us to calculate the molecular weights from a few simple measurements. If a reaction involves only a small number of molecules — since the numbers of molecules are proportional to the volumes — it follows that small numbers of volumes will be involved also.

The liter is not the most convenient volume for making molecular and atomic weight determinations. If a unit of 22.4 liters is chosen, those substances with only one atom of hydrogen per molecule will contain almost exactly one gram of hydrogen. This volume will also contain the atomic weights in grams of all other atoms occurring one per molecule, and some multiple of this when there is more than one atom per molecule.

To determine the molecular weight it is only necessary to get the weight of 22.4 liters of the substance. Atomic weights are more difficult to determine. These involve studying many of the compounds of the element, weighing 22.4 liters of the gaseous substance, decomposing it, and weighing the particular element alone. If a given weight is found in some of the experiments, twice that weight in others, three times in still others, etc., the lowest weight is the atomic weight.

To obtain the formula for a compound, its molecular weight is first found. Next the compound is decomposed into its elements and the weight of each element in 22.4 liters determined. The weight of each element divided by its atomic weight gives the number of its atoms in the molecule.

CHAPTER 10

THE CHLORINE FAMILY OF ELEMENTS

Introduction. The study of chemistry is somewhat simplified by the fact that we may group together certain elements that are very similar. The elements fluorine, chlorine, bromine, and iodine form such a family of similar elements. They are often called the *chlorine family* and sometimes the *halogens*, which means the *salt formers*. Chlorine is a constituent of table salt, and all the others form quite similar salts.

Family or group relationships. The order in which the members of the family were named—fluorine, chlorine, bromine, and iodine—is significant. All the physical and all chemical properties vary in this same order, which is also

BROMINE AND IODINE. Bromine is a heavy red liquid; iodine is a dark violet solid.

the order of increasing atomic weights with fluorine 19, chlorine 35.5, bromine 80, and iodine 127. The intensity of color varies from yellow fluorine, through green chlorine and red bromine, to purple iodine. The density increases in the same way. Fluorine is a light gas; chlorine, a denser gas; bromine, a liquid; and iodine, a solid under ordinary temperatures and pressures. The boiling points and melting points of these elements are in the same order. These elements are all very reactive, but this property, too, varies in degree in the same order. Fluorine is the most reactive substance known. Chlorine is very reactive but not so reactive as fluorine. Bromine is less reactive than chlorine but more reactive than iodine, the last in the list. However, iodine is considered a reactive substance.

Carl Wilhelm
Scheele

1742–1786

Scheele was apprenticed to an apothecary at the age of four-teen. This pharmacist made his own medicines from the crude drugs, a procedure which indicates he was well versed in chemistry. Young Scheele became much interested in his chemical books, and repeated most of the experiments described in them. His memory was so good that he remembered all of the chemical facts in a book after one reading.

In 1773 Scheele accepted a position in Lokk's pharmacy at Upsala. One day Lokk observed that potassium nitrate which had been fused for some time gave off red fumes when treated with vinegar, which other potassium nitrate did not do. The professors at the University of Upsala could not explain this curious phenomenon. Scheele, however, explained it by assuming that heating the potassium nitrate caused it to lose oxygen and become the salt of nitrous acid, it being the nitrous acid which showed the unusual properties. Carrying on research at the pharmacy, Scheele discovered oxygen independently of Priestley.

The research for which Scheele is best known is his discovery of the element chlorine. He allowed hydrochloric acid to stand in contact with finely powdered pyrolusite (crude manganese dioxide). He observed that the mixture acquired a suffocating odor. The substance responsible for the suffocating odor he named "dephlogisticated muriatic acid." Later, however, Davy concluded that this substance is an element. Scheele found that chlorine dissolves slightly in water, giving it an acid taste. It also bleaches flowers and green leaves and attacks most metals.

He also paved the way for the discovery of tungsten by preparing tungstic acid from a white Swedish mineral called tungsten.

Finally, he also discovered nitrogen independently by absorbing the oxygen of the air by a mixture of sulfur and iron filings.

Occurrence of the halogens. Because of the great activity of the elements of the chlorine family, they occur only in their compounds. Chlorine, by far the most abundant member of the family, occurs in common salt, or sodium chloride. There is enough salt in the ocean to cover the land surface many feet deep; hence, the ocean is an inexhaustible source of both chlorine and sodium. Besides the salt that is found in the ocean and other salty bodies of water, such as the Great Salt

Courtesy of Leslie Salt Co

TABLE SALT FROM OCEAN WATER. Although most of our salt comes from inland deposits, on the west coast there is a thriving industry of recovering salt from sea water. Notice the conveyor system by which the salt is removed from the bottom of the evaporating pond to the pile in the background.

Lake and the Dead Sea, there are large beds of salt in various parts of the world.

Sodium bromide, a salt quite like table salt, occurs in comparatively small amounts in sea water and certain salt deposits in Europe. It is the principal source of bromine.

Fluorine occurs principally as calcium fluoride (CaF_2) and a mineral called cryolite (Na_3AlF_6), which is a double fluoride of sodium and aluminum. Fluorine occurs in somewhat the same amounts as bromine.

Iodine is very scarce, occurring in sea water only as a trace. The seaweed needs it and concentrates it in its tissues. For a long time the seaweed was our chief source of iodine. At the present time the bulk of our iodine comes from Chile, where it is deposited along with nitrates.

As a consequence of Chile's practical monopoly, until quite recently iodine cost several dollars per pound, a cost which limited its use for many purposes for which it was suitable. Recently, however, it has been discovered that the brine at the bottom of the oil wells in southern California is quite rich in iodine compounds. Competition from this new iodine industry in the United States has resulted in lowering the price of iodine almost to one dollar per pound wholesale. As a consequence of this price reduction, its use in the future is likely to increase.

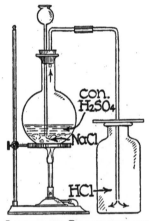

LABORATORY PREPARATION OF HYDROGEN CHLORIDE. When concentrated sulfuric acid and sodium chloride are heated together, the sodium and hydrogen trade places. The resulting hydrogen chloride, being a gas, is driven off by the heat into the collecting bottle.

Hydrogen chloride (hydrochloric acid). A little before the French Revolution, a man by the name of Leblanc invented a method of making soda. Hydrogen chloride was a waste product of the process, which escaped into the atmosphere. Its strong acid action annoyed the neighbors, rusted their iron implements, and destroyed their garden plants. The government made Leblanc move the plant out into the country. Out in the country, it killed the trees and crops of the farmers, who raised such a fuss that the factory had to do something with the hydrogen chloride other than allow it to escape into the atmosphere. It was run into the water of a stream in the hope that this would end the nuisance. However, it killed the fish and made the water unfit for the livestock to drink. At this stage of the process there began to be a need for a strong acid. When the gaseous HCl was passed into water, it was found to be very soluble and to make a concentrated solution of a very strong acid. Soon the acid became very valuable, as more uses for it were learned.   In time the hydrogen chloride, which had previously been a nuisance, became more profitable than the soda.   Today hydrochloric acid is one of our necessary chemicals — one which is used by the thousands of tons, and one which is needed in every chemical laboratory.

Preparing hydrogen halides. Hydrogen chloride is prepared from table salt by the same general method by which nitric acid was obtained — that is, by heating the salt with sulfuric acid.

$$2\,NaCl + H_2SO_4 \longrightarrow Na_2SO_4 + 2\,HCl$$

The HCl, being a gas, separates from the other materials in the generating flask and escapes into the receiving flask.   Any of the other hydrogen halides can be made by the same method if the corresponding fluoride, bromide, or iodide is used with the sulfuric acid.

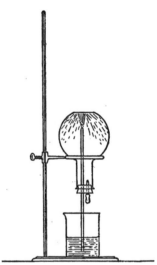

VACUUM FOUNTAIN. If a flask of hydrogen chloride is connected with a beaker of water as shown, the gas dissolves so rapidly in the water that the pressure in the flask is reduced and atmospheric pressure forces the water to spray into the flask.

The hydrogen compounds of the halogens. All of the halogens form hydrogen compounds similar to HCl.   They are all invisible gases very soluble in water, forming strong acids.   These hydrogen compounds are not equally stable.   The fluorine end of the series is very stable, but the stability gradually decreases from member to member until HI slowly decomposes in its solutions until they turn dark because of free iodine.

Hydrogen fluoride. Hydrogen fluoride has two very unusual properties which make it useful.   It dissolves glass, which property makes it useful for marking glassware, such as measuring glasses.   The glass to be marked is covered with a protecting coating of paraffin.   The figures and lines to be

etched on the glass are scratched through the wax so the glass
is exposed.   The glassware is now put into a lead room in

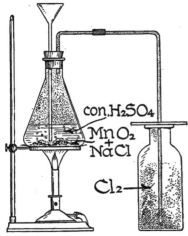

which the hydrogen fluoride
is generated.   It attacks the
glass and makes the mark.
As a laboratory reagent, hydrofluoric .acid, the water
solution of hydrogen fluoride,
must be kept in a paraffin
bottle, as it would dissolve
the glass in a glass bottle and
thereby both exhaust the
reagent and destroy the
bottle.

Glass dissolves because
it contains silicon.   Hydrofluoric acid reacts with silicon compounds to form gaseous silicon fluorides.   This
fact makes hydrofluoric acid
a useful reagent in chemical

LABORATORY PREPARATION OF CHLORINE.
If hydrogen chloride is prepared in the
presence of an oxidizing agent, such as
manganese dioxide, the oxidizing agent removes the hydrogen, leaving free chlorine.

analysis.   Silicon compounds are very difficult to remove from
samples of ores.   The silicon must be removed, however, before a successful analysis can be made.   Hydrofluoric acid

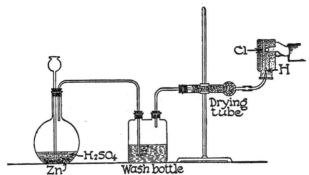

CHLORINE REACTS WITH HYDROGEN.   Hydrogen burns in an atmosphere of
chlorine very much as it does in oxygen.

GLASS ETCHED WITH HF. Hydrogen fluoride reacts with the silicon of the glass, forming gaseous silicon fluoride, which escapes and leaves a mark where the glass was previously.

changes the silicon to its gaseous compound, which can be driven off by heat.

$$SiO_2 + 4\,HF \longrightarrow SiF_4 + 2\,H_2O$$

Hydrofluoric acid will also dissolve iron rust. The laundryman uses it to remove rust spots from clothes.

Salts of the halogens. Most of the other compounds of the halogens are called salts. There are a large number of salts, the chlorides, the bromides, the fluorides, and the iodides. Most of the salts are white substances, which look like table salt but taste very differently.

Uses of the salts of the halogens. Sodium chloride as a flavoring and food preservative tops the list of the halogen salts in usefulness. It is used by the carload in many industries. The livestock industry uses it for salting the animals and for preserving meats. The laundry industry uses it to regenerate its water softeners when they become filled with lime from the hard water. The soap-making industry uses salt to separate the soap from

PARAFFIN BOTTLE. Because hydrofluoric acid attacks glass, it must be kept in a p'raff'n bottle.

the liquor in which it is cooked. Common salt is the source of all chlorine gas and the chlorine in most chlorine compounds, as well as the source of all sodium metal and the sodium in most sodium compounds.

The bromides, or salts of bromine, are used in medicine as sedatives. They are also the source of what free bromine is needed in industry.

The iodides also are used in medicine. In certain localities they are added to the drinking water to prevent goiter. They are the source of the free (uncombined) iodine. Because of their scarcity, iodine and iodides are very expensive, which limits their use in the industries. Chlorides, bromides, and iodides are all used in photography.

QUESTIONS OF FACT

1. Name the halogens in the order of: (a) increasing atomic weights; (b) increasing density; (c) increasing boiling points; (d) decreasing tendency to react; (e) decreasing stability of the hydrogen compound. Are the arrangements all in the same order?

2. Which of the following statements are true and which are false?

(a) Halogen means salt former.

(b) Chlorine is the most abundant halide.

(c) The halogens usually occur free.

(d) Iodine compounds are more expensive than those of the other halogens.

3. Name two compounds of fluorine.

4. Give two uses of hydrogen fluoride.

5. Tell the story of hydrochloric acid in connection with the soda works of Leblanc.

6. Give three uses of sodium chloride.

7. Give one use of bromides and one of iodides.

8. Discuss the occurrence of sodium chloride.

9. Why is iodine so much more expensive than the other halogens?

10. Give the formulas of two minerals containing fluorine.

11. What discovery greatly lowered the price of iodine in the United States?

12. How are the hydrogen halides prepared from their salts?

References for Supplementary Reading

1. Clarke, Beverly L. *Marvels of Modern Chemistry*. Harper and Brothers, New York, 1932. Chap. XIII, pp. 178–184, "The Red-Headed Halogens."
    (1) Name the element with which fluorine does not react. (p. 179)
    (2) Relate the account of one potassium chlorate explosion. (p. 180)
    (3) Discuss the relative stability of the oxygen compounds of the halogens. (pp. 179–180)
    (4) Where does fluorine occur in the human body? (p. 181)
    (5) How fast is fluorine thought to be escaping from the ground in the Valley of Ten Thousand Smokes in Alaska? (p. 181)
    (6) What is the principal source of bromine in the United States?
2. Foster, William. *The Romance of Chemistry*. D. Appleton-Century Company, New York, 1936. Chap. VIII, pp. 111–128, "A Natural Family of Elements, the Halogens."
    (1) Are the members of the chlorine family metallic or nonmetallic elements? (p. 111)
    (2) What are the simple salts of these elements called: (*a*) of fluorine? (*b*) of chlorine? (*c*) of bromine? (*d*) of iodine? (p. 112)
    (3) How thick is the rock salt at Sperenberg, Germany? (p. 112)
    (4) Give the dimensions of the salt deposit in Galicia. (p. 112)
    (5) What did the name bromine originally mean? (p. 118)
    (6) What per cent of sodium iodide should be in the salt used by people drinking water from melting snow? (p. 124)
    (7) Explain the relation of chlorine to smoke screens. (pp. 125–126)

How the halogens are obtained free from chemical combination. One method of freeing an element from other elements is by the use of electricity. This method will be explained in a later chapter. The common method is first to prepare the hydrogen compound, from which the hydrogen can then be removed by oxidation.

$$4 \, HCl + MnO_2 \longrightarrow MnCl_2 + Cl_2 + 2 \, H_2O$$

Bromine and iodine can be prepared by similar reactions.

Uses of the halogens in the free state. The halogens when uncombined are very reactive. Chlorine gas escaping from tanks of liquid chlorine was used as a war gas in the World War. The effects on the soldiers before gas masks were used were horrible beyond description. Chlorine, however, has a legitimate peacetime use for sterilizing water in swimming pools.

Enough of the substance can be dissolved in the water to kill the germs and still be tolerated by the swimmers. The germ-killing ability of chlorine is remarkable. It can even be used to sterilize drinking water and is so used by many cities. Water for household purposes naturally should not contain enough chlorine to taste. Even in dilutions that do not spoil the taste, chlorine will kill practically all of the dangerous bacteria.

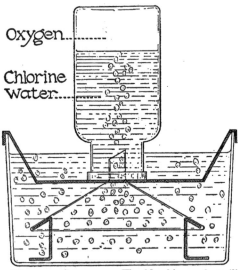

Oxygen

Chlorine
Water

CHLORINE AND WATER IN SUNLIGHT. The bleaching and sterilizing action of chlorine is due to the liberation of oxygen. When sunlight falls on chlorine dissolved in water, oxygen soon separates out and may be collected as shown in the drawing.

Liquid chlorine is used by laundries in preparing their bleaching solutions. For this purpose, the chlorine is passed into soda solution, forming sodium hypochlorite (NaClO) solution. This solution readily parts with its oxygen, which is very reactive for an instant, while in the nascent, or atomic, condition. This active oxygen bleaches soiled clothes readily. Sodium hypochlorite solution is sold under the trade names of "Javelle Water" and "Clorox" for home use. It is a good antiseptic for flesh wounds. For this purpose, it goes under the name of *Carrel-Dakin solution*. Chlorine is used in water solution as *chlorine water*, which is an oxidizing and chlorinating reagent.   Chlorine finds considerable use in mak-

ing organic chlorine compounds such as chloroform, methyl chloride, and carbon tetrachloride. Chloroform ($CHCl_3$) is an anesthetic and a solvent.   Carbon tetrachloride (CCl_4) is used in fire extinguishers and in dry cleaning.

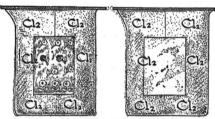

Free bromine is used in preparing certain drugs, dyes, and chemicals.

BLEACHING WITH CHLORINE. A great deal of chlorine is used in laundries and the textile industry in bleaching cloth. It is really oxygen that does the bleaching, since the cloth will not bleach if perfectly dry. The chlorine reacts with water to liberate oxygen, which does the bleaching. Note the difference between the square of dry cloth in the beaker at the left and that of the moist cloth in the beaker at the right.

Free iodine in alcohol solution is called tincture of iodine and finds common use as an antiseptic for cuts and skin wounds.

CHLORINE REACTS WITH ANTIMONY. Chlorine like other members of the family is an active element. It reacts violently with antimony to form antimony chloride.

Chemical properties of the members of the chlorine family. Fluorine, chlorine, bromine, and iodine are very reactive elements. They combine directly with many metals. Antimony reacts with them violently enough to produce light and heat. Hydrogen may be ignited and burned in an atmosphere of chlorine in much the same way as in air.

Tests for the halogens. A test in chemistry is a chemical reaction that distinguishes a substance from other substances. There are very few tests that are specific, that is, given by one substance only.   However, it is often possible to add some reagent which will keep all but one substance from giving the test, which makes the test practically specific.   A reaction that gives a colored product or a gas with a characteristic odor is always considered a good test.   A white

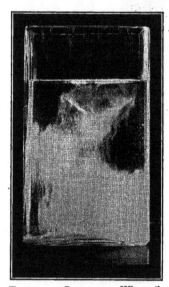

precipitate (under such conditions that all the substances present but one are soluble) is one of the most common tests, and, although not so satisfactory a test as a colored product, it serves very well.

Fluorine is so reactive that it is not obtained free, except for a short time under very special conditions.   The other three members of this family are easily identified in the elementary condition: chlorine an irritating green gas, bromine a red irritating liquid, and iodine a solid that *sublimes* — that is, changes directly to the vapor state.   Its vapor has a beautiful purple color, which is not duplicated in appearance in any other gas.

TEST FOR A CHLORIDE.   When silver nitrate is added to a nitric acid solution of any chloride, a white precipitate results.   This reaction is the usual test for a chloride.

In the combined condition, the halogens are frequently tested for as fluorides, chlorides, bromides, and iodides.

The fluoride is identified by liberating the hydrogen compound with sulfuric acid and seeing that it will etch glass.

The other three halides can be identified by other tests.   If it is merely a question of one of the three — chloride, bromide, or iodide — the substance may be dissolved and a little nitric acid added.   Now if silver nitrate solution is added, a white or yellow insoluble precipitate shows the presence of chloride, bromide, or iodide.   It is not always easy to tell them apart.

Silver chloride is white, while the other two are yellowish.   If an iodide is suspected, free bromine may be added and the solution shaken up with chloroform.   If it is an iodide, the bromine will release the iodine, which will form a purple solution in the chloroform; otherwise, the bromine itself will turn the solution red.   If the unknown halide is to be tested for bromide, chlorine may be used to set free the bromine, which will color the chloroform red.

Since the chloride is white when precipitated as silver chloride, silver nitrate is said to be the test for the chloride.   A typical reaction would be:

$$NaCl + AgNO_3 \longrightarrow AgCl + NaNO_3$$

Suppose the test were made on $AlCl_3$.

$$AlCl_3 + 3 AgNO_3 \longrightarrow 3 AgCl + Al(NO_3)_3$$

The replacing of one halide by another again emphasizes their arrangement in a sequence according to their atomic weights — Fl, Cl, Br, I.   Any one will replace all of those that follow it in this sequence.

The naming of salts.   There are many oxygen salts of the halogens. The table below gives typical ones of chlorine and tells what each name implies with respect to oxygen content.

TEST FOR BROMIDE.   If chlorine gas is passed into a solution of a bromide, the chlorine enters the solution, forcing the bromine out.   The bromine is soluble in chloroform, with the formation of the characteristic reddish-brown color.

COMPOUNDS OF SODIUM, CHLORINE, AND OXYGEN

FORMULA	NAME	IMPLICATION OF NAME
$NaClO_4$	sodium *perchlorate*	contains most oxygen
$NaClO_3$	sodium *chlorate*	contains much oxygen
$NaClO_2$	sodium *chlorite*	contains less oxygen
NaClO	sodium *hypochlorite*	contains least oxygen
NaCl	sodium *chloride*	contains no oxygen

Thus we see that a salt with the prefix *per* and the ending *ate* contains the most oxygen. The ending *ate* without the prefix *per* means much oxygen in the salt. The ending *ite* means less oxygen than the *ate*. The prefix *hypo* and the ending *ite* really mean less oxygen than the *ite* substance, yet some oxygen. The *ide* ending shows that there is no oxygen present.

Sublimation. Under the preceding topic, iodine was said to sublime. There are a few substances which when heated change directly to the gaseous state without becoming liquids first. When cooled, these substances change directly back again to solids. Dry ice is dry because it sublimes and goes into the colorless gaseous condition without becoming liquid. Naphthalene, from which ordinary moth balls are made, also sublimes. This fact makes naphthalene suitable for protecting clothes. It evaporates fast enough to produce a concentration of odorous vapor which keeps away the moths yet never produces any liquid which might stain the cloth. Substances that sublime can be purified from nonvolatile impurities by sublimation and then condensing the vapor by cooling.

REVIEW QUESTIONS

1. Give three uses for free chlorine.
2. What is Carrel-Dakin solution?
3. What do we mean when we say a substance sublimes?
4. Which element in the chlorine family sublimes?
5. Which halogen in the free state was used as a war gas?
6. What chlorine compound is an anesthetic?
7. What chlorine compound is used in fire extinguishers?
8. Give three uses of free bromine.
9. What is tincture of iodine?
10. What may serve as a test for chloride?
11. How could a salt be proved to be a fluoride?
12. How can chlorides, bromides, and iodides be distinguished from one another?
13. In naming the oxygen salts of the halogens, what combination is applied to those containing the most oxygen? Which to those with least oxygen?
14. What is meant by the endings *ate* and *ite?*

REFERENCES FOR SUPPLEMENTARY READING

1. Foster, William. *The Romance of Chemistry.* D. Appleton-Century
   Company, New York, 1936. Chap. VIII, pp. 111–128, "A Natural
   Family of Elements, the Halogens."
   (1) How much chlorine can be prepared daily by the plant at the Edg-
       wood Arsenal in Maryland? (p. 113)
   (2) Compare the rate of reaction of hydrogen and chlorine in darkness
       and in the light. (p. 114)
   (3) What chlorine reactions must be catalyzed by water? (p. 115)
   (4) Explain how the use of tetraethyl lead determines largely the price
       of bromine in the United States. (p. 119)
2. Weeks, Mary Elvira. *The Discovery of the Elements,* Journal of Chemical
   Education, Easton, Pa., 1934. Chap. XVII, pp. 253–277, "The
   Halogen Family."
   (1) By whom was chlorine discovered? (p. 253)
   (2) What did he call it? (p. 253)
   (3) What did Berthollet think chlorine was? (p. 255)
   (4) Who proved chlorine is an element and not a compound? (p. 256)
   (5) When was iodine discovered and by whom? (p. 257)
   (6) Relate the incident of the discovery of iodine. (pp. 258–259)
   (7) What did Gay-Lussac discover about iodine? (p. 262)
   (8) Relate the story of how a student practically discovered bromine.
       (p. 262)
   (9) Who is credited with the discovery? (p. 262)
   (10) Relate Ballard's discovery. (p. 264)
   (11) Relate how Liebig just missed discovering bromine. (p. 266)
   (12) Which halogen compound poisoned several chemists? (p. 267)
   (13) Name two men who predicted the discovery of fluorine. (p. 267)
   (14) Name several substances which are attacked by fluorine. (p. 268)
   (15) Who discovered fluorine and when? (p. 268)
   (16) How was this element finally obtained? (pp. 270–273)

SUMMARY

Fluorine, chlorine, bromine, and iodine have similar chemical
properties. They constitute a chemical family called the halogens. These elements are too active to occur uncombined,
but their salts are found in the ocean, in other bodies of salt
water, and in a few deposits resulting from the drying up of
salt lakes. The chlorides are by far the most abundant halogen compounds.

The hydrogen halides are colorless gases which are readily

soluble in water, forming strong acids. Hydrofluoric acid has the uncommon property of etching or dissolving glass. These substances are prepared by heating the salt with concentrated sulfuric acid.

$$NaBr + H_2SO_4 \longrightarrow NaHSO_4 + HBr$$

Large quantities of sodium chloride are used in the laundry, soap, and food industries.

With the exception of fluorine the free halogens can be obtained from their acids by removing the hydrogen with an oxidizing substance.

$$4\,HBr + MnO_2 \longrightarrow MnBr_2 + Br_2 + 2\,H_2O$$

In the free state chlorine is used for sterilizing water, bleaching, preparing bleaching solutions, and synthesizing fire-extinguishing and dry-cleaning liquids. Free bromine and iodine are used in synthesizing medicinal chemicals. Tincture of iodine is a valuable antiseptic.

The halogen salts in solution form precipitates with silver nitrate, precipitates which do not dissolve in nitric acid. This is considered a test. To distinguish between chlorine, bromine, and iodine, chlorine is passed into the solution. If the substance is a chloride, nothing happens; if it is a bromide, bromine is liberated, which can be extracted in chloroform to form a red solution; while if it is an iodide, the chloroform will become purple.

The halogens form a series of oxygen salts, with one oxygen atom in the hypochlorite, two oxygen atoms in the chlorite, three oxygen atoms in the chlorate, and four oxygen atoms in the perchlorate molecule.

Iodine sublimes — that is, changes directly from the solid to the gaseous state.

CHAPTER 11

CHEMISTRY AND ELECTRICITY

Two kinds of electricity. If we take a piece of hard rubber and rub it with a cat skin, it becomes electrified. It shows this electrification by attracting and picking up small shreds of paper or ravelings of cloth or any very light substance.

If we change our materials, using a glass rod instead of hard rubber, and a silk cloth instead of the cat skin, we get the glass rod electrified also. This electricity, like that on the rubber

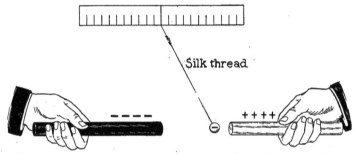

Hard rubber rubbed with wool   Glass rubbed with silk

TWO KINDS OF ELECTRICITY. Glass rubbed with silk produces positive electricity; rubber rubbed with wool produces negative electricity. The two kinds of electricity behave differently toward a charged pith ball.

rod, will pick up small shreds of any substance. However, it can be shown that there is a difference between the electricity on the two rods. Suppose we suspend a pith ball by a silk thread and touch it with the charged rubber rod. Now if the rubber rod is again charged and brought near the charged pith ball, the latter is repelled and pushed away long before the rod can touch it. If the charged glass rod is brought near it, however, the charged pith ball is attracted and pulled towards the charged glass. This difference in behavior of the charged pith ball shows that there is a fundamental difference

between the electricity on the rubber rod and that on the glass
rod.

Two kinds of electricity can neutralize each other. Suppose we touch one suspended pith ball with the charged rubber
rod and another one near by with the charged glass rod. The
two pith balls will attract each other and bump together. If
the amounts of electricity are equivalent, neither of the pith
balls will show any electrification after touching. The two
kinds of electricity will have neutralized each other. Because
the two kinds of electricity neutralize each other, and because
of other properties of electrification, one kind of electricity has

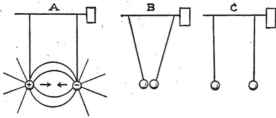

UNLIKE CHARGES NEUTRALIZE EACH OTHER. If two equally but oppositely
charged pith balls touch each other, the charges are neutralized, leaving both
pith balls uncharged.

been called positive (+) and the other negative (−). The
electricity on the glass was arbitrarily chosen as positive, and
that on the rubber as negative.

Electrical nature of matter. We wonder where the electricity comes from when we electrify the rubber rod. Modern
scientists believe that all substances are made entirely of positive and negative electricity. When there is an equal amount
of both kinds, they neutralize each other, and the substance
shows no electrification. If one kind is in excess, the substance
is electrified, or charged. When the cat skin is rubbed on the
rubber rod, the friction pulls apart equal amounts of positive
and negative electricity. The positive electricity is held by
the cat skin, while the negative electricity remains in excess on
the rubber rod. In the case of the silk and the glass, the silk
takes away the negative electricity and leaves the positive on
the glass.

Benjamin Franklin

1706–1790

Benjamin Franklin, although not primarily a chemist, was a great scientist and inventor. His work in connection with electricity laid a solid foundation, upon which many of our important chemical laws have been built and without which much of our chemical knowledge and many of our great chemical industries never could have come into existence.

At a time when science was stumbling over rival theories as to the nature of electricity, Franklin evolved a clear-cut conception, which has stood the test of time and is still held. He believed that electricity was composed of very small particles, so small that they can pass between the atoms of matter. He believed that there were two kinds of electrical particles, which he called positive and negative particles because the opposite kinds attract each other and then neutralize each other on coming in contact.

Franklin proved that lightning is a flash of electricity of the same kind that is obtained from a battery. This he did by bringing the electricity down the string of a kite which was flying in a thunderstorm. He found that electricity from a cloud would charge a condenser exactly as does the electricity from a battery. When Franklin announced his discovery, European scientists laughed at the idea. However, when they repeated his experiments, they found he was right, after which he was recognized as a real scientist. The invention of the lightning rod was the immediate practical result of Franklin's experiments with the lightning.

One discovery of Franklin's was essentially chemical, and that was that breathing the air in a room removes oxygen so that the air in a closed room eventually becomes unfit for breathing. In other words, he discovered the need for ventilation and preached the doctrine of fresh air.

Units of electricity. Electricity, both positive and negative, is thought to be made of very small particles, chunks, packages, or charges. These fundamental separate negative charges of electricity are called *electrons*, while the corresponding positive charges are called *protons*. The electron and the proton are electrically equivalent, but very different in mass. The proton weighs about 1845 times as much as the electron. The most recent research seems to indicate that it is possible to separate the proton into a small mass equal to the mass of the hydrogen atom, and a positive charge of electricity of practically the same mass as that of the electron. This electrical charge is called the *positron*. However, since in all common chemical processes the proton seems to be the positive unit, we will use it in our electrical discussions.

There has also been recently discovered a particle of mass the same as that of the proton but without an electrical charge. This particle, called the *neutron*, is thought to be a proton which has an electron joined to it to make it neutral. There is also thought to be another neutral particle made by removing a positron from a proton. This is called the *neutrino*. Since these neutral particles are rarely detected and since the positron remains separated only a small fraction of a second, we will consider in this chapter only the proton and the electron, which are the only ones involved ordinarily in chemical reactions.

The chemist often speaks of a plus charge of electricity, meaning the charge of the proton. He often speaks of a minus charge of electricity, meaning the charge of the electron. The physicist has defined a larger quantity of electricity as the unit charge. However, in this text we shall use the chemist's understanding of the charge rather than the physicist's.

Law of electrical charges. It will be recalled that a negatively charged pith ball is attracted by a positively charged glass rod but repelled by a negatively charged rubber one. Similarly, a positively charged pith ball is attracted by the negative electricity on the rubber and repelled by the positive charge on the glass. These facts may be generalized into one statement, or law: *Like charges repel and unlike charges attract.*

Conductor and insulator. A copper wire allows electricity to pass through it readily. Such a substance is called a *conductor* of electricity. Many metals and some solutions are good conductors of electricity. A substance like glass or porcelain that stops the flow of electricity is called an *insulator* of electricity. Silk, cotton, rubber, distilled water, and certain liquids and solutions are insulators. Many substances are intermediate between these extremes and may be classed as poor conductors and poor insulators. Since the positive charge has the mass attached to it, and since it does not seem probable that the mass can travel through the metal of a wire, we think that the current in a copper wire consists of a stream of electrons only.

LAW OF ELECTRICAL CHARGES. This cartoon illustrates the principle: Like charges repel; unlike charges attract. The hens (top) represent negative charges, which repel each other. The roosters (middle) represent positive charges, which also repel each other. The hen and the rooster (below) are pictured as attracting each other.

Electrolytes and nonelectrolytes. The passage of electricity through certain conducting water solutions has led to some very interesting and profitable results. Several men have made fortunes from this study.

The first thing we learn when we try to pass electricity through solutions is that, in many cases, it cannot go. It cannot pass through any pure liquids. Pure distilled water offers great resistance to the flow of electricity. In fact, it is an excellent insulator.

If we dip the electrodes (ends of the wires made wide so as to have good contact with the solution) from a battery into a water solution of any of the following substances, no current

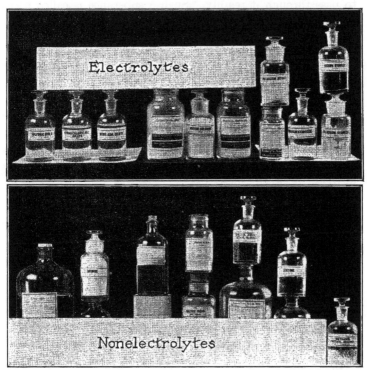

ELECTROLYTES AND NONELECTROLYTES. The upper illustration shows (left to right): sulfuric acid, hydrochloric acid, nitric acid, calcium chloride, sodium chloride, magnesium sulfate, potassium nitrate (above), lead nitrate, ammonium hydroxide, sodium hydroxide (above), and potassium hydroxide. The lower illustration shows xylene, sucrose, pyridine, thymol (above), maltose, ethanol (above), carbon tetrachloride, acetone, and methanol. Acids, bases, and salts are electrolytes; all other substances, nonelectrolytes.

will flow: sugar, glucose, ether, alcohols, esters, and benzene. However, if we dissolve in the water an acid, a base, or a salt, the current will flow. These substances — acids, bases, and salts — that, when dissolved in water make it a conductor of electricity, are called *electrolytes*. The rest of the substances, typified by sugar, which even in concentrated solutions does not carry the current, are called *nonelectrolytes*.

What is an electric current? In order to understand how an electrolyte makes water a conductor of electricity, we must

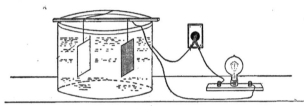

TESTER FOR ELECTROLYTES. To determine whether or not a substance is an electrolyte, it is dissolved in water in the cell and connected as shown in the drawing. If it is an electrolyte, the globe will light up; if not, no light will appear.

know the nature of an electric current. An electric current may be either a stream of negative electrons moving in one direction or a stream of positive protons moving in the opposite direction or both streams moving simultaneously.

Since the current cannot flow through pure water, it is evident the electrons and protons from the battery cannot dive in and swim across the water. Current does flow, however, when an electrolyte is dissolved in the water. The natural conclusion would be that the electrolyte supplies positive and negative charges in the water, charges which might move and become a current.

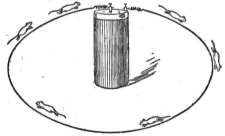

Theory of ionization. To explain how an electrolyte causes a current

NATURE OF AN ELECTRIC CURRENT IN A WIRE. The cartoon represents the electrons as individual mice moving along the wire. The electrons themselves are of course invisible.

to flow through water, the chemist Arrhenius invented the *theory of electrolytic dissociation* or *theory of ionization*. According to the theory, when the electrolyte dissolves in water, it separates into positively charged particles and negatively charged particles, called *ions*. Sometimes there are an equal number of each kind. This, however, is true only when the ions carry equal charges. The total number of positive charges always equals the total number of negative charges. When an atom or radical becomes an ion, it carries the same kind and number of charges as it has positive or negative valences. Representative substances ionize as follows:

$$NaCl \longrightarrow \overset{+}{Na} + \overset{-}{Cl}$$
$$K_2SO_4 \longrightarrow 2\overset{+}{K} + \overset{=}{SO_4}$$
$$AlBr_3 \longrightarrow \overset{+++}{Al} + 3\overset{-}{Br}$$

The separation takes place as soon as the electrolyte dissolves. No battery is needed to produce it. If the water should be evaporated, the ions would be forced to unite again into the neutral substance.

It may be wondered where the charges on the ions came from. According to the theory, the atoms and radicals are held together in chemical union by electrical forces — that is, the attraction between positive and negative charges. In discussing valence, it was stated that some of the elements and radicals formed the positive parts of the compounds, while others formed the negative parts. Those elements or radicals forming the positive parts of compounds (which have the positive valences) also form the positive ions. The negative parts of the compounds (which have the negative valences) form the negative ions. The ionizing solvent is thought to weaken the forces which hold the positive and negative parts of a compound together so that these parts can separate and circulate by themselves in the solution. A negative ion is an atom or a radical with one or more unneutralized electrons, as $\overset{-}{Cl}$, $\overset{-}{Br}$, $\overset{-}{OH}$, $\overset{-}{NO_3}$, $\overset{=}{SO_4}$, and $\overset{\equiv}{PO_4}$, in which the minus charges represent the number of unneutralized electrons. A positive ion is an element or radical with one or more unneutralized positive charges, such as Na^+, K^+, NH_4^+, Ca^{++}, Al^{+++} and Fe^{+++}, in which the number of positive charges represents the number of unneutralized protons. Another way of considering posi-

Svante Arrhenius

1859–1927

While attending the University of Upsala, Arrhenius became interested in the conductivity of electricity through solutions. The phenomenon of electrolysis had been known for almost a century, yet no one had ever been able to explain what took place in the solutions during electrolysis. When ready for his doctorate, Arrhenius went to the University at Stockholm to study the problem of his interest. Day after day and far into the night he studied the electrolysis of different substances in solution.

Finally there came to him a picture of what was happening in solutions. He saw the dissolved molecules separating into independent particles, some of which carried positive charges and some negative charges. When the electrodes connected to a battery are placed in such a solution, the positive ions are attracted towards the negative electrode, while the negative ions are pulled towards the positive electrode. When these ions arrive and have their charges neutralized, they undergo the reactions occurring there.

When he finished his thesis containing the new radical theory, Arrhenius returned to Upsala to take his examination for his Doctor's degree. The custom was to appoint someone to attack the thesis of an applicant and find all manner of fault with it. The applicant then had to explain and defend his thesis.

Arrhenius defended his thesis carefully but conservatively. He said afterward that if he had expressed the full meaning of his theory, he would not have been granted his degree. As it was, the thesis was given such a low rating that the degree was more of a humiliation than an honor.

Today the theory of electrolytic dissociation has gained such universal acceptance that it is taught in colleges and high schools the world over.

tive ions is to think of them as having lost one or more electrons, thereby acquiring the excess positive charges.

How the theory explains carrying of current by electrolytes. When a spoonful of the electrolyte dissolves in a quart of water, its molecules separate into an enormous number of charged particles, which are distributed throughout all parts of the solution. The negative electrode, receiving a strong negative charge from the battery, pulls on all of the positive ions and pushes against all the negative ions. The positive electrode, on the other hand, pulls on all the negative ions and pushes on all the positive ones. Both electrodes working together keep the ions moving through the solution. We have already learned that a current may be a stream of negative charges moving in one direction, and a stream of positive charges moving simultaneously in the opposite direction. This accounts for the current in the solution; positive charges are moving in one direction, and negative charges in the opposite direction.

The magnitude of the current may depend upon the number of moving charges and the speed at which they travel. The mass of the ion upon which the charge is riding makes little difference in the current except that the heavy ions tend to move more slowly than light ones.

Ions may have more than one charge. Some ions carry one charge, such as the positive ions:

$$Na^+, K^+, NH_4^+, H^+, Ag^+, Cu^+, Hg^+,$$

and the negative ions:

$$Cl^-, Br^-, I^-, OH^-, NO_3^-, HCO_3^-, HSO_4^-.$$

Other ions carry two charges. The commoner ones are:

$$Zn^{++}, Ca^{++}, Mg^{++}, Fe^{++}, Sn^{++}, Ba^{++}, Sr^{++},$$
$$Pb^{++}, S^=, SO_3^=, SO_4^=, O^=, SO_3^=.$$

Ions with three charges are not so common. The following should become familiar to the student:

$$Fe^{+++}, Al^{+++}, P^{+++}, PO_4^\equiv.$$

A few ions carry four and five charges, but no common ion carries more than five charges. Those of four and five charges are :

$$Sn^{++++}, Si^{++++}, P^{+++++}.$$

COMPARISON OF ATOMS AND IONS. Although atoms and ions differ only in electric charge, their properties are quite unlike. For example, copper atoms (in metallic copper) are copper colored, while copper ions (in solution of copper salts) are blue. Likewise, lead atoms are metallic looking, while lead ions are colorless; zinc atoms are bluish-gray, while zinc ions are without color; and iron atoms are different from the yellow iron ions.

It will be noticed that the number of charges carried by an ion is the same as its valence or holding power for other ions.

The difference between an ion and an atom. *An ion is an atom or a radical carrying an electric charge.* Its properties as an ion are often quite different from its properties as an atom. For instance, the silver ion (Ag$^+$) is invisible, while silver atoms (Ag) in clusters form silver, black in color. The silver ion is soluble in water, while silver metal is not. The hydrogen ion is sour in taste, while the hydrogen atom has no taste. The cupric ion (Cu^{++}) is blue, while the copper atom is copper colored. In every case, the ion has some property different from the atom or group of atoms it becomes when it loses its electric charge.

QUESTIONS OF FACT

1. How can an object be charged with negative electricity?

2. How will a charged body act towards very light objects, such as small bits of paper?

3. How can a body be given a charge of positive electricity?

4. How will two negatively charged objects act towards each other?

5. How will two positively charged particles act towards each other?

6. How will a positively charged particle act towards a negatively charged particle?

7. State the law of electrical charges.

8. What is the nature of the charge of a proton?

9. What is the nature of the charge of the electron?

10. Name four substances which are insulators.

11. Name four substances which are conductors of electricity.

12. In which class is distilled water?

13. How does ordinary water compare with distilled water as a conductor of electricity?

14. Describe an experimental method of proving there are two kinds of electricity.

15. What is the difference between an electrolyte and a nonelectrolyte?

16. Name four substances that are nonelectrolytes.

17. Name three kinds of substances that are electrolytes.

18. What is an electric current?

19. Name seven ions with one plus charge on each.

20. Name seven ions with one minus charge on each.

21. Name eight ions with two plus charges on each.

22. Name five ions with two minus charges on each.

23. Name three ions with three plus charges and one with three minus charges on each.

24. Name those ions with four and five charges.

25. Who first thought of the idea of substances separating into ions?

QUESTIONS OF UNDERSTANDING

1. What is meant by + electricity and − electricity neutralizing each other?

2. Where does the electricity come from when a rubber rod is rubbed with a cat skin?

3. Does an unelectrified body contain electricity? Explain.

4. What is the difference between an insulator and a conductor?

5. Define electrolyte and nonelectrolyte.

6. State the theory of electrolytic dissociation.

7. What is an electrode?

8. State three different ways an electric current may be constituted.

9. What is the particle formed by a proton uniting with an electron called? Is it electrified?

10. Compare the positron with the electron as to (a) mass; (b) magnitude of charge; (c) nature of charge.

11. What would have to be removed from a proton in order to get a neutrino?

12. How does the theory of electrolytic dissociation explain how an electrolyte carries current through water?

13. State the difference between the ion and the atom in the following cases: (a) Ag^+ and Ag; (b) H^+ and H; (c) Cu^{++} and Cu.

REFERENCES FOR SUPPLEMENTARY READING

1. Foster, William. *The Romance of Chemistry.* D. Appleton-Century Co., New York, 1936. Chap. X, pp. 145–160, "Electricity in the Service of Chemistry."
   (1) How was the first man-made electrical effect produced? (p. 145)
   (2) What property came to amber when it was rubbed by cloth? (p. 146)

(3) What additional facts were learned about electricity by Gilbert, A.D. 1600? (p. 146).

(4) How was another kind of electricity discovered? (p. 146)

(5) Who named the two kinds of electricity? (p. 146)

(6) What great discovery did Benjamin Franklin make regarding electricity?

(7) Define a positively charged body and a negatively charged body. (p. 146)

(8) How is it proved that Cu^{++} carries two charges? (p. 153)

2. Jaffe, Bernard. *Crucibles.* Simon and Schuster, New York, 1930. Chap. XII, pp. 219–241, "Biography of Svante Arrhenius."

(1) Write a detailed biography of Svante Arrhenius.

(2) Did Arrhenius have any better opportunity of becoming a great scientist than you have?

Other evidence in favor of the ionic theory. There is another set of facts that also demands the ionic theory for its explanation. This has to do with the *freezing points* of solutions. Pure water freezes at.0° C. If some sugar is dissolved in the water, it has to be cooled below zero before it will freeze. The more sugar that is dissolved in the water, the lower the freezing point becomes. When the molecular weight of the sugar in grams is dissolved in 1000 grams water, the solution freezes at − 1.86° C. The gram-molecular weight of a substance is called a *mole.* It makes no difference which nonelectrolyte is dissolved in the water, the freezing point of the molal solution is always the same distance below zero. Its molecules may be either large or small, it may be sweet or tasteless, it may differ from other nonelectrolytes in a variety of properties; but so long as it is a nonelectrolyte, it will produce the same freezing point lowering when it is in the same concentration in moles per 1000 grams of water. The only way we can explain why all molal solutions lower the freezing point by the same amount is to assume that the lowering is dependent only on the number of particles. The 22.4 liters of different substances studied in a previous chapter always gave a mole of the substance, and, according to Avogadro's hypothesis, there are the same number of molecules in a mole of any substance that has not dissociated or separated into simpler substances, that is, into ions.

When any electrolyte is made up into a molal solution in water, the freezing point is abnormally low.   For instance, for HCl it is lowered almost $2 \times 1.86°$ in contrast to the $1.86°$ for nonelectrolytes.   This could be explained on the basis of each molecule having separated into two particles, the ions.

$$HCl \longrightarrow H^+ + Cl^-$$

If the electrolyte $CaCl_2$ is used, the freezing point lowering is almost 3 times the usual $1.86°$.

$$CaCl_2 \longrightarrow Ca^{++} + 2\ Cl^-$$

If $AlCl_3$ is chosen, the lowering will be nearly $4 \times 1.86°$.

$$AlCl_3 \longrightarrow Al^{+++} + 3\ Cl^-$$

These cases also can be explained on the basis of the molecule separating into its ions.

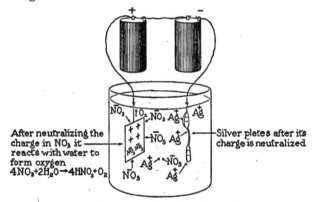

After neutralizing the charge in NO_3 it reacts with water to form oxygen $4NO_3 + 2H_2O \rightarrow 4HNO_3 + O_2$

Silver plates after its charge is neutralized

ELECTROLYSIS OF SILVER NITRATE.   The electrolysis of silver nitrate results in the deposition of silver at the negative electrode and the escape of oxygen at the positive electrode.   As the negative nitrate ions are neutralized at the positive plate, they react with the water of the solution to free oxygen and form nitric acid.

Thus we see that the theory that the molecules of electrolytes dissociate into ions while those of the nonelectrolytes do not explains the abnormal lowering of the freezing point by electrolytes.   The fact that this new line of thought leads to the same conclusion as the conductivity of solutions gives us faith

that the theory of ionization is true although it is not susceptible of direct proof and must still be called a theory.

Electrolysis. The part of this subject that has made fortunes
for many chemists and built up great industries is electrolysis.
By electrolysis we mean the chemical reactions that take place
when the ions reach the electrodes and have their charges
neutralized. If we know the
properties of the atom or radical that is formed when the
ionic charge is removed, we
can predict what will happen.

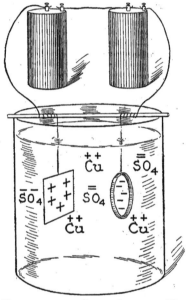

Plating metals. Let us
consider silver plating when
silver nitrate solution is electrolyzed. When the Ag^+
loses its charge, it becomes
silver metal; hence the negative electrode becomes coated
with silver. If we wish to
silver-plate a watch fob, we
have to make it part of the
negative electrode. When
the NO_3^- ion loses its charge,
it reacts with the water to
form nitric acid, and oxygen
is given off:

ELECTROLYSIS OF COPPER SULFATE. The
copper plates out on the negative electrode as the positive ions arrive. The
sulfate ions have their negative charges
neutralized at the positive pole, and the
resulting sulfate radicals react with water
to form hydrogen gas and sulfuric acid.

$$4\,NO_3 + 2\,H_2O \longrightarrow 2\,HNO_3 + O_2$$

This process "kills three birds
with one stone." It deposits
silver, generates oxygen, and synthesizes nitric acid. Practically, silver plating is the only part of the process that is
made use of. The oxygen is allowed to escape and the nitric
acid is not recovered, as these are not formed in sufficient
amounts to make it worth while to collect them.

Gold, copper, and other metals can be deposited by a similar
process. In practical plating, the process is not quite so

simple as this suggests. As outlined, the silver would deposit too fast and be soft. Sodium cyanide is added to the solution to cut down the number of silver ions, so the plate will be firm. Cleaning the metal for plating is difficult in some cases. There are also problems of current density and voltage.

Electrolysis of salt. In the electrolysis of sodium chloride solution, the chlorine ion has its charge neutralized and becomes chlorine gas, most of it bubbling out of solution. When the

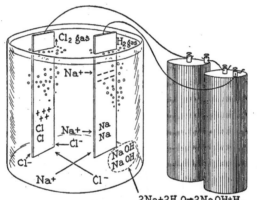

$$2Na+2H_2O \rightarrow 2NaOH+H_2$$

Reaction of sodium after its charge has been neutralized

ELECTROLYSIS OF SALT. The electrolysis of salt produces three useful substances, hydrogen gas, chlorine gas, and sodium hydroxide. The negative electrode attracts the sodium ions and their charges are neutralized. The resulting sodium atoms react with water to form hydrogen and sodium hydroxide. The hydrogen bubbles away at the electrode, but the sodium hydroxide accumulates in the solution. The chlorine ions are neutralized at the other electrode, from which chlorine gas escapes.

sodium loses its charge and becomes sodium metal, it reacts with water to liberate hydrogen and leaves sodium hydroxide in the solution.

$$2\,Na + 2\,H_2O \longrightarrow 2\,NaOH + H_2$$

Here again three useful substances are formed from salt, one of the cheapest chemical substances. Most of the chlorine and sodium hydroxide used in industry is made in this way.

Electrolysis of water. In the chapter on hydrogen, the electrolysis of water was mentioned. When the hydrogen ion loses its charge by having it neutralized, it escapes as hydrogen gas. The sulfate radical reacts with water to free oxygen gas and re-form the sulfuric acid. Since the process does not use up H_2SO_4 but replaces it, and since water is consumed, it is essentially the electrolysis of water.

$$2\,SO_4 + 2\,H_2O \longrightarrow 2\,H_2SO_4 + O_2$$

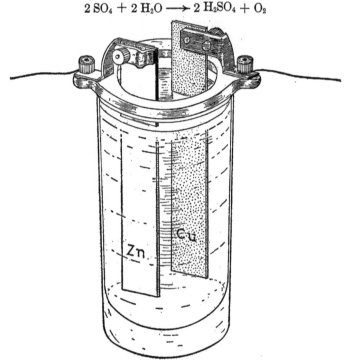

SIMPLE CELL. When a zinc electrode (Zn) and a copper electrode (Cu) are dipped into dilute sulfuric acid and connected into a circuit with a current-measuring instrument, electricity is found to be flowing in the circuit.

In this process only two useful substances result. However, in every electrolysis there is one chemical reaction at each electrode, so there must be at least two substances formed. If one of the products reacts with water, there will be three new substances.

Plating metals with copper, silver, gold, nickel, and chromium, and the manufacture of chlorine, hydrogen, oxygen, sodium hydroxide, and potassium hydroxide are only a few of the industrial applications of electrolysis.

How chemistry produces electricity. If a rod of pure zinc and a rod of copper are dipped into a solution of sulfuric acid, nothing seems to happen. However, if these metals are connected through a galvanometer, an electric current will be shown to be flowing. We can change the metals at will and change the electrolyte in solution, but as long as the two poles (as we call the pieces of metal) are different, more or less voltage will be created. The amount of the voltage will depend upon the metals used. This arrangement for creating an electric current is called a *primary cell*.

EXPERIMENT

The primary cell. Into a glass of water containing a little dissolved salt, place a piece of zinc and a piece of copper connected through a switch to a sensitive galvanometer. Close the switch for an instant and note the galvanometer needle swinging. This shows the presence of current.

Copper sulfate cells. Several different primary cells are used commercially. One is called the gravity cell because gravity keeps the solutions apart. The copper positive pole is placed at the bottom of a glass jar with a rubber-covered wire leading up over the side of the jar. The negative zinc pole is shaped something like a man's hand and has a hook over the top of the jar. In the bottom of the jar is a saturated copper-sulfate solution. Above this is a dilute solution of zinc sulfate. In this cell, the zinc dissolves at the top, and copper plates out on the copper pole below. Although the gravity cell has a low voltage, it has one advantage over other cells for continuous use. The peculiar advantage of this cell is that it does not produce any hydrogen. When hydrogen results from cell action, it causes trouble in two ways: First, it tends to hold the solution away from the pole; and secondly, it sets up an opposing voltage which causes the cell to weaken in long use. Instead of hydrogen, this cell deposits copper. The freshly deposited copper helps lower the resistance of the cell instead of hindering the current.

The *Daniell cell* uses the same substances as the gravity cell. In it, however, the zinc pole and the zinc sulfate solution are put inside a porous earthenware jar, which keeps this solution separate from the copper sulfate solution around the copper pole, which is a flat sheet of metal circling the porous jar.

Ammonium-chloride cells. Another type of cell has won more favor commercially than those just described because it gives a higher voltage. In this cell, zinc is the fuel as usual, but carbon or a composition material forms the positive pole. Ammonium chloride, called "sal ammoniac" by the trade, is the electrolyte. The reactions of the cell are complicated. Some provision must be made to offset the effects of hydrogen. Frequently an oxidizing substance is included in the composition of the positive pole to react with and remove the hydrogen.

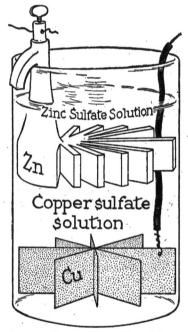

GRAVITY CELL. When this cell is on closed circuit, zinc dissolves in the upper part of the solution and copper plates out of the lower part.

In the so-called "dry cell" (which is not dry), the electrolyte is a paste. Black manganese dioxide is the oxidizing agent. In spite of this oxidizing agent, a dry cell on heavy duty will soon drop in voltage because of the effect of hydrogen. When the cell rests, the manganese dioxide catches up with the hydrogen, and the cell recovers its voltage. The dry cell has become by far the most popular primary cell in use. This is because of its portability and freedom from spilled solution. The zinc pole forms a can which is closed with sealing wax at the top.

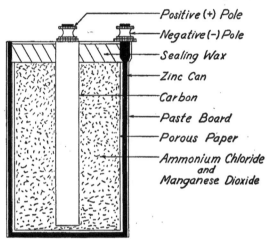

Positive (+) Pole
Negative (−) Pole
Sealing Wax
Zinc Can
Carbon
Paste Board
Porous Paper
Ammonium Chloride
and
Manganese Dioxide

DRY CELL. The so-called dry cell is not really dry, as it uses a moist paste of ammonium chloride for the electrolyte. It is one of the most useful cells because of its portability.

QUESTIONS OF FACT

1. What is the freezing point of pure water?

2. What is the effect on the freezing point of water of dissolving something in it?

3. Is the lowering of the freezing point proportional to the weight of substance dissolved?

4. How do the electrolytes compare with the nonelectrolytes in regard to the freezing point lowering for the same molecular concentration?

5. What is the freezing point of a molal solution?

6. Name several useful electrolyses processes.

7. What will the following ions become after their electric charges are neutralized in electrolysis: (a) Cu^{++}? (b) Ag^+? (c) Cr^{+++}? (d) Au^+?

8. Is it possible to plate such metals as sodium and potassium, which react with water in a water solution?

9. What becomes of the metal if such a process is attempted?

10. If an article is to be plated with a metal, to which electrode must it be attached?

11. Give the reaction that follows if Na^+ has its charge neutralized in water solution.

12. What is meant by electrolysis?

13. Discuss the following electrolyses as to (1) products formed, (2) action of the ions when they arrive at the electrodes:

(a) Electrolysis of silver nitrate.

(b) Electrolysis of table salt, NaCl.

14. What are the essential constituents of a primary cell?

QUESTIONS OF UNDERSTANDING

1. What is defined as a mole of some substance?

2. Calculate the weight of a mole of each of the following compounds: (a) $NaCl$; (b) KNO_3; (c) C_2H_5OH; (d) $C_6H_{12}O_6$; (e) $C_{12}H_{22}O_{11}$. (Look up the atomic weights in the back of the book.)

3. Compare the number of molecules in a mole of all nonelectrolytes.

4. Explain how your conclusion in ex. 3 follows from Avogadro's hypothesis. (See page 170.)

5. Define a molal solution.

6. Why do molal solutions of all nonelectrolytes freeze at the same temperature?

7. Is this statement true? The lowering of the freezing point is dependent only on the number of dissolved particles per given weight of water.

8. If one solution has a lower freezing point than another, are we reasonably sure that it has a greater concentration of dissolved particles than the other?

9. Is this statement true? More freezing-point lowering for equal weights of water means a greater number of dissolved particles, or a greater number of dissolved particles will produce a greater lowering.

10. What is the meaning of the fact that solutions of electrolytes lower the freezing point more than solutions of nonelectrolytes of equal molal concentration?

11. If table salt ionizes completely according to the equation $NaCl \longrightarrow Na^+ + Cl^-$, what would be the freezing point of a molal solution?

12. If ferrous chloride ionizes as follows, $FeCl_2 \longrightarrow Fe^{++} + 2\ Cl^-$, what is the lowest freezing point a molal solution of this substance could have?

13. If the actual freezing point is above your answer for ex. 12, what does this mean?

14. How many times would the freezing-point lowering of stannic chloride be that of a nonelectrolyte of equal molal concentration if it were to ionize completely according to the equation

$$SnCl_4 \longrightarrow Sn^{++++} + 4\ Cl^-?$$

15. In what sense is the electrolysis of water wrongly named?

Reference for Supplementary Reading

1. Findlay, Alexander. *Chemistry in the Service of Man.* Longmans, Green and Co., New York, 1931. Chap. XII, pp. 242–263, "Electricity and Chemistry."
   (1) When was the birth of electrochemistry? (p. 242)
   (2) Who made the first discovery in this field? (p. 242)
   (3) How did Volta interpret Galvani's discovery? (p. 243)
   (4) What invention resulted from Volta's interpretation? (p. 243)
   (5) What two metals did Davy discover by means of the electric battery? (p. 243)
   (6) Describe an experiment which proves that ions move through the solution of an electrolyte during electrolysis. (pp. 249–251)

Electropotential series. All of the metals can be arranged in a series according to the amount of voltage that is created when they are paired off into a cell. If a cell could be made using the metal at the top of the list and the one at the bottom of the list, the cell would have the highest voltage of any cell. If two adjacent metals were used, the cell would have a very low voltage. This arrangement of metals is called the *electropotential series.*

Electropotential Series

1. Cesium	8. Aluminum	15. Nickel	22. Antimony
2. Rubidium	9. Manganese	16. Tin	23. Mercury
3. Potassium	10. Zinc	17. Lead	24. Silver
4. Sodium	11. Chromium	18. *Hydrogen*	25. Palladium
5. Lithium	12. Cadmium	19. Arsenic	26. Platinum
6. Calcium	13. Iron	20. Copper	27. Gold
7. Magnesium	14. Cobalt	21. Bismuth	28. Osmium

Unfortunately, the metals at the top of the series react with water, which makes it impossible to use them in a practical cell, while those at the lower end of the list are too expensive to use in a practical cell.   It is theoretically possible to make a cell of nearly 4.5 volts.   However, because of the objections mentioned, no practical cell has over 2.5 volts; most of them range from 1 to 1.5 volts.   Zinc has turned out to be the most feasible metal for the negative pole, and carbon has been found to be the cheapest material to use for the positive pole.   Ammonium chloride is the most practical electrolyte.

Where electricity comes from.   It is a little difficult to explain where the electricity comes from in the primary cells.   We think the material of the electrodes, as in all uncharged matter, is composed of equal numbers of positive and negative charges. The zinc electrode of the cell dissolves in the electrolyte.   Since each zinc ion carries two positive charges, the two negative charges with which these positive charges were formerly paired must remain behind on the electrode, making it negatively charged.   The two positive charges on the ion on entering the solution give the solution an excess of positive charges.   Two other positively charged ions, usually $\overset{+}{H}$, leave the solution by the other electrode, making it positive.   These charges meet each other through the external circuit and neutralize again. Moving charges constitute the current.

Zinc does not change to hydrogen.   When a piece of impure zinc is put into acid solution, hydrogen begins to form and escape from the surface of the zinc.   Students sometimes draw the incorrect conclusion that the zinc is changing into hydrogen. What is really happening is that the hydrogen is separating on the small particles of impurities, which are the positive poles of very small electric cells.   These cells are short-circuited through the metal so they continue to run down and the zinc continues to dissolve.   Chemically pure zinc will not dissolve in the same way.

The storage cell.   We speak of the storage battery as if it stored electricity.   This is not true.   When a storage battery is being charged, it is simply used as a cell for electrolysis.   The electrolysis produces lead dioxide at the positive pole. In making the storage cell, the positive pole is made by putting PbO_2 paste into a gridlike arrangement. When the cell runs down by producing electricity similarly to the primary cell, the PbO_2 is reduced to PbO, which forms $PbSO_4$ with the acid in the solution.

$$PbO + H_2SO_4 \longrightarrow PbSO_4 + H_2O$$

Charging the cell consists of applying a stronger voltage to the cell and forcing the current to flow in the direction opposite to

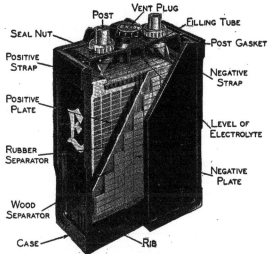

POST
VENT PLUG
FILLING TUBE
SEAL NUT
POST GASKET
POSITIVE STRAP
NEGATIVE STRAP
POSITIVE PLATE
LEVEL OF ELECTROLYTE
RUBBER SEPARATOR
NEGATIVE PLATE
WOOD SEPARATOR
CASE
RIB

Courtesy of Electric Storage Battery Co.

THE STORAGE CELL. This cell has a relative high voltage. After it has run down, it may be renewed (within limits) by connecting it with a direct current of the proper voltage. The positive pole of this cell is PbO_2, and the negative pole is lead. When the cell is producing current, the PbO_2 is reduced to PbO, and the lead is oxidized to PbO. Both electrodes of the exhausted cell react with the sulfuric-acid electrolyte to form $PbSO_4$.

which it flows when the cell is working. The electrolysis oxidizes the lead sulfate back to the original PbO_2. Thus we see that the storage battery does not store electricity as such.

In reality a reaction is taking place at both poles while the storage cell is running, and also a different reaction at each pole when the cell is being charged. However, the complete explanation of what takes place in these four cases belongs to a more advanced course in chemistry.

EXPERIMENT

The storage cell. (Apparatus: Tumbler of dilute sulfuric acid, two lead plates, voltmeter with 2-volt range, electric bell, emery paper, connecting wires, storage battery or dry cells.)

Clean the surface of the lead with emery paper. Connect the lead electrodes with the storage battery and place them in the acid and leave for one-half minute. Turn off the battery and show the formation of reddish lead dioxide on the positive electrode.

Now connect the electrodes in the acid to the voltmeter with the positive connection connected to the electrode with the reddish deposit and show the electromotive force.

Charge again by connecting to the battery for a minute or more. Disconnect the battery and connect it to the electric bell. If the bell is in good shape, it will ring.

Acids give hydrogen ions. All acids have certain properties in common, such as tasting sour and turning blue litmus paper pink. We naturally look for a common cause of these properties. An examination of the formulas of the acids in the fol-

HYDROGEN IONS ARE ACID. The properties of acids are due to the presence of hydrogen (H^+) ions.

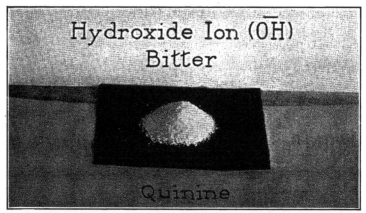

HYDROXYL IONS ARE BASIC. Basic properties are due to the presence of hydroxyl (OH⁻) ions. The bitterness of quinine may not all be due to basic properties, although quinine is a base and typical bases are bitter.

lowing table will show that hydrogen is the only element common to all the acids.

ACIDS

HCl	$HC_2H_3O_2$	HBr	HI
HNO_3	H_2SO_4	$H_2C_2O_4$	H_2SO_3

However, hydrogen occurs in thousands of substances that are not acids, so hydrogen alone does not suffice for acid properties. Acids are all electrolytes. When we study them to see what ions are formed, we find that the hydrogen ion ($\overset{+}{H}$) is always formed. We sometimes write the formation of ions as a chemical reaction:

$$(a) \quad HCl \longrightarrow H^+ + Cl^-$$
$$(b) \quad HNO_3 \longrightarrow H^+ + NO_3^-$$
$$(c) \quad H_2SO_4 \longrightarrow 2\,H^+ + SO_4^{--}$$
$$(d) \quad H_2C_2O_4 \longrightarrow 2\,H^+ + C_2O_4^{--}$$

The hydrogen ion, being the only thing common to all acids, must be responsible for the acid properties. We could say, then, that *an acid is a substance that forms hydrogen ions in water solution.*

The ions of bases. An examination of the formulas of a large number of bases shows us that the OH radical is common to all of them. This radical is sometimes called the *hydroxyl* radical. Probably a better name is hydroxide radical, as we call the compounds of it hydroxides.

Bases

NaOH	$Ba(OH)_2$	$Mg(OH)_2$	NH_4OH
KOH	$Ca(OH)_2$	KOH	$Fe(OH)_3$

However, thousands of compounds contain hydroxyl radicals but are not bases. Among these are the alcohols and the sugars. In order to understand why the bases are different from the other hydroxyl compounds, we need to study the ions of the bases and remember that the other hydroxyl compounds are not electrolytes:

$$(a) \quad NaOH \longrightarrow Na^+ + OH^-$$
$$(b) \quad KOH \longrightarrow K^+ + OH^-$$
$$(c) \quad Ba(OH)_2 \longrightarrow Ba^{++} + 2\,OH^-$$
$$(d) \quad Ca(OH)_2 \longrightarrow Ca^{++} + 2\,OH^-$$

The bases all have the hydroxyl ion (OH^-) as their negative ion. This must make the difference between the bases and the other hydroxyl bodies, and all properties characteristic of bases must be the properties of the OH^-. Its properties, then, must be an alkaline taste and the ability to turn pink litmus blue and to turn colorless phenolphthalein pink.

Neutralization explained by ions. We learned in a previous chapter that if we mix an acid solution with an equivalent amount of basic solution, neutralization takes place; that is, the acid destroys the basic properties and the base destroys the acid properties. The solution is still a conductor of electricity, which proves that an electrolyte is still present. Neutralization can be represented by an ionic equation.

$$(a) \quad Na^+OH^- + H^+Cl^- \longrightarrow H_2O + Na^+Cl^-$$
$$(b) \quad K^+OH^- + H^+NO_3^- \longrightarrow H_2O + K^+NO_3^-$$
$$(c) \quad Mg^{++}(OH)_2^- + H_2^+SO_4^{--} \longrightarrow 2\,H_2O + Mg^{++}SO_4^{--}$$

When we study these equations, we see that what really happens during neutralization is that the H^+ of the acid has united with the OH^- of the base to form un-ionized water.

Sodium was an ion in the base before reaction, and it is still an ion in the salt solution. The Cl^- in the acid is still an ion in the salt.

New definition of a salt. We can now define a salt in terms of ions. Let us look at the ions from several salts.

SALTS

Na^+Cl^-	$K^+NO_3^-$	$Al^{+++}PO_4^{---}$	$Ca^{++}Br_2^-$
K^+Cl^-	$Al^{+++}(NO_3^-)_3$	$K_2^+S^{--}$	$P^{+++}I_3^-$
$Mg^{++}Br_2^-$	$Fe^{++}SO_4^{--}$	$Ba^{++}I_2^-$	$P^{+++++}Cl_5^-$

We find no common ions in the salts. Acids, bases, and salts are the only electrolytes. Acids all give the hydrogen ion. Bases all give the hydroxyl ion. This gives us a chance to define salts as *those electrolytes that give neither hydrogen ions nor hydroxyl ions in solution.* This is the simplest definition of a salt.

Strong and weak acids and bases. Many acids and bases do not separate completely into their ions. The more water we put them into, the more completely they will ionize; but with some acids and bases, it is not possible to give them room enough to ionize completely. Those acids that ionize to a large extent in practical dilutions are called *strong* acids, while those that ionize to only a small extent are called *weak* acids. Likewise highly ionized bases are *strong* bases, and slightly ionized bases, *weak* bases.

IONIZATION OF ACIDS AND BASES IN DILUTE SOLUTIONS

ACID	IONIZED	BASE	IONIZED
HCl	92 per cent	NaOH	91 per cent
HNO_3	92 per cent	KOH	91 per cent
H_2SO_4	61 per cent	$Ba(OH)_2$ . . .	77 per cent
Acetic	1.3 per cent	NH_4OH . . .	1.3 per cent

We may have concentrated acids or dilute acids, and strong acids or weak acids. A concentrated acid means a large amount of acid compared to the water. A dilute acid means little acid compared to the water. "Weak" or "strong" says nothing about the concentration, but has reference to the proportion of the molecules that ionize at any working concentration. There is no exact dividing line between a concentrated acid and a dilute acid. What is dilute for one purpose might be concentrated for another. Likewise there is no recognized dividing line between strong acids and weak acids. Some acids, such as the mineral acids, are so strong one would not dare to take them into the mouth unless extremely dilute. Acetic acid, in vinegar, may be eaten at about 5 per cent strength. The very weak acid, boric acid, can be put into the delicate eye without, injury to the latter.

The same four terms — concentrated, dilute, strong, and weak — may be applied to the bases with the same meanings. Sodium hydroxide and potassium hydroxide are the strongest bases. Ammonium hydroxide is a moderately weak base — weak enough that it will not harm cloth when used to neutralize acids.

QUESTIONS OF FACT

1. What determines the arrangement of the elements in the electropotential series?

2. Classify the following metals as to whether they are in the first, second, third, or fourth quarter of the series: gold, calcium, silver, iron, aluminum, copper, lead, zinc, and platinum.

3. Why are the first six metals not used in primary cells? The last six?

4. Does the storage cell store electricity?

5. What happens when a storage cell is charged?

6. What substance is used up when the storage cell is charged?

7. What substance is used as the positive pole of many cells because of its cheapness?

8. What substance is the negative pole of most primary cells?

9. What is the electrolyte in the so-called "dry cell"?

10. How is this electrolyte kept from drying out?

11. Do the chemical reactions in the dry cell produce hydrogen?

12. How is this hydrogen disposed of?

13. What is made the positive pole of a lead storage cell?

14. What happens to the PbO_2 as the cell runs down?

15. Are acids electrolytes or nonelectrolytes?

16. What ion is responsible for acid properties?

QUESTIONS OF UNDERSTANDING

1. Define a primary cell as opposed to a storage cell.

2. What determines the voltage of a primary cell?

3. In what two ways does the production of hydrogen by the reactions in a cell reduce its efficiency?

4. In what sense is the dry cell not dry?

5. Why cannot the dry cell be kept on a continuous load?

6. What ion tastes sour?

7. What ion is responsible for the properties of bases?

8. Define an acid in terms of ions.

9. Define a base in terms of ions.

10. Define a salt in terms of ions.

11. Can any other substance be defined in terms of ions?

12. Explain the difference between concentrated, dilute, weak, and strong acids and bases.

13. Name one of each of the following:

(a) a strong acid   (d) a strong base

(b) a moderately weak acid   (e) a moderately weak base

(c) a very weak acid

14. What is the only reaction that actually takes place during neutralization?

15. Name the following ions:

(a) one that tastes sour   (c) two that are colorless

(b) one that tastes bitter   (d) two that exist in the gravity cell

Select all of the correct responses for exercises 16 to 20.

16. The hydrogen ion is united with the hydroxyl ion:

(a) during neutralization   (c) during precipitation

(b) during electrolysis   (d) during ionization

17. A strong acid in water solution (a) will have no action on litmus, (b) will largely ionize.

18. The hydrogen ion (a) turns litmus pink, (b) unites with the hydroxyl to form water, (c) is the characteristic ion of an acid, (d) tastes sour.

19. A base (a) is a hydroxide, (b) is responsible for hydrogen ions, (c) reacts with an acid to form a salt, (d) produces hydroxyl ions in solution, (e) will turn pink litmus blue in solution, (f) will neutralize acids, (g) is any hydroxide.

20. The hydroxyl radical (a) is present in all bases, (b) tastes sour, (c) is a negatively charged ion, (d) is colorless in solution, (e) turns phenolphthalein pink.

REFERENCES FOR SUPPLEMENTARY READING

1. Findlay, Alexander. *Chemistry in the Service of Man*. Longmans, Green and Co., New York, 1931. Chap. XII, pp. 242–263, "Electricity and Chemistry."
   (1) Explain the action of the ammonium chloride cell. (pp. 251–253)
   (2) Name a useful substance produced in an electric furnace. (pp. 259–261)
2. Foster, William. *The Romance of Chemistry*. D. Appleton-Century Company, New York, 1936. Chap. X, pp. 145–160, "Electricity in the Service of Chemistry."
   (1) Name several products of the electric furnace. (p. 159)
   (2) What was the first price of carborundum, a product of the electric furnace? (p. 160)

Ionic equilibrium. A weak acid like the acetic acid in vinegar is only partly ionized, yet when we neutralize it with alkali, it is finally neutralized as completely as if it were all in the form of ions. We think there is a dynamic equilibrium between the ions and the un-ionized molecules. We can represent this equilibrium by an equation with two arrows pointing in opposite directions, meaning that two reactions in opposite directions are going on at the same time.

$$NH_4OH \rightleftharpoons NH_4^+ + OH^-$$

We might try to illustrate what we mean by a dynamic equilibrium — that is, two reactions going on at the same time in opposite directions. Suppose two children had the task of amusing the baby. The baby is given a box of building blocks. It begins to throw the blocks all over the room. This represents the reaction in one direction, corresponding to the molecule breaking into the ions. The two children begin gathering up the blocks and putting them back into the box. This corresponds to the opposite reaction, that of the ions combining into the molecules. Try as hard as they can, the children cannot gather up the blocks faster than the baby can throw them out. Suppose two thirds of the blocks are always in the box and one third are scattered on the floor. This distribution of blocks remains constant, so we call it a situation of equilibrium. Blocks are on the move in both directions, so we call it a dynamic equilibrium.

Suppose that instead of putting the blocks back into the box, the children removed them to another room; the box would soon be emptied. This corresponds to what takes place when we neutralize the ions in solutions. The reaction in the opposite direction keeps on. However, the ions are removed so the backward or reverse reaction is stopped and the acid is all neutralized. Any process that removes one of the reacting parts eventually gets it all.

Other equilibrium reactions. Equilibrium reactions are very common, but they are not so useful as the reactions that go to completion in one direction.

If KCl and NaNO$_3$ are mixed in solution, we might expect the following reaction:

$$KCl + NaNO_3 \longrightarrow KNO_3 + NaCl$$

The above reaction does take place to some extent, but not entirely. Besides the expected products, KNO$_3$ and NaCl, there still remain some of the original KCl and NaNO$_3$.

On the other hand, if we begin by putting the substances on the right side of the above equation into the solution first, we get the same mixture of all four substances. This shows that the reaction goes in the reverse direction.

$$KCl + NaNO_3 \longleftarrow KNO_3 + NaCl$$

In the final condition in either case it is reasonable to think that both reactions are taking place at the same time and that we have a condition of dynamic equilibrium.

$$KCl + NaNO_3 \rightleftharpoons KNO_3 + NaCl$$

As it stands, the reaction in either direction is not useful because of the difficulty of getting any of the substances out of such a complex mixture.

Conditions that make reactions go to completion. It has just been pointed out that if one of the products of a double decomposition reaction between ionized substances is un-ionized it cannot take part in the reaction in the reverse direction and the reaction goes to completion in the direction of the un-ionized product.

$$\overset{+}{H}\overset{-}{Cl} + \overset{+}{Na}\overset{-}{OH} \longrightarrow \overset{+}{Na}\overset{-}{Cl} + H_2O \text{ (un-ionized)}$$

A second condition that causes reactions to go to an end is when one of the products of the reaction is a gas and escapes from the reaction vessel, often hurried by heating.

$$2 KBr + H_2SO_4 \longrightarrow 2 HBr \text{ (gas)} + K_2SO_4$$

The product of a reaction, itself, may not be volatile, but it may break up into gaseous substances. If H_2CO_3 is one of the products of a reaction, it will decompose into H_2O and the gas CO_2 and affect the reaction just as if it were a gas.

The third condition which makes a reaction go to completion is when one of the products leaves the solution because of insolubility.

$$\overset{+}{K}\overset{-}{Cl} + \overset{+}{Ag}\overset{-}{NO_3} \longrightarrow AgCl \text{ (insoluble)} + \overset{+}{K}\overset{-}{NO_3}$$

Using equilibrium reactions. Some equilibrium reactions can be used. One way of using them is to change the conditions of the reaction so it is no longer an equilibrium between two opposing reactions. In NH_4OH solution we have a condition of equilibrium as:

$$NH_4OH \rightleftharpoons NH_3 + H_2O$$

If we boil this solution in an open vessel, the volatile NH_3 is driven out of the solution so the reverse reaction can no longer take place. In other words, the reaction goes to an end, NH_3 escaping into the air and hot water remaining in the vessel.

Mass action. Another way of using an equilibrium reaction to advantage is by adding a large mass of the substances on the side of the reaction having the more soluble constituents. This may produce enough of the substances on the other side to reach the saturation point of the least soluble constituent. From then on as more is added, this least soluble constituent will separate from the solution as crystals or a precipitate. Let us apply this to an equilibrium described in a previous paragraph.

$$KCl + NaNO_3 \rightleftharpoons KNO_3 + NaCl$$

NaCl is the least soluble of the four constituents of the equilibrium mixture. The addition of more KCl and $NaNO_3$ will eventually result in NaCl crystallizing out and the solution getting richer in KNO_3. Finally the solution gets so rich in KNO_3 that this substance can be obtained by evaporating off the water. This is the method used to obtain KNO_3 from the naturally occurring KCl and $NaNO_3$.

The most useful reactions, however, are the ones that go to completion because of the formation of a gaseous product, or an insoluble precipitate, or an un-ionized substance.

QUESTIONS OF FACT

1. Do all substances ionize completely?

2. Do all of the reactants and reaction products of an equilibrium reaction exist in solution at the same time?

3. Which is of commoner occurrence, reactions that go to completion, or reactions that remain in equilibrium?

4. State three conditions which will cause a reaction to go to completion instead of remaining in equilibrium.

5. Define and illustrate mass action.

6. Write one equilibrium reaction.

QUESTIONS OF UNDERSTANDING

1. When an electrolyte is not completely ionized, what condition is thought to exist?

2. Describe the illustration used to illustrate a dynamic equilibrium.

3. In the equilibrium reaction, $NH_4OH \rightleftharpoons \overset{+}{N}H_4 + \overset{-}{O}H$, which exists in solution to the largest extent, the ionized part or the un-ionized part?

4. Which of the three possible conditions which cause reactions to go to completion operate in each of the following reactions?

(a) the production of nitric acid. (p. 144)

(b) the production of HCl. (p. 189)

(c) the test for chlorides. (p. 197)

(d) the formation of sodium hydroxide. (p. 261)

(e) the production of hydrogen. (p. 73)

(f) the neutralization of acids and bases. (p. 228)

5. Explain how an equilibrium reaction can be used to produce a useful substance.

ADDITIONAL EXERCISES FOR SUPERIOR STUDENTS

1. What two equilibrium reactions are going on in a bottle of ammonium hydroxide?

2. What equilibrium reaction is going on in a bottle of vinegar (solution of $HC_2H_3O_2$)?

3. Mention several reactions that go to an end (a) because of volatility of one of the products, (b) because of the insolubility of one of the products.

SUMMARY

Matter is composed essentially of two kinds of electricity, called positive and negative, respectively, because of their ability to neutralize each other. Negative electricity can be produced by rubbing a rubber rod with a cat skin, and positive electricity can be obtained by rubbing a glass rod with silk.

The respective units of negative and positive electricity are the electron and the proton. Each of these is extremely small, but the latter weighs 1845 times as much as the former. There is some evidence for the existence of a positron, or positive charge equal in mass and charge to the electron. Another small particle is the neutron. It is without charge and is probably a proton which has captured an electron.

Like charges of electricity repel each other, while unlike charges attract.

A *conductor* is a substance through which electricity passes readily, while an *insulator* is one through which it does not. Distilled water is an insulator. When acids, bases, and salts are dissolved in distilled water, the solution becomes a conductor. This is not the case when other substances are dissolved in water. Those substances which make water conduct electricity are called *electrolytes*, while those which do not are called *nonelectrolytes*.

To explain how electrolytes carry the current through water, it is assumed that the neutral molecules dissociate — that is, separate in positively and negatively charged parts called ions.

An electric current is electric charges in motion. When the pull of the charges from the battery is applied to the ions, they begin to drift, the negative ions towards the positive electrode and the positive ions towards the negative electrode. The moving ions constitute the current.

The positive ions are formed from the positive parts of the compound and have as many charges as the radical or atom has positive valences. Negative ions come from the negative parts of the compound, with negative charges equal to the negative valences.

The theory of ionization also explains the fact that all solutions of electrolytes have abnormally low freezing points as compared to the nonelectrolytes, which all have the same freezing point for the same molecular concentration. Those electrolytes which separate into two ions lower the freezing point nearly twice as much as the nonelectrolytes; those which form three ions per molecule lower it almost three times the expected amount, etc., which suggests that ionization increases the number of dissolved particles.

Electrolysis is the use of electricity to produce chemical reactions in solutions, one of the most important being electroplating. In all cases of electrolysis the reactions can be predicted if we know the chemical properties of the elements or radicals formed when the charges of the ions are neutralized.

Electricity is produced from chemical action in electric cells. The conditions necessary for a primary cell are two pieces of different metals projecting into the solution of an electrolyte and connected externally. The farther apart the metals are in the electropotential series, the greater the voltage the cell will have. Zinc is the most practical metal for the negative pole, and ammonium chloride the most practical electrolyte. Before the chemical action dissolves the zinc, the electricity is in the zinc.

The lead storage cell does not store electricity as such but produces lead dioxide by electrolysis as one of its reactions during charging. As the cell runs down, this lead dioxide reduces to lead monoxide, which reacts with sulfuric acid, removing it from solution. At the same time, the other electrode is oxidized from metallic lead to lead monoxide, which in turn removes more sulfuric acid from solution.

Acids are definable as those substances which form hydrogen ions in solution. The characteristic acid properties are simply the properties of this ion.

Bases are those substances which form hydroxyl ions in solution, while salts are those electrolytes whose ions include neither $\overset{+}{H}$ or $\overset{-}{OH}$ ions. Neutralization may be thought of as the combination of hydrogen ions with hydroxyl ions to form undissociated water.

Not all acids and bases ionize to the same extent. Those whose molecules mostly separate into ions are said to be strong acids and bases, while those which ionize to only a small degree are said to be weak.

Where ionization is not complete, it is thought to be in a state of dynamic equilibrium — that is, continually separating into ions and combining back into undissociated molecules so that both ions and molecules exist together. Many other chemical reactions are equilibrium reactions not going to completion in either direction. For a reaction to go to completion, one of the reactants must leave the reaction because it becomes a gas, is insoluble, or is un-ionized.

Sometimes an equilibrium reaction can be made useful by *mass action*. Some constituent is added until one of those on the other side of the reaction reaches the saturation point, after which it will separate from solution.

CHAPTER 12

⇄

THE METALS

In studying the different elements when they are separated from chemical union with other elements, we are impressed with the general similarity of many of them, such as gold, silver, copper, iron, lead, zinc, and tin. We classify these elements and many others not so well known as metals.

The metals include some of our most useful substances. Without iron our great bridges, skyscrapers, fast-working steam shovels, railroads, large transatlantic steamships, automobiles, and most manufacturing machinery would be impossible. There would be no stoves, no axes, no shovels, and no knives. Without tin the great canned-food industries would be badly handicapped. Without silver, photography as we know it today would be impossible. This would mean no movies, and very few pictures of any kind. Without copper for electric wiring, electric lighting, electrical machinery, electric streetcars, the telegraph, the telephone, and the radio would probably never have been developed to where they are today. Without aluminum we would not have the fine sanitary cooking utensils and light airplane parts so essential to aviation. In fact, without metals we would probably still be cooking in baskets with hot rocks as did the Indians. Without gold and silver the nations would have difficulty in establishing an acceptable measure of value, and we might have to use beads and shells for a medium of exchange as did the Indians, or resort to barter as do the savages of Africa.

Without metals, modern surgery never could have developed. When we look about us in our daily lives and think how much we are dependent on metals, we realize to some extent how important they are to us. Nations have gone to war over a deposit of useful metals. Search is going on in nearly every country to discover new deposits of the metals.

A CONCRETE STEEL BUILDING.   In addition to the steel in the framework, rust-
proof steel is widely used for trim.

Appearance of the metals. Although gold is yellow, copper reddish, and other metals are various shades of gray, it is easy to recognize a metal by its luster or shininess. The metallic luster of the metals is hard to describe, yet easy to see. There is little chance of confusing the metals with the nonmetals, such as sulfur, chlorine, bromine, oxygen, or nitrogen. The

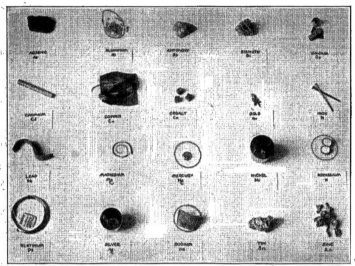

Photo by Frederick King

COMMON METALS. Top row, left to right: arsenic, aluminum, antimony, bismuth, and calcium. Second row: cadmium, copper, cobalt, gold, and iron. Third row: lead, magnesium, mercury, nickel, and potassium. Fourth row: platinum, silver, sodium, tin, and zinc.

luster of a metallic surface depends to a large extent upon its smoothness and freedom from tarnish. A piece of lead pipe that has lain out in the rain for several seasons may not show any luster because of a coating of tarnish. A piece of iron gathered from the creek bed may show no luster for the same reason. However, all metals whose surfaces are freshly cut, scraped, or polished smooth show the characteristic luster.

Other physical properties common to the metals. Besides luster, the metals all have certain other physical similarities.

They are all good conductors of heat and electricity. We use iron stoves and stovepipes because iron conducts heat readily. We use aluminum cooking utensils because aluminum conducts heat readily. For the same reason, we often find fault with the aluminum cups in our camping outfits, which conduct heat too rapidly to our lips when we drink hot beverages.

Courtesy California Bridge Authority

A MODERN STEEL STRUCTURE. The metallurgy of iron has made possible such fine bridges as the one shown above.

Aluminum and (especially) copper are used for electric wiring because of their great conductivity of electricity. Metals which can be drawn out into fine wires are said to be *ductile*. Sheet iron, tin foil, aluminum foil, and gold leaf all depend for their existence on the fact that these metals are *malleable* — that is, can be rolled out into strong, thin sheets. The usefulness of metals as wires and sheets is dependent not only on their ductility and malleability but also on their tensile strength. We marvel at the ability of a wire cable to hold up the enormous weight of a great suspension bridge.

The metals show a wide range of variation in the common physical properties, such as density and hardness. Lithium, the lightest of them all, weighs only .53 gram per cubic centimeter and readily floats on water. On the other hand, a cubic centimeter of the rare metal osmium weighs 22.5 grams. In comparison with the common rocks, whose densities range only from 2.5 to 7.5 grams per cubic centimeter, these metals are truly remarkable. In hardness, the metals range from the wax-like sodium, which is easily cut with a knife, to chromium, which will scratch glass. One of the metals, mercury, is a liquid under ordinary temperatures.

The chemical properties of the metals. In their chemical reactions the metals show marked differences. Some, such as gold and platinum, are so resistant to chemical reaction that they are obtained free, or uncombined, from the earth. Others, like sodium and potassium, are so reactive that a fresh surface will tarnish in a few seconds when exposed to the atmosphere. Because of their great reactivity, these two metals can only be kept under kerosene. Even then the surface is always corroded. With some metals the surface tarnishes readily, but the tarnish serves as a protecting coat to prevent further tarnish. Old copper kettles, boiler bottoms, and wire soon become covered with a dark-colored tarnish, which often has a greenish cast. Often an old boiler in the junk pile will have all rusted away except the copper bottom. Lead also gets a protecting coat of tarnish. Zinc, too, loses its luster, yet a zinc drain on a roof will serve for a hundred years and still be about as good as ever.

Iron, on the other hand, will rust completely away. Rust begets rust. If a sportsman wishes to keep the inside of his gun barrel from getting pitted with rust, he must keep out the first trace of rust. When once started, the rust acts as a catalyst to speed up the rusting process. If the farmer knew how rust encourages rust, he would not leave his plow out in the field during autumn or spring rains. Millions of dollars' worth of machinery are lost every year through rust.

Action towards acids. The metals show even greater differences in their ability to withstand acids. Platinum and gold will not dissolve in any single acid no matter how concentrated or strong. Silver will dissolve only in a concentrated solution of a strong oxidizing acid like nitric acid. Zinc and tin, on the other hand, dissolve readily in all strong acids in even quite dilute solution. Sodium and potassium do not even need an acid to dissolve them, but dissolve readily in water.

Actions towards bases. In general, the bases do not dissolve metals as readily as do the acids. However, a few metals, such as aluminum, dissolve readily in bases. For this reason the housewife must take care not to get alkalies in her aluminum cooking utensils. Zinc also will dissolve in strong basic solution.
$$2\,Al + 6\,NaOH \longrightarrow 2\,Na_3AlO_3 + 3\,H_2$$

The positive part of a compound. In writing formulas, we learned that part of the compound is positive and part negative. Also when an electrolyte forms ions in water solution, part of the ions carry positive charges and part negative charges. Both in compounds and in solutions the metals nearly always form the positive part. $\overset{+}{H}$ and $\overset{+}{NH_4}$ are the only two common positive ions that are not metals. Some of the common metallic ions are $\overset{+}{Na}$, $\overset{+}{K}$, $\overset{+}{Li}$, $\overset{++}{Ca}$, $\overset{++}{Sr}$, $\overset{++}{Ba}$, $\overset{++}{Mg}$, $\overset{+++}{Al}$, $\overset{++}{Zn}$, $\overset{++}{Cu}$, $\overset{+}{Ag}$, $\overset{+++}{Au}$.

Occurrence of the metals. Gold, platinum, and a few other rare metals occur usually in the uncombined, or free, state. Copper and silver may also occur free, but more frequently in their compounds. Other familiar metals occur only in combination. Metallic ores (naturally occurring minerals from which metals are recovered) are usually carbonates, oxides, and sulfides. Iron is usually found as the oxide; copper may occur as oxide, sulfide, or carbonate; and mercury occurs almost entirely as the sulfide.

It is an important fact that metals such as gold, silver, copper, lead, and tin, which form only a small fraction of one per cent of the earth's crust, are so concentrated that thousands of tons of their compounds occur in one small locality. Moreover, in one mine the mineral will be mostly copper ore; in another, a tin compound; and in a third, lead sulfide. All the factors that have determined this spotted concentration are not known. In general, the heavy metals occur on the tops of *volcanic intrusions*. During volcanic activity the rock material far underground becomes melted. Sometimes it breaks through the outer crust and boils out as lava. This seldom results in any concentration of the heavy metallic compounds. In other places, however, the melted rock never reaches the surface. However, this intrusion, as the geologist calls it, often pushes up the horizontal sedimentary rock layer, producing many cracks. The compounds of the heavy metals, being more volatile than the general rock substance, distill up into these cracks and there are sought for mining. But why one mine should produce silver while another near by yields copper is not explained.

QUESTIONS OF FACT

1. Take a sheet of paper and make a list of all the useful things you can think of that are mostly metals.

2. Make a list of things in which metal forms only a small part but which would be useless without the metallic parts.

3. Name two metals that are colored other than light colored.

4. Name one metal that is a liquid at ordinary temperatures.

5. Name three metals that are ductile.

6. Name the lightest metal and the densest metal.

7. Name two metals very resistant to chemical action.

8. Which two metals cannot be dissolved by any single acid?

9. Name two metals which dissolve in bases.

10. What three classes of compounds are the common ores of the metals?

QUESTIONS OF UNDERSTANDING

1. What is meant by metallic luster?

2. Explain why it is hard to drink hot liquids in aluminum cups.

3. What metal is very malleable? Name two others that are malleable.

4. What is meant by ductility?

5. Why are most metals not found uncombined in nature?

6. Why must sodium and potassium be kept under kerosene?

7. Name three metals whose tarnish serves as a protective coating to prevent further tarnish.

8. What kind of solution is quite injurious to aluminum?

9. Give the reaction for aluminum dissolving in sodium hydroxide.

10. What is the charge of the characteristic ions of the metals?

11. What is a volcanic intrusion?

12. How have volcanic intrusions concentrated the metals?

REFERENCES FOR SUPPLEMENTARY READING

1. Foster, William. *The Romance of Chemistry.* D. Appleton-Century Co., New York, 1936. Chap. XVII, pp. 261–286, "Minerals and Metals."
   (1) Define the term mineral. (p. 261)
   (2) About how many minerals are known? (p. 262)
   (3) Which are more common, single minerals or aggregates of minerals? (p. 262)
   (4) What are the most abundant minerals? (p. 263)
   (5) Give the evidence in favor of the hypothesis that the center of the earth is mostly iron. (p. 263)
   (6). What elements are in the feldspars? (pp. 264–265)
   (7) What is our most abundant iron ore? (p. 267)
   (8) Where is uncombined copper found? (p. 267)
   (9) Name six metals known to the ancients. (p. 272)
   (10) How thin has gold been hammered? (p. 273)
   (11) How long a wire can be made from one ounce of gold? (p. 274)
   (12) Give the extremes of the melting points of the metals. (p. 274)
   (13) How much copper is produced annually in the world? (p. 275)
   (14) Where does tin occur? (p. 275)
2. Sadtler, Samuel Schmucher. *Chemistry of Familiar Things.* J. B. Lippincott Co., Philadelphia, 1924. Chap. IX, pp. 113–132, "Metals."
   (1) What relation does carbon dioxide have to the rusting of iron? (p. 119)
   (2) Why do cast iron and steel rust more easily than wrought iron? (p. 120)
   (3) What is the effect of neutral salts on rusting? (p. 120)
   (4) What is the effect of alkalies on the corrosion of iron? (p. 120)
   (5) What is necessary besides acid for corrosion of iron? (p. 120)
3. Whitlock, Herbert P. *The Story of Minerals.* American Museum Press, New York, 1932.
   (1) Describe a geode. (p. 53)

(2) What mineral is lead sulfide? (p. 71)

(3) Name the sulfide of zinc.

(4) What is the mineralogical name of rock salt? (p. 78)

(5) Name a mineral containing fluorine. (p. 78)

(6) Name five minerals that are carbonates. (pp. 91–101)

(7) Name four semiprecious stone minerals which are silicates. (pp. 112–122)

Metallurgy by furnace heat. Metallurgy is the science of extracting metals from their ores. The processes employed in separating metals from their compounds vary from one mineral to another. Mercury, for instance, can be broken loose

METALLURGY OF IRON. The principal reaction in the metallurgy of iron is:
ore + coke ⟶ metal + carbon monoxide.

from the sulfur in cinnabar, its common ore, simply by heat. Some other metals are separated with great difficulty. Although in some cases there are short cuts in the metallurgy of a metal and in other cases some special process must be used, most metals are separated by the same general set of processes.

(1) The ore is roasted to the oxide. Roasting means heating or burning in the presence of air. If the ore is already the oxide, as is often the case with iron, roasting as such is not needed. The carbonates and sulfides require roasting. The chemistry of roasting is quite simple, as the following reactions show:

$$(a) \ 2\,ZnS + 3\,O_2 \longrightarrow 2\,ZnO + 2\,SO_2$$
$$(b) \qquad ZnCO_3 \longrightarrow ZnO + CO_2$$

The process (b) in which no oxygen enters the reaction is sometimes called *calcining*.

(2) Reduction of the oxide with carbon. Coke is the form in which carbon is most frequently used for reducing oxides. When heated in the furnace with the oxide of the metal, the coke gets hold of the oxygen to form CO. The gaseous carbon monoxide escapes from the hot furnace and leaves the metal behind.

$$ZnO + C \longrightarrow Zn + CO$$

Metallurgy is not usually so simple as would appear from the two reactions just described, because the ore often has a great deal of other rock mixed with it. Sometimes the ore

FORMATION OF SLAG. The foreign rocks in an ore are called gangue. Another rock is added to react with the gangue to form an easily fusible slag. The second rock is called a flux. Often quartz (SiO_2) is the gangue, and limestone ($CaCO_3$) is used as the flux. Occasionally limestone is the gangue, and quartz is used as the flux. The reaction is the same in either case:

$$gangue + flux \longrightarrow slag.$$

can be concentrated by removing some of this useless rock. In furnace metallurgy a *flux* must usually be added to react with and remove this foreign rock. The substance resulting from the reaction between the flux and the impurity in the ore is called the *slag*. Almost invariably the slag is lighter than the melted metal; hence it can be drawn off in the form of a scum floating on the metal. If the impurity is quartz, limestone is the flux. The reactions are:

(1) Calcining the limestone by furnace heat:

$$CaCO_3 \longrightarrow CaO + CO_2$$

(2) The quartz unites with the lime flux to form slag — the scum above the metal.

$$SiO_2 + CaO \longrightarrow CaSiO_3$$

If the impurity is already limestone, as is often the case, quartz is' added, with the formation of slag as before:

$$CaO + SiO_2 \longrightarrow CaSiO_3$$

Metallurgy by electrolysis. Some metals cannot be obtained by the processes outlined in the previous paragraphs, especially the more reactive ones that burn readily. Several of these have been successfully obtained by electrolysis. Sodium and potassium are obtained by the electrolysis of melted NaOH or KOH in a cell that shuts out all air. If oxygen could reach these metals while heated, they would readily burn, which would undo everything accomplished by the electrolysis. The metal aluminum was too expensive for extensive use until Dr. Hall perfected a method of getting it out of its ores by electrolysis. This process has developed into one of our great industries, producing aluminum cooking utensils, wire, automobile parts, and a thousand other articles. The Hall process electrolyzes the oxide of aluminum dissolved in molten cryolite as the solvent. The process makes use of a fact previously unknown; namely, that a melted mineral may serve as a solvent for a substance that is not soluble in water. The aluminum oxide (Al_2O_3) seems to ionize in the melted cryolite into $\overset{+++}{Al}$ ions and $\overset{=}{O}$ ions.

Metallurgy by high temperature replacement. Some metals replace others from their compounds at high temperatures, usually with the liberation of heat. This type of reaction is not very practical because of the high cost of the replacing metal. The first aluminum metal was obtained by heating aluminum chloride with metallic potassium, but the high cost of potassium made the aluminum cost about $60 per pound. The metals chromium, uranium, and manganese are very difficult to free from their compounds. If the oxide of one of these metals is mixed with powdered aluminum metal and ignited, a reaction takes place by which aluminum takes the oxygen away from the other metal. This type of reaction is called the Goldschmidt process after the inventor. Because of the high temperature of the reaction, the Goldschmidt process is used to weld together broken rails or castings. A mixture of powdered aluminum and iron oxide is placed in a crucible over the crack in the steel rail. The mixture (thermit) is ignited by means of a lighted magnesium ribbon. A temperature of over 3000° C. is obtained by the reaction.

$$Fe_2O_3 + 2 Al \longrightarrow Al_2O_3 + 2 Fe$$

Courtesy of Metal and Thermit Corp.

WELDING WITH THERMIT. When a mixture of powdered aluminum and iron oxide are ignited above a break in an iron or steel casting, the iron from the reaction runs into the break and fuses with the metal of the casting to unite the parts.

The iron from the reaction sinks into the crack, fusing the broken parts into one piece. Oxyacetylene welding has largely replaced thermit in practice.

Metallurgy by metallic replacement in solution. Some metals replace others in water solution. Although this process is not used very extensively as a method of getting metals from their ores, it has some practical use. The photographer recovers silver from his waste solutions by adding iron nails. Some copper mines recover copper from the mine waters by the use of scrap iron.

$$CuSO_4 + Fe \longrightarrow Cu + FeSO_4$$
$$2 AgNO_3 + Fe \longrightarrow 2 Ag + Fe(NO_3)_2$$

One can predict which metals will replace another metal by reviewing its position in the electropotential series of Chapter 11, which is also often called the *replacement series*. Any metal will replace those below it in this series from their compounds.

Likewise this series tells us which metals dissolve in dilute acids to form salts by replacement of hydrogen.   Although hydrogen is not a metal, it does form the positive part of many compounds.   In the acids, especially, it forms the positive part; and when they ionize, it forms the positive ion. Since the simple dissolving of many metals in acids is really a replacement reaction similar to the replacement reactions of the metals, and since hydrogen forms an electric cell with any metal as was mentioned in the previous chapter, it is desirable to give hydrogen its appropriate position in the series of the metals. From its position we can

Courtesy of Metal and Thermit Corp.

A RAIL WELDED WITH THERMIT.   A piece of metal welded in this manner is as strong as before the break.

predict which metals will dissolve in dilute acids and which will not.   All of those metals above hydrogen dissolve in dilute acid, while those below it do not.

EXPERIMENT

Deposition of lead on zinc.   Cover a zinc rod with one layer of asbestos paper and immerse it in a 10 per cent solution of lead nitrate until the next class period.   Crystals of lead will appear on the outside of the paper.   What does this experiment indicate as to the relative positions of lead and zinc in the replacement series?

Salts of the metals.   Thousands of useful compounds are salts of the metals.   There are several general methods of preparing salts, most of which we have already studied.

(1) Neutralization of a base with an acid.

$$Ca(OH)_2 + H_2SO_4 \longrightarrow CaSO_4 + 2 H_2O$$

(2) Dissolving a metal in an acid.

$$Zn + H_2SO_4 \longrightarrow ZnSO_4 + H_2$$

(3) Dissolving the oxide of a metal in an acid,

$$PbO + 2\,HNO_3 \longrightarrow Pb(NO_3)_2 + H_2O$$

(4) Heating a nonvolatile acid with the salt of a volatile acid.

$$H_2SO_4 + 2\,NaCl \longrightarrow \underset{\text{new salt}}{Na_2SO_4} + \underset{\text{volatile acid}}{2\,HCl}$$

(5) Dissolving a carbonate in an acid.

$$CaCO_3 + 2\,HCl \longrightarrow CaCl_2 + H_2O + CO_2$$

(6) A double decomposition or trading of ions.

$$NaCl + AgNO_3 \longrightarrow AgCl + NaNO_2$$

(7) Direct combination (in a few cases where the salt contains only two elements.)

$$2\,Na + Cl_2 \longrightarrow 2\,NaCl$$

(8) Replacement of one metal by another.

$$Fe + CuSO_4 \longrightarrow FeSO_4 + Cu$$

Types of chemical reactions. In this chapter we have illustrated most of the possible types of chemical reaction.

Decomposition. The simplest type of chemical reaction is called decomposition. In the metallurgy of the metals which occur as carbonates, we learned that the carbonate is decomposed into two parts by the furnace heat, one of these being the oxide of the metal and the other carbon dioxide.

$$FeCO_3 \longrightarrow FeO + CO_2$$

Combination. The opposite type of reaction in which two substances unite into a single substance is called *combination.* Sometimes combination may take place between free elements and sometimes between compounds, as the following reactions show.

(a) $\quad 2\,K + Cl_2 \longrightarrow 2\,KCl$

(b) $\quad NH_3 + HCl \longrightarrow NH_4Cl$

Replacement. Another type reaction is illustrated by the formation of hydrogen gas by zinc dissolving in acid or by one metal replacing another from its salts.

$$Zn + 2\,HCl \longrightarrow ZnCl_2 + H_2$$
$$Sn + 2\,AgNO_3 \longrightarrow 2\,Ag + Sn(NO_3)_2$$

Double decomposition. A reaction in which two metallic ions trade partners is called a double decomposition.

$$CaCl_2 + 2\,AgNO_3 \longrightarrow 2\,AgCl + Ca(NO_3)_2$$

QUESTIONS OF FACT

1. What is metallurgy?
2. What metal has the simplest metallurgy?
3. What is roasting?
4. What kind of ore would not require roasting?
5. Define calcining.
6. Why is coke the form of carbon used for the reduction of most metals?
7. How did Hall's process for obtaining aluminum from its oxide differ from any metallurgical process up to that time?
8. What is thermit?
9. Of what use is thermit?
10. Name eight different ways of preparing salts.
11. Name four types of chemical reactions.

QUESTIONS OF UNDERSTANDING

1. Why is a flux needed in metallurgy?
2. If SiO_2 is in a metallic ore, what would be the suitable flux?
3. If $CaCO_3$ were the foreign substance, what should be the flux?
4. Explain the term slag.
5. Outline the electrometallurgy of sodium and potassium.
6. Explain how thermit is used to weld steel rails.

Looking at the replacement series, or electropotential series, on page 223 answer the following questions:

7. Name four metals that will dissolve in dilute acids.
8. Name four metals that will not dissolve in dilute acids.
9. Name two metals that will replace mercury from its solutions readily.

10. Name one metal which will just barely replace mercury from its solutions.

11. Illustrate each of the four types of chemical reactions.

REFERENCE FÓR SUPPLEMENTARY READING

1. Foster, William. *The Romance of Chemistry.* D. Appleton-Century Co., New York, 1936. Chap. XVII, pp. 261–286, " Minerals and Metals."
   (1) How are many ores concentrated? (p. 280)
   (2) Explain the flotation process. (p. 281)
   (3) What is the chemical reaction of the reduction of the ores? (p. 282)
   (4) What is electrometallurgy? (pp. 284–285)
   (5) Describe the electrometallurgy of copper. (p. 285)

SUMMARY

Although differing in color and other physical properties, the metals all have a characteristic shininess or luster. They are more ductile, more malleable, and of greater tensile strength than the nonmetals.

In their chemical reactions, some metals are very active, others very inert. Sodium and related metals react with water, while gold will not dissolve in any single acid. Aluminum and zinc dissolve in bases as well as acids.

A few of the nonreactive metals are found uncombined. Most metals occur as oxides, sulfides, and carbonates. Iron occurs usually as an oxide, lead as a sulfide, and calcium as a carbonate.

In the metallurgy of metals, the ore is first roasted, or heated with air. This changes the other compounds into the oxides. The oxides now react with coke to form CO and set the metal free. In the metallurgy of iron and similar metals it is necessary to remove other substances mixed with the ore. A flux is added, which unites with the rock to form a molten scum, which floats on the heavier metal and is drawn off as slag. Some metals which cannot be freed from their compounds by furnace reactions can be freed by electrolysis. Aluminum, sodium, and potassium are obtained in this way. Certain metals difficult to obtain by other methods can be freed by heating their oxides with aluminum metal. Metals can often be obtained from their soluble compounds by replacement by metals above them in the replacement series. Copper and silver are often recovered from solution by adding iron.

Salts of the metals can be prepared by: (1) neutralization of a base with an acid; (2) dissolving a metal in an acid; (3) dissolving the oxide in an acid; (4) heating a nonvolatile acid with the salt of a volatile acid; (5) dissolving a carbonate in an acid; (6) double decomposition; (7) direct addition, and (8) replacement of one metal by another.

The general types of chemical reaction are decomposition, combination, replacement, and double decomposition.

CHAPTER 13

THE ALKALI METALS

Introduction. Lithium, sodium, potassium, rubidium, and cesium constitute a family of metallic elements called the alkali metals, all of which have a valence of one in their compounds. These five metals are grouped into a family because their physical and chemical properties are very similar. All of their compounds look and act almost alike. Like a human family, the individuals have many similarities and a few differences. Their family name "alkali metals" comes from the fact that their hydroxides are the strongest bases known. The terms alkali and base do not mean exactly the same. Some alkalies are not bases but are substances which react with water to form bases. If the reader does not fully remember what is meant by a base, he should refer back to Chapter 7. The agriculturist often uses the term alkali to include table salt and other salts that do not react with water to form bases, but this is too broad a use for the term.

The alkali metals as a family. As a family, the alkali metals — like all other typical families of elements — all have the same chemical properties, differing only in the speed with which they react. Their physical properties are similar also, variations being most noticeable between elements of very different atomic weights. The relation of atomic weights to physical and chemical properties is shown in the following table.

PROPERTIES AND ATOMIC WEIGHTS

Element	Li	Na	K	Rb	Cs
Atomic weight	7	23	39	85	133
Melting point	186° C.	97.5° C.	62.3° C.	38° C.	26° C.
Boiling point	1400° C.	880° C.	760° C.	696° C.	670° C.
Speed of reaction	least	next	next	next	greatest
Electrons by light	fewest	next	next	next	most

Although most of the variations shown in the table are just the reverse of what one might expect, they are consistent. The

Courtesy of International Salt Co

INSIDE A SALT MINE. Besides the vast accumulations of salt in the oceans and the salt lakes, there are extensive underground deposits of salt in various parts of the world.

usual situation as exhibited by the halogens is that boiling points and melting points increase with increasing atomic weights, while the tendency to react decreases. For some unknown reason the members of this family change in the reverse order from the halogens; that is, melting points and boiling points decrease as the atomic weights increase, while the reactivity increases.

Lithium, rubidium, and cesium. Lithium is so scarce that it has little commercial use. Aside from a limited use of the carbonate and bromide in medicine, its uses are hardly worth mentioning. Rubidium and cesium are still scarcer. Although expensive because of scarcity, rubidium and cesium find some use in photo cells, as they seem to give off electrons more readily than other metals when struck by light. Their greater efficiency more than makes up for their difference in cost above other metals.

Sodium and potassium. Sodium and potassium are quite abundant, and many of their compounds are used extensively. The earth's crust, as we know it, is 2.64 per cent sodium and 2.40 per cent potassium. Although the amount of potassium is almost as great as that of sodium, potassium compounds are relatively scarce and sodium compounds are quite common and cheap. This strange situation is explained by the fact that potassium is needed in large quantities by plants, while very little sodium is required. Rain falls on the land and works its way towards the ocean, leaching out the soluble potassium and sodium compounds. Plants all along the way, and even in the ocean itself, take up the potassium and allow the sodium compounds to pass on. The result is that sodium compounds are concentrated in the waters of the ocean and certain inland lakes, such as the Great Salt Lake and the Dead Sea.

Salt. The sodium compound that is most concentrated in the ocean is sodium chloride ($NaCl$) or table salt. There are also fabulous amounts of this salt in the Great Salt Lake, the Dead Sea, and in many dried-up lake beds. This substance is the cheapest source of the sodium used in other sodium compounds and the chlorine in all chlorine compounds. The supply is inexhaustible and easy to obtain. A large part of our table salt is obtained by the evaporation of sea water in shallow ponds. One may see piles of salt as large as houses around these salt works. Salt is so much in excess of the other compounds in the sea water that it reaches saturation and begins to crystallize long before they do. Salt for table use must be recrystallized several times to make it conform to Government standards.

EXPERIMENT

Evaporation of water rich in dissolved substances. If the ocean or alkaline ponds are near, evaporate some of the water to dryness. Examine the resultant crystals with a high-powered magnifier. Taste the residue. Try the flame test on the materials. (See p. 266.)

Sodium chloride is found in large deposits, from which it is obtained as impure rock salt, sometimes used for salting livestock without any further purification.

Salt is easily soluble in cold water but, unlike most substances, is only slightly more soluble in hot water. When the water of solution is evaporated, the salt crystallizes in cubical crystals, which when perfect are true cubes; but it is seldom that the faces are evenly developed.

Sodium carbonate. There are "soda" lakes in the western part of the United States, but the greatest natural deposit of sodium carbonate is Lake Magadi, British East Africa. This lake is literally a solid mass of crystalline sodium carbonate. However, most of this substance used commercially is prepared from salt. The process used in the more modern plants is called the Solvay process. This process is rather simple to carry out but quite hard to explain. In practice, carbon dioxide gas and ammonia gas are passed into a concentrated solution of sodium chloride until $NaHCO_3$ precipitates.

$$CO_2 + H_2O \rightleftharpoons H_2CO_3 \text{ (carbonic acid)}$$
$$NH_3 + H_2O \rightleftharpoons NH_4OH \text{ (ammonium hydroxide)}$$
$$NH_4OH + H_2CO_3 \rightleftharpoons NH_4HCO_3 + H_2O \text{ (neutralization)}$$
$$NaCl + NH_4HCO_3 \rightleftharpoons NaHCO_3 + NH_4Cl$$

The reason the $NaHCO_3$ separates out is that it is the least soluble of all the possible products in this complicated solution. When the sodium bicarbonate is heated, the following reaction takes place:

$$2\,NaHCO_3 \longrightarrow Na_2CO_3 + H_2O + CO_2$$

Sodium carbonate, Na_2CO_3, is sold to the housewife as washing soda (sal soda) to use in her washing machine and to the laundries as "soda ash" for the same purpose. Large quantities of it are also used in glass making, petroleum refining, water softening, cleansers, textile industries, soap making, and the manufacture of paper.

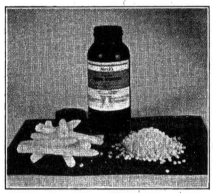

SODIUM HYDROXIDE. Sodium hydroxide is the cheapest of the strong bases. A great deal of it is used in making soap, rayon, and paper and in refining oils.

Baking soda. The first product of the Solvay process is baking soda ($NaHCO_3$). The principal use of baking soda is in the manufacture of baking powder. Practically all of the baking powders have baking soda as one of the constituents. Baking soda serves as the source of the carbon dioxide that makes the bread to rise.

Sodium hydroxide. Sodium hydroxide, another important sodium compound, is sometimes called caustic soda. A caustic substance is one that will dissolve the skin. If one handles solid NaOH or dips the hands into strong solutions of it, the fingers soon become raw. In fruit canneries, peaches are lye peeled. Lye is a mixture of NaOH and KOH, both of which are caustic. The KOH is known as caustic potash. The lye causes the skin of the peach to slip off easily. Cannery workers have to wear rubber gloves to protect their fingers. The greatest use for sodium hydroxide is in soap making; large amounts of it are also used in the manufacture of rayon and paper. Sodium hydroxide was formerly made from sodium carbonate and calcium hydroxide according to the following reaction:

$$Na_2CO_3 + Ca(OH)_2 \longrightarrow CaCO_3 + 2\,NaOH$$

This reaction goes to completion because the $CaCO_3$ is insoluble and leaves the reaction as a precipitate. An electrolysis process has largely replaced the chemical one just described. If a solution of table salt is electrolyzed, chlorine escapes from the positive electrode, and NaOH accumulates in the solution around the negative electrode, as was explained on page 217.

QUESTIONS OF FACT

1. Name the alkali metals.
2. Compare the properties of the alkali metals when arranged according to increasing atomic weights.
3. Give one use of lithium compounds and one of rubidium and cesium compounds.
4. What are the percentages of sodium and potassium in the earth's crust?
5. Discuss the abundance of sodium chloride.
6. Give three common names of sodium carbonate.
7. What are the active constituents of lye?
8. How is NaOH made from NaCl commercially?
9. What is the shape of NaCl crystals?
10. What substance is called caustic soda?
11. State three uses of NaOH.
12. Give the reaction when $NaHCO_3$ is heated.

QUESTIONS OF UNDERSTANDING

1. What does the chemist mean by alkali?
2. What is meant by a family of elements?
3. Account for the low concentration of potassium salts in the ocean as compared with sodium salts.
4. Name seven possible compounds in the mixture in the Solvay process.
5. Name four substances that are put into the reacting vessel of the same.
6. In the Solvay process, of the six possible substances besides water, which is the least soluble?

Sir Humphry Davy

1778–1828

After the death of his father, which left his mother a widow with five children, Davy apprenticed himself to a surgeon apothecary at Penzance. The one of his premedical studies that most aroused his interest was chemistry, especially a treatise written by Lavoisier.

In 1798 Davy was installed as superintendent of a recently established institute for research on the medical properties of gases. The next year his investigation of the properties of nitrous oxide established his reputation. When inhaled, this gas produces a sort of intoxication, which gave it the name laughing gas.

In 1800 Davy was hired as assistant lecturer and laboratory director of the Royal Institute in London. He did so well in this position that he was made lecturer and full professor two years later.

After studying and writing on tanning and agriculture for several years, Davy turned to experimentation with electrolysis, which resulted in two outstanding discoveries, the metals sodium and potassium. Soon after this he proved that "oxymuriatic acid" was in reality not a compound but an element, which he named chlorine.

Honors began to come to Davy rapidly now. In 1810 he was given the honorary degree of LL.B. by Trinity College, and in 1812 he was made a knight, hence the title Sir Humphry.

Two more discoveries followed in the next few years; namely, that iodine is an element instead of a compound and that diamond is nothing but carbon. Following a series of serious explosions in coal mines, Davy was asked (about 1815) to see if something could be done to reduce the number of such explosions. After a careful study of the principles of combustion and the travel of flames, he produced the Davy safety lamp, which is still used in coal mines, and which has reduced explosions to a minimum.

7. If too much baking soda is put into soda biscuits, why do they taste bitter?

8. What is meant by a substance being caustic?

9. Give the chemical reaction for preparing NaOH from Na_2CO_3 and tell why the reaction goes to completion.

10. Give four equilibrium reactions in the Solvay soda process.

ADDITIONAL EXERCISES FOR SUPERIOR STUDENTS

1. Write a hundred-word essay on sodium chloride.

2. Assume sea water to be 2.8 per cent salt. Look up "ocean" in the encyclopedia and get the estimated volume of all the sea water of the earth. One cu. m. weighs 1.13 tons. Calculate the weight of the salt.

3. Which would neutralize the most acid, a pound of KOH, a pound of LiOH, or a pound of NaOH? Look up the atomic weights before answering.

4. Which could produce the greatest weight of chloride, an ounce of sodium or an ounce of potassium?

5. If a cheap source of cesium should be discovered, is it likely that more uses would be found for this substance?

REFERENCES FOR SUPPLEMENTARY READING

1. Clarke, Beverly L. *Marvels of Modern Chemistry.* Harper and Brothers, New York, 1932. Chap. XVI, pp. 204–213, "The Alkali Metals."
   (1) Describe the appearance of the alkali metals. (p. 204)
   (2) Make a list of sodium compounds and their uses. (p. 206)
   (3) Where and when did Solvay live? (p. 207)
   (4) What process did Solvay invent? (p. 207)

Potassium compounds. Potassium compounds are more expensive than sodium compounds, since nature has not left us many deposits of concentrated potassium salts. Sodium compounds can be used instead of the potassium compounds in most cases. Potassium nitrate is a constituent of black gunpowder. In this case sodium nitrate cannot be used instead of the potassium compound because sodium nitrate is hygroscopic and absorbs moisture from the air. The moisture keeps the powder from exploding. Potassium nitrate is usually called saltpeter.

Metallic sodium. Metallic sodium was first prepared by Sir Humphry Davy in 1807 by passing electricity through melted NaOH. It is now usually prepared by the electrolysis of sodium chloride. This metal when uncombined is so active that it has to be kept under kerosene or sealed in an airtight can. Even then it does not keep very well, because oxygen dissolves in the kerosene and gets into the can while it is being sealed. Metallic sodium is used in making tetraethyl lead, a

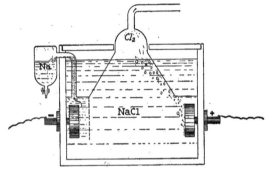

PREPARATION OF METALLIC SODIUM. The diagram shows a tank of molten salt with an electrode at either end. The salt partly ionizes into Na⁺ and Cl⁻ ions. These ions have their charges neutralized at the electrodes of opposite sign, chlorine gas being collected at the positive electrode and metallic sodium being formed at the negative electrode. Since sodium is lighter than the salt, it rises to the top and is drawn off into a receiver.

substance which greatly increases the efficiency of gasoline as a motor fuel.

All of the other metals of this family can be obtained in the free state by processes similar to the one used for sodium (see illustration). They too must be kept out of the atmosphere to reduce loss by reactions with other substances.

Potassium nitrate. Potassium nitrate, aside from its chemical uses, is most valuable as a constituent of fertilizers. It contains two of the elements most likely to be needed by plants, potassium and nitrogen. As a necessary constituent of black gunpowder, it was even more essential to nations a century ago than it is now. There are no important natural deposits of this compound. However, there are large deposits of $NaNO_3$ in Chile. Until about 1920, Chile supplied nitrate for the entire world. Today, however, a large part of the nitrate used is made by nitrogen fixation, and each year Chile finds her markets shrinking. The method of preparing KNO_3 from $NaNO_3$ and KCl, the cheapest potassium compound, was given to illustrate the idea of "reversible chemical reactions" and "mass action" in Chapter 11. Sodium nitrate and potassium chloride are both quite soluble. When they are dissolved together, the possible products of the reaction, KNO_3 and NaCl, are also soluble; hence no precipitate forms, and there remains in solution a mixture of the four salts.

$$\overset{+}{Na}\overset{-}{NO_3} + \overset{+}{K}\overset{-}{Cl} \longrightarrow$$
$$\overset{+}{K}\overset{-}{NO_3} + \overset{+}{Na}\overset{-}{Cl}$$

Presumably both reactions are going on simultaneously in opposite directions so neither one gets anywhere. NaCl is the least soluble of the four. If a large mass of

POTASSIUM AND PLANT GROWTH. Potassium is one of the three elements (nitrogen and phosphorus being the other two) that are most likely to be deficient in the soil. The plant at the left grew in soil containing plenty of potassium; the one at the right failed to grow well because the soil lacked sufficient potassium.

$NaNO_3$ is dumped into the solution, it soon becomes saturated with NaCl, which begins to separate out. This process of adding an excessive mass of a more soluble constituent is, you remember, called mass action. As the salt is continuously leaving the solution, the latter is getting richer in KNO_3. Sodium ions and chlorine ions are being removed in the salt. Finally the solution can be evaporated, and the potassium nitrate made to crystallize out. This process is helped by the fact that sodium chloride is not much more soluble in hot water than in cold, while KNO_3 is much more soluble in hot water.

The flame test for metals. Sodium, lithium, and potassium can be detected in their compounds by the color they impart to a colorless flame. A flame is colored when there are incandescent particles floating in it. Any sodium compound imparts the characteristic sodium yellow to the flame. Potassium compounds give a lavender color to it. Lithium gives a beautiful crimson color. The flame tests, in general, are very sensitive; that is, only a trace of the substance is necessary for the color. The method of applying the test is as follows: Dissolve the substance in hydrochloric acid. Then dip a platinum wire into this solution and introduce it into the colorless flame of a Bunsen burner.

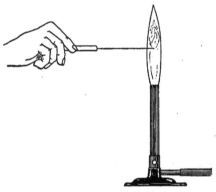

THE FLAME TEST. The metal or metallic compound is first dissolved in hydrochloric acid. A platinum wire is dipped into this solution and then held in a colorless Bunsen flame.

There is often difficulty in recognizing the lavender of potassium in the presence of the yellow of sodium when compounds of this element are also present. This is usually the case, for the sodium compounds are widely distributed and hard to separate completely from the similar potassium compounds. The difficulty may be overcome by observing the flame through a blue glass plate, which completely absorbs the yellow light from the sodium but transmits the lavender light from the potassium.

Several other elements give characteristic flame tests. Boron gives a green color. Copper gives a green flame test in some of its compounds and a blue one in others. Calcium gives an orange flame test. Strontium colors the flame scarlet. Flame tests are the quickest ways to test for any of these elements. Spectacular effects in colored flames in the fireplace can be obtained by soaking pieces of firewood in solutions of the salts described in this topic and allowing them to dry thoroughly. The wood so treated will burn with brilliant colored flames.

EXPERIMENT

Flame tests. Materials needed : Platinum wire sealed into a glass tube, hydrochloric acid, small beaker, LiCl, KCl, NaCl, $Cu(NO_3)_2$, $Sr(NO_3)_2$, potassioscope — sheet of blue glass. Note the flame tests. After each test dip the wire into the acid and heat until the previous color has ceased.

Ammonium compounds and the alkali metals. The ammonium compounds are often compared with those of the alkali metals because of their great similarities. Ammonium hydroxide is a strong base, which although somewhat weaker is comparable with the hydroxides of the alkali metals. Ammonium chloride is similar to the alkali chlorides. Likewise, every other ammonium salt is practically as much like the corresponding salt of one of the alkali metals as the corresponding salts of two alkali metals are like each other. The ammonium radical is of course not a metal; hence we must consider the similarity of its compounds to those of the alkali metals as little more than a coincidence. This similarity may in part be caused by the fact that this radical has a valence of one, as have all the members of this family.

Percentage of an element in a compound. When the formula and atomic weights are known, it is quite simple to calculate the percentages of the elements which make up a compound. For simplicity, the procedure can be set up in a formula: $p = \dfrac{100\,na}{m}$, where p is the percentage, n the number of times the atom under consideration occurs in the formula, a its atomic weight, and m molecular weight, which can be calculated. The percentage of any element in a compound is useful, since all that is necessary in order to find the weight of the element in a given weight of compound is to multiply the latter by the percentage.

Problem: Find the percentage of oxygen in H_2SO_4.

$$\frac{4 \times 16}{98} \times 100 = \frac{6400}{98} = 65.31\%$$

QUESTIONS OF FACT

1. What substance is called saltpeter?
2. Why cannot $NaNO_3$ replace KNO_3 in gunpowder?
3. Who first prepared metallic sodium?
4. How is metallic sodium kept from the air?
5. Describe the flame test for potassium compounds in the presence of sodium compounds.
6. What color do each of the following impart to a flame?

    (a) sodium   (d) copper
    (b) lithium   (e) strontium
    (c) boron

7. How is metallic sodium prepared?
8. What are the valences of the alkali metals?
9. What is the valence of the ammonium radical?

QUESTIONS OF UNDERSTANDING

1. Why is the ammonium radical often studied with the alkali metals?
2. What is meant by saying ammonium hydroxide is a weaker base than NaOH?
3. Why is potassium nitrate more valuable as a constituent of plant fertilizer than sodium nitrate?
4. Why does not sodium or potassium metal kept under kerosene show a bright surface?
5. Tell how KNO_3 is prepared from $NaNO_3$ and KCl.
6. What other substance is prepared at the same time as the KNO_3 by the process in ex. 5?

PROBLEMS

1. Calculate the percentage of copper in $Cu(NO_3)_2$.
2. Calculate the percentage of silver in $AgNO_3$.
3. Calculate the percentage of gold in $AuCl_3$.

REFERENCE FOR SUPPLEMENTARY READING

1. Clarke, Beverly L. *Marvels of Modern Chemistry.* Harper and Brothers, New York, 1932. Chap. XVI, pp. 204–213, "The Alkali Metals."
   (1) How much potassium occurs in feldspar? (p. 208)
   (2) What pressure and temperature are developed inside a gun? (p. 212)

SUMMARY

Lithium, sodium, potassium, rubidium, and cesium form a family of metals known as the alkali metals, which are all very reactive, form similar compounds, and have similar physical properties. Within the family, their properties vary somewhat in the order of their increasing atomic weights. Because of their scarcity, lithium, rubidium, and cesium are very limited in use. Although sodium amounts to but 2.64 per cent and potassium to but 2.40 per cent of the earth's crust, the compounds of both these metals are very important. Due to the fact that potassium is needed in large quantities as food by plants, it has never accumulated in the ocean or salt lakes in anything like the quantities of the sodium compounds. Of the sodium compounds sodium chloride is by far the most abundant.

Starting with the sodium in table salt, other useful sodium compounds are prepared. The carbonates are prepared by the Solvay process of passing ammonia and carbon dioxide into strong brine until baking soda separates out. Sodium carbonate, or washing soda, is obtained by heating the baking soda. Sodium hydroxide, another sodium compound, is used in large quantities in making soap, rayon, and paper. It is obtained by the electrolysis of a salt solution.

Potassium compounds are in great demand for uses the sodium compounds cannot serve, as in making liquid soaps and black powder, and as a constituent of fertilizers. Because of the great need for fertilizers, the Unites States has with some success fostered the domestic production of potassium compounds. (See Chapter 37.)

The uncombined metals of this family are obtained by the electrolysis of the melted hydroxides.

Although potassium nitrate does not occur in nature to any extent, it can be prepared from the naturally occurring sodium nitrate. This process illustrates mass action. Sodium nitrate and potassium chloride are added to a solution until it becomes saturated with sodium chloride, which thereafter separates out on the addition of more of either of the initial substances, allowing potassium nitrate to accumulate in the solution, from which it can be recovered by evaporation.

The common alkali metals can be identified in their compounds by their characteristic flame tests: sodium, yellow; potassium, lavender; and lithium, crimson. Copper with a green or blue flame test and strontium with scarlet can be identified in the same way.

CHAPTER 14

⇄

THE ALKALINE-EARTH METALS

Introduction. The typical members of this family are calcium, strontium, and barium. Beryllium and magnesium are related metals that will be studied at this time. Radium, another related metal, has such unusual properties that a special chapter is given to it later in the text. The family name "alkaline earth" results from the fact that the properties of these elements lie between those of the alkali metals on one side and those of a group of elements called the earth metals on the other side. Like sodium and the other alkali metals, the alkaline-earth metals all react with water to liberate hydrogen.

$$Ca + 2H_2O \longrightarrow Ca(OH)_2 + H_2$$

The first two, beryllium and magnesium, require hot water to start the reaction; the others react with cold water, but with less violence than the alkali metals. Since these metals are less reactive than the alkali metals, they can be kept fairly well in a bottle without kerosene. Their hydroxides are not so caustic as sodium hydroxide or potassium hydroxide; but, although not very soluble, they are quite strong bases. All of the metals of this family have the same valence, two.

Courtesy of Amer. Museum of Nat. Hist
CRYSTAL OF BERYL

Beryllium. Beryllium occurs chiefly as the mineral beryl, a beryllium aluminum silicate. This mineral often occurs in very large crystals. One crystal found in Maine was 14 feet long and 3 feet in diameter. One of our most precious stones, the

271

emerald, is transparent beryl tinted green. Beryllium, although about as abundant as zinc or lead, is very expensive because of the difficulty of getting the metal out of its minerals. Chemical research has reduced the price from $500 per lb. to $50 per lb. At the latter price it is beginning to have industrial possibilities. This metal has certain properties that will make it increasingly useful as the price is reduced.

Courtesy of Signal Corps U. S. Army

MAGNESIUM FLAME. Because of its brilliant flame, magnesium is used in flares and fireworks.

The low density (1.8) of beryllium as compared with that of the metals now used in airplanes makes it a promising material for future developments in airplane structure. Aluminum (density 2.7) cannot compete with it in lightness. Magnesium (density 1.74) is the only metal which compares favorably with it in strength and weight. However, beryllium is far superior to magnesium for engine parts, since beryllium melts at 1350° C. while magnesium melts at only 651° C. Even aluminum (melting point 658.7° C.) is at a disadvantage in this respect. Its elasticity is greater than that of any other known metal, which property will probably earn it many new uses. Its electrical conductivity is high. When alloyed with copper (melting point 1083° C.), it forms a valuable material for the electrical industries, being a good conductor of electricity and less likely than copper to melt when overheated. Millions of dollars are invested in beryllium mines by the electrical companies. This copper-beryllium alloy ought to make good automobile pistons. The alloy of beryllium with aluminum is lighter, stronger, and of higher melting point than aluminum, yet can be worked equally well.

Magnesium. Magnesium is the sixth metal in abundance in the earth's crust. In the metallic form it has several uses.

Because it burns with such brilliance, it is used in fireworks and for military illumination. Its strong tendency to unite with oxygen makes it valuable as a "scavenging metal" in the manufacture of steel. Small air bubbles and particles of iron oxide greatly weaken steel. A little magnesium metal added to a batch of steel will find all the iron oxide and take its oxygen. Since magnesium unites with both oxygen and nitrogen, it removes air bubbles as well. Mixed with aluminum, magnesium is used in the light alloy, magnalium, employed in airplane and dirigible construction. The principal advantage of magnalium over pure aluminum is the ease with which it is worked. Pure aluminum tends to stick to the cutting tools, while magnalium does not.

MAGNESIUM METAL. Magnesium is usually sold in the ribbon or the powder form.

EXPERIMENT

Magnesium flame. Burn a piece of magnesium ribbon and note the intense light. Scratch the metal and see if there is tarnish on it. Cut the ribbon with a knife. Is it softer than iron?

Compounds of magnesium. Magnesium is one of the very few elements which unite directly with nitrogen.

$$3 \, Mg + N_2 \longrightarrow Mg_3N_2$$

The resulting compound is called magnesium nitride. This reaction has never been used as a commercial method of nitrogen fixation because of the cost of magnesium.

Magnesium compounds are all laxative in their physiological action. Common ones used for this purpose are: magnesium hydroxide, called milk of magnesia because it looks like milk when mixed with water; magnesium sulfate, known as Epsom salts; and magnesium citrate. Talc, naturally occurring magnesium silicate, is a constituent of talcum powder. Small amounts of magnesium are necessary to plant life, being found in the green chlorophyll.

QUESTIONS OF FACT

1. Name the alkaline-earth metals.
2. What is the reason for the name?
3. Give the similarities between the alkali metals and the alkaline-earth metals.
4. What substance is an emerald?
5. Why is the metallic beryllium so expensive?
6. Name the properties of metallic beryllium that will make it useful.
7. Describe two uses of metallic magnesium.
8. Name three magnesium compounds used in medicine.
9. What is talc?
10. What is magnalium, and what advantage does it have over aluminum?

QUESTIONS OF UNDERSTANDING

1. Discuss $Mg(OH)_2$ as a base.
2. If cheap metallic beryllium were available, what would it be used for?
3. If beryllium occurred as the oxide or carbonate, would it be cheaper? Give reasons for answer.
4. What is meant by a scavenging metal?
5. What property of beryllium is extreme?
6. Why has magnesium nitride not been made the basis of a nitrogen fixation process?

REFERENCES FOR SUPPLEMENTARY READING

1. Weeks, Mary Elvira. *The Discovery of the Elements.* Journal of Chemical Education, Easton, Pa., 1934. Pp. 153–156, "Discovery of Beryllium," and pp. 133–135, "Discovery of Magnesium."
    (1) In the discovery of what element was geometry an aid? (p. 153)
    (2) In what way did it aid? (p. 153)

(3) How do the compounds of beryllium differ from those of aluminum with which it occurs? (p. 154)

(4) How and by whom was metallic beryllium prepared? (p. 155)

(5) What is the first mention of magnesium compounds? (p. 133)

(6) Who first isolated some magnesium metal? (p. 135)

(7) Describe Bussy's method of preparing magnesium. (p. 135)

(8) When did Bussy develop his method? (p. 134)

2. Howard, Joseph W. "Emeralds." *Journal of Chemical Education,* Vol. XI, No. 6, pp. 323–327, June 1934.

(1) Give a short discussion of the emerald.

Courtesy of Indiana Limestone Quarryman's Assn.

LIMESTONE QUARRY. Limestone for buildings and monuments is removed in large blocks from quarries like the one above. The waste stone is burned into quicklime.

Calcium. Calcium, the fifth metal in abundance, like magnesium is also needed by plants and animals. The mineral part of the bones is principally calcium phosphate. Calcium metal is of little commercial value as yet, largely because of its cost.

Calcium compounds. The calcium compounds, because of their comparative cheapness and great importance, play quite a role in the industries and arts. Calcium compounds occur everywhere. Large bodies of calcium carbonate (limestone) are found in various portions of the globe. Limestone under-lies large areas in the Mis-sissippi Valley. The Mam-moth Cave in Kentucky is dissolved out of limestone. Crystallized limestone, commonly called marble, finds use as building stone and tombstones. Large amounts of calcium sul-fate, or gypsum, are used in making plaster of Paris, and in counteracting alkali in the soil.

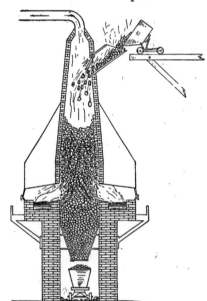

Lime. The oxide of cal-cium formed by heating the carbonate is called *quicklime.*

$$CaCO_3 \longrightarrow CaO + CO_2$$

Quicklime is used in making mortar for cement-ing bricks together and in

LIME KILN. Quicklime (CaO) is made by heating limestone (CaCO₃) in a furnace like the one shown above.

making the lime-sulfur washes used for dipping sheep and spraying fruit trees. When quicklime is moistened, the hy-droxide is formed, with the evolution of much heat. This formation of hydroxide by action of an oxide in water illus-trates a *general method for making any hydroxide.* The hydroxide of calcium is called *slaked lime.*

$$CaO + H_2O \longrightarrow Ca(OH)_2$$

When quicklime stands exposed to dry air, it reacts with car-bon dioxide according to the equation:

$$CaO + CO_2 \longrightarrow CaCO_3.$$

All lime that has been standing exposed to the air will contain both the hydroxide and carbonate. Air-slaked lime is largely carbonate and in time will become entirely the carbonate.

Heat by slaking lime. Place several lumps of quicklime in an evaporating dish and moisten with hot water. Ignite a match by placing the head in one of the crevices as the lime crumbles.

Calcium oxide as drying agent. Moisten the inside of two bell jars. Turn one over a few lumps of calcium oxide and the other over on the table. In a short time the moisture inside the one containing the quicklime disappears.

Mortar. Mortar, the substance used in laying brick, consists of a mixture of slaked lime, sand, and water. When the mortar hardens, or sets, a reaction takes place between the lime and the carbon dioxide of the air, and limestone is formed.

$$Ca(OH)_2 + CO_2 \longrightarrow CaCO_3 + H_2O$$

TILE HELD TOGETHER BY MORTAR. Perhaps during the construction of a new building you have noticed a man stirring a mixture with a hoe. This mixture is made of lime, sand, and water. It is called mortar and is used for laying brick and tile.

This reaction explains why mortar is so long in drying; water is formed as the reaction progresses. The hardening begins on the outside and slowly works inward.

Hard water. Although calcium and magnesium carbonates are very insoluble and the sulfates quite so, the hydrogen carbonates or bicarbonates dissolve to a larger extent. Small amounts of the latter get into the drainage waters of limestone countries. The carbon dioxide of the air is somewhat soluble;

hence, it is taken up by the rain water and carried into the ground. When carbon dioxide is dissolved in water, carbonic acid is formed. The latter reacts with the limestone it meets on its way, to form soluble acid calcium carbonate.

$$H_2O + CO_2 \longrightarrow H_2CO_3$$
$$CaCO_3 + H_2CO_3 \longrightarrow$$
$$Ca(HCO_3)_2$$

The second reaction is the one by which limestone caves are formed. An *acid salt* is part acid and part salt; that is, the hydrogen has not all been replaced by the metallic element.

STALACTITES AND STALAGMITES.

Acid calcium carbonate is such a salt. The other calcium carbonate is called the *normal carbonate*. A compound half salt and half base, such as Ca(OH)Cl, is called a *basic salt.* A basic salt may be formed by partly neutralizing a base with an acid. Acid calcium carbonate is more soluble when the water is under pressure. As water flows underground, it is under more or less pressure and absorbs more limestone. In many limestone countries this water finds its way to caves, where the pressure is suddenly released and the excess limestone separates out, forming long projections from the roof, called stalactites. Other projections on the floor, caused by the impregnated water dripping continuously in one spot, are called stalagmites. Stalactites and stalagmites often meet to form columns and produce fantastic scenes in limestone caves. Water containing dissolved calcium and magnesium salts is called *hard water*.

$$Ca(HCO_3)_2 \longrightarrow CaCO_3 + H_2O + CO_2 \text{ (stalactite formation)}$$

Temporary hardness. If the salt dissolved in water is acid calcium carbonate, or acid magnesium carbonate, the resulting hardness is called temporary hardness because it can be removed by boiling.

$$Ca(HCO_3)_2 \longrightarrow CaCO_3 + CO_2 + H_2O$$

Such water, however, cannot be used in boilers because the limestone forms a thick "boiler scale" inside, which wastes much of the heat. Temporarily hard water can be used for laundry purposes after it has been boiled. Hard water is very bad for most industrial uses. In the

BOILER SCALE. Hard water, especially temporarily hard water, forms a deposit inside the boiler, which soon renders the boiler useless.

laundry, for example, hard water precipitates the soap, not only wasting it but also producing little gray spots in the clothes.

EXPERIMENT

Boiler scale. Examine some boiler scale or teakettle scale. Test it with HCl for evolution of CO_2. Boil water with temporary hardness to show the scum resulting.

Permanent hardness. The dissolved salts of calcium and magnesium, other than the bicarbonates, cause *permanent hardness* in water. Permanently hard water cannot be used for laundry purposes, because the calcium and magnesium precipitate out and waste the soap. In such cases washing soda (Na_2CO_3) or other similar substance is added. As soon as this

soluble carbonate is added, the highly insoluble calcium and magnesium carbonates separate out and do not then interfere with washing. Removing the dissolved calcium and magnesium is called softening the water.

BOILING SOFTENS TEMPORARILY HARD WATER. If water containing magnesium and calcium acid carbonates is boiled, the carbonates separate out. The beaker at the right shows such water before boiling; the one at the left, after boiling.

$$CaCl_2 + Na_2CO_3 \longrightarrow CaCO_3 + 2\,NaCl$$

$$MgSO_4 + Na_2CO_3 \longrightarrow MgCO_3 + Na_2SO_4$$

Drying agent. Calcium chloride ($CaCl_2$) is used as a dehydrating or drying agent because of the great avidity with which it reacts with water or water-vapor to form $CaCl_2 \cdot 6\,H_2O$. It is used for keeping the atmosphere dry in desiccators and for drying liquids such as ether.

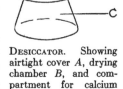

$$CaCl_2 + 6\,H_2O \longrightarrow CaCl_2 \cdot 6\,H_2O$$

It will be noted that the product contains water of hydration.

Calcium hydroxide. Calcium hydroxide is a base of limited solubility. The clear solution is called lime water. Lime water is added to baby's milk for two reasons: (1) to remove all traces of acid; and (2) to furnish calcium needed for bone building. If a large excess of undissolved $Ca(OH)_2$ is left in lime water, it is called "milk of lime," because it looks somewhat like milk. This mixture is used as a cheap base to neutralize acids. It is

DESICCATOR. Showing airtight cover A, drying chamber B, and compartment for calcium chloride C.

DIFFERENT FORMS OF CALCIUM CARBONATE. Left to right: marble, precipitated chalk, calcite, dark limestone, and light limestone.

sometimes sprayed on old buildings as "whitewash" to protect the wood and, in the case of chicken houses, to act as a disinfectant. The hydroxide of the whitewash soon reacts with the carbon dioxide of the air, so in the end the building is covered with limestone instead of hydroxide.

$$Ca(OH)_2 + CO_3 \longrightarrow CaCO_3 + H_2O$$

Calcium hydroxide as water-slaked lime has already been mentioned as a constituent of mortar.

Calcium carbonate. Calcium carbonate ($CaCO_3$) in its various forms is a substance that serves many uses in industry. As the crude limestone, it is used to make quicklime, to neutralize acid soils, and in the metallurgy of iron. Other names for commercial calcium carbonate are marble, chalk, and calcite. Marble, the beautiful building stone and material for tombstones and statuary, is limestone that has been heated under pressure by volcanic action and crystallized in small crystals. Small crystals of calcium carbonate look white in the same way that the small crystals of ice in snow look white. In a few localities the carbonate is found crystallized in large transparent crystals of calcite. Those calcite crystals which are perfectly clear are called Iceland spar. Iceland spar has

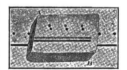

ICELAND SPAR. Notice that the positions of the line and the dots appear to have shifted and that they seem to be doubled.

some very unusual properties that make it valuable for certain optical instruments. One of these properties is that of double refraction. A single line seen through a crystal appears as two lines. A pearl is essentially calcium carbonate which has been deposited in layers by an oyster around some irritant that has got inside its shell. The Japanese produce artificial pearls by inserting sand into an oyster, which they keep until a pearl develops. Oysters are now sold with a guarantee that each contains a pearl. However, not many such pearls are of a size and quality to give them much value.

Coral also is primarily calcium carbonate. It is the bony structure of a small saltwater organism. Entire islands in the Pacific Ocean are made of coral. .

QUESTIONS OF FACT

1. Discuss the distribution of limestone.
2. What compound is called quicklime?
3. How can quicklime be prepared?
4. Give a general method of preparing a hydroxide.
5. When quicklime is exposed to the atmosphere, what changes take place?
6. What is mortar?
7. Where does calcium occur principally in the human body?
8. Hard water is (a) ice, (b) rose water, (c) water containing dissolved calcium and magnesium, (d) water containing black alkali.

Which item (a, b, c, or d) correctly completes the foregoing statement?
9. Define an acid salt. Illustrate.
10. Define a basic salt. Illustrate.
11. Those salts which are neither acid salts nor basic salts are called what?
12. Give the objections to hard water for (a) boilers; (b) laundry purposes; (c) hot-water heaters.
13. Name three substances which remove hardness from water.
14. What unusual properties are exhibited by Iceland spar?
15. . What substance is the pearl?

QUESTIONS OF UNDERSTANDING

1. What happens in the air slaking of lime?
2. Write the reaction for the water slaking of lime.
3. Write the reaction for the hardening of mortar.

4. Explain the formation of (a) limestone caves; (b) stalactites.
5. Give the formulas of the following substances:

(a) milk of lime   (e) cause of temporary hardness
(b) limestone   (f) marble
(c) whitewash   (g) calcite
(d) chalk   (h) a pearl

6. What causes temporary hardness of water, and how may it be removed?
7. What causes permanent hardness of water, and how is it removed?
8. How is marble produced in nature?
9. Give peculiar properties and use of Iceland spar.
10. Why did these reactions go to an end?

(a) $\qquad CaCO_3 \longrightarrow CaO + CO_2$ (furnace)
(b) $Mg(NO_3)_2 + Na_2CO_3 \longrightarrow MgCO_3 + 2\,NaNO_3$
(c) $Mg + 2\,H_2O \longrightarrow Mg(OH)_2 + H_2$

11. Given the equilibrium reaction:

$$H_2CO_3 + CaCO_3 \rightleftharpoons Ca(HCO_3)_2$$

How can it be made to go to completion towards the left?

12. Why is calcium metal not used as much as its properties should warrant?

13. Pick out the (a) acid salts, (b) basic salts, (c) normal salts, from the list given below:

$CaCl_2$	$BaHPO_4$	$Sr_2(PO_4)_3$	$Ca(OH)Br$
$NaHSO_4$	$Ba(NO_3)_2$	$Ba(OH)Cl$	$SrSO_4$

REFERENCES FOR SUPPLEMENTARY READING

1. Clarke, Beverly L. *Marvels of Modern Chemistry.* Harper and Brothers, New York, 1932. Chap. XVIII, pp. 225–237, "Lime and Magnesia."
   (1) What per cent of the earth's crust is calcium? (p. 225)
   (2) Discuss the hardness of calcium. (p. 225)
   (3) How much water vapor is absorbed by 100 grams of anhydrous calcium chloride? (p. 227)
   (4) What calcium compound is luminous in the dark? (p. 229)
2. Weeks, Mary Elvira. *The Discovery of the Elements.* Journal of Chemical Education. Easton, Pa., 1934. Pp. 128–132, "The Discovery of Calcium."
   (1) Who first decided that lime is an oxide and not an element? (p. 128)
   (2) Who first produced calcium metal? (p. 129)
   (3) Describe the method. (pp. 129–130)

Calcium sulfate. Crystallized calcium sulfate, or gypsum, has been used in making plaster for 4000 years. On the Egyptian pyramids we find plaster work that has outlasted the very rock on which it was laid. Gypsum ($CaSO_4 \cdot 2 H_2O$) often occurs in layers more than fifty feet thick and extending for miles. When pure, gypsum is pure white. The alabaster vessels mentioned in the Scriptures and also those found in Tutankhamen's tomb were carved from transparent gypsum. They are still in good condition. Gypsum is applied to land

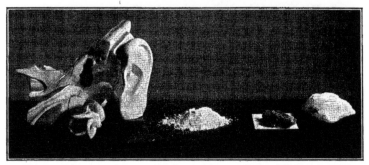

PLASTER OF PARIS AND GYPSUM. Left to right: a plaster of Paris model of the parts of the ear, powdered plaster of Paris, impure gypsum, and pure gypsum (alabaster).

containing "black alkali," which interferes with the growth of plants. Black alkali is sodium carbonate, which reacts with calcium sulfate as follows:

$$CaSO_4 + Na_2CO_3 \longrightarrow CaCO_3 + Na_2SO_4$$

The insolubility of calcium carbonate causes this reaction to go to an end. The resulting sodium sulfate is less injurious to plant life than is sodium carbonate.

When gypsum is heated to 125° C. it loses three-fourths of its water of crystallization, forming plaster of Paris.

$$\underset{\text{gypsum}}{2 CaSO_4 \cdot 2 H_2O} \longrightarrow \underset{\text{plaster of Paris}}{(CaSO_4)_2 \cdot H_2O} + 3 H_2O$$

Plaster of Paris will set into a solid when mixed with water. It recovers its water of crystallization and becomes gypsum again. The physician uses powdered plaster of Paris to make casts for broken bones. Statues and death masks are made with it also. It expands slightly on setting making clean, sharp impressions.

$$(CaSO_4)_2 \cdot H_2O + 3\ H_2O \longrightarrow 2\ CaSO_4 \cdot 2\ H_2O$$

Plaster is this gypsum product mixed with sand and wood fiber or hair. When plaster is put on the outside of a house, it is called stucco. Stucco contains no fiber nor hair binder. The setting of plaster or stucco is essentially the same reaction as the setting of plaster of Paris.

If plaster of Paris is overheated and all of the water of crystallization driven off, it strangely loses the property of setting into a solid.

EXPERIMENT

Plaster of Paris cast. Mix plaster of Paris and water to form a thick paste. Press some object such as a thimble into the paste and allow one-half hour for the paste to harden.

Bleaching powder. Bleaching powder $(CaCl \cdot ClO)$ is a calcium compound of considerable importance. It is a mixed salt, part chloride and part hypochlorite; that is, the calcium is attached by one valence to Cl and by the other to ClO. The hypochlorite part readily loses its oxygen and is useful for bleaching purposes and for disinfecting.

Calcium phosphates. There are three calcium phosphates, the normal phosphate, $Ca_3(PO_4)_2$, and two acid phosphates, $CaHPO_4$ and $Ca(H_2PO_4)_2$. These are the source of the phosphate used in fertilizer. Calcium phosphate, $Ca_3(PO_4)_2$,

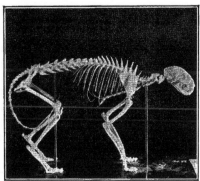

BONES. The stiffness of bones is due to calcium phosphate.

is so insoluble that it only becomes available to plants as they can secrete acid to dissolve it. When the agriculturist has soil deficient in phosphorus, he wants the phosphorus in a more soluble form. The acid phosphate with the two hydrogen atoms per phosphate radical is readily soluble and is sold to the farmer under the name "superphosphate." This soluble phosphate is made by treating the normal phosphate with sulfuric acid.

$$Ca_3(PO_4)_2 + 2\ H_2SO_4 \longrightarrow Ca(H_2PO_4)_2 + 2\ CaSO_4$$

$Ca(H_2PO_4)_2$ is also used in the phosphate baking powders. The other phosphate $(CaHPO_4)$ is used in medicine.

Strontium and barium. As compared with calcium, strontium is of little importance, due largely to its scarceness. However, this element has one commercially important property — that of giving a scarlet color to a flame, when its salts are volatilized. Strontium compounds are used in red fire and fireworks.

Because of their opaqueness to X rays, barium compounds are given to a patient when the physician wishes to take pictures of the alimentary canal. Barium sulfate is used as a filler in cheap paints and in rubber. Other uses of barium are quite limited. The oxide of barium was formerly used in one process of getting oxygen from the air.

EXPERIMENT

Red fire. Mix the following substances in finely powdered condition. (Under no circumstances should potassium chlorate be ground with any other substance. If it must be powdered, keep everything else out.) 1 g. potassium chlorate, 11 g. strontium nitrate, 120 g. of flowers of sulfur, and .5 g. lampblack. Place in the hood on a piece of asbestos paper or dish in a pile. Stick a few inches of magnesium ribbon in the top. Ignite the end of the ribbon and stand back. The powder will burn with a red flame.

EXPERIMENT

Green fire. Mix finely powdered potassium chlorate 3 g., barium nitrate 8 g., flowers of sulfur 3 g. Ignite on asbestos, where it burns with a green flame.

QUESTIONS OF FACT

1. Give the reaction by which plaster of Paris is made from gypsum.

2. What is the chemical change when plaster of Paris sets?

3. What are plaster and stucco?

4. What is the objection to overheating the gypsum in preparing plaster?

5. Give formulas: (a) bleaching powder; (b) normal calcium phosphate; (c) calcium superphosphate.

6. What is alabaster?

7. What compounds are used to contribute redness to red fire?

8. What element of the alkaline-earth metals causes a flame to be green?

QUESTIONS OF UNDERSTANDING

In the first three questions select the most suitable response.

1. Plaster of Paris is made by heating (a) chalk, (b) gypsum, (c) marble, (d) calcite, (e) lime.

2. One use of strontium compounds is (a) to make plaster, (b) to whitewash houses, (c) to make red fire, (d) as a medicine.

3. Barium compounds are used (a) to make X rays effective, (b) in medicine, (c) in bleaching.

4. Why is soluble acid phosphate usually preferred to the normal phosphate in fertilizers?

5. What property of barium compounds causes them to be used with X rays?

REFERENCE FOR SUPPLEMENTARY READING

1. Weeks, Mary Elvira. *The Discovery of the Elements.* Journal of Chemical Education, Easton, Pa., 1934. Pp. 132–133, "The Discovery of Barium and Strontium."
    (1) When were barium compounds shown to differ from the corresponding calcium compounds? (p. 132)
    (2) Who first formed metallic barium? When? (p. 132)
    (3) When were strontium compounds first recognized to be different from the corresponding barium compounds? (p. 133)
    (4) Who isolated the metal strontium? When? (p. 133)

SUMMARY

Beryllium, magnesium, calcium, strontium, barium, and radium constitute a family of elements called the alkaline-earth metals because their properties are intermediate between the alkali metals and the earth metals.

Beryllium occurs in the mineral beryl and the precious stone emerald. Due to the difficulty of getting it out of its compounds, beryllium is too expensive for many of the uses for which it is suited. Among its properties are resistance to corrosion, great elasticity, lightness, and electrical conductivity.

Magnesium finds use in pyrotechnics and as a scavenging metal in steel. Because of its lightness it is used in magnalium, which is superior to aluminum for machining.

Magnesium hydroxide as milk of magnesia, magnesium sulfate as Epsom salts, and magnesium citrate are used as laxatives.

Calcium compounds, because of their abundance and cheapness, are very important in industry. Lime, calcium oxide formed by calcining limestone, is used in making mortar. Coral, chalk, pearls, limestone, marble, calcite, and Iceland spar are all forms of calcium carbonate in various degrees of purity and crystallization.

Hard water is water containing dissolved calcium and magnesium compounds, the bicarbonates causing temporary hardness and all the others permanent hardness. An acid salt is part acid and part salt, while the basic salt is part salt and part base.

Calcium hydroxide mixed with water is called lime water and whitewash. Crystallized calcium sulfate (gypsum) is used to improve the condition of the soil. In its purest form, gypsum is called alabaster and is made into beautiful vases and similar articles. When heated until it loses part of its water of crystallization, gypsum becomes plaster of Paris, which sets into a solid when mixed with water. Stucco consists of plaster of Paris and sand, while plaster is made from the same substances and hair or wood fiber.

Calcium phosphate is the principal hard constituent of bones. Phosphate rock is important as a fertilizer.

Strontium compounds are used to make red fire. Barium salts are used internally for taking X-ray pictures and in a few chemical reactions, but as compared to calcium, the uses of barium are few. Radium will be studied in a separate chapter.

CHAPTER 15

THE MAGNETIC METALS

Introduction. The three metals iron, cobalt, and nickel are strongly attracted by a magnet. No other elements possess this peculiar property in any comparable degree. These three metals have other resemblances, which make it convenient to study them as a family.

Importance of iron. Iron was one of the few elements known to the alchemists. In all the intervening centuries this metal has increased in use and importance. There is hardly a thing we use in our complex civilization today that does not depend di-

MAGNETIC PROPERTIES OF NICKEL AND COBALT. The magnet on the left is picking up cubes of cobalt, while that on the right is attracting nickel powder.

rectly or indirectly upon iron in some form or other. This is often called the Iron Age or Steel Age, and rightly so. Of the ninety-two known elements, iron ranks fourth in abundance in the earth's crust, being exceeded only by oxygen, silicon, and aluminum. It is found nearly everywhere in rock and soil. Iron is one of the elements essential to plant growth, yet practically all soils contain sufficient iron for plant needs, something which cannot be said of some of the other necessary elements. Red soil, especially, is rich in iron compounds. The green chlorophyll of plants and the red corpuscles of the blood of animals contain iron.

Iron is seldom found in the uncombined or metallic condition. Such iron is found in the iron meteorites that drop onto the

earth from outer space.   Most of the commercial iron and steel
is obtained from the oxides or carbonates of iron.   A rich ore
contains over 50 per cent iron, while one with less than 25 per
cent does not pay to work.   Iron ore is mined in fifteen states
of the Union and many other parts of the world.   The most

OPEN PIT IRON MINE.   *Keystone View*

valuable ores in the United States come from Michigan, Wis-
consin, and Minnesota.   The Mesabi Range in Minnesota
yields more iron ore than any other place in the world, about
three-fifths of the country's entire output.   In this district the
ore is near the surface and is hauled by cars out of enormous
open pits.   The principal ore in the United States is Fe_2O_3,
called hematite.

Making cast iron. Cast iron, the most impure commercial
iron product, is produced from the iron ore in a blast furnace —

so called because the oxygen for combustion is supplied by a blast of hot air. Coke, a strong, porous form of carbon, is the usual fuel. It also reduces the iron oxide to metal.

$$Fe_2O_3 + 3\,C \longrightarrow 2\,Fe + 3\,CO$$

All iron ore contains more or less foreign material called "gangue," usually refractory silicates. Limestone is added

Photo by Galloway

IRON BY THE TRAINLOAD. Iron from the Lake Superior region is often taken from open pits by steam shovels and loaded onto cars, which are hauled to ports on the lake. From them it is carried by ore steamer to the great centers of the iron industry on Lakes Michigan and Erie.

with the ore as a "flux." It reacts with the gangue to form an easily fused slag, which separates from the molten iron and floats on top of it. Blast furnace slag is frequently called "cinder." The reactions leading to the formation of slag are as follows:

$$CaCO_3 \longrightarrow CaO + CO_2$$
limestone   lime carbon dioxide

$$CaO + SiO_2 \longrightarrow CaSiO_3$$
flux   gangue   slag

The blast furnace is a tall, round tower of heavy steel plates over 100 feet high and 20 feet in diameter, lined with firebrick. About every four hours, the molten iron and slag are tapped off into separate ladles. From 80 to 100 tons of iron flow out in a fiery, hissing stream into 75-ton ladles to be carried to other furnaces for further refining. If the iron is to be used as the impure cast iron, it is often cast in large chunks called "pigs."

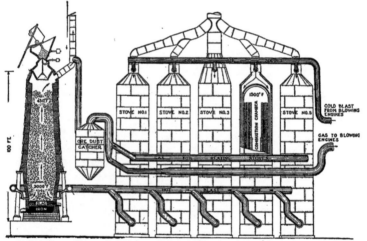

Simplified diagram, with gas-purifiers and pipes omitted, showing internal construction of the blast furnace and hot-blast stoves.

Courtesy Amer. Steel and Wire Co.

IRON BLAST FURNACE. The furnace proper (left) is built of iron plates lined with firebrick. The lower part, which becomes the hottest, is cooled by the circulation of water through bronze jackets. Ore, fuel (coke), and flux (limestone) are introduced at the top. Molten iron and slag (cinder) are tapped off at the bottom. Oxygen for combustion is supplied by a hot blast from the stoves, which are heated by waste gases from the furnace.

Cast iron. Cast iron as it comes from the blast furnace contains five impurities, as the following representative analysis shows:

Carbon . . . . . 3.50 per cent	Manganese . . . 2.00 per cent
Sulfur . . . . . .04 per cent	Phosphorus . . . .30 per cent
Silicon . . . . . 1.25 per cent	

These foreign substances make cast iron too weak and brittle for most uses. Only where weight and bulk rigidity alone are needed can cast iron be used. It is the least useful form of iron.

Wrought iron. The purest form of iron made by any furnace process is wrought iron. The properties of wrought iron are quite different from those of cast iron. It is tough and can be drawn out into rods and wires. It can be bent in any way when hot, and to a considerable degree when cold. Wrought iron can be welded; that is, two pieces can be heated and hammered until the two parts stick together practically as one piece. The blacksmith uses wrought iron for wagon tires, bolts, rods, and horseshoes.

Photo by F. H. King

WORKING WROUGHT IRON. Wrought iron, the purest form in which iron is obtained by any metallurgical process, can be worked up in various forms, such as horseshoes and bolts.

Steel. Steel was formerly considered to be the substance formed when the impurities of iron are burned out and from .5 to 2.0 per cent carbon added, but today the term "steel" includes iron products containing, besides carbon, one or more of the other metals that contribute certain useful properties to iron.

The addition of carbon to iron gives steel some very useful properties, principally elasticity and hardness. High carbon steel, especially, can be tempered; that is, its hardness can be varied over a considerable range by heat treatments. If the steel is heated to one temperature and then chilled suddenly, it has a definite degree of hardness. If it is heated to a higher temperature before chilling, the steel is harder and more brittle. If a blacksmith tempers a plowshare too hard, it may break if it strikes a rock; if it is too soft, it will not stay sharp long. A razor must be tempered hard, but if a pick is made very hard, it breaks too easily.

Tempering. It seems strange that tempering should give steel such a variety of properties. A careful study shows that iron exists in several allotropic forms, two of which are pretty well known. These are called (from the Greek letters) alpha and gamma iron. Alpha iron is soft and magnetic. It does

not dissolve carbon. When heated above 900 degrees, alpha iron changes into gamma iron, which is not magnetic and will dissolve carbon. Gamma iron is very hard. Tempering steel which has been heated to high temperatures fixes a large part of the iron in the gamma form. Some of the carbon in steel is present as iron carbide (Fe_3C). This also tends to increase the hardness. If the steel is not tempered but cools slowly, all of the iron changes into the soft alpha form, and most of the carbon becomes small particles of soft graphite distributed throughout the iron.

Photo by F. H. King

TEMPERING STEEL. Steel is iron alloyed with carbon and often with various amounts of metals other than iron. The hardness of steel can be varied through a wide range by chilling hot steel at different temperatures. The chilling process is called tempering.

The Bessemer process of making steel. About 1847 an American named Kelly observed that a draft of air caused molten iron in his Kentucky iron furnace to seethe and boil. This strange phenomenon of a draft of cold air heating up molten iron was also noticed by an Englishman named Sir Henry Bessemer. He reasoned that the oxygen of the air must be burning up the fuellike impurities in the iron. This observation led to a rapid process of making low-grade steel known as the Bessemer process.

The molten cast iron from the blast furnace is poured into a pear-shaped furnace equipped with a number of compressed-air nozzles in its base. Air is blown through the molten iron for 10 to 15 minutes. The air oxidizes most of the impurities, the sulfur and carbon escaping as gaseous oxides, the manganese and silicon oxides rising to the surface as slag. High-carbon iron is added to bring up the carbon content.

Bessemer steel is good enough for rails and building construction. For most purposes a better quality steel is needed.

Courtesy of U S Steel Corp.

THE BESSEMER CONVERTER. The Bessemer converter is used in the rapid production of cheap steel by blowing hot air through molten iron to burn out the impurities. After the impurities are removed, carbon is added to the iron to form steel.

This is made by the slower, better-controlled open-hearth process.

Open-hearth steel. The open-hearth furnace is a large oven 35 feet long, 13 feet wide, and 2 feet deep, heated by gases burned above it. Greater heat is obtained by heating the gases before they are burned. For about half an hour, gas and air enter through the valves on one side and burn in the furnace. The hot gas produced by the combustion escapes through a brickwork on the other side, heating it. Then the valves are reversed, and the gas and air are admitted through the heated chambers on the opposite side. This periodic reversal produces a continually rising heat until the 75-ton charge is ready. This usually takes about 12 hours — almost as many hours as the Bessemer process takes minutes.

OPEN-HEARTH FURNACE. This furnace is used in the production of high-quality steel. Molten iron from the blast furnace and limestone — together with special minerals for the special steels — are placed in the hearth. Fuel gas and hot air burn above the molten iron and keep it hot. In about twelve hours the impurities have all been driven off or are floating on the steel as slag. The air and gas enter the furnace through separate flues, in which they are heated by a hot brick checkerwork. The waste gases escape through a similar checkerwork, which is thereby heated, and through which the incoming gases will later be routed.

Limestone is dumped into the furnace ahead of the iron. It quickly changes to quicklime. When the molten iron is added, the lime rises up through the iron and removes the phosphorus as calcium phosphate, the sulfur as calcium sulfide, and silica as calcium silicate. Samples are tested from time to time to determine when the lot is pure enough. The character of the product is modified by the addition of carbon and nonferrous metals in the form of iron alloys rich in these elements.

QUESTIONS OF FACT

1. Name the three metals that a magnet will attract.
2. Why is this sometimes called the Iron Age?
3. Name the common source of uncombined iron.
4. Give the name and formula for the principal iron ore in the United States.
5. A study of the blast furnace.
(a) What is the origin of the name of this furnace?
(b) What part does coke play in the process?
(c) Why is a flux added?
(d) What is the chemical nature of the slag?
(e) Define gangue.
(f) What use is made of the CO which issues from the blast furnace?
6. Name the five principal impurities in cast iron.
7. Name the uses for cast iron.
8. What is wrought iron?
9. Explain welding of wrought iron.
10. Give three articles made from wrought iron.
11. What is steel?
12. What is meant by tempering steel?
13. Describe the Bessemer process for making steel.
14. What are the uses of Bessemer steel?
15. Give the differences between the two steel processes.
16. In preparing open-hearth steel how are the following impurities removed?   (a) phosphorus   (b) sulfur   (c) silica.
17. What happens if steel is tempered (a) too hard? (b) too soft?

QUESTIONS OF UNDERSTANDING

1. Questions pertaining to the blast furnace:
(a) What kind of gangue is often found in iron ore?
(b) What becomes of the oxygen of the air blown into the blast furnace?
(c) What becomes of the nitrogen of this air?
(d) Why is the air heated before being blown into the furnace?
(e) Does any of the coke added to the blast furnace burn?
(f) Why do large quantities of CO_2 not come from this furnace?
2. Write the reactions which take place in the blast furnace.
(a) Formation of slag.

(b) Reaction which produces most of the heat.

(c) Reaction which frees the iron.

(d) Action of heat on $CaCO_3$.

3. If SiO_2 is the gangue and CaO the flux, what will be the slag?

4. If $CaCO_3$ is added with iron ore as the flux, what will be its first reaction? What is such a reaction called? To which type reaction does it belong?

5. Compare the properties of alpha and gamma iron.

6. What do we call the different forms of an element?

7. What other element previously studied exists in two allotropic forms?

8. Carbon in steel may exist in what iron compound?

9. How is steel tempered to be soft?

REFERENCES FOR SUPPLEMENTARY READING

1. Clarke, Beverly L. *Marvels of Modern Chemistry.* Harper and Brothers, New York, 1932. Chap. XIX, pp. 238–251, "Iron and Steel."
   (1) What kind of iron comes to the earth from outer space? (p. 238)
   (2) What substance in the animal body contains iron? (p. 239)
   (3) What is a use of slag from a blast furnace? (p. 240)
   (4) What is the economy of the Gayley process of removing moisture from the air blast? (p. 241)
   (5) What is metallography? (p. 243)
   (6) Who invented the open-hearth furnace? (p. 246)
   (7) What is the variety in the color of Fe_2O_3? (p. 250)

2. Glover and Cornell. *The Development of American Industries,* Prentice-Hall Co., Inc., 1933. Chap. XVII, pp. 357–379, "The Iron and Steel Industry."
   (1) When was the wedge of iron in the great Pyramid of Cheops supposed to have been put there? (p. 357)
   (2) Tell about the iron pillar of Delhi, India. (p. 357)
   (3) How widely distributed are iron ores? (pp. 357–358)
   (4) What form of iron was first manufactured? (p. 358)
   (5) Describe the first iron furnaces. (p. 358)
   (6) When did Pittsburgh begin to be a center of iron production? (p. 359).
   (7) What part does the chemist have in keeping the quality of steel uniform? (pp. 363–364)
   (8) What is the capacity of the open-hearth furnace? (p. 365)
   (9) What is the total capacity for steel production by all the companies in the United States? (p. 378)
   (10) Discuss the future of the steel industry. (pp. 378–379)

3. Howe, H. E. *Chemistry in Industry*, Vol. 1.   The Chemical Foundation,
   New York, 1925.   Chap. XI, pp. 147–156, "The Elements of Iron and
   Steel Manufacture" by A. E. White.
   (1) What is the earliest record of iron implements?   (p. 147)
   (2) What was the first substance used to reduce iron from its ores?
       (p. 148)
   (3) What changes have taken place in the size of the furnace since the
       early days of the industry?   (p. 148)
   (4) When was Bessemer steel invented?   (p. 148)
   (5) When was high-speed steel first developed?   (p. 149)
   (6) What is called malleable iron?   (p. 152)

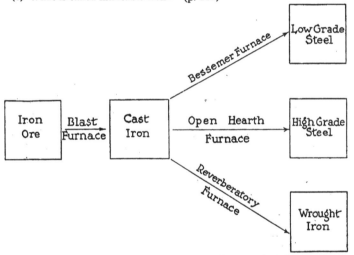

Summary of Iron and Steel Manufacture. Iron ore is fed into the blast
furnace, where it is made into cast iron. The cast iron is transferred: (1) to a
Bessemer furnace, where it is made into cheap steel; (2) to an open-hearth
furnace, where it is made into high-grade steel; or (3) to a reverberatory furnace,
where it is made into wrought iron.

Making wrought iron. In making wrought iron a reverberatory furnace is employed.   This is essentially a small open-
hearth furnace lined with Fe_2O_3.   In this furnace all impurities are removed from the iron, except small traces of slag,
which helps in welding.   The impurities are more combustible
than the iron itself — especially the carbon, sulfur, and phosphorus.   The fact that the sulfur and carbon burn to gaseous
oxides makes them easy to eliminate.   Although the oxides of

the other three impurities (manganese, silicon, and phosphorus) are solids, they are acidic oxides, which react under furnace heat with the basic oxide (Fe_2O_3) lining of the furnace. Having reacted with this oxide, they become part of the lining itself, which must be renewed from time to time. These reactions are similar to the formation of slag in the blast furnace.

$$Fe_2O_3 + 3 SiO_2 \longrightarrow Fe_2(SiO_3)_3$$
$$Fe_2O_3 + P_2O_5 \longrightarrow 2 FePO_4$$
$$Fe_2O_3 + 3 Mn_2O_7 \longrightarrow 2 Fe(MnO_4)_3$$

Alloy steels. Certain properties of steel can be improved by the addition of various amounts of other metals. These steels are often called alloy steels. Alloy steels are very important. Automobiles would cost twice as much as they do if it were not for high-speed machine tools made from molybdenum and tungsten steels. In the early days of the industry when a cutting tool got hot, it became soft and useless. The alloy steels just mentioned may be used as cutting tools at high speed and still retain their hardness. This means that hundreds of automobile parts can now be made in the time formerly required for one. Bulletproof vests, burglarproof safes, and auto axles that will not break are a few of the new steel products. To these may be added stainless steel knives and acid-resistant pipes and tanks. Since any number of metals may be added together in any proportion, it is possible to make an alloy steel of many desirable properties. The following table gives the characteristics contributed by various alloy metals and some important uses of the resulting steels.

ALLOY STEELS

STEEL	PROPERTIES	USES
Chromium	Hardness; resistance to shock and corrosion	Jaws of rock crushers; weatherproof and stainless steel
Manganese	Strength; hardness	Safes; steam shovel lips
Molybdenum	Hardness; heat resistance	High-speed tools
Nickel	Strength; toughness	Armor plate
Silicon	Hardness; resistance to corrosion; brittleness	Acidproof steels; electrodes
Titanium	Hardness	Rails for curves; frogs

Oxides of iron. Iron exists in both bivalent and trivalent forms. When it has the valence 2, its compounds are called ferrous compounds; and when the valence is 3, they are called ferric. The two sets of compounds are quite different in their properties. For instance, ferrous hydroxide is bluish white, while ferric hydroxide is rusty red; ferrous oxide is black, while ferric oxide is the yellow of iron rust or the red of rouge. When the blacksmith heats a piece of iron too hot, black scales of brittle ferrous oxide (FeO) fall from it. Iron exposed to moist air rusts readily. It is estimated that a billion dollars' worth of iron each year is lost by rusting away into the ferric oxide. The natural oxide magnetite (Fe_3O_4) contains both ferric and ferrous iron. When magnetic, this oxide is known as lodestone. It is the principal constituent in black sand. If a magnet is run through dark sand, many black particles of magnetite will be found clinging to it.

Tests for iron. Besides the usual compounds of the metals, iron forms two complex substances containing iron in the negative part. These are the ferrocyanides (*e.g.*, $K_4FeC_6N_6$) and the ferricyanides (*e.g.*, $K_3FeC_6N_6$). The ferrocyanides and ferricyanides serve as tests for both ferrous and ferric iron. The solution of any ferrocyanide forms a deep-blue substance with the ferric ion but not with the ferrous ion. On the other hand, the ferricyanide ion forms the deep blue with the ferrous ion and not with the ferric; hence we have a test for each ion in the presence of the other. Ferric ferrocyanide, $Fe_4(Fe(CN)_6)_3$, Prussian blue, is sometimes used as bluing.

Blueprints. Blueprints for house plans and mechanical drawings of all kinds owe their color to the formation of ferrous ferricyanide.

When a solution containing ferric chloride and a reducing agent, such as oxalic acid, is exposed to the sunlight, the ferric salt is reduced to the ferrous salt.

$$2\ FeCl_3 + H_2C_2O_4 \longrightarrow 2\ CO_2 + 2\ HCl + 2\ FeCl_2$$

When a sheet of paper is coated with such a mixture in a darkened room, dried, and then exposed under a negative to the sunlight, the greatest reduction will take place where the light is brightest. On covering the exposed paper with a solution of potassium ferricyanide, the blue will develop wherever ferrous iron exists, and the depth of color will be proportional to the amount of ferrous salt present. In other words, the potassium ferricyanide acts as a developer. Where the paper has been protected from the light, the materials are unchanged. The picture can be fixed by washing away the ferric chloride and the excess of potassium ferricyanide.

In making commercial blueprint paper a single compound, ammonium ferric citrate, serves both as the ferric salt and the reducing agent. The paper is coated with a mixture of this salt and the developer, potassium ferricyanide. Such a paper, after exposure, is both developed and fixed by simply washing in water.

QUESTIONS OF FACT

1. Exercises on the removal of impurities of cast iron in the reverberatory furnace.

(a) Write the reaction when sulfur burns.

(b) Account for the fact that the burning eliminates the sulfur.

(c) Will carbon be eliminated by the same process?

(d) Write reactions for the formations of the oxides of Si; of Mn_2O_7; of P_2O_5.

(e) How are the solid oxides removed from the iron?

(f) Write reactions of SiO_2, P_2O_5, and Mn_2O_7 with the oxides of iron in the lining of the furnace.

2. What is the basic lining of the reverberatory furnace?

3. Name the properties contributed to steel by the addition of each of the following metals: (a) chromium; (b) molybdenum; (c) nickel; (d) silicon; (e) titanium.

4. What two sets of iron compounds are there?

5. Compare the colors of ferrous hydroxide and ferric hydroxide; of ferrous oxide and ferric oxide.

6. Which oxide of iron is magnetic?

7. What iron compound is used for bluing?

QUESTIONS OF UNDERSTANDING

1. What black substance is taken out of sand by a magnet?

2. What is the difference between the formulas of potassium ferrocyanide and potassium ferricyanide?

3. Describe the tests for ferric ions and ferrous ions.

4. Name the substances on blueprint paper, and tell what each chemical does.

5. What takes place when the blueprint is exposed?

ADDITIONAL EXERCISES FOR SUPERIOR STUDENTS

1. Tell which of the following are insoluble and blue and which are tests for something and for what.

$$(a)\ \overset{++}{Fe_3}(\overset{\equiv}{FeC_6N_6})_2 \qquad (c)\ \overset{++}{Fe_2}\overset{\equiv}{FeC_6N_6}$$

$$(b)\ \overset{+++}{Fe}\overset{\equiv}{FeC_6N_6} \qquad (d)\ \overset{+++}{Fe_4}(\overset{\equiv}{FeC_6N_6})_3$$

2. Which of the following are true and which false?

(a) Cast iron can be tempered.

(b) Steel is the purest form of iron.

(c) Prussian blue is $Fe_4(FeC_6N_6)_3$.

(d) The ferricyanide ion is a test for the ferrous ion.

(e) An anemic person has too much iron in the blood.

Inks. Many inks are simply solutions of the aniline dyes. They are satisfactory if the ink is intended to last only a few years. After the lapse of centuries, the aniline dyes will fade out and can never be restored. The old-fashioned iron black ink is better if the ink is expected to last a century or more. Ordinary black ink is a rather complicated mixture. It contains tannic acid, gallic acid, ferrous salt, dilute mineral acid, phenol, and indigo. The indigo is added to give a blue color to serve until the slowly forming black of the ferric gallate and ferric tannate can form. It will be noticed that such ink turns from blue to black as it remains exposed to the air on the paper. The mineral acid prevents the ferrous salt from oxidizing in the bottle and precipitating the tannate and gallate. The phenol is a preservative, which keeps the tannate from fermenting. After the ink is spread on the paper, the alkali in the sizing of the paper neutralizes the mineral acid. The oxygen of the air now oxidizes the ferrous iron to ferric iron, which forms insoluble black ferric tannate and ferric gallate. Even if the organic part of black ink has faded with age, the iron can be made to take on color again.

Preparing ink. Ink·may be prepared for school use. Into 800 cc. of water dissolve 23.4 g. tannic acid; 7.7 g. gallic acid; 30 g. ferrous sulfate; 25 cc. dilute hydrochloric acid; 1 g. phenol; 2.2 g. of any water-soluble blue dye. Dilute the solution to 1000 cc.

Cobalt. Cobalt, the second member of the magnetic family of metals, has little use. Small amounts of the metal are added to glass to color it blue. Some of it is used in making magnets for telephone coils. The chloride has the peculiar property of changing color when exposed to air of varying humidity. These changes are explained by the fact that the substance takes on varying amounts of water of crystallization and passes readily from one form to another. The less hydrated forms are blue or lavender, while the more hydrated forms are red. Heated or exposed to dry air, the red salt loses water of hydration and is changed to a blue, less hydrated form. Advantage is taken of this fact to make sympathetic ink, which is invisible until heated, and simple apparatus for indicating the amount of moisture in the air. Cobalt nitrate is used as a test for aluminum and zinc.

Change of color of cobalt chloride with moisture. Cut two paper dolls out of filter paper and moisten with a saturated solution of cobalt chloride and allow to dry in the air. Suspend these inside of two bell jars. Turn one bell jar over a plate with a dish of water on it. Turn the other bell jar with its paper doll over another plate with a dish of anhydrous calcium chloride. Notice the difference in the color of the dolls. Reverse the jars and see that the color reverses. Keep the dolls to determine the condition of the air from day to day.

Borax bead tests. The easiest way of testing for the presence of cobalt in its compounds is by means of the borax bead.

If a piece of platinum wire with a small loop in the end is heated and dipped into borax and then strongly heated in the flame of a Bunsen burner, the borax melts and forms a bead of clear borax glass in the loop of wire. If now this bead is touched to some cobalt compound or solution containing cobalt and reheated, the glass is colored a deep blue. This is a quick and delicate test for this element.

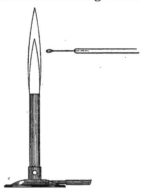

The borax bead test can also be applied to the elements iron, chromium, nickel, and manganese. The test for these is not so easy, and the results are not so positive as in the case of cobalt. One difficulty with the test for these elements is that one color may be obtained in the oxidizing flame and a different color in the reducing flame. For

THE BORAX BEAD TEST.

instance, iron will show greenish in the reducing flame, and rusty color in the oxidizing part. Manganese produces purple in the oxidizing part, and brown in the reducing flame.

Structure of flames. This mention of the oxidizing and reducing flames requires a further study of the structure of the Bunsen flame. A glance at the lighted burner will show a bluish inner cone and a partially luminous covering over it. The inner cone consists of the unburned mixture of gas with air, which has entered through the holes at the bottom of the burner. The reducing flame is the region just above the inner cone. Here there is an excess of hydrogen and carbon from the decomposing hydrocarbon of the gas. The nascent hydrogen and carbon tend to reduce any element that is capable of changing to a lower valence. Thus ferric iron changes to the ferrous.

The oxidizing flame is the region near the outer part of the outer cone. Here the oxygen of the air is coming in from the outside and is in excess of hydrogen and carbon, which have mostly already reacted. In the oxidizing part of the flame all metals capable of increasing in valence are changed to the higher valent condition.

Nickel. Early in the eighteenth century a copper-silver mine in Saxony yielded an ore which would not respond to the ordinary methods of metallurgy. In disgust the miners called it "kupfernickel," the first word being the German for copper and the second from Old Nick, who they thought had bewitched the ore. In 1751 Cronstedt isolated a new metal from the ore, which he named nickel.

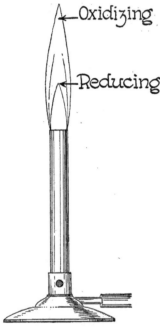

The uses for nickel have increased from then up to the present. Nickel plating was once an important means of protecting iron objects, but chromium plating has almost superseded nickel for this use. The nickel coin is 75 per cent copper, so it is not attracted by a magnet as is the pure metal. The World War created an enormous demand for nickel to make tough crankshafts for the Liberty Motors. Since the War, the use of nickel in resistant steel has increased. The chromium-nickel steels are strong, tough, and resistant to oxidation in acids. Stainless steel of about 8 per cent nickel is used in automobile parts, especially axles, pinions, gears, crankshafts, clutch plates, and steering knuckles. Nickel-chromium wire has very high electrical resistance, high melting point, and high resistance to oxidation. All electric heating devices, such as electric stoves, heaters, toasters, and coffee pots, use this wire. Permalloy of 30–80 per cent nickel has such exceptional magnetic properties

STRUCTURE OF THE BUNSEN FLAME. The Bunsen flame consists of two distinct parts, an inner cool, blue cone of unburnt gas and a glowing outer cone.

that transformers using it as a core have smaller coils than others. Monel metal is an alloy of 2 parts nickel to 1 part copper. Its silvery brightness and resistance to rust and corrosion make it valuable for kitchen sinks and washing machines.

QUESTIONS OF FACT

1. What iron compound is added to ink?
2. Give two uses for cobalt.
3. What is one use for cobalt nitrate?
4. What is a borax bead?
5. What color does cobalt give to glass?
6. Describe the color given by iron to the borax bead: (a) in the oxidizing flame; (b) in the reducing flame.
7. Give the composition of a nickel coin.
8. What are the properties of nickel steel?
9. What are some of the uses of steel containing 8 per cent nickel?
10. What is the composition of permalloy?
11. What is monel metal?
12. What are some uses of monel metal?

QUESTIONS OF UNDERSTANDING

1. What would go wrong if ferric sulfate were added to ink instead of the ferrous sulfate?
2. When the ink is spread on the paper, what happens to the ferrous iron?
3. Why does not the acid in ink prevent oxidation when the ink is used?
4. Name the substances put into iron ink, and give the reason for each constituent.
5. Why is a nickel coin not attracted by a magnet?
6. What is the relation of nickel to the automobile industry?
7. Give three other useful nickel alloys.

REFERENCES FOR SUPPLEMENTARY READING

1. Howe, H. E. *Chemistry in Industry*, Vol. II. The Chemical Foundation, New York, 1926. Chap. XI, pp. 190–203, "The Chemistry of Inks."
   (1) What is the date of the first writing in ink? (p. 190)
   (2) What was the composition of this first ink? (p. 190)
   (3) When did the iron inks come into use? (p. 191)

(4) What else besides iron was in these inks? (p. 191)

(5) What is used in modern colored inks? (pp. 192–193)

(6) Describe the composition of the inks which cannot be erased. (p. 193)

(7) Discuss printing inks. (p. 195)

(8) In lithographic printing from flat surfaces, of what is the ink composed? (p. 195)

(9) Name the white pigments in the order of their transparency. (p. 197)

(10) What is an ink drier? Name several. (p. 198)

2. Weeks, Mary Elvira. *The Discovery of the Elements*. Journal of Chemical Education, Easton, Pa., 1934. Pp. 21–24, "The Discovery of Cobalt and Nickel."

(1) What were the first uses of cobalt compounds? (p. 21)

(2) What did Kobald mean in German mythology? (p. 22)

(3) When did the Swedish chemist Brandt discover cobalt? (p. 22)

(4) What was the origin of the name cobalt? (p. 22)

(5) What was the early use of nickel ores? (p. 22)

(6) Relate Cronstedt's experiment which surprised him. (p. 23)

(7) When did Cronstedt separate metallic nickel? (p. 23)

(8) What was the nationality of Cronstedt? (p. 22)

SUMMARY

The magnetic metals iron, cobalt, and nickel have the unusual property of being attracted by a magnet. Iron, the fourth element in abundance in the earth's crust, is used in larger quantities than any other metal. The ores of iron to be worth working must run above 25 per cent of the metal. The workable ores are usually the oxides.

The metallurgy of iron is carried out first in the blast furnace, from which comes the molten cast iron. Owing to the limited uses of cast iron, it is rarely allowed to solidify but is transferred to a second furnace for further purification into wrought iron or steel. In the blast furnace, the coke removes the oxygen from the iron, while the flux and gangue unite to form slag.

In the reverberatory furnace the carbon, manganese, sulfur, silicon, and phosphorus are removed from the impure cast iron to form wrought iron.

In the Bessemer steel process most of the impurities are burned out by blasts of hot air in a few minutes of heating.

To make high-quality steel for automobile parts, these impurities must be more completely removed in the open-hearth furnace by reaction with the oxygen of the air and the CaO lining of the furnace. Since steel contains carbon, this element must be returned by the addition of high-carbon iron. Other metals are added to make steels for special purposes — molybdenum and tungsten for high-speed tools, chromium for stainless steel, nickel for armor plate, and silicon for acid-resistant containers.

Iron forms three oxides — the black ferrous, the yellow and red ferric, and the magnetic oxide of both ferrous and ferric. In solution, the ferrous ion can be detected by the addition of the ferricyanide ion, and the ferric ion by the addition of the ferrocyanide ion. Mixing the other way does not produce a deep blue.

Blueprints result from the formation of ferrous ferricyanide. Ferric iron, potassium ferricyanide, and a reducing agent are on the paper originally. Exposure to light causes the reducing agent to change the ferric iron to ferrous, which now forms the color with the ferricyanide.

Iron ink contains ferrous ions, gallic acid, indigo, sulfuric acid, and phenol. The phenol prevents fermenting organisms destroying the gallic acid, which makes the ultimate black ferric gallate when the iron is oxidized to the ferric condition on the paper. The acid prevents oxidation and consequent precipitation in the ink bottle.

Cobalt metal is used to some extent in magnets; its salts find use in novelties where their property of changing color with varying humidity is an advantage.

Beads of borax glass in the flame of the burner can be used to test for cobalt and a few other metals. Cobalt always colors the glass blue. Iron, however, colors it green when heated in the reducing flame but reddish yellow in the oxidizing part of the flame.

Nickel is a constituent of permalloy and monel metal, of corrosion-resistant steels, and of resistance wire for electric heaters.

CHAPTER 16

THREE COMMON PRECIOUS METALS

Introduction. All of the metals may be grouped into families for study, but the differences are often more marked than the similarities. In this and the following chapters the metals will be studied as individuals, rather than as members of groups. Gold, platinum, and silver will be considered in the present chapter. They are all familiar precious metals.

Gold. Gold is the most beautiful of the metals and is the one that has held the attention of the world for centuries. It is the standard of value over most of the globe. No adventure is more alluring than the quest for gold, and no substance is more desired and sought for than this metal. Gold usually occurs in the free, or uncombined, state. This is due to its chemical inactivity as is shown by its position near the bottom of the replacement series. All other metals replace it from its compounds, and many nonmetals also reduce it to the metallic state.

Occurrence of gold. Gold is found embedded in quartz and a few other minerals, and as particles of various size in the sand or gravel beds of streams. Recovering the metal from sand or gravel is called placer mining. Placer mining may be done with a rocker or a sluice. In either case the lighter rock is washed out and the heavy gold caught in the riffles in the bottom. Gold, with a specific gravity of 19.3, easily sinks to the bottom in a stream that will wash away ordinary rock whose average density is about 2.75.

Sometimes strong jets of water are used to tear up the gravel and wash it through the sluice. This method is called hydraulic mining. At other times an electric dredge is used to scoop up the gravel and sift out the gold.

Quartz mining. The gold occurs originally in deposits of quartz or related rock, sometimes as microscopic specks and sometimes as quite large pieces. The quartz is pounded into a powder in a stamp mill. This gold-bearing powder is washed over brass plates covered with mercury. The mercury collects and holds the gold but allows the rock dust to pass by. Sometimes gold occurs in pockets in the earth but not in a ledge. The gold in both the ledges and pockets originally came up from the interior of the earth through cracks caused by mountain building. In the case of the pocket, there was just a small hole through which the melted quartz with its gold in solution came. The iron in the soil probably precipitated the gold

HYDRAULIC MINING. A stream of water is used to cut through hills and carry the gold-bearing gravel through the sluice boxes. The stream of water is about 150 yards long. It is shot from a nozzle 12 inches in diameter and 20 feet long. The nozzle is directed by a man who sits astride it and controls it with an electric mechanism.

and deposited it in the pocket. Streams in time cut through ledges and pockets and scattered the gold along their channels, where the placer miner finds it.

Occasionally the gold must be recovered from the ore by a solvent process. Potassium cyanide, one solvent for gold, is the basis of one process, and chlorine of the other. In the cyanide process, a complex cyanide of gold is formed. The gold is released from the cyanide solution by the addition of zinc.

Recovery of gold. The cyanide process for recovering gold is used when the gold is in a very finely divided condition. Near Jackson, California, a mining company dumped the powdered rock from its stamp mills on some vacant land.

There accumulated a pile of powdered quartz covering several acres. With the recent rise in gold prices an analysis was made on this material, which resembled a huge pile of ashes. The analysis showed considerable values in gold, which in terms of the estimated weight of the pile would amount to several millions of dollars. Research showed that the cyanide process was suitable for the recovery of this valuable metal. About 20 large tanks, 30 feet high and 20 feet in diameter, were

AN ELECTRIC DREDGE. The electric dredge floats on the water of a pond which it digs itself — and then fills up. It eats its way through the gravel and soil, removes the gold in the mill within it, and dumps the waste rock at its rear. It is not unusual for a dredge to recover a thousand dollars' worth of gold a day.

erected for cyaniding. At present $500 worth of gold per day is being recovered.

The chlorination process is similar to the cyanide process in principle. The free chlorine reacts with the finely divided gold to form gold chloride.

$$2\,Au + 3\,Cl_2 \longrightarrow 2\,AuCl_3$$

As in the cyanide process, the gold is recovered by adding a metal high in the replacement series. This metal takes the place of the gold in the solution, and throws it down as an insoluble sludge.

Recovering gold from jewelry. If an old high-school pin or piece of broken locket or other broken jewelry can be obtained, treat with concentrated nitric acid to remove the other metal and leave the gold.

Uses of gold. Of the many uses of gold, coinage is by far the most important, although, strange as it may sound, practically every nation has withdrawn all its gold coins from circulation. Nevertheless, the gold is deposited in well-guarded underground vaults, where it remains a basis for issuing paper money. Gold coins are usually made with 10 per cent copper to reduce loss by wear, since pure gold is quite soft.

Because of its softness and malleability, gold is used commercially for gold lettering. The gold leaf can be made so extremely thin that the amount necessary to do considerable lettering does not cost a great deal. The lettering, moreover, is both beautiful and durable. For jewelry, gold is alloyed with other metals. The composition of these gold alloys is expressed in carats. By carats is meant twenty-fourths. Pure gold is 24 carats fine or $\frac{24}{24}$ pure, while 18 carats fine is 18 parts gold and 6 parts of something else. The white gold and green gold used in jewelry are gold alloyed with some other metals in such proportions that the appearance of gold is largely obscured.

Gold has certain properties that make it highly desirable as a filling for teeth; it is malleable cold, is highly resistant to chemical action, and is stronger than other materials available for this purpose. Unfortunately, gold used for filling teeth is usually permanently lost to the world's supply when it is buried with the individual. Gold used in jewelry eventually finds its way back into the trade for use again.

If gold jewelry were limited to substances which contain a considerable percentage of gold, we could not afford to enjoy this beautiful metal as much as we do. The electroplating of gold on a cheaper metal makes available inexpensive ornaments, which are identical in appearance with those made of pure gold. Rolled gold, which is often used for jewelry and watch cases, is made by mechanically covering a base metal with a thin layer of gold. It is more durable than gold plate.

Platinum. Platinum is a metal resembling silver or nickel in appearance, but more resistant to all chemical action. Gold and silver get part of their value because of coinage. Platinum, however, stands purely on its merits, and in normal times commands a price somewhat higher than that of gold. It comes principally from Russia, Canada, Columbia, and South

PLATINUM DISHES. Platinum is used where a metal is needed that may be heated without corroding and that will withstand strong acids.

Africa. In northern California platinum occurs with the gold. In the early days of mining in that region, the purchaser of the gold dust deducted from the price for the weight of the platinum. Since then the uses of platinum have been learned, and it is now worth more than the gold.

The properties of platinum which make it so useful are resistance to chemical action and catalytic activity. Because of its strength and resistance to tarnish, it is used to mount diamonds in jewelry. The chemist uses platinum crucibles in his accurate analyses to prevent contamination of his solutions. As a catalyst, platinum is exceptional. For use as a catalyst, it must be precipitated as finely divided platinum black, usually on a piece of asbestos. Platinum is a standard catalyst for the contact process of making sulfuric acid. The catalytic action of platinum can be illustrated by turning on the gas in a Bunsen burner and holding a small piece of platinized asbestos in the stream of gas. It catalyzes the oxidation of the gas until the accumulated heat ignites it.

Other members of the platinum family. Whether found in the alluvial sands of the Ural Mountains or recovered from the nickel ores of Ontario, crude platinum is always alloyed with 15 to 40 per cent of five other similar metals — ruthenium, rhodium, palladium, osmium, and iridium. Being very similar to platinum in properties, these metals are classified in the same family. The individuals of the platinum family have some very unusual properties which would make them very useful if they could be obtained in larger quantities.

Among these unusual properties are the resistance of rhodium and iridium to attack by aqua regia and the resistance of iridium to attack by chlorine. Iridium alloyed with platinum increases the latter's resistance to corrosion, which gives rise to platinum-iridium pen points. Palladium will absorb a greater volume of gas than any other metal, 1 volume of palladium absorbing 900 volumes of gas. Its ability to concentrate gaseous molecules into an active region makes this metal an excellent catalyst of gaseous reactions. Palladium may be substituted for platinum for some purposes, and is used for soldering platinum. Osmium is the heaviest substance. It also enters into compounds with the highest known valence, 8. Osmium tetroxide (OsO_4) is a white, crystalline poison with a burning taste and an irritating odor. It is used in solution for staining and hardening tissues for microscopic study.

Silver. Silver is more abundant and much less valuable than gold. Although used for coins, this use has been limited in the United States since 1873. One presidential campaign was run on the platform of free and unlimited coinage of silver at a ratio of 16 to 1 with gold; that is, one ounce of gold was to be considered as worth 16 ounces of silver. Ordinarily the ratio between the prices of gold and silver varies. In recent years the United States has purchased considerable quantities of silver to add to her metallic reserves as a basis for issuing more paper money.

Occurrence and metallurgy of silver. Although silver sometimes occurs in the free state, it is usually found in its compounds, especially as the sulfide. Silver sulfide is usually associated in small amounts with lead sulfide, or galena. The special blast furnace or mechanical hearth used in the metallurgy of this silver-bearing lead produces a mixture of the two metals. A very clever process called the Parkes process is used

SILVERWARE. Silver turns black on long standing, the sulfur compounds of the air reacting with the silver to form silver sulfide.

to separate the lead and silver. This process depends upon the facts that molten zinc does not mix with molten lead and that silver is much more soluble in molten zinc than in molten lead. Small quantities of zinc are melted with the silver-lead mixture. Soon the zinc containing most of the silver will be floating on top of the heavier lead. The zinc crust can be lifted off and distilled away from the silver, ready to use again.

Properties and uses of silver and silver compounds. The beauty of silver is the basis of many of its uses in jewelry, dishes, and ornaments. This metal, being a little too soft to wear well, like gold is also alloyed with some harder metal, usually copper. Sterling silver articles are 7½ per cent copper; the American silver dollar, 10 per cent. Much silverware is only silver plated. By using the silver as a thin layer plated over some cheaper metal, articles can be made with all the beauty of silver at a much cheaper cost. Silver tarnishes quite readily when sulfur compounds are in the atmosphere. Although the tarnish is black, it serves as a protective coating, and the metal does not all become corroded as in the case with rusting iron. Silver nitrate, sometimes called lunar caustic, is corrosive. It is cast into sticks and used by the physician to burn out sores or to remove abnormal growths.

QUESTIONS OF FACT

1. What property of gold is the basis of its separation from sand and gravel in placer mining?
2. Describe quartz mining and the use of a stamp mill.
3. Give the composition of (a) gold coins; (b) white gold.
4. What is meant by 14 carats fine?
5. Give the properties and uses of platinum.
6. Describe the Parkes process.
7. What is sterling silver?
8. In what condition must platinum be when it is used as a catalyst?
9. What unusual property have rhodium and iridium?
10. Give some interesting facts about the other metals of the platinum family.

QUESTIONS OF UNDERSTANDING

1. Why does gold usually occur free or uncombined?
2. What is the action of mercury in the quartz mill?
3. How can gold be recovered from its solutions?
4. What is supposed to be the origin of gold pockets?
5. How can people afford to use a metal as valuable as gold for lettering on windows?
6. Why is iridium alloyed with platinum in pen points?
7. What is the valence of osmium as shown by the following formula, OsO_4?
8. What compound is called lunar caustic?
9. Devise a cheap method of recovering silver from waste photographic solutions.

ADDITIONAL EXERCISES FOR SUPERIOR STUDENTS

1. Calculate the percentage of gold in each of the following: (a) an 18-carat pin; (b) a 12-carat pin.

2. One Troy ounce weighs approximately 31.1 grams. If gold is quoted at $35 per ounce, what would it be per gram?

3. If a piece of jewelry weighs 20 grams and is marked 18 carat, calculate the value of the gold it contains if worth $35 per ounce. (Use the answer to problem 2.)

4. If one gram of gold chloride ($AuCl_3$) costs $2.50, what is the price of gold per gram?

5. If silver sulfide ore is treated in a furnace, write the reactions for: (a) roasting; (b) reduction with coke.

REFERENCES FOR SUPPLEMENTARY READING

1. Clarke, Beverly L. *Marvels of Modern Chemistry.* Harper & Brothers, New York, 1932. Chap. XXII, pp. 271–277, "Some of the Rarer Metals."
   (1) Which is the best conductor of heat and electricity among the metals? (p. 271)
   (2) What ores of silver usually occur? (p. 271)
   (3) With what other metal does silver usually occur? (p. 271)
   (4) Why does silverware often develop dark stains? (p. 271)
   (5) Compare the United States silver coins with sterling silver in the percentage of silver. (p. 272)
   (6) Discuss the silvering of mirrors. (p. 273)
   (7) Name the six platinum metals. (p. 277)
2. Sadtler, Samuel Schmucker. *Chemistry of Familiar Things.* J. B. Lippincott Company, Philadelphia, 1924. Chap. X, pp. 133–140, "Gold and Silver."
   The subject matter of this reference is very similar to that of the text.
3. Weeks, Mary Elvira. *The Discovery of the Elements.* Mack Printing Co., Easton, Pa., 1934. Chap. VIII, pp. 99–115, "Discovery of the Platinum Metals."
   (1) Who first discovered platinum in the Old World? (p. 99)
   (2) Relate the story of how a substance now more valuable than gold was once used to adulterate gold. (p. 100)
   (3) What metal did the Spaniards of South America call "little silver of the river Pinto"? (p. 100)
   (4) What did Dr. Wollaston learn about platinum that made the metal more useful? (p. 101)
   (5) When did Dr. Wollaston discover rhodium and palladium? (p. 104)

(6) How did he get the platinum away from the palladium? (p. 104)
(7) What property of rhodium compounds gave it its name? (p. 105)
(8) Write a biography of Tennant. (pp. 106–110)
(9) When was ruthenium discovered? (p. 110)

Silver in photography. Certain silver compounds are sensitive to light. Strong light will reduce these salts to the black, finely divided silver. Even short exposures to light, although making no visible change, so affect the silver salt that it is easily reduced by a mild reducing agent, which would not reduce a salt untouched by light. This sensitiveness to light has given rise to the photographic industry in all its various fields : moving pictures, snapshots of all kinds, and the new photography in natural colors.

Silver bromide. Silver bromide prepared in darkness and not exposed to the light is not affected by mild organic reducing agents. However, if it is

Courtesy of Eastman Kodak Co.

DEVELOPING PHOTOGRAPHIC FILM. Photographic film has to be developed in a dark room, illuminated only by a red light, to which the chemicals on the film are not sensitive.

exposed to white light for even a fractional part of a second, it will be reduced to black metallic silver by these same reducing agents (developers). Since these reducing agents have complex formulas which require a knowledge of organic chemistry to understand, in the following reaction we will think of them as simply supplying hydrogen, which does the reducing.

$$AgBr + H \longrightarrow Ag + HBr$$

A photographic dry plate consists of an emulsion of gelatin and silver bromide on a glass plate. A film is very similar except that the emulsion is on a celluloid film instead of on a glass

Louis Jacques Mande Daguerre

1789–1851

The first successful commercial photographic process was invented by Daguerre in France and published in 1839. During the next twelve years this process was highly popular.

A daguerreotype photograph is a positive image formed by mercury vapor upon a silver-coated copper plate. The process includes cleaning and polishing the silver plate, sensitizing the surface, exposing in the camera, developing, fixing, and finishing.

The silvered surface had to be polished perfectly, first with pumice powder and olive oil and finally with velvet. Next the smooth silver surface was sensitized by exposing it to the fumes of iodine in a covered box over a small dish containing iodine crystals. Other investigators also used bromine and chlorine vapors, which shortened the necessary exposure by about four-fifths.

The first exposures were from five to thirty minutes. In the modified forms the exposures dropped to from five to thirty seconds. To develop the image, the exposed plate was put into another box over some mercury which could be heated through the bottom of the box by an alcohol lamp. In about 20 minutes the image was developed. To fix the image so light could make no further changes it was treated with a solution of salt or hypo, which removed any unchanged silver salt.

Daguerre was born at Corneilles, not far from Paris, in 1789. His first job was as a revenue officer, but later he changed to scene painter for the opera. At this work he developed a remarkable power of representing light and shade. To help get the first sketch Daguerre used the camera obscura, the predecessor of our modern camera. The hopelessness of trying to reproduce the beauty of the natural image caused Daguerre to think of the possibility of some method of automatic production of the image.

plate. The sensitive film prepared in the darkness is kept out of the light until it is exposed through the lens of the camera for a short time when taking the picture. White objects in the subject photographed reflect relatively much light, and dark objects correspondingly less. The light from white objects affects the sensitive silver bromide the most, the part of the plate opposite dark objects is relatively unaffected, and objects of intermediate shades affect the plate proportionally. No change in the silver bromide is outwardly observable; but when it is developed in some good reducing agent, the image soon appears in black and white. All the white silver-bromide grains touched by light are reduced in the developer to black, finely divided silver. Although a silver spoon or coin is shiny white, finely powdered silver is black. After a short rinse in cool water the film is put into the hypo fixing bath, which is a solution of a sodium-oxygen-sulfur compound that dissolves all the unchanged silver bromide and leaves only the metallic silver. The film is now called a *negative*, since all the white parts of the subject appear black, and all the dark parts look clear.

Until some years ago photographic emulsions were sensitive only to blue, violet, and ultraviolet light. Anything green, yellow, or red looked black in the finished print. A few decades ago it was learned that certain organic dyes added to the emulsion extended the sensitiveness to other colors of light. Not many films of the earlier type are used today. When the sensitiveness includes green light in addition to these mentioned, as in the case of most of the kodak film sold today, the film is called orthochromatic. However, when the film is sensitive to all the colors, it is said to be panchromatic film. The dyes that make the film panchromatic also increase its speed. Further treatment of panchromatic film with ammonia produces the supersensitive film, which is fast enough to photograph basketball games at night.

Developing papers. Photographic printing consists in letting the light pass through the negative onto a paper covered with another silver emulsion a little less sensitive than the one originally on the film.   This exposed paper is developed, fixed, and washed the same as the negative.   The print, however, differs from the negative in that it is a positive.   By this is meant that the white of the original subject shows white on the print, and the black shows black.

Photographic papers. are made in different degrees of smoothness and in special textures to represent linen and

NEGATIVE AND POSITIVE.   The action of light on silver salts causes a developed film to show light objects dark and dark objects light, hence the name negative. By passing light through such a film to a sensitized paper, the process is again reversed, and a positive print results.

tapestry.   Buff and cream-colored paper stock are often used to produce the so-called warm effects.   The papers also differ in speed of printing.   For contact printing, where the light is not being spread out in enlarging the image, the silver salt used is silver chloride, which is much slower or less responsive to light than silver bromide.   Even in these chloride papers there is some variety in the printing speed due to the fact that the crystals of silver salt can be made in different sizes.   In general, large crystals are more sensitive to light than small ones. This difference in the speed of contact papers is of no advantage in itself, but it is accompanied by a difference in contrast, which is desirable. If the negative that is being printed is rather flat, a hard, or contrastive, paper will produce a more brilliant print than a soft paper.

When the image is to be enlarged, the light coming through the negative is spread out over more area and hence weakened in intensity, which increases the printing time. If silver bromide is used on the paper emulsion, due to its much greater speed than the silver chloride, the printing time may not need to be increased. In general, the bromide paper is about fifty times as fast as the chloride paper. All the advantage is not with the bromide paper, however. Chloride paper has a longer scale; that is, it can print jet black for a dark part of the print and still leave the white objects white. In order to get brilliant blacks with the bromide paper it is sometimes necessary to overprint the whites and make them muddy.

Because of their longer scale, the chloride papers are sometimes used for enlarging in spite of the long printing time. Among the best enlarging papers are the chloro-bromide papers — those containing both silver chloride and silver bromide in the emulsion and having speeds intermediate between the chloride and the bromide papers.

Printing-out papers. Printing-out papers are covered with the less sensitive silver chloride in its least sensitive form. They are printed in direct sunlight instead of by artificial light as are the developing papers. The sunlight partially breaks down the silver chloride to produce the image without the necessity of development.

These printing-out papers are useful to the portrait photographer for making proofs. These can be made without the trouble of development, fixing, washing, and drying. Although they will not last long unless given further treatment, and although they have a dingy reddish color, they enable the purchaser to choose which of several exposures he wishes made. If it should be desired to improve the tone of these prints and make them permanent, it can be done — usually by immersion in a gold solution. It will be recalled that gold is below silver in the replacement series, which means silver will replace gold from its solutions. When the gold in solution touches the silver image, the silver and gold change places; that is, the gold stays on the image, and the silver goes into the solution. The picture then becomes a picture in gold, which has different pleasing tints according to the thickness of the gold layer. It is fixed to remove the unchanged silver chloride.

Any metal below silver in the replacement series can be used for toning. Platinum toning gives a very permanent and beautiful black print.

$$3 \text{ Ag} + \text{AuCl}_3 \longrightarrow \text{Au} + 3 \text{ AgCl}$$

Sepia toning.   Prints may be toned in many colors by changing the silver metal into some colored silver salt or colored salt of some other metal. Brown or sepia tones are very pleasing. In sepia toning the silver print is dipped into a potassium ferricyanide solution, which changes the black silver image to colorless silver ferricyanide. The photographer calls this the bleaching solution. The bleached print is now dipped into a dilute sodium sulfide solution to change the image into brown silver sulfide.

$$3 \text{ Ag} + \text{K}_3\text{FeC}_6\text{N}_6 \longrightarrow \text{Ag}_3\text{FeC}_6\text{N}_6 + 3 \text{ K (bleaching)}$$
$$2 \text{ Ag}_3\text{FeC}_6\text{N}_6 + 3 \text{ Na}_2\text{S} \longrightarrow 3 \text{ Ag}_2\text{S} + 2 \text{ Na}_3\text{FeC}_6\text{N}_6 \text{ (toning)}$$

QUESTIONS OF FACT

1. What is the difference in the behavior of grains of silver salt that have been touched by light from those which have not?

2. What is the process called which causes the photographic image to become visible?

3. What do we call the chemical which makes it become visible?

4. What is the substance out of which the image in the negative is made?

5. In what state is silver black?

6. What would happen to a negative that was developed and not fixed?

7. What substance is removed by fixing?

8. What chemical does the fixing?

9. What property of silver salts makes them so important in photography? •

10. Describe the making of a negative, telling what is done by the developer and by the fixing bath.

11. Write the reaction for toning a silver image with $AuCl_3$.

QUESTIONS OF UNDERSTANDING

1. What is meant by a negative?

2. Explain how the print becomes a positive.

3. How can printing papers using the same silver salt have different printing speeds?

4. Recalling the arrangement of the halogens according to increasing atomic weights, would you expect silver iodide to be more sensitive or less sensitive to light than silver bromide?

5. With which of the following will metallic silver react?

(a) hypo　　　(b) $K_3FeC_6N_6$　　　(c) developer　　　(d) Na_2S

6. Write the two reactions for sepia toning of the silver image.

7. What do we call the chemical reaction which changes AgBr to Ag?

8. Which silver salt is the more sensitive to light, AgCl or AgBr?

9. In sepia toning what substance forms the final image?

10. What do we mean by saying a substance is sensitive to light?

11. Which print would be the most permanent, one whose image is silver, platinum, or gold?

REFERENCES FOR SUPPLEMENTARY READING

1. Howe, H. E. *Chemistry in Industry*, Vol. I. The Chemical Foundation, New York, 1925. Chap. XVIII, pp. 312–328, "Photography" by S. E. Sheppard.

(1) When was it first noticed that light darkens silver nitrate? (p. 312)

(2) When was the darkening of silver nitrate by light first used to take profiles? (p. 313)

(3) What two great drawbacks were there to this process? (p. 313)

(4) Describe Talbot's method of making the silver salts more sensitive to light and also his method of fixing the print. (p. 314)

(5) Relate the invention and history of the daguerreotype. (p. 315)

(6) Describe Fox Talbot's discovery in 1841. (p. 316)

(7) How much did Talbot's discovery increase sensitivity? (p. 316)

(8) Describe the wet-plate process of photography. (p. 319)

(9) What were the first developers of wet plates? (p. 319)

(10) How are films made sensitive to all colors of light? (p. 323)

2. Eastman Kodak Company. *Elementary Photographic Chemistry.* Rochester, New York, 1936.

(1) In an elon-hydroquinine developer what purpose does each of the following constituents serve? (a) elon; (b) sodium sulfite; (c) hydroquinine; (d) sodium carbonate; (e) potassium bromide. (pp. 17–26)

(2) Give the purpose of each constituent in the acid-hardening fixing bath: (a) hypo; (b) sodium sulfite; (c) acetic acid; (d) alum. (pp. 27–32)

(3) Give the formula for hypo. (p. 27)

(The following exercises should be assigned to the more competent students.)

(4) Give a report on intensification and reduction of negatives. (pp. 36–41)

(5) How does the term reduction in photography differ from the chemical meaning of the term?

(6) Give a demonstration of the toning of prints. (pp. 42–48, and formulas, pp. 82–92)

Summary

As would be expected from metals so near the bottom of the replacement series, gold and platinum usually occur uncombined, while silver occurs occasionally in this form. When gold occurs in gravels, it is mined by either hydraulic mining, placer mining, or dredging. When it occurs in quartz, the rock is crushed to a powder and washed over amalgamated brass plates where the mercury catches the gold but permits the powdered quartz to go by.

Most of the gold in the world is held by the governments as a reserve to back their paper money. However, considerable quantities of it get into the trade for jewelry, gold leaf, and fillings for teeth. Gold, being too soft for most uses, is alloyed with copper and other metals to make it more resistant to wear. Its composition is usually expressed in carats (twenty-fourths) gold.

Platinum is a rare metal more expensive than gold. Because of its great efficiency as a catalyst, its high melting point, and its great resistance to chemical reaction, it finds use in jewelry and in chemical crucibles, pans, electrodes, and test wires.

Silver is usually separated from the lead with which it occurs by the Parkes process of extraction with molten zinc, in which the silver is much more soluble than it is in lead. Silver is used in making coins and in manufacturing tableware, but one of its greatest uses is in photography.

The value of silver in photography lies in the fact that the silver salts are sensitive to light. Strong light will decompose them to produce the silver metal. Even light too weak to produce any noticeable effect makes the silver salts reducible to the metal by organic reducing agents called developers.

The silver salts are prepared in darkness and spread in a thin gelatin layer on film or paper. After the film is exposed in the camera, it is developed in the reducing agent, which makes the image visible in black, finely divided silver. After a short rinse the film is put into a fixing bath of hypo, which dissolves out all of the silver bromide that was not reached by the light from the lens, so there will be no further change when this negative is taken into the light after washing and drying. The printing paper is made with an emulsion similar to but of slower speed than the film emulsion. By printing through the negative, light and dark parts are reversed, so the finished print will be a positive. The print too must be developed, rinsed, fixed, washed, and dried.

Printing-out proof papers have the image brought out by a strong light. Without further treatment, they soon fade. However, they can be made permanent by toning in a gold solution and fixing. Sepia prints are made by changing the black silver image into brown silver sulfide.

⇆

CHAPTER 17

SOME COMMON USEFUL METALS

Aluminum. In 1827 a German chemist, Wöhler, produced some metallic aluminum by heating aluminum chloride with metallic potassium.

$$3 \text{ K} + \text{AlCl}_3 \longrightarrow 3 \text{ KCl} + \text{Al}$$

This was man's first experience with the pure element. We know now that aluminum is the commonest of all metals, forming nearly 8 per cent of the earth's crust. Only oxygen and silicon exceed it. Although aluminum is so abundant, it is so difficult to recover from its compounds by Wöhler's process that it would cost $160 per pound.

In France, under the patronage of Napoleon III, the chemist Deville took up the attempt to find a cheaper way of freeing aluminum from its compounds. By using sodium instead of potassium, Deville reduced the cost of the new metal to about $20 per pound, still too expensive for industrial use. The next step in the cheapening of aluminum came indirectly through cheapening of sodium by a process invented by Castner. This brought the price down to $4 per pound, still too costly for extensive commercial use.

One of Wöhler's pupils, an American named Jewett, was a chemistry professor at Oberlin College in 1885. In one of his lectures he made the statement that a fortune awaited anyone who could discover a cheap way of obtaining aluminum from its compounds. In that class there was a student by the name of Charles Martin Hall. At the above statement by the professor, Hall poked his neighbor in the ribs and said, "I am going after that metal."

True to his word, Hall set up a laboratory in his father's woodshed. On February 23, 1886, he succeeded in making some aluminum pellets by a new process, which was destined to create a great industry and make a fortune of forty million dollars for the inventor. The Hall process has reduced the cost of aluminum from $4 to 20 cents a pound.

The Hall process. Hall's process involved an entirely new principle, that of electrolyzing the oxide in melted rock. The melted mineral serves as the solvent; and the oxide, the electrolyte. There had been electrolytic processes where a soluble

Courtesy of Aluminum Company of America

A BAUXITE MINE. Bauxite is the crude aluminum oxide from which all commercial aluminum is obtained.

salt was electrolyzed to produce metals, but this was the first use of a mineral solvent for an oxide. The solvent used by Hall was melted cryolite (Na_3AlF_6) and the electrolyte Al_2O_3.

Aluminum ore. Although all earth contains aluminum, only the oxide ore pays to work. This ore is called bauxite. The percentage of aluminum in the pure oxide is 53, but the crude ore often does not contain over 35 per cent of the metal. Before the electrolysis, however, the aluminum oxide must be separated in a pure white form from the more or less colored foreign gangue.   First the ore is heated with sodium carbonate to form soluble sodium aluminate, which can be dissolved away from the gangue.

$$Al_2O_3 + 3 Na_2CO_3 \longrightarrow 2 Na_3AlO_3 + 3 CO_2$$

Next carbon dioxide is passed into the solution to precipitate the aluminum as the hydroxide.

$$CO_2 + H_2O \longrightarrow H_2CO_3$$
$$2 Na_3AlO_3 + 3 H_2CO_3 \longrightarrow 2 Al(OH)_3 + 3 Na_2CO_3$$

The pure hydroxide is separated from the water and heated to form the oxide.

$$2 Al(OH)_3 \longrightarrow Al_2O_3 + 3 H_2O$$

Uses of aluminum.   The uses of aluminum depend largely on its beauty, lightness, conductivity, and resistance to tarnish. It is made into electric wires, cooking utensils, light alloys, aluminum foil, and aluminum paint.   Automobile engines are made of duralumin — 95 per cent aluminum with a little copper, magnesium, and manganese.   Another alloy, magnalium, which is aluminum with 5–30 per cent magnesium, is used largely to form the metal structure of dirigible balloons.   Automobile parts, such as pistons and crank cases, also use a lot of these light alloys.   Even furniture is being made of aluminum.

Aluminum cooking utensils.   A century ago most cooking utensils were made of iron.   These, with possibly a copper kettle and some "tin" pans, were all the housewife had.   The iron utensils were rough and black.   If food stood in them, it often dissolved enough iron to give it an objectionable taste. Copper was expensive and dangerous to the extent that copper compounds are poisonous.   Certain foods standing in copper dissolve the metal.

Today, most of our cooking utensils are of aluminum — beautiful, clean, efficient, and absolutely nonpoisonous.   There are three types of aluminum cooking utensils.   Heavy utensils are often made of cast aluminum, which tends to be brittle and is more likely to pit because of specks of impurities.   Another form of aluminum ware is spun aluminum.   Thin sheets of the metal are pressed around spinning forms of wood until they have the correct shape. Spun aluminum ware dents quite readily. The best utensils are the pressed or rolled ware. The sheets of metal are considerably thicker than the final pan should be. These sheets are now rolled under great pressure, and the utensil stamped from the compressed and hardened sheets. The pressed utensils are quite resistant to denting or bending.

Solubility of aluminum metal. Aluminum metal does not seem to rust or corrode. A careful examination, however, will show that

ALUMINUM COOKING UTENSILS. Aluminum cooking utensils are popular because of their durability, beauty, and heat conductivity.

the surface is slightly dulled by a thin layer of aluminum oxide that protects the metal from further oxidation. Aluminum will dissolve in hydrochloric acid but not in the weak acids in fruit and other foods. Concentrated nitric acid attacks silver and mercury, which will not dissolve in hydrochloric acid. Strangely, aluminum will not dissolve in concentrated nitric acid. The protecting oxide that forms on the surface keeps it from dissolving. The oxide of aluminum is one of the most resistant substances we know. Aluminum, however, will dissolve in strong bases, such as NaOH and KOH. The housewife must be careful not to put strongly alkaline soap powders in her aluminum kettles. Lye will dissolve aluminum very rapidly.

$$2\,Al + 6\,NaOH \longrightarrow 2\,Na_3AlO_3 + 3\,H_2$$

Aluminum hydroxide. Aluminum hydroxide is unusual in its chemical reactions. Ordinarily the hydroxides of metals are bases which are dissolved readily by acids but not by other bases. Aluminum hydroxide is dissolved by both acids and bases. Such "two-faced" substances are said to be *amphoteric*. This means that aluminum hydroxide acts as a base in the

presence of a strong acid, while in the presence of a strong base
it acts as an acid.

$$Al(OH)_3 + 3\ HCl \longrightarrow AlCl_3 + 3\ H_2O$$
$$\text{base} \qquad \text{acid} \qquad \text{salt} \qquad \text{water}$$

$$Al(OH)_3 + 3\ NaOH \longrightarrow Na_3AlO_3 + 3\ H_2O$$
$$\text{acid} \qquad \text{base} \qquad \text{salt} \qquad \text{water}$$

The second salt is called sodium aluminate.

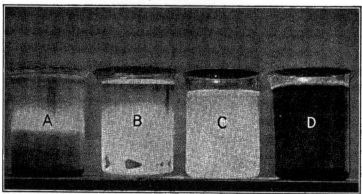

CLARIFYING WATER WITH ALUMINUM SULFATE. Beaker *D* contains muddy
water. Beaker *A* shows the mud settling out after the addition of some aluminum
sulfate. Beaker *C* is full of ordinary clear water. Beaker *B* shows the result of
adding sulfate to *C*.

In an early chapter it was mentioned that aluminum compounds are used to settle sediment suspended in water. Aluminum hydroxide is the compound that actually removes the mud.

$$Al_2(SO_4)_3 + 6\ NH_4OH \longrightarrow 2\ Al(OH)_3 + 3\ (NH_4)_2SO_4$$

QUESTIONS OF FACT

1. What remark did Professor Jewett once make to a class which
contained Martin Hall?
2. What effect did this remark have upon Hall?
3. How was the first metallic aluminum obtained?
4. Tell of Hall's discovery of a better process of obtaining the
metal.
5. What is the commercial ore for obtaining aluminum?
6. How is the aluminum oxide dissolved away from the gangue?
7. How is pure oxide then formed?

8. Why is aluminum suitable for cooking utensils?

9. Discuss aluminum metal as to: (a) protective coating; (b) solubility in HCl; (c) solubility in conc. HNO_3; (d) solubility in alkalies.

10. What special name is given to a hydroxide that will dissolve in either an acid or a base?

11. What elements are in the mineral cryolite?

12. Give the composition and uses of two aluminum alloys.

QUESTIONS OF UNDERSTANDING

1. Give the new features of Hall's process.

2. Write reactions for: (a) $Al(OH)_3$ dissolves in HCl; (b) $Al(OH)_3$ in NaOH.

3. In what sense did Napoleon's aluminum card tray cost $100,000?

4. How did Castner's new sodium process affect the aluminum industry?

5. Give the reaction when Al_2O_3 is heated with Na_2CO_3.

6. What product is formed when aluminum hydroxide is calcined?

7. Is $Al(OH)_3$ soluble or insoluble in water?

8. Why is cast aluminum likely to develop pits?

9. Which type of aluminum cooking utensils is the most durable?

10. Describe the preparation of rolled aluminum ware.

11. Why does aluminum resist being dissolved in concentrated nitric acid, whereas it dissolves readily in hydrochloric acid?

12. Give the formula of sodium aluminate.

13. What ions would sodium aluminate form in water solution?

REFERENCE FOR SUPPLEMENTARY READING

1. Weeks, Mary Elvira. *The Discovery of the Elements*. Journal of Chemical Education, Easton, Pa., 1934. Pp. 163–177, "The Discovery of Aluminum."

(1) Who first suspected alum contained a new element? About when? (p. 163)

(2) What two famous chemists tried to separate it electrically and failed? (p. 163)

(3) Who first prepared pure aluminum? When? (p. 173)

(4) What noble act did Deville do for Wöhler when troublemakers tried to get him to claim the discovery for himself? (p. 174)

Alums. An alum is one of a whole set of compounds with very similar formulas. Ordinary alum, sometimes called potassium alum, is $K_2SO_4 \cdot Al_2(SO_4)_3 \cdot 24\,H_2O$. It is a double sulfate of potassium and aluminum with 24 molecules of water of crystallization. The other alums differ from potassium alum by having other elements in place of potassium and aluminum. Either ammonium, rubidium, cesium, thallium, or almost any other univalent metallic element may be in the place of potassium. Either ferric iron, trivalent chromium, trivalent manganese, trivalent thallium, or almost any other trivalent metallic element may be in the place of aluminum. These compounds are all called alums. All alums form crystals of the same shape. In fact, a crystal may be built up of different layers of these different-colored alums by first allowing the crystals to remain in the saturated solution of one alum, then in another, and so on. When alum is heated, it loses the water of crystallization, forming *burnt alum*. Alum is used in fireproofing draperies, in "sizing" papers, and in dying cloth.

Aluminum sulfate. Aluminum sulfate is used in calico printing and in sizing paper. Paper is sized to prevent the absorption and consequent spreading of the ink. For writing paper, gelatin solution is often employed. In making printing papers, rosin soap is mixed with the pulp; and aluminum sulfate is added. The rosin and aluminum hydroxide are precipitated in the pulp, perhaps in feeble combination, and pressing between hot rollers afterwards melts the former and gives a surface to the paper. Clay, from which bricks, pottery, and porcelain are made, is an aluminum compound.

Aluminum chloride. A useful compound of aluminum is the chloride ($AlCl_3$). This substance is very reactive and fumes when exposed to the moisture of the air. What happens is that it reacts with moisture to form HCl gas, which fumes in the air. Aluminum chloride has long been known to catalyze certain reactions in organic chemistry. However, these reactions were limited in their usefulness because aluminum chloride cost about $1.50 per pound. It was then discovered that aluminum chloride was useful in cracking oils for gasoline, but the price made the process prohibitive. The Texas Oil Company, therefore, set about finding a method of producing the aluminum chloride at a price which would enable it to be used. The most promising method seemed to be that of heating aluminum oxide (obtained by roasting bauxite, the common ore of aluminum) with chlorine. However, a difficulty arose. The aluminum attacked the firebrick lining of the furnace, destroying it in such a short time that the product was still expensive. Finally, after several years of research, involving the expenditure of three million dollars, the process was perfected and the cost of aluminum chloride reduced to 5 cents a pound in carload lots.

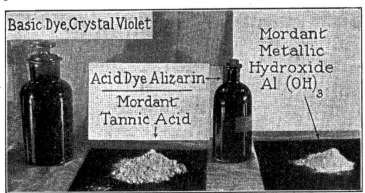

DYES AND MORDANTS. A mordant is the chemical substance necessary to fix the dye as an insoluble precipitate on cotton, linen, or rayon fibers.

Dyeing as related to aluminum compounds. The problem of the dyer is to give a desirable permanent color to a fabric, usually of cotton, linen, rayon, wool, or silk. The color should be "fast" — that is, unaffected by the action of soap or moderate sunlight. The problem is complicated by the fact that the fabrics differ markedly in chemical composition. Cotton, linen, and rayon are cellulose $(C_6H_{10}O_5)_y$, which does not react readily with organic dyes. Wool and silk, on the other hand, contain a considerable percentage of nitrogen, the compounds of which are much more reactive. These two fabrics may, therefore, be easily and permanently stained with organic compounds. Since these materials contain both acid and basic groups, they will combine *directly* with either basic or acid dyes.

For some unknown reason, a few of the *direct* dyes also color cotton successfully. In general, however, cotton, linen, and rayon require different treatment. The inability of cellulose to react with dyes makes it necessary to precipitate the color within the fabric as an insoluble salt. This necessitates the use of a separate chemical to bring about the precipitation. If the dye is an acid compound, it will form insoluble salts with certain hydroxides, such as $Al(OH)_3$, $Fe(OH)_3$, and $Sn(OH)_2$. These hydroxides are called *mordants*. If the dye itself is a base, the mordant must be an acid, usually tannic acid. Although the precipitated dye salts are insoluble, they are probably attached to the fiber only by physical means; hence dyed cotton does not hold color so well as the animal fibers, silk and wool. Colored precipitates are also prepared for use like pigments by artists and others. They are then called *lakes*.

QUESTIONS OF FACT

1. What kind of element can replace potassium in an alum?
2. What kind of metallic element can replace the aluminum in an alum?
3. Give three uses of ordinary alum.
4. How does burnt alum differ in composition from crystallized alum?
5. What is aluminum chloride used for?
6. What kind of dyes need aluminum sulfate?
7. Give the use of lakes made from dyes.
8. What enables silk and wool to hold the dyes without a mordant?

QUESTIONS OF UNDERSTANDING

1. What similarity must elements have in order to replace one another in alums?
2. Alum is said to be a double sulfate of potassium and aluminum. What does this mean?
3. Explain what is meant by the sizing of paper.
4. Do bricks contain aluminum? Give the reason for your answer.
5. Why was it a difficult task to prepare cheap $AlCl_3$?

6. Write formulas for: (a) common alum in crystalline form; (b) sodium-aluminum alum; (c) ammonium-iron alum; (d) potassium-chromium alum.

7. What is a mordant?

8. Why is a mordant needed in dyeing cotton?

9. Define a lake as related to a dye.

10. How many atoms of hydrogen are in a molecule of crystallized alum? How many atoms of oxygen?

11. Calculate the molecular weight of a molecule of crystallized alum.

12. What percentage of crystallized alum is water?

ADDITIONAL EXERCISES FOR SUPERIOR STUDENTS

1. If the water of crystallization of a mol of potassium alum were vaporized, how many liters would it occupy under standard temperature and pressure?

2. What would a mol of burnt alum weigh?

3. What percentage of burnt alum is potassium?

4. Write the history of the discovery and production of aluminum.

Occurrence of copper. Copper was one of the first metals to be used by man. It is found in the metallic state in a few places. There is one such locality in the Upper Peninsula of Michigan. One mass discovered there weighed over 500 tons of 99.92 per cent copper. The Indians formerly mined free copper in this Lake Superior region. The ores of copper are usually the sulfides, oxides, and carbonates. At Bingham Canyon, Utah, is a large open mine, from which enough material has been removed to make a small mountain. There are other large copper deposits in Montana and Arizona. Copper exists in minute quantities in bird feathers, oysters, lobsters, and chocolate.

Metallurgy of copper. The metallurgy of copper involves more operations than are necessary for most of the common metals. The principal operations are as follows: (1) roasting the ore to remove antimony, arsenic, and part of the sulfur; (2) smelting the roasted material to iron-copper sulfide ("matte") in a reverberatory or blast furnace; (3) "converting" the matte to impure "blister" copper — this is similar to the manufacture of steel by the Bessemer process — and (4) refining the blister copper by electrolysis. These operations are necessary because the principal copper ores are sulfides, which are harder to work than oxides or carbonates, and because copper must be purified much more than most metals. Its chief use, for electrical wire, requires a very pure form of the metal, since traces of impurities increase the electrical resistance greatly.

COPPER MINER AT WORK. While in a few localities copper ore lies near enough to the surface that it may be removed with steam shovels, most of the richer workings are deep under ground. The photograph shows a miner operating a pneumatic drill in the face of a copper mine.

The first step in the metallurgy of copper has caused the smelters a great deal of trouble. Large volumes of sulfur dioxide (SO_2), from the roasting sulfides, are discharged from the smokestacks. This SO_2 reacts with the moisture in living plants to form sulfurous acid, which soon kills the vegetation. For miles around the large smelters, there is not a living plant. Some smelters have installed equipment for turning waste SO_2 into useful H_2SO_4 instead of allowing it to escape into the air to destroy vegetation and invite lawsuits from farmers.

Recovery of dissolved copper. Water from copper mines often carries copper in solution. This fact was discovered quite accidentally. At one large copper mine the miners' houses were built along a stream running out of the mine. The housewives used to throw empty tin cans into the creek. Someone noticed that the cans seemed to be turning to copper. What was happening was that the iron was replacing the copper in solution and leaving the copper metal on the can.

$$Fe + CuSO_4 \longrightarrow Cu + FeSO_4$$

Ewing Galloway

COPPER SMELTER. Notice how bare the landscape is. All vegetation has been destroyed by sulfur dioxide from the roasting ores.

The mines immediately profited by this method of recovering copper that would otherwise have been lost. One company recovered $1000 worth of copper per day by dumping scrap iron into the water running out of the mines.

EXPERIMENT

Replacement of copper by iron. Dip a clean knife blade into some copper sulfate solution and show the deposited copper. Drop some iron nails into the copper sulfate solution and allow to stand until the color leaves the solution.

Concentration of ores by flotation. The compounds of the more valuable metals often constitute such a small percentage of their ores that it is necessary to concentrate the metallic compounds before treatment for the metal. One method of concentration often used on copper ores is called the flotation process. The underlying principle of this process was discovered by a washerwoman in a mining town, although she did not profit financially by the discovery. She noticed that when she washed the dirty overalls of the miners the compounds of the heavy metals would float in the froth of the soapsuds while ordinary rock would not. The flotation process developed from this discovery. Oils like coal-tar oil and pine oil are used to make the froth in the practical process.

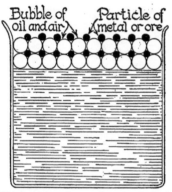

FLOTATION OF ORES. Certain oils froth with water. When powdered rock and ore are mixed with the froth, the rock sinks to the bottom, but the ore is held up by the froth.

Uses of copper metal. Copper is a very common substance. This is an electrical age, and copper wire is almost everywhere. The uses of copper are increasing and are likely to continue to increase with the development of power projects, the electrification of railroads, and the coming of television. Gold, silver, and nickel coins are all alloyed with copper. Gold jewelry is usually an alloy of gold and copper. Other alloys containing copper are brass (copper and zinc), special brasses (containing lead, tin, aluminum, iron, or manganese in addition), German silver (copper, nickel, and zinc), ordinary bronze (copper and tin), and special bronzes (containing also such other elements as phosphorus, lead, and zinc).

EXPERIMENT

Color of copper metal. Pour 1 cc. of ethyl alcohol into a test tube. A copper spiral is heated and introduced into the test tube of alcohol. The copper oxide on the surface of the wire is reduced to a fresh surface of copper.

Copper compounds. The copper compounds are all colored, usually blue or greenish blue. There are two sets of these compounds, the cuprous with a valence of one and the cupric with a valence of two. Copper sulfate (bluestone or blue vitriol) is somewhat poisonous; hence it is used in insecticides and for preventing rust, a disease of wheat.

QUESTIONS OF FACT

1. What is supposed to have been the first metal known to man?
2. Give the location of two large copper mines in the United States.
3. Name several substances which contain small quantities of copper.
4. What difficulty have the copper smelters experienced with the agricultural interests?
5. Name five copper alloys and name the other metals in each case.
6. Give the common name and uses of copper sulfate.
7. What is the color of the cupric compounds?

QUESTIONS OF UNDERSTANDING

1. Give two reasons why the metallurgy of copper is difficult.
2. Describe the recovery of copper from mine waters.
3. Describe the flotation process of concentrating ores.
4. Write formulas for: (a) cuprous oxide; (b) cuprous chloride; (c) cupric nitrate; (d) cuprous sulfate.
5. Why does copper have to be obtained in a state of greater purity than most metals?

ADDITIONAL EXERCISES FOR SUPERIOR STUDENTS

1. If one mole of crystallized copper sulfate, $CuSO_4 \cdot 5\,H_2O$, is heated until the water is driven out, what volume will the water vapor occupy under standard conditions?
2. What weight of copper could be obtained from the copper sulfate in ex. 1?

REFERENCE FOR SUPPLEMENTARY READING

1. Glover, John George, and Bouck, Cornell William. *The Development of American Industries.* Prentice-Hall, Inc., 1933. Chap. XVIII, pp. 381–394, "The Copper Industry."

(1) When did the copper industry begin in the United States? (p. 382)
(2) How low a percentage of copper ore has been worked? (p. 383)
(3) How does the production of copper in the United States compare with that elsewhere? (p. 383)
(4) Where is the largest copper mine in the world? (p. 383)
(5) Explain the three different processes of copper mining. (pp. 384–385)
(6) Describe the making of copper wire. (pp. 387–388)

Discovery of zinc. Zinc is interesting historically because of the fact that centuries before it was known as a metal, an alloy (brass) was made from its ores. The reason the metal itself was not discovered sooner is that it is very volatile when heated; hence it escaped from the furnace into the air. The

GALVANIZING IRON. The figure shows the iron wires being drawn through the tank of molten zinc.

earliest recorded recognition of zinc as a metal is that of Mada-napila, a Hindu king, about 1374. Some two and a half centuries later a Chinese book was written which describes in detail the metallurgy and uses of zinc. The first notice of this metal in Europe was in the cracks in the walls of a lead furnace in Prussia. Because of its durability and its use in electric batteries, zinc is one of our important metals.

Metallurgy of zinc. The principal ore of zinc is the sulfide. In theory the metallurgy of zinc is simpler than that of most metals; in practice the percentage of recovery is decidedly lower. Zinc ore is first roasted to remove the sulfur. The resulting oxide is then mixed with coal and placed in a clay receptacle (retort), which is heated by a gas or oil flame. The carbon removes the oxygen, and the metallic zinc (spelter) distills off into a condenser. The escape of zinc vapor into the air and the loss of metallic zinc in the retort walls reduce the efficiency of the process.

Uses. The principal use of zinc is for protecting (galvanizing) iron. Zinc is very resistant to atmospheric action, due to the formation of a protecting coat of zinc oxide. The zinc is applied to the iron in three ways: (1) by dipping the iron in molten zinc; (2) by heating the iron electrically in zinc dust; and (3) by spraying the iron with molten zinc. As long as the galvanized iron is completely covered with zinc, it is safe from rust; but if the zinc covering gets broken or is damaged by fire so that some of the iron is exposed, the iron will rust and dissolve much faster than if there were no zinc upon it. This is because of electrochemical action. The iron and zinc form an electric cell with rain water as the solution. Since the cell is short-circuited, the action is rapid and continuous, the zinc disappearing in solution.

EXPERIMENT

Galvanized iron. Melt some metallic zinc in a small porcelain crucible with a porcelain cover, over a hot burner. When the zinc is melted, stick the end of a cleaned iron nail into the molten zinc. The iron will become roughly galvanized.

The next largest use of zinc is in making brass. About one-fourth of the world's output is used in this way. Sheet zinc finds considerable use where a sheet metal is exposed to the weather. Other uses are in dry cells, die castings, and as zinc dust.

Zinc compounds. The only compound of zinc of much industrial importance is zinc oxide. Added to rubber, it improves the quality of automobile tires. Zinc oxide is also used as a white paint pigment under the name of zinc white. Zinc oxide has a rival in another white pigment, white lead, but zinc white has one advantage over white lead. When there are sulfur compounds in the air, both substances tend to form the sulfides of the metals on their exposed surfaces. Lead sulfide is black; hence the lead paint gets dingy on standing. Zinc sulfide, on the other hand, is white like the oxide; hence the zinc paint does not get dingy. In the manufacture of paint, zinc oxide is frequently used in connection with basic lead sulfate, another white pigment.

BLOWPIPE TEST FOR ZINC.

The blowpipe test for zinc and other metals. Zinc is a hard element to identify by chemical reaction. The most nearly specific reaction or test for zinc is a flame reaction at the temperature of the blowpipe. The blowpipe is a tapering brass pipe with a fine pinhole at one end, which is curved at right angles to its length. If the flame of a Bunsen burner is turned down low and blown against a piece of charcoal by the fine stream of air from the blowpipe, there is obtained a quiet flame hot enough to produce high temperature reactions. A small hole is scooped out of the charcoal block. The substance to be tested is mixed with sodium carbonate in this hole and heated for a short time with the blowpipe. Then a little cobalt nitrate is added to the mixture, and it is heated again. If zinc is present, a green-colored substance is formed. This green substance is $CoZnO_2$.

The same blowpipe test on aluminum compounds gives a parallel substance blue in color. Magnesium compounds give a pink color with this test. The blowpipe test is the best one for the three metals zinc, aluminum, and magnesium in their compounds.

QUESTIONS OF FACT

1. Where was metallic zinc first noticed in Europe?
2. What is the principal zinc ore?
3. Discuss the metallurgy of zinc.
4. What is galvanized iron?
5. Name three ways of applying zinc in galvanizing.
6. What happens when the iron of galvanized iron becomes uncovered in spots?
7. In what respect is zinc oxide superior to white lead as a pigment?
8. In what sense is the metallurgy of zinc simpler than that of most metals?
9. Describe the blowpipe test for zinc.
10. Give the colors in the blowpipe tests for magnesium and aluminum.

QUESTIONS OF UNDERSTANDING

1. Why was metallic zinc hard to discover?
2. In what form will a metal leave the reduction furnace if its temperature of volatility is lower than the temperature of reduction?
3. In what country was metallic zinc first obtained?
4. What kind of a substance do the following reactions show zinc to be?

$$Zn(OH)_2 + 2\ HCl \longrightarrow ZnCl_2 + 2\ H_2O$$
$$Zn(OH)_2 + 2\ NaOH \longrightarrow Na_2ZnO_2 + 2\ H_2O$$

5. From the fact that the zinc instead of the iron dissolves when the galvanizing surface of iron is cracked, which would you say is higher in the replacement series, zinc or iron?
6. Zinc and aluminum, both quite high in the replacement series, and consequently quite active, seem to last indefinitely in the atmosphere. Explain.

REFERENCES FOR SUPPLEMENTARY READING

1. Glover, John George, and Cornell, William Bouck. *The Development of American Industries.* Prentice-Hall, Inc., New York, 1933. Chap. XX, pp. 405–418, "The Zinc Industry."
   (1) What did the Romans call zinc? (p. 405)
   (2) What substance containing zinc was made during the Dark Ages without knowing that it contained zinc? (p. 405)
   (3) What people produced zinc about the seventeenth century? (p. 405)

(4) When was the smelting of zinc started in Europe? In the United States?. (p. 405)
(5) What percentage of the world's production of zinc is produced in the United States? (p. 405)
(6) How is zinc ore concentrated? (p. 407)
(7) Describe the electrolytic method of producing zinc. (p. 410)
(8) What substance is a by-product of zinc production? (p. 411)

2. Weeks, Mary Elvira. *The Discovery of the Elements.* Journal of Chemical Education, Mack Printing Co., Easton, Pa., 1934. Pp. 20–21, "The Discovery of Zinc."

(1) Where was an idol found that was 87.5 per cent zinc? (p. 20)
(2) When was the metallurgy of zinc described in China? (p. 20)
(3) Who isolated the metal in 1746? (p. 21)

Summary

Aluminum, the most abundant element on the earth, for a long time was one of the hardest to get out of its compounds. Deville obtained some of the metal by heating its chloride with metallic potassium or sodium. Not until Hall produced it by the electrolysis of aluminum oxide in melted cryolite, was this metal made cheap enough for commercial use. The crude aluminum oxide (bauxite) is purified by heating with Na_2CO_3 to form Na_3AlO_3. This is changed to $Al(OH)_3$ by passing in CO_2. Heating changes the hydroxide to the pure white Al_2O_3.

Aluminum is used in electric wires, cooking utensils, aluminum foil, aluminum paint, and in light alloys. Because of the formation of a protecting coating of aluminum oxide, the metal does not tarnish badly and does not dissolve in oxidizing acids. It does dissolve in other strong acids and bases.

Aluminum hydroxide, being able to act the part of either an acid or a base, is said to be amphoteric. Alum is one of a group of double sulfates of a monovalent and a trivalent metal.

Aluminum sulfate is used in sizing paper and as a mordant in dyeing.

Aluminum chloride is useful in the oil industry. As a mordant, it fixes the acid dyes on cotton cloth.

Copper, which often occurs as the sulfide and which must be purified to a greater degree than iron, has a complicated metallurgy. Roasting, the first step, liberates large quantities of

SO_2, which kills surrounding vegetation. Dissolved copper can be recovered from mine waters by the use of scrap iron. In some regions copper ores are concentrated by the flotation process. The last step in the preparation of pure copper is a purification by electrolysis. The chief use of copper metal is in the electrical industries. Some is used also in the copper alloys.

The essential steps in metallurgy of zinc are roasting, reduction, and distillation. Because it is protected by a coat of oxide, zinc is used in making galvanized wire, sheet metal, and fencing. Other uses of zinc are in dry cells and in making the white pigment, zinc oxide. The best test for both zinc and aluminum is the blowpipe test — heating the zinc compound with sodium carbonate and cobalt nitrate to form green $CoZnO_2$. Aluminum under the same test gives a blue aluminate, and magnesium a pink.

$$\xleftarrow{\quad}\atop\xrightarrow{\quad}$$

THE OTHER METALS

Introduction. Of the 92 elements, three-fourths are metals. Most of the important nonmetals have already been studied, as well as most of the most important metals. It is impossible, in a beginning course, to consider all of the metals; hence we limit our study to those that are either very widely distributed or of great practical or theoretical importance. Because of their great theoretical significance, the radioactive elements will be given a chapter to themselves. In this chapter is given a list of metals which are of considerable practical importance in some limited field. The length of treatment in each case is approximately in proportion to the number of uses of the metal. Elements, such as lead, tin, and mercury, are given considerable space, while cerium and thorium, for which there is only one commercial use, are given only a paragraph. With a grouping based on usefulness only, we can expect few similarities.

Lead. Lead is one of the six metals known before modern times, as its symbol, Pb (Latin *plumbum*), would suggest. This metal has been found in the tombs of Egyptian kings.

The common ore of lead is the sulfide (galena), which occurs in bluish-black cubical crystals. Low-grade lead ores are usually first roasted to the oxide and then smelted to metal in a blast furnace. High-grade concentrates, on the other hand, are smelted directly in an ore hearth without preliminary roasting. In principle the ore hearth is merely a fire on a shallow basin of lead. A low blast keeps the fire alive. From time to time the fire is stirred and charged with ore, coal, and lime. The following complicated oxidizing and reducing reactions take place simultaneously in different parts of the fire.

$$2\,PbS + 3\,O_2 \longrightarrow 2\,PbO + 2\,SO_2$$
$$PbS + 2\,O_2 \longrightarrow PbSO_4$$
$$2\,PbO + C \longrightarrow 2\,Pb + CO_2$$
$$PbS + 2\,PbO \longrightarrow 3\,Pb + SO_2$$
$$PbS + PbSO_4 \longrightarrow 2\,Pb + 2\,SO_2$$

Uses. Because of its cheapness, weight, and ease of working, lead has many uses. Lead is used for making shot, bullets, and weights because of its great density — it is 11.3 times as heavy as water. Because of softness and low melting point, it is used, by the plumber for joints and short lengths of pipe. If two iron pipes are to be joined together and are not quite in line, it is almost impossible to make the union with iron. By joining them with lead, which can be bent in any direction, the connection is quite easy. Some alloys of lead are common necessities, such as solder, a low-melting alloy of lead with tin. Type metal and pewter are other useful combinations. The

USES OF LEAD. The uses of lead depend upon its weight, its softness, and its resistance to chemical action.

former is 70 per cent lead, 10 per cent tin, 18 per cent antimony, and 2 per cent copper, while the latter does not vary much from these percentages. Lead has three common oxides: litharge (PbO), a buff-colored compound; lead dioxide (PbO$_2$), a chocolate-brown substance; and red lead (Pb$_3$O$_4$), a bright red powder.

Lead compounds are all poisonous. What is worse, small amounts of lead taken from time to time stay in the human system and their poisonous action is cumulative. Painters are often made ill by the small amounts of lead absorbed through the skin during their daily labor. A lead pipe for drinking water might seem dangerous. The danger, however, is remote because lead sulfate is insoluble. There is a great deal of soluble sulfate in most water. The lead pipe soon becomes coated with insoluble sulfate so that lead ceases to get into solution.

Lead pigments. Many lead compounds are used as pigments, or coloring materials of paints. Paint is an emulsion or mixture of pigment in a "drying oil," which rapidly hardens and holds the pigment in a tough film. Red lead is one of the commonest and most substantial red paints for rough work. Lead chromate, or chrome yellow ($PbCrO_4$), is a brilliant yellow.

LEAD PIGMENTS. The principal lead pigments are: chrome yellow, lead chromate (left); red lead, Pb_3O_4 (center); and white lead, basic lead carbonate (right).

White lead, the basic carbonate $Pb_3(OH)_2(CO_3)_2$, is used as a white pigment.

In the preceding chapter it was pointed out that zinc oxide has an advantage over white lead in that zinc sulfide is white, so the zinc pigment does not darken with age. However, the zinc pigment does not have all the good features. White lead has a greater spreading power than zinc oxide; that is, it goes farther. This is due to the fact that white lead paint is more opaque than zinc oxide paint. Zinc oxide also shows a tendency to crack, which the lead compound does not. Red lead is the best paint pigment for protecting iron and steel. Any steel bridge or building is always given a preliminary coating of red-lead paint. Lead dioxide is made by dissolving red lead in nitric acid. It is used for the positive plate of the lead storage cell.

_____

Precipitation of chrome yellow. Mix solutions of lead nitrate and sodium chromate. Yellow pigment precipitates. Discuss the reaction from the point of view of what causes the reaction to go to completion.

_____

Tin. Tin is another of the prehistoric metals. Its symbol (Sn) comes from the Latin name, *stannum*. This element occurs as the oxide. It is mined extensively in Wales, the Malay Peninsula, and Bolivia. Although the United States is the greatest consumer of tin, it produces very little. The oxide is easily reduced by the usual metallurgical processes.

$$SnO_2 + 2 C \longrightarrow Sn + 2 CO$$

_____

Deposition of tin. (a) A small porous earthenware jar half filled with dilute sulfuric acid is set into a beaker containing concentrated stannous chloride. A zinc rod with a copper wire attached to it is put into the dilute acid and the copper wire bent over until it reaches about 1 cm. below the surface of the stannous chloride solution. After standing under a bell jar for 24 hours, tin crystals appear outside the porous jar.

(b) Place a zinc rod in a dilute solution of stannous chloride. Tin crystals form.

_____

Uses of tin. Although soluble in acids, tin is not attacked by the atmosphere. This property of atmospheric resistance makes tin useful for tin cans and pans for certain uses. The tin serves as a protective coating for the iron. As is the case with galvanized iron, tinned iron with an exposed iron surface rusts faster than if the tin were not there. Here again electrical action causes it to dissolve. Tin plate is made by dipping the iron into molten tin.   Tin is readily rolled into tinfoil for wrapping purposes.   Solder, babbitt bearing metal, and the bronzes are all alloys of tin.

Tin compounds.   Tin has two sets of compounds, the stannous compounds, in which it has a valence of two, and the stannic compounds, in which it has a valence of four.   Stannous

TINWARE.   A generation ago tin cups, pans, and buckets were common.   Today this use of tin is limited almost entirely to the manufacture of cans for the preservation of food.

chloride is used as a mordant in dyeing and as a reducing agent. Stannic sulfide is a bright-yellow substance.   The compounds of tin, however, are not so important as the metal.

EXPERIMENT

Preparation of mosaic gold (stannic sulfide).   Into a small porcelain crucible put equal volumes of powdered tin and flowers of sulfur. After thoroughly mixing, cover with a 2 mm. layer of ammonium chloride.   Heat for some time in a Bunsen flame with the crucible covered.   Stannic sulfide sublimes as a brilliant yellow crystalline deposit on the crucible lid.

QUESTIONS OF FACT

1. Give the uses of lead that depend on its density.
2. Why is there little danger of poisoning from lead water pipes?
3. Name three alloys containing lead.
4. Discuss the lead paint pigments.
5. What lead compound is used in the lead storage cell?

6. What property of tin is the basis of its use in the tin can?

7. Name three alloys containing tin.

8. What is the common lead ore?

9. Describe crystals of lead sulfide.

10. Give the colors of each of the following compounds:

   (a) PbO_2 (c) $Pb_3(OH)_2(CO_3)_2$

   (b) Pb_3O_4 (d) $PbCrO_4$

11. How is PbO_2 made?

12. What is the common tin ore?

13. Where does tin occur?

14. Give two uses of stannous chloride.

QUESTIONS OF UNDERSTANDING

1. The common valence of lead is 2; what is the valence of this metal in PbO_2?

2. The formula for red lead is Pb_3O_4. Can all of the lead atoms have either a valence of 2, 4, or any other valence? Explain.

3. Can Pb_3O_4 be satisfactorily explained on the basis of some of the lead having a valence of 2 and some a valence of 4? How many atoms per molecule should there be of each kind?

4. Name two possible oxidation products of PbS.

5. Write the reactions for the oxidation of lead sulfide to form the two oxidation products of ex. 4.

6. What becomes of the sulfur when PbS reacts with PbO?

7. Write the reaction between PbS and $PbSO_4$.

8. Why is less coke needed in the metallurgy of lead than in the metallurgy of iron?

9. Why is lead poisoning especially serious?

10. What properties of lead make it useful for (a) bullets? (b) plumb bobs? (c) pipe joints? (d) sinkers for fishing? (e) solder? (f) type metal?

11. What is the formula for white lead?

12. Compare white lead with zinc oxide as a white paint pigment.

13. Name the following tin compounds: (a) $SnBr_4$; (b) SnO_2; (c) $SnBr_2$; (d) SnS.

REFERENCES FOR SUPPLEMENTARY READING

1. Clarke, Beverly L. *Marvels of Modern Chemistry*. Harper and Brothers, New York, 1932. Pp. 259–260, "Tin"; pp. 261–265, "Lead"; pp. 266–268.

    (1) What is "tin cry" and what is its cause? (p. 259)
    (2) What is "tin disease"? (p. 259)
    (3) Give the formula for stannic acid. (p. 260)
    (4) What great use of lead came because of electric communication?
        (p. 261)
    (5) Give a use of galena in radio. (p. 261)
    (6) What is "sugar of lead"? (p. 262)
    (7) How many oxides of lead are there? (p. 262)
    (8) Describe the Dutch process of making white lead. (p. 264)
2. Glover, John George, and Cornell, William Bouck. *The Development of American Industries.* Prentice-Hall, Inc., New York, 1933. Chap. XIX, pp. 395–403, "The Lead Industry."
    (1) Name four metals mentioned in the Bible as the riches of Tarshish.
        (p. 395)
    (2) How long have lead pipes been in use at Bath, England? (p. 395)
    (3) Name some early mention of lead compounds. (p. 395)
    (4) What part of the world's production of lead is in the United States?
        (p. 395)
    (5) Make a list of the uses of lead. (p. 397)
    (6) What valuable impurities occur in lead ore? (p. 399)

Mercury. Mercury, the only liquid metal, is sometimes called quicksilver. Mercury also was known before the time of modern chemistry, although the people of that time did not call it a metal. The chief ore of mercury is the sulfide (HgS), called cinnabar. This ore is found in Spain and in California. The metallurgy of mercury is the simplest of all the metals, it being so volatile that it is not difficult to get it free from its relatively unstable compounds.

Uses of mercury. Because mercury is a liquid, it finds use in thermometers and many automatic electrical instruments, such as controls for constant temperature baths. Mercury dissolves other metals to form alloys called *amalgams.* The so-called "silver filling" of the teeth is an alloy of silver, mercury, and other metals. This amalgam is soft and easily molded into the shape of the cavity. In a short time it becomes hard and crystalline. Gold and silver amalgamate readily. Advantage of this fact is taken in the amalgamation process of recovering gold from ore that has been crushed to powder in the stamp mill. The surface of brass plates is heavily amalgamated with mercury. As the gold washes over the plates, it also is amalgamated by the mercury and held fast. The placer miner also sometimes puts mercury in the riffles of his sluice boxes to catch the fine gold.

Properties of mercury. Lift a bottle of mercury to feel its weight. Note that a small drop is spherical. Clean a strip of copper with concentrated nitric acid and see how a drop of mercury will amalgamate it when rubbed.

Mercury compounds. The mercury compounds are all poisonous. Several are used in medicine. Mercurous chloride is called calomel. A generation ago calomel was a common purgative, but it has been replaced by less poisonous substances. Mercuric chloride was used in dilute solution as an antiseptic wash, under the name of *corrosive sublimate*. Mercurochrome is a very complex organic compound containing mercury. It is not particularly poisonous to man but is highly poisonous to disease germs, which makes it a very good antiseptic.

Oxidation and reduction. When mercuric chloride and stannous chloride are mixed together, there results the reaction represented by the following equation:

$$2\,HgCl_2 + SnCl_2 \longrightarrow 2\,HgCl + SnCl_4$$

In this reaction the valence of tin has increased from two to four. At the same time the valence of mercury has dropped from two to one. Such a reaction is called an oxidation-reduction reaction. The terms oxidation and reduction are not new. In Chapter 2 oxidation was defined as the addition of oxygen to an element, while in Chapter 4 reduction was defined as the removal of oxygen from an element. Both of these definitions are true as far as they go, but are too limited to include similar phenomena. Suppose that PbO took on another oxygen atom to form PbO_2. This would be oxidation in the restricted sense. At the same time the lead added the oxygen, the valence of the lead increased from two to four.

As far as the lead is concerned, the oxidation consists in having its valence increased. Now oxidation will be defined as a chemical reaction in which one or more elements have their valences increased.

If adding oxygen or its equivalent increases the valence of some element, the removal of oxygen or other similar elements will reduce the valence of some element. Therefore, reduction is the lowering of the valence of an element. A reaction in which one element is oxidized always has some other element simultaneously reduced; hence it is called an oxidation-reduction reaction. The substance containing the element whose valence is reduced is called the oxidizing agent, while the substance containing the element whose valence is increased is called the reducing agent. Another way of wording the definitions of oxidation would be: *Oxidation is increasing the number of active valences of a positive element or decreasing the number of active valences of a negative element.* Reduction, on the other hand, would be *decreasing the active valence of a positive element or increasing the number of negative valences on a negative element.*

TANTALUM. Metallic tantalum is used in large-size radio tubes.

Tantalum. Tantalum is a metal of unusual properties. It is the most resistant of all metals to chemical action. No single acid will dissolve either gold or platinum, but a mixture of three parts concentrated hydrochloric acid and one part concentrated nitric acid will dissolve them. This mixture was named aqua regia by the alchemists. (Aqua regia is Latin for royal water, water being the term used by the ancients for all liquids and solutions.) Aqua regia, however, will not dissolve tantalum. Only hot concentrated sodium hydroxide and hydrofluoric acid can attack tantalum. Tantalum is scarcer than gold. This limits its use for most purposes. It is used for dental instruments, surgical tools, pen points, hypodermic needles, and acid-proof pumps.

Gallium. Gallium is another metal that is as yet too expensive for a great variety of uses. In recent years its price has been reduced from $200 per gram to $3.60 per gram, which is still much more expensive than gold. Gallium melts at body heat but does not boil until 2000° C. is reached. When gallium is used in radio tubes, the tube will work at lower temperatures than when any other metal is used. Gallium is nonpoisonous and may be used in tooth fillings.

Indium. Several times in the last century some element previously only a rarity has been developed until it could be used commercially. Such an element is indium. In 1924 an order was placed with a New York firm for a considerable quantity of this rare metal. After long correspondence with all the chemical supply companies of the world, one gram of this substance was purchased for $10, this being the entire supply in the world. Eight years later there was displayed in New York 3000 grams of this metal, which was valued at more than $20,000. Today this same quantity of indium can be purchased for less than $3000. A drop in price of from $10 to $1 per gram in eight years shows considerable progress. At this price, however, it is almost as expensive as gold and cannot be used as much as it might at a cheaper price.

The story of this metal is interesting. In 1924 a Mr. Murray learned that indium has the property of making metals more resistant and lasting. This fact was of little value, however, since the metal was not on the market and was considered by mineralogists as unobtainable. Mr. Murray began a search for indium in as many different ores as he could locate. His method of search was by the spectroscope, since indium shows a characteristic blue line. Finally he found a sample with the blue line strong enough to show promise. In time he located a mine with ore yielding 1.93 ounces of indium per ton. It was still necessary to work out a satisfactory process for recovering the metal. A Mr. Gray was called in, and in time the problem was solved quite satisfactorily.

Indium shows promise of being useful in medicine. It may also be used to replace gold in filling teeth and in alloys.

Vanadium. Vanadium is used primarily in high-speed tool steel. Auto-spring steels contain up to 47 per cent. High-speed cutting tools, such as drills, saws, and oil-well drills, may contain up to 12 per cent. In Peru the highest commercial mine in the world is a vanadium mine. Vanadium compounds, especially the silver vanadate, are good catalysts in the contact sulfuric-acid process. This vanadium catalyst does not become poisoned as easily as the platinum catalyst.

QUESTIONS OF FACT

1. What metal is liquid at ordinary temperatures?
2. What is the principal ore of mercury?
3. In what sense is the metallurgy of mercury simple?
4. Tell how mercury is used in gold mining.
5. What property of mercury compounds is the basis of their medicinal use?
6. Write the chemical formulas of (a) calomel; (b) corrosive sublimate.
7. Discuss the resistance of tantalum to chemical action.
8. Name some useful articles made from tantalum.
9. What is interesting about the melting point of gallium?
10. Relate the story of the search for indium in the United States.
11. What properties of indium make it useful?
12. Discuss vanadium on the following points: (a) relation to high-speed tools; (b) automobile springs; (c) catalyst in the sulfuric acid process; (d) highest commercial mine in the world.

QUESTIONS OF UNDERSTANDING

1. What is an amalgam?
2. Mercurochrome is an antiseptic compound of what metal?

3. Extend the definitions of oxidation and reduction so they include changes in the valence of elements without necessarily involving oxygen.

4. Give a reaction illustrating question 3.

5. What unit is used as a quantitative measure of oxidation and reduction?

6. If one element is oxidized, is another one reduced at the same time?

7. If in an oxidation-reduction reaction one metal is reduced two valences, how many valences must some other element be oxidized?

8. Do you think a tantalum mine would be a valuable thing to own? Give reasons for your answer.

9. How in your judgment does the range of temperatures at which gallium exists as a liquid compare with that of other substances?

REFERENCES FOR SUPPLEMENTARY READING

1. French, Sidney. "A Story of Indium." *Journal of Chemical Education*, Vol. II, No. 5 (May 1934), pp. 270–272.
   (1) Relate the history of the indium industry in the United States.

2. Weeks, Mary Elvira. *The Discovery of the Elements*. Journal of Chemical Education, Easton, Pa., 1934.
   (1) Relate the account of the discovery of tantalum. (pp. 81–87)
   (2) Tell of the discovery of gallium. (pp. 215–219)
   (3) Write the story of the discovery of indium. (pp. 196–199)
   (4) Write a report of the discovery of vanadium. (pp. 87–96)

TUNGSTEN FILAMENT. Because of its high melting point (3000° C.) tungsten is especially suitable for the filaments of electric-light bulbs. As it can be heated to a higher temperature than any other filament, it gives a more brilliant and whiter light.

Tungsten. The symbol for tungsten is W, taken from the German name, wolfram. Tungsten has the highest melting point of all the elements except carbon. The high melting point has made tungsten useful for the filament of electric-light globes. Tungsten, as a constituent of steel, makes it hard and heat resistive. Tungsten steel in lathe tools can get red hot

without loss of hardness.   Tungsten carbide (carboloy) is the
hardest cutting-tool material known.

Arsenic.   Arsenic metal has no very desirable properties and
is little used.   A little arsenic is used to harden lead in what
the duck hunter knows as "chilled shot."   The uses for arsenic
compounds can all be summed up in the word *poisons*.   Many
ant pastes and insecticides are arsenic compounds.   Arsenious
oxide is the white powder often sold as "white arsenic" or

Courtesy of Case Co.

SPRAYING FRUIT WITH SODIUM ARSENATE.   All arsenic compounds are more
or less poisonous, hence their use in ant pastes, insecticides, and germicides.

sometimes merely as "arsenic."   Paris green is a copper-
arsenic compound.   As both copper and arsenic are poisonous,
this insecticide is very poisonous.   Pears and vegetables are
often sprayed with lead arsenate, $Pb_3(AsO_4)_2$.   All pears and
vegetables should be thoroughly washed before being eaten.
Every once in a while the newspapers tell of someone being
poisoned from the vegetable spray.   The German chemist Paul
Ehrlich made use of the fact that all arsenic compounds are
poisonous to develop an important remedy for a severe con-
tagious disease of man.   He found that the complex organic
compounds of arsenic are not equally poisonous to all living

organisms. He succeeded in synthesizing arsenic compounds that are highly fatal to certain animal germs without being fatal to human beings. His 606th compound of arsenic showed this selective poisonous action enough so it could be used in the treatment of the disease which is called syphilis.

Chromium. The metal chromium has come to the front in importance rapidly in the last few years. Because of its beauty and great resistance to tarnish, it has almost replaced nickel as a protective coating of iron. Chromium-plated articles are much in demand now. Chromium-nickel alloys resist oxidation, and chromium steels are being used more and more. Stainless steel is likewise a chromium steel. Nickel-chromium resistance wire for electrical heating has already been mentioned in connection with the study of nickel.

CHROMIUM PLATE. Chromium plating has largely replaced nickel plating, which was very common a few years back.

Compounds of chromium. Lead chromate ($PbCrO_4$) is a good yellow pigment for paints. The soluble chromates, such as Na_2CrO_4 and the dichromate $Na_2Cr_2O_7$, are among the cheap oxidizing agents.

Cerium and thorium. The oxides of cerium and thorium are used in gas mantles to produce high luminosity in flames not naturally very luminous. When gas was used for lighting purposes, these mantles were much in demand. Such mantles are still necessary in gasoline lamps. The mantles, which consist of 99 per cent thorium oxide and 1 per cent cerium oxide,

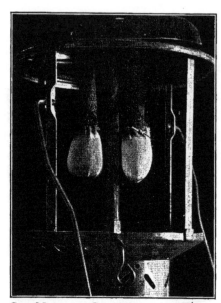

GAS MANTLES. Gas and gasoline flames, which are too nearly colorless for illuminating purposes, can be made highly luminous by means of a gauze mantle of cerium and thorium oxides. These mantles are used on all gasoline lamps and lanterns.

glow with a brilliant white flame when heated to a high temperature.

Titanium. Titanium is an abundant metal, but widely scattered. Titanium oxide (TiO_2) is an important white pigment for paint.

Boron. The element boron is valued for two of its compounds. One of these, borax ($Na_2B_4O_7$), occurs naturally in the desert regions in or surrounding Death Valley. Borax crystallizes with 10 molecules of water of crystallization. The chief uses of borax are in enamel and as a mild alkali. Enamel consists of iron with a borax glass surface. Bath tubs, sinks, and cooking utensils are the common uses of enamel.

The other useful boron compound, boric acid, is usually made from borax by treating it with sulfuric acid.

$$Na_2B_4O_7 + 5 H_2O + H_2SO_4 \longrightarrow Na_2SO_4 + 4 H_3BO_3$$

Boric acid is a very weak acid and also a mild antiseptic. If the doctor needs to use an acid on some delicate tissue, such as the eye, he chooses boric acid.

Minor metals. There are a number of additional metals whose uses are extremely limited, either because the metals themselves are scarce or because they are lacking in outstanding properties. A few of the more important of these minor metals are shown in the table at the top of the following page.

MINOR METALS

METAL	USE OF METAL	USEFUL COMPOUNDS
Antimony	Low-melting alloys for fuses, type metal, babbitt, pewter. Minor use in medicine.	Sulfides used in matches.
Bismuth	Same as antimony	
Cadmium	Low-melting alloys; electric fuses; safety plugs in steam boilers; dental alloys.	CdS, a yellow pigment for paint.
Manganese	Very hard, wear-resisting steel 6–15 per cent manganese; used in wire for resistance boxes.	MnO_2, cheap oxidizing agent; $KMnO_4$, a vigorous oxidizing agent.
Molybdenum	Molybdenum steel: resists twisting; auto axles and rods. Used with tungsten in electric light filaments.	

Hydrolysis. In the last paragraph, we spoke of borax as being a mild alkali. The term *alkali* is rather flexible in meaning when used by different persons. The agriculturist uses it in a much wider sense than the chemist. The former considers it to mean all chemicals in the soil that hinder plant growth, while the latter limits the term to bases and substances that form bases in solution. It will be recalled that bases are those hydroxides that ionize in

BORAX. This lump of borax was mined near Kramer in the Mojave Desert.

solution to give hydroxyl ions ($O\overline{H}$), which are characterized by turning pink litmus paper blue.

Borax ($Na_2B_4O_7$) and washing soda (Na_2CO_3), whose formulas show no hydroxide radicals, both turn pink litmus blue in solution. This shows us that there are $O\overline{H}$ ions in these solutions. The only way we can account for their presence is to assume that the salts have reacted with the water to produce hydroxides that ionize. This type of reaction of a salt with water is called *hydrolysis*.

There are other salts which hydrolyze to produce an acid litmus reaction. Copper sulfate ($CuSO_4$) and aluminum sulfate ($Al_2(SO_4)_3$) do this. Here again, since neither salt contains any hydrogen, the only way we can account for the acid hydrogen ion ($\overset{+}{H}$) is to assume the salt hydrolyzed to form acid.

Certain questions immediately come to our minds: How can some salts hydrolyze to give a basic reaction, while others hydrolyze to give acid? Can we predict whether a salt will hydrolyze or not, as some salts such as $NaCl$ and KNO_3 do not hydrolyze? Among those salts that do hydrolyze, can we predict which will form an acid reaction and which a basic reaction?

To answer the first question, let us look at the equations of these reactions:

$$Na_2B_4O_7 + 2\,H_2O \longrightarrow H_2B_4O_7 + 2\,NaOH$$
$$Na_2CO_3 + 2\,H_2O \longrightarrow H_2CO_3 + 2\,NaOH$$

The products of the first reaction are tetraboric acid and sodium hydroxide. Tetraboric acid is a weak acid, while its companion substance, sodium hydroxide, is a very strong base. This means that the acid puts only a few hydrogen ions into the solution, while the sodium hydroxide puts in a large number of hydroxyl ions. The many hydroxyl ions from the sodium hydroxide predominate over the few hydrogen ions from the acid, and the solution as a whole is basic to litmus. The same situation holds for the second reaction, in which the weak un-ionized H_2CO_3 is obscured by the strong, highly ionized $NaOH$. The conclusion suggested by these reactions is that if the salt hydrolyzes into a weak acid and a strong base, the solution will be alkaline. Another way of stating it is that those salts that are obtained by the neutralization of a strong base and a weak acid hydrolyze to give a basic reaction.

An examination of the equations for the hydrolysis of copper sulfate and aluminum sulfate shows the opposite situation.

$$CuSO_4 + 2\,H_2O \longrightarrow Cu(OH)_2 + H_2SO_4$$
$$Al_2(SO_4)_3 + 6\,H_2O \longrightarrow 2\,Al(OH)_3 + 3\,H_2SO_4$$

In these equations the resulting acid is the stronger of the two as both $Cu(OH)_2$ and $Al(OH)_3$ are weak and mostly unionized. Naturally, the strong, highly ionized sulfuric acid predominates over the weak base and the solution is acid.

Let us now examine two salts that do not hydrolyze. NaCl can be obtained by neutralizing the strong base NaOH with the strong acid HCl. In a similar way KNO_3 can be obtained by neutralizing the strong KOH with the strong HNO_3. Therefore, if the salt can be obtained by the neutralization of a strong base with a strong acid, it will not hydrolyze; but if either the acid or the base is weak, the salt will hydrolyze.

It will be seen that hydrolysis is the opposite of neutralization. The facts are that in the case of a strong base and weak acid or vice versa, the neutralization is not complete, and the reaction goes to equilibrium. The reactions should be written with the double arrows, as :

$$H_2SO_4 + Cu(OH)_2 \rightleftharpoons CuSO_4 + 2 H_2O$$
$$3 H_2SO_4 + 2 Al(OH)_3 \rightleftharpoons Al_2(SO_4)_3 + 3 H_2O$$
$$H_2CO_3 + 2 NaOH \rightleftharpoons Na_2CO_3 + 2 H_2O$$
$$H_2B_4O_7 + 2 NaOH \rightleftharpoons Na_2B_4O_7 + 2 H_2O$$

where the symbol $\cdots\rightarrow$ stands for neutralization, and the symbol $\longleftarrow$ stands for hydrolysis.

QUESTIONS OF FACT

1. Why is tungsten useful in incandescent bulbs?
2. Give the properties and uses of tungsten carbide.
3. What is "white arsenic"?
4. What word could describe the uses of arsenic compounds?
5. Why has chromium plating largely replaced nickel plating?
6. What is the purpose of a gas mantle, and what two oxides does it contain?
7. What are the uses of borax and boric acid?
8. From the table at the end of the chapter name : (a) a compound that is a pigment; (b) a cheap oxidizing agent; (c) an element that forms a hard alloy with platinum; (d) an element whose steel resists twisting.
9. Give the formula of a chromium compound which is : (a) a pigment; (b) of two which are cheap oxidizing agents.

10. What titanium compound is a commercial white pigment?

11. Name two substances that are alkalies.

QUESTIONS OF UNDERSTANDING

1. Explain why tungsten has the symbol W.

2. In what property does tungsten excel all other metals?

3. Discuss Ehrlich's research with arsenic compounds.

4. Where in electrical devices are chromium-alloy wires used?

5. Borax in solution turns pink litmus paper blue. What is proved to be present?

6. Could this ion possibly come from $Na_2B_4O_7$?

7. What, then, is the only source of the $\overline{OH}$?

8. What do we call the reaction necessary to produce the $\overline{OH}$?

9. How does an alkali differ from a base?

10. In what way does copper sulfate hydrolyze?

11. Can the presence of $\overset{+}{H}$ be accounted for from $CuSO_4$?

12. What type of salt does not hydrolyze?

13. What type of salt hydrolyzes to give a basic reaction?

14. What type of salt hydrolyzes to give an acid reaction?

15. Explain the relation between hydrolysis and neutralization.

16. H_3BO_3 is a slightly soluble weak acid. Explain "slightly soluble" and "weak."

17. Which of the following will hydrolyze and which not? Select those salts that hydrolyze acidic and those that hydrolyze basic.

(a) KNO_3 (c) Na_2CO_3 (e) $NaCl$

(b) $CuSO_4$ (d) $Al_2(SO_4)_3$ (f) $NaBr$

REFERENCE FOR SUPPLEMENTARY READING

1. Weeks, Mary Elvira. *The Discovery of the Elements.* Journal of Chemical Education, Easton, Pa., 1934.
   (1) Report on the discovery of one or more of the following metals:
      (a) pp. 50–52, "Discovery of Tungsten";
      (b) pp. 10–11, "Discovery of Arsenic";
      (c) pp. 58–63, "Discovery of Chromium";
      (d) pp. 146–149, "Discovery of Cerium";
      (e) pp. 149–151, "Discovery of Thorium";
      (f) pp. 142–146, "Discovery of Titanium";
      (g) pp. 156–161, "Discovery of Boron."

SUMMARY

Because of its cheapness, weight, softness, and resistance to solution, lead has many uses. The metallurgy of lead differs from that of other metals in that oxidation and reduction may take place simultaneously according to a number of different reactions.

The following lead compounds are well-known pigments: red lead (Pb_3O_4), chrome yellow ($PbCrO_4$), and white lead, $Pb_3(OH)_2(CO_3)_2$.

Tin, a very resistant metal, is used in making tin cans. Alloyed with other metals, tin forms solder, babbitt, and bronzes.

Mercury, the only liquid metal, is used in thermometers and automatic heating devices. Mercury is easily freed from its sulfide ore by heat. Mercurous chloride (calomel), mercuric chloride (a deadly poison), and mercurochrome are mercury compounds used in medicine or surgery.

When mercuric chloride is mixed with stannous chloride, both the mercury and the tin change their valences. The valence of mercury drops from two to one, while that of tin rises from two to four. Such a reaction is called an oxidation-reduction reaction. The substance containing the element whose valence is reduced is called the oxidizing agent, while the substance containing the element whose valence is increased is called the reducing agent.

The metal tantalum, one of the most resistant of all metals, which will not even dissolve in aqua regia, is used in making dental instruments, surgical tools, pen points, and acid-proof pumps.

Gallium, a metal which melts at body heat, is used in radio tubes and tooth fillings. Vanadium is valuable primarily in high-speed tool steel, and in auto springs. Tungsten, having the highest melting point of any metal, is used as the filaments of light globes. Tungsten carbide (carboloy) is the hardest cutting-tool material. Arsenic, besides some use in hardening shot, is used in its compounds to poison insects. Indium, a new metal, has a very interesting story, but it is still as expensive as gold.

Chromium is very popular as a plating metal. Nickel-chromium is very resistant to acids. The metal is also used in nickel-chromium wire for electric heaters. Cerium and thorium increase the brilliance of gas mantles. Boron in its compounds borax and boric acid is used in washing powders, enamel, and borax glass.

Some salts, such as borax or sodium carbonate, react basic in solution. On the other hand, copper sulfate in solution reacts acid. These facts are explained by the salt reacting with the water to form the base and the acid in each case. Such a reaction is called hydrolysis. If a salt is obtained by neutralizing a strong base with a strong acid, it does not hydrolyze. However, if either the acid or base is weak, the salt will hydrolyze, giving the reaction of the stronger.

⇄

CHAPTER 19

THE RADIOACTIVE ELEMENTS

What radioactivity is.   Certain of the heavy elements possess remarkable properties, which have come to be known as *radioactivity*.   Something of the nature of these properties is indicated in the following paragraphs.

(1) If a small amount of radioactive material is allowed to remain for some time above a key lying on a protected photographic plate, a shadow picture of the key will appear when the plate is developed.   Some form of energy evidently passed right through the opaque covering but was stopped by the heavy metal key.   We speak of this energy as *radioactive* rays.

(2) A radioactive substance held near the top of a charged electroscope (see figure, p. 370) causes the charge to disappear quickly and the foil leaves to drop together.   The explanation of this is that the radioactive rays ionize the air. Those ions whose charges are opposite to the charge on the

Courtesy of Bureau of Mines

SHADOWGRAPH TAKEN WITH RADIUM. If thick metallic objects, such as coins and keys, are laid on a plate holder containing a sensitive photographic plate, and a tube of radium held above them for a short time, the plate is exposed through the closed plate holder, and shadow pictures of the metallic objects appear when the plate is developed.

electroscope quickly neutralize the latter.   A charged electroscope often is used to hunt radioactive minerals.   In a certain hospital the radium that was being used to treat cancer got mislaid.   A charged electroscope located the radium in the furnace.   The small tube containing the radium had got mixed with waste gauze and bandages and was thrown into the refuse to be burned. By extracting the ashes of the furnace, most of the valuable radium was recovered.

(3) Radioactive materials cause certain substances to fluoresce, or emit a faint light. This property has been put to practical use in the luminous dials of watches and clocks. A very small amount of radioactive substance is mixed with zinc sulfide, a cheap compound that fluoresces under the action of the rays. This mixture is painted on the watch dial. In daylight, one observes nothing unusual about the paint, but in total darkness it can be seen to emit light, making the hands and figures visible.

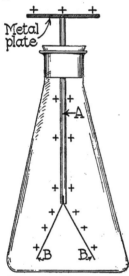

CHARGED ELECTROSCOPE. When an electroscope is charged, the leaves B stand apart from each other. When radium is held near a charged electroscope, the charge is neutralized so that the leaves drop down until they touch each other.

(4) A radioactive substance attached to the bulb of a thermometer causes the thermometer to remain a little above the prevailing temperature, thus showing that radioactivity is a continuous source of heat. It has been determined that a gram of radium, one of the most vigorous radioactive elements, yields 132 calories of heat per hour and that half the gram will still remain after this activity has been going on for sixteen centuries.

(5) Before the properties of radium were well understood, a college professor got a severe burn by carrying a small glass tube of radium in his vest pocket. This property of killing living cells is the basis of radiumtherapy, a method of treating cancer. The cancer cells are much more susceptible to the radioactive rays than are most normal cells; hence a short treatment may be fatal to the cancer cells without killing many of the healthy cells.

Danger from radioactive substances. One sometimes sees radioactive water advertised as having wonderful curative properties. If such water actually contains radioactive substances, as is sometimes the case, it is positively dangerous. Several years ago a watch factory, where luminous-dial watches were being made, employed girls to paint the dials. The girls would point the tips of the paint brushes by putting them into their mouths. After a while several of the

Photo by Calvin Coover

LUMINOUS WATCH DIAL. When a small amount of radioactive substance is mixed with certain materials, the latter are made to fluoresce, or emit a feeble light which is visible in darkness.

girls began to feel ill. Their physician, not knowing the cause of the trouble, treated them for ordinary anemia, but with little success. One night one of the girls happened to glance into the mirror before turning on the light. To her

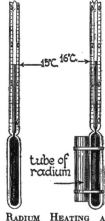

amazement and horror she observed herself emitting a fluorescent light. This incident gave the doctors a clue as to the cause of the anemia. The radioactive substance was collecting in the bones and there destroying the blood cells where they are formed. The doctors know no way of removing this radioactive poison, so it was feared that these young women were doomed. The watch company tried to do what it could for them and indemnified each with $10,000. One of the leading magazines came out with the heading: "Ten Thousand Dollars and Only One

RADIUM HEATING A THERMOMETER. Radium continually gives out energy, day and night, summer and winter, year after year. The thermometer with the radium salts attached to it shows a higher temperature than the other thermometer without radium.

Marie
Sklodowska
Curie

1867–1934

Marie Sklodowska, a young Polish girl, was much interested in chemistry. Because of lack of opportunity in Poland, she moved to Paris. While attending the University she supported herself meagerly by washing glassware and similar tasks. At the home of a friend in Paris she met young Pierre Curie. In time the young couple married and set up housekeeping on a very modest scale, as Pierre was still a graduate student. About this time the chemist Becquerel discovered that uranium ore would affect a photographic plate through the plate holder, and the amount of change seemed to be in proportion to the percentage of uranium. Soon, however, he experimented with uranium ore from another locality in Austria, which showed this property many times as great as would be expected from the quantity of uranium. This made him think that this very active ore contained something else more active than uranium. At this point he turned the problem over to Madame Curie, who began a thorough search for the new element. Her husband now dropped the work he was doing on crystals and both plunged enthusiastically into the new problem. After nearly three years of almost superhuman effort, during which a daughter was born to the couple, they announced the discovery of a new element, polonium, and shortly, a still more active element, radium.

Fame now began to come to the Curies. Pierre was made Professor of Physics at the Sorbonne, while Marie was put in charge of physics lectures at the Higher Normal School for Girls near Paris. After the death of her husband Madame Curie was given the professorship vacated by him, something entirely without precedent in French universities. Continuing her researches with the aid of a fund donated by Andrew Carnegie, she finally separated the metal radium out of its salts by electrolysis, her crowning achievement.

Year to Live; What Would You Do?" However, most of the afflicted women lived more than the year; some of them lived several years. No radioactive substance should ever be taken internally.

The discovery of radium. Radioactivity was discovered in 1896 by Henri Becquerel in connection with the mineral pitchblende, the oxide of uranium (U_3O_8). Pierre Curie and his wife began to study radioactivity. They found that the naturally occurring oxide of uranium was four times as radioactive as the same oxide when prepared in the laboratory. This suggested that possibly there was a more radioactive substance than uranium in the pitchblende. By working over tons of the ore, the Curies were able to separate a small amount of the chloride of a metal a million times as radioactive as uranium. This new element was named radium. In 1906 Monsieur Curie was killed in a street accident, but his wife continued the work. She found the new element to have an atomic weight 226 and chemical properties similar to those of barium.

Quantity and cost of radium. The first source of radium was the pitchblende of Austria. Next, carnotite, an ore from Colorado and Utah, was found to contain a higher concentration of radium than pitchblende. Even so, it still took 500 tons of ore to yield 1 gram of radium salt. The vast amount of work necessary to produce radium made it cost about $120,000 per gram. In 1922 a very rich source of radium was found in Belgian Congo. Only 70 tons of mineral have to be worked in order to obtain a gram of radium. The price dropped to around $50,000 per gram and might have gone much lower except for the fact that a monopoly kept the price up. In 1932 another deposit was discovered in Northern Canada, which may break the Belgian monopoly and bring the price down where every hospital can afford it. The radium treatment for cancer is one of the best, but it takes at least $10,000 to get enough for the treatment. Many hospitals cannot now afford to give this type of treatment, necessary as it is.

REVIEW EXERCISES

1. Tell five things a radioactive substance can do.
2. Is radioactive water safe to drink?
3. In what mineral was radium discovered?
4. Why is radium so costly?
5. Compare the yield from the Belgian Congo ore with that of carnotite from Colorado.
6. Describe a charged electroscope and tell what happens when it is brought near a radioactive substance.
7. Explain why the electroscope loses its charge.
8. How can radioactive substances be located?
9. Discuss radiumtherapy.
10. Relate the story of the women in the watch factory who became anemic because of radioactivity.
11. What is the ore pitchblende?

REFERENCES FOR SUPPLEMENTARY READING

1. Foster, William. *The Romance of Chemistry*. D. Appleton-Century Co., New York, 1936. Chap. XV, pp. 218–238, "The Wonders of Radium."
   (1) What discovery led eventually to radioactivity? (p. 218)
   (2) What are cathode rays and what can they do? (pp. 218–219)
   (3) Describe the Becquerel rays discovered in 1896. (p. 219)
   (4) What did Madame Curie discover? (p. 219–224)
   (5) How did she prepare pure radium? (p. 224)
   (6) Describe the spinthariscope and tell what it shows. (p. 225)
2. Holmes, Harry N. *Out of the Test Tube*. Ray Long & Richard R. Smith, New York, 1934. Chap. XII, pp. 137–158, "The Fall of the House of Uranium."
   (1) Describe Crookes's experiment. (p. 138)
   (2) Describe accidental experiment of Röntgen. (pp. 139–140)
   (3) What did Röntgen's discovery suggest to Becquerel? (p. 140)
   (4) How did Madame Curie become interested in the uranium minerals? (p. 142)
   (5) What did Madame Curie name the first radioactive element she discovered? (pp. 142–143)
   (6) How much more radioactive is radium than uranium? (p. 143)
   (7) What great honors came to Madame Curie? (p. 143)
   (8) What assumption in the atomic theory was disproved by radioactivity? (pp. 144–146)
3. Jaffe, Bernard. *Crucibles*. Simon and Schuster, New York, 1930. Chap. XIII, pp. 242–264, "Curie."

(1) Did Marie Sklodowska have any better chance of becoming a famous scientist than any modern high-school girl?

(2) Have you ever thought of being somebody in the world or are you satisfied just to live and die?

(3) Write a biography of Madame Curie.

4. Weeks, Mary Elvira. *The Discovery of the Elements.* Journal of Chemical Education, Easton, Pa., 1934. Chap. XIX, pp. 292–309, "The Radioactive Elements."

(1) Describe the appearance of radioactive substances in the dark. (p. 297)

(2) What does radium do to the air? (p. 300)

(3) Who first explained why the air becomes radioactive? (p. 300)

Radioactive rays. If a small amount of a radium compound is put into a thick lead box with a tiny hole in the top and a zinc-sulfide screen held above it in a dark room, a spot of fluorescent light is observed on the zinc-sulfide screen. If now a positively charged plate is held on one side of this beam and a negatively charged plate on the opposite side, there will be three spots of light instead of one. One spot is over toward the side of the negatively charged plate.

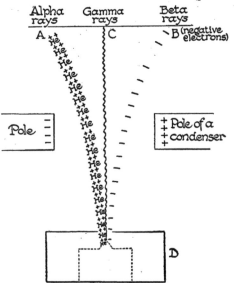

RADIOACTIVE RAYS. A little radium salt is placed in the lead box D, which permits the escape of radium rays only through the small hole at the top. If in a darkened room a photographic plate is placed along A C B, the narrow beam of rays will cause a spot to develop at C. If, however, an electric field is applied as shown in the drawing, three spots will develop. At A the alpha rays will produce one spot; at B the beta rays will produce another spot; and at C the gamma rays still strike the old spot. The alpha rays are helium atoms carrying positive charges; hence they are pulled out of line by the negative plate of the condenser. The beta rays are negative electrons; hence they are pulled towards the positive plate of the condenser. The gamma rays are similar to light and remain unaffected by the electric field.

This is interpreted to mean that the spot is caused by bombardment with high-speed projectiles carrying positive charges, since a negatively charged plate can attract only positive charges. This beam of rays is called the alpha rays and has been proved to be a stream of high-speed helium atoms, carrying two unit positive charges each. On the opposite side, the positive plate has pulled out a stream of very high-speed negative electrons. This stream of electrons is called the beta rays. Part of the rays still strike the original spot. These are called the gamma rays. They are like very short wave X rays and hence are not attracted by the charged plates. The gamma rays are thought to be caused by the other rays striking the

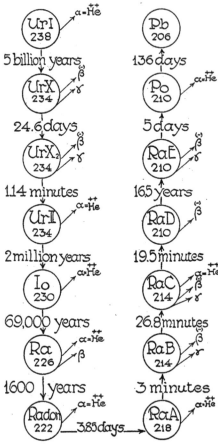

RADIOACTIVE TRANSFORMATIONS. The diagram represents the series of changes by which uranium eventually becomes lead. The time that it takes half the given mass of substance to decompose is indicated by the figure cutting across the arrow. An alpha particle is an atom of helium carrying two positive charges. Every time one of these particles is emitted, four units of atomic weight are lost. The beta particles are high-speed electrons, weighing only $\frac{1}{1845}$ of an atomic unit; hence no loss of weight is indicated when a beta particle is emitted. The gamma ray is like a very short light ray and is caused by an electron striking an atom of a radioactive substance.

unchanged atoms. . Thus the rays of fundamental importance are the alpha and the beta rays.

Nature of radioactive transformations. The proof that alpha rays are helium atoms carrying positive electric charges introduced a new conception of the nature of the atoms. Here was radium, an element of atomic weight 226, throwing out a helium atom of atomic weight 4 and leaving a new element (radon) of atomic weight 222. The rate at which this change takes place is not influenced by anything scientists have been able to do up to the present. No catalyst will speed it up or slow it down. A certain proportion of the atoms seem to explode during each interval of time.

The half-life of a radioactive substance is the time it takes for half of any sample to change. The half-life of radium is 1600 years. This means that if one gram of radium were purchased today, in 1600 years one-half gram would be left, in spite of the fact that it had been emitting rays all the time, and heat at the rate of 132 calories per hour.

Radon, however, has a half-life of only 3.825 days. It runs down quite rapidly. In the treatment of cancer, radon is usually used instead of radium itself. There are two reasons for this: First, it is a gas, and small amounts can be sealed in a glass tube; and secondly, it is not likely to overdo the treatment.

When a beta particle is expelled, there is very little change in mass. The negative electron weighs only $\frac{1}{1845}$ of the weight of the hydrogen atom. A long series of changes has been worked out to show how uranium can begin a series that is eventually supposed to end in lead. Some of the changes require very long times and are quite hypothetical. Others requiring short times are very well established. Thorium also is thought to go through a similar series of transformations.

Dalton's atomic theory changed. Dalton, in 1808, built his atomic hypothesis on the supposedly established fact that the atoms of each element are simple units of matter and unchangeable. This idea of the indestructible and unchangeable atom held unchallenged sway from 1808 to about 1900, when radioactivity began to be understood.

Dalton's theory was built upon the experience of all scientists up to his time. In the days of the alchemists, when facts were being learned which later developed into the science of chemistry, it was thought that one element might be changed into another. Above all, it was hoped that the cheaper metals might be changed into gold. There began a long series of attempts by the alchemists to make gold from other substances, and a few of the pioneer investigators stumbled onto alloys having some resemblance to gold. They thought they had actually succeeded in their quest and left written formulas for making gold. Most of the alchemists, however, soon got mixed up with magic and mysticism and tried to hoodwink others into thinking that they had succeeded in the quest of a method for making gold.

Time eventually weeded out the hocus-pocus and sifted all the evidence down to the bare fact that not one had succeeded in synthesizing gold from something that did not already contain gold. Out of centuries of effort came the conclusion that it is impossible to change one element into another. After this theory of the unchangeable nature of the atoms had held unchallenged sway for a century, radioactivity proved it false.

Atoms are complex. Studies of radioactivity have given us a new understanding of the nature of the atom. Dalton's idea of the atom as a single piece of a simple substance has been replaced by the picture of a complex combination of positive and negative charges of electricity. In the long series of radioactive transformations following the disintegration of radium (see figure, p. 376) five helium atoms, each consisting of four positive charges and two negative charges, are expelled. Besides these in the beta rays, a series of electrons also come out of the radioactive atoms. The atomic weight of lead, the final product of one radioactive series, is 206; and lead is probably itself quite complex. In going from radium, atomic weight 226, to lead, atomic weight 206, many charges have escaped. If 20 positive charges and an equal number of negative charges can account for the 20 units of the atomic weight of radium which are lost in radioactive transformations, there is every reason to suspect that positive and negative charges might also account for the rest of its atomic weight and also for the weights of all the other atoms. Recently atoms of smaller atomic weight, such as sodium, have been artificially made radioactive, which supports the supposition that other atoms than those of the naturally radioactive elements are complex. We now believe that the atoms of all the elements are more or less complex and are composed of positive and negative charges of electricity.

REVIEW QUESTIONS

1. Describe how it is proved that radium emits three kinds of rays.

2. What are the three rays?

3. What happens to radium when it loses an alpha particle?

4. Can the rate of radioactive change be speeded up or slowed down?

5. What is supposed to be the final decomposition product of radium?

6. Why do we now think atoms are complex in structure?

7. How is it proved that the particles in the alpha and beta rays carry electrical charges?

8. Explain what is meant by the half-life of a radioactive substance?

9. Does radioactivity reverse the conclusion reached by the alchemists and the early chemists that one element cannot be changed into another?

10. Compare the mass of an alpha particle and that of the hydrogen atom.

11. Compare the mass of a beta particle and that of a hydrogen atom.

12. Compare the charge carried by an alpha particle with that carried by a hydrogen ion.

13. Compare the charge carried by a beta particle with that carried by a chloride ion. (pp. 204, 208)

14. Compare the charge carried by a beta particle and each of the following: (a) proton; (b) electron; (c) positron. (p. 204)

Summary

The properties shown by radium, such as exposing photographic plates through the plate holders, discharging electroscopes, killing cancerous cells, heating a thermometer, penetrating metals, and causing minerals to fluoresce, are called radioactivity. Except under a doctor's supervision, radioactive substances should not be taken into the human body.

There are three rays given off from radioactive substances — alpha rays, positively charged helium atoms; beta rays, negative electrons; and gamma rays, which are short X rays.

Radium of atomic weight 226 expels an alpha particle and becomes radon of atomic weight 222. Similar transformations are believed to take place until the final product becomes lead, of atomic weight 206.

In a limited way the alchemists' dream of transmuting one element into another has been fulfilled. Dalton's theory of the unchangeable atom has also been overthrown. This conception of complex atoms has resulted in entirely new conceptions of the fundamental nature of matter.

CHAPTER 20

THE PERIODIC LAW OF THE ELEMENTS

Introduction. The study of the elements has been greatly simplified by the fact that they can be collected into groups or families having similar chemical properties. Although the chemical reactions are similar for the members of any family of elements, such as the halogens, the alkali metals, or the alkaline-earth metals, there is considerable variation within the family in the readiness or speed with which the reactions take place. Family likenesses and differences are also characteristic of the other properties of the elements.

Dobereiner's triads. Back in 1829 Dobereiner noticed that in many of the families of closely related elements the variation in the properties was in the same order as the variation in their atomic weights. He studied several groups of three elements, and found that the element whose atomic weight was between the other two had properties that were between those of the other two. Take the halogen family, for example. Chlorine has an atomic weight of 35.5, bromine of 80, and iodine of 127. Chlorine is the most reactive, then bromine, while iodine is least. The hydrogen compound of chlorine is most stable, that of bromine intermediate, and that of iodine least stable. The physical properties of these elements also vary in the same way. Chlorine is the least dense and is gaseous at ordinary temperatures. Bromine is next and is a liquid. Iodine is the most dense, hence ordinarily a solid. Intensity of color varies also in the same order as the atomic weights, from pale green through red to deep violet. In the case of the alkali metals all of their chemical and physical properties vary in the same order as their atomic weights — lithium 7, sodium 23, and potassium 39. This gave scientists the first hint that perhaps in some way the atomic weight of an element determines its properties.

Newlands's octave arrangement of the elements. The idea of the members within a family of elements varying as the atomic weight increased seemed quite natural. About 1864, however, an English chemist named Newlands discovered among the elements of the different families a significant variation that also seemed to be determined by their atomic weights. He took the following group of elements as illustrating his discovery :

Li 7   Be 9   B 11   C 12   N 14   O 16   F 19

These elements are arranged in the order of increasing atomic weights. They include all of the elements whose atomic weights are neither less than 7 nor more than 19. Their chemical differences and the differences in their compounds can be arranged in the same order. For instance, lithium hydroxide is a very strong base, beryllium hydroxide is a weak base, boron hydroxide a weak acid, and the hydroxides of the rest, consecutively stronger acids. Valence towards oxygen is least (one) with lithium and increases one bond at a time until fluorine has a valence of seven.

The next remarkable set of facts discovered by Newlands, however, is that sodium, whose atomic weight (23) is slightly above that of fluorine (19), has properties almost identical with those of lithium (7).; and magnesium, the element of next higher atomic weight, closely resembles beryllium. The next five elements complete the second octave, which is quite similar to the first, each element in the second group having properties parallel to those of the element immediately above it. Newlands was able to arrange three such octaves.

Li	7	Be	9	B 11	C 12	N 14	O 16	F 19
Na 23		Mg 24		Al 27	Si 28	P 31	S 32	Cl 35
K	39	Ca	40	Sc 45	Ti 48	V 51	Cr 52	

It will be noticed that such an arrangement places the elements of the same family one above the other.

The periodic law. About 1870 the Russian chemist Mendeléeff extended Newlands's idea to include all the elements in a much more complete scheme which we now call the *periodic*

Dmitri
Ivanovitch
Mendeléeff

1834–1907

Dmitri Mendeléeff, the young-
est of seventeen children, was
born in Tobolsk, Siberia. Shortly
after his birth his father, who
was director of the local high
school, became blind and soon
died. His mother opened a glass
factory to support the family. After the glass factory burned, his
mother went to St. Petersburg, where she put young Dmitri in the
Science Department of the Pedagogical Institute, where he specialized in mathematics, physics, and chemistry.

After graduating at the head of his class, Mendeléeff went to
South Russia for his health. He taught a few years and then went
to France and Germany for advanced study. After a year abroad
he returned to Russia and married. At this time he wrote an
organic chemistry text, which won him a prize. Soon he won his
doctor's degree, and at the age of thirty-two was made full professor
at the University of St. Petersburg.

In 1869, after twenty years spent studying the sixty-three elements then known and cataloguing their properties, he announced
the periodic law and predicted three new elements, whose properties
he described and which he named eka-aluminum, ekasilicon, and
ekaboron. The scientific world was skeptical of the periodic
arrangement, and the idea of describing the properties of elements
still undiscovered seemed preposterous. Nevertheless, in 1875 a
Frenchman named Boisbaudran discovered an element (gallium) with
exactly the properties predicted for eka-aluminum. Two years later
Nilsen in Scandinavia discovered another element (scandium) with
the properties of ekaboron. Finally Winkler in Germany found a
third element (germanium) with the properties Mendeléeff had given
to ekasilicon. By this time, the scientists were thoroughly convinced that Mendeléeff's periodic table and law were really scientific.

arrangement of the elements or the *periodic table*. He was able to put the facts in a concise form, which is now called the *periodic law*: *The properties of the elements are periodic functions of their atomic weights.*

This means that within a period the properties of the elements vary in the same order as their atomic weights. At the end of a period, however, there is an abrupt change in properties, so that the next succession of elements repeats the period. After all of the periods are placed one below another, each vertical row contains chemical families of similar properties.

At the time that the periodic table was first worked out by Mendeléeff, there were several vacant places in the series; that is, the elements had not yet been discovered. Mendeléeff noticed that there was a vacant space in the vertical column under boron and aluminum. From the properties of aluminum above the missing element (which he called eka-aluminum), of zinc on the horizontal row to the left, and of arsenic on the right, he was able to estimate the properties of the unknown element. Since then a metal has been discovered with properties remarkably like those predicted for eka-aluminum. The estimated atomic weight was 69; so was the real atomic weight. The predicted density was 5.9; the actual density was 5.93. The new metal was called gallium. Its predicted chemical properties agreed almost as closely with those found as did the physical properties just mentioned.

Two other metals, scandium and germanium, were described quite closely before they were discovered, by the use of the periodic arrangement of the elements.

The periodic chart also serves as a rough check upon the atomic weights of the elements. If the atomic weight of any element throws it into a column where its properties are not between those of its immediate neighbors, both horizontally and vertically, there is a suspicion that its atomic weight is not correct, due usually to the fact that the element has not been obtained pure. A few of the irregular atomic weights have been corrected. However, there are still two or three

PERIODIC ARRANGEMENT OF THE ELEMENTS ACCORDING TO ATOMIC NUMBERS [1]

Group	0	I	II	III	IV	V	VI	VII	VIII
Type of Hydride		RH	RH_2	RH_3	RH_4	RH_3	RH_2	RH	
Type of Oxide		R_2O	RO	R_2O_3	RO_2	R_2O_5	RO_3	R_2O_7	RO_4
Series 1	He 2 — 4.002								
2	Ne 10 — 20.183	Li 3 — 6.940	Be 4 — 9.02	B 5 — 10.82	C 6 — 12.00	N 7 — 14.008	O 8 — 16.0000	F 9 — 19.00	
3	A 18 — 39.944	Na 11 — 22.997	Mg 12 — 24.32	Al 13 — 26.97	Si 14 — 28.06	P 15 — 31.02	S 16 — 32.06	Cl 17 — 35.457	
4		K 19 — 39.096	Ca 20 — 40.08	Sc 21 — 45.10	Ti 22 — 47.90	V 23 — 50.95	Cr 24 — 52.01	Mn 25 — 54.93	Fe 26 55.84 · Co 27 58.94 · Ni 28 58.69
5	Kr 36 — 83.7	Cu 29 — 63.57	Zn 30 — 65.38	Ga 31 — 69.72	Ge 32 — 72.60	As 33 — 74.91	Se 34 — 78.96	Br 35 — 79.916	
6		Rb 37 — 85.44	Sr 38 — 87.63	Y 39 — 88.92	Zr 40 — 91.22	Cb 41 — 92.91	Mo 42 — 96.0	Ma 43 — ?	Ru 44 101.7 · Rh 45 102.91 · Pd 46 106.7
7	Xe 54 — 131.3	Ag 47 — 107.880	Cd 48 — 112.41	In 49 — 114.76	Sn 50 — 118.70	Sb 51 — 121.76	Te 52 — 127.61	I 53 — 126.92	
8		Cs 55 — 132.91	Ba 56 — 137.36	Rare Earths 57-71	Hf 72 — 178.6	Ta 73 — 181.4	W 74 — 184.0	Re 75 — 186.31	Os 76 191.5 · Ir 77 193.1 · Pt 78 195.23
9	Rn 86 — 222	Au 79 — 197.2	Hg 80 — 200.61	Tl 81 — 204.39	Pb 82 — 207.22	Bi 83 — 209.00	Po 84 — 210	? 85 — ?	
		? 87 — ?	Ra 88 — 225.97	Ac 89 — ?	Th 90 — 232.12	Pa 91 — 231	U 92 — 238.14		

[1] This arrangement does not provide for hydrogen (H 1 1.0078) nor for the rare-earth elements (57-71). The discovery of numbers 85 and 87 has been reported but awaits confirmation.

385

elements whose atomic weights refuse to be corrected so as to make them consistent with the periodic table.

The periodic arrangement has served its purpose as regards the prediction and discovery of new elements. Most of the elements have been discovered and quite thoroughly studied.

QUESTIONS OF FACT

1. What fact has greatly simplified the study of the elements?

2. Who first noticed that in groups of related elements the variation in individual properties is in the same order as the magnitudes of their atomic weights?

3. When did Dobereiner discover the facts mentioned in question 2?

4. Arrange the related elements chlorine, iodine, and bromine in the order of increasing atomic weights.

5. Compare the following properties of the elements chlorine, bromine, and iodine in the order of increasing atomic weights: (a) tendency to react; (b) stability of the hydrogen compound; (c) density.

6. Arrange the alkali metals sodium, potassium, and lithium in the order of increasing atomic weights. (a) Arranged thus, lithium is least reactive; which is most reactive? (b) Potassium is the densest of the three metals; which is the least dense?

7. When did Newlands make a further discovery as to the relationship between the properties of the elements and their atomic weights?

8. Did Newlands's arrangement of elements extend beyond the ordinary families of elements?

9. In Newlands's arrangement do the properties of the elements vary in the same order as magnitude of atomic weight?

10. Do the properties at the end of the octave abruptly jump back to those of the beginning of the octave so that the second octave almost duplicates the first?

11. When did Mendeléeff invent a complete periodic arrangement of the then known elements?

12. State the periodic law.

13. Describe two services performed by the periodic arrangement of the elements.

14. Compare the predicted properties of eka-aluminum with the actual properties of gallium.

Henry G. J.
Moseley

1887–1915

At the age of thirteen Henry Moseley entered Eton with a King's scholarship. Five years later he entered Trinity College, Oxford, and began studying under Rutherford. After working upon a few short problems in radioactivity, he began experimenting with X rays upon crystals. Out of this study came one of the most astounding discoveries of this century.

When a target is bombarded with high-speed electrons, it emits a beam of X rays which is characteristic of the substance of which the target is made. When such X rays from various elements are passed through a crystal, there is a simple yet wonderful relation between the frequencies of the X rays and the atomic numbers of the elements.

Moseley's X-ray spectra arranged the elements in almost the same order as Mendeléeff's table but with no irregularities and a place for every element — including the rare earth metals, which did not fit the other system. Moseley's system also told definitely how many elements remained to be discovered.

Shortly after this discovery, Moseley enlisted in the World War and was killed at Gallipoli. Dr. Millikan said of Moseley: "In a research which is destined to rank as one of the dozen most brilliant in conception, skillful in execution, and illuminating in results in the history of science, a young man twenty-six years old threw open the windows through which we can glimpse the subatomic world with a definiteness and certainty never dreamed of before. Had the European War had no other result than the snuffing out of this young life, that alone would make it one of the most hideous and most irreparable crimes in history."

Even though Moseley's scientific career was of the shortest duration of that of any famous scientist, his fame is nevertheless secure as the result of one of the most significant discoveries of all time.

QUESTIONS OF UNDERSTANDING

1. From your study of the periodic table, which element or elements would you say are most like each of the following:

(a) F 9	(c) C 6	(e) Na 11	(g) Ir 77
(b) Li 3	(d) N 7	(f) Ca 27	(h) Ne 10

2. Calculate the average of the atomic weights of calcium and titanium and compare it with that of scandium.

3. Compare Newlands's octaves with Mendeléeff's series.

4. How does a family of elements occur in the arrangement of Mendeléeff?

5. Do the facts discovered by Dobereiner fit according to Mendeléeff's arrangement?

REFERENCE FOR SUPPLEMENTARY READING

1. Jaffe, Bernard. *Crucibles.* Simon and Schuster, New York, 1930. Chap. XI, pp. 199–218, "Mendeléeff."
   (1) Write a fuller biography of Mendeléeff than that given in the text.

Present uses of the periodic system. The periodic system is still useful. It helps the student of chemistry to correlate, compare, and remember the properties of the elements. To remember with exactness all the many properties of all the elements is a task which no one person would care to try. However, if he knows the position of a given element in the periodic table and is familiar with several neighboring elements, he can usually estimate the properties of the one he wishes to learn. The periodic system also has enabled the chemist to find compounds with certain definite desirable properties. For instance, Thomas Midgely, Jr., in his search for an antiknock compound for gasoline was able to predict which compounds would be the best antiknock materials. He predicted that tetraethyl lead would be the best of all the possible compounds before it had been tried. All his predictions were based on a study of the periodic arrangement. Again, in their search for an ideal liquid refrigerant for electric refrigerators, Midgely and his assistants were able in three days of intensive study of the progressive variation of the compounds of elements arranged in periodic order to settle upon dichloro-difluoro-

B. Smith Hopkins

1873–

Professor Hopkins began his college training at Albion College, at which he got his Bachelor of Arts degree in 1896 and his Master's degree the next year. Continuing his training at Johns Hopkins University, he obtained the Doctor's degree in 1906. He began teaching at Wesleyan University, then changed to Carrol College, and finally to the University of Illinois, where he still is, with the rank of full professor, having rapidly climbed the ladder of advancement from instructor through assistant professor, and associate professor.

Doctor Hopkins' research on osmotic pressure and the properties of a group of rare elements called the rare earths has attracted much attention. Of these he determined the atomic weights of yttrium, erbium, dysprosium, and holmium. In addition to these he studied the properties of beryllium, selenium, and tellurium.

Another subject of his researches is luminescence, or the formation of cold light by chemical action.

The work for which Hopkins is most famous is his discovery of a new element which he named illinium, in honor of his university and state, Illinois. This achievement makes him the first American to discover a new element.

Illinium was discovered in monasite sand in which others of the rare earth metals occur. By painstaking fractional crystallization of the rare earths, Hopkins was able to concentrate the new element in one fraction. Moseley's X-ray spectra had showed a vacant place which proved that there was another element in the the rare earth group with properties intermediate between those of neodymium and samarium. Because of this, several chemists were looking for this element. Hopkins won the race because his element gave the new spectral lines of the nature expected and an X-ray spectra which filled the gap in Moseley's series.

methane, a nonpoisonous, noninflammable, easily vaporized liquid.

Defects of the periodic classification of the elements. The periodic table is far from perfect. It has no place for hydrogen nor for the fifteen related elements known as the rare earth metals. It also throws elements together that are not so similar as one would expect. For instance, copper, silver, and gold form one branch of Group I, of which lithium, sodium, and potassium form the other branch. In reality, the differences between these two groups of metals are more numerous than the similarities. Group VIII also is a sort of appendage and does not fit into the system. Finally, a strict arrangement according to atomic weights would interchange a few pairs of elements, putting them in families where they do not belong according to their atomic weights. In such instances it has been the policy to follow properties rather than atomic weights in placing the elements. Argon and potassium are an illustration of such an *inversion*. There have been dozens of attempts to work out a perfect periodic table, but they have all failed. Mendeléeff's scheme is about as good as any of them and is better known; hence it is retained with all its imperfections for want of a better system.

Atomic numbers. A young Englishman named Moseley invented an X-ray spectroscope. With it he made a fundamental discovery about the elements. Each element when bombarded with high-speed electrons was found to emit X rays of a pattern characteristic of the element — that is, different from that of any other element. When the spectral lines were plotted in a certain way, it was found that each element gave characteristic lines in a definite place in the complete spectral field. It was also found that the position of these spectral lines arranged the first eighteen elements in the same order as the atomic weights. Among the elements of higher atomic weight the new classification agreed with the atomic weights except in the case of the inversions noted in the preceding paragraph. In all such cases the new arrangement agreed with the properties rather than with the atomic weights. The elements are now numbered according to their position in this row of spectral lines, beginning with H = 1, He = 2, Li = 3, etc. There are no irregularities in this arrangement. The position of each element, or number designating its position in the line, is called its *atomic number*. Atomic number is thus of more importance in any periodic classification than the atomic weight.

Moseley's arrangements told us exactly how many elements there were to be discovered by the gaps in the lines. Below the element uranium of highest known atomic weight there were only about five gaps. Since Moseley's discovery, all of these missing elements have definitely been discovered except elements of atomic numbers 85 and 87, and these have been reported spectroscopically.

MOSELEY'S X-RAY SPECTRA. The atomic numbers of the elements are determined by the positions of their X-ray spectra. Moseley found that the gaps in his series of lines corresponded to elements not yet discovered.

Isotopes. A large number of atomic weights are practically whole numbers; for instance, carbon is 12, nitrogen 14, oxygen 16, and fluorine 19. Others, such as chlorine 35.5, are definitely not whole numbers. The fact that so many of them are whole numbers has led chemists for a long time to suspect that there may be some fundamental reason why all atomic weights should be exact whole numbers. Experiment seemed to be against them, however, in those elements like chlorine, which when carefully purified refused to give an integral atomic weight.

The English scientist Aston worked out the solution of this problem. He discovered that the elements whose atomic weights are not whole numbers are mixtures of atoms of the element having different atomic weights but the same chemical properties. For instance, chlorine is a mixture of chlorine atoms of atomic weight 35 and 37, there being just enough of the heavier atoms to make the mixture have an atomic weight of 35.5.   Hydrogen could never be purified so its atomic weight was exactly 1 when compared with oxygen as 16.   Recently it has been discovered'that there are some hydrogen atoms of atomic weight 2 and a few of atomic weight 3.   Where an element possesses atoms of different atomic weights, the different atoms are called isotopes. Many of the elements exist in isotopes. Water made of the heavy hydrogen isotope is called deuterium oxide.   Deuterium oxide has been found in a great many natural waters. Scientists have been able to concentrate deuterium oxide (heavy water) and have found it quite different from ordinary water. Many types of life do not do well when deuterium oxide is substituted for

SUPPOSED POISONOUS ACTION OF DEUTERIUM OXIDE.   The preliminary experiments seemed to indicate that heavy water is very poisonous to goldfish and other forms of life.   Later experiments indicate that the poisonous properties of the oxide were exaggerated.

ordinary water.   Preliminary experiments seem to show that it is somewhat poisonous to goldfish.

QUESTIONS OF FACT

1. In what way is the periodic arrangement of the elements of value to chemical students today?

2. Relate two instances where the periodic system of the elements enabled a chemist to prepare substances with specified desirable properties.

3. State the faults of this arrangement.

4. How did Moseley show how many elements there are below uranium?

5. What are atomic numbers?

6. What are isotopes?

7. What do isotopes reveal about fractional atomic weights?

Harold
Clayton Urey

1893–

Professor Harold Clayton Urey, now teaching at Columbia University, received his Bachelor of Science degree at the University of Montana in 1917. Taking graduate work at the University of California, he achieved his Doctor's degree in 1923. After this he taught two years at the University of Montana, his first alma mater. Then, getting an American-Scandinavian Foundation Fellowship, he studied a year at the University of Copenhagen. Returning to the United States, he taught until 1929 at Johns Hopkins University, after which he came to Columbia.

Professor Urey's research work began with a study of gases, then extended to the study of atomic structure, and then to the study of the absorption of light and its meaning in terms of the structures of the molecules.

The discovery which has made Urey famous is heavy hydrogen, a hydrogen isotope with twice the usual atomic weight. . This heavy hydrogen has been named deuterium. In this connection he discovered heavy water, or deuterium oxide. Deuterium oxide is a molecule of water each of the hydrogen atoms of which has been replaced by a deuterium atom. As a consequence of the fact that deuterium is twice as heavy as hydrogen, the oxide has a greater density than ordinary water and other differences in properties.

Urey's method involved evaporating large quantities of liquid hydrogen. The lighter isotope escaped first, leaving the heavier one. Its difference from ordinary hydrogen was shown by its spectrum, obtained by making the gas glow under an electrical discharge.

The discovery of deuterium started a large number of research projects for preparing other compounds of deuterium by replacing hydrogen atoms with deuterium atoms. The discovery of deuterium brought Urey the Nobel prize for Chemistry.

8. What is the name of the isotope of hydrogen with atomic weight 2?

9. What is the meaning of the term "heavy water," and what is its chemical name?

QUESTIONS OF UNDERSTANDING

1. Locate two places in the periodic system where the elements are not arranged according to ascending atomic weights.

2. Pick out the element in the following most like aluminum:

A   N   B   P   Cl   Zn

3. Rearrange the following elements into a period:

S   Ne   Mg   Na   Cl   Si   Al   S

4. Tell how Moseley's atomic series is superior to the periodic arrangement of Mendeléeff in the following respects: (a) elements not in the expected order; (b) elements without a position in the system.

5. How do the chemical reactions of the isotopes of an element compare?

QUESTIONS FOR SUPERIOR STUDENTS

1. What is the regular valence of the elements in the family number 1? (See periodic table, p. 385.)

2. What two elements in Group I have other valences than 1?

3. The rare metal masurium (Ma) will be most like what common element?

4. Name some marked differences between the first branch of Group I (Li, Na, K, Rb, Cs) and the second branch (Cu, Ag, Au).

5. Study one of the other periodic arrangements of the elements and note any advantages over the Mendeléeff arrangement.

References: (1) *Journal of Chemical Education*, vol. 14, no. 5, pp. 232–235. (2) Any college chemistry.

SUMMARY

In 1829 it was noticed that within a chemical family whose members were arranged in the order of increasing atomic weights the properties of any intermediate element were between those of the preceding and the succeeding elements. In 1864 Newlands noticed that there was a gradual relationship between the properties of the different chemical families which also varied with the order of increasing atomic weights and that after a period of several elements there is an abrupt change in properties to the beginning of a new period. In 1870 Mendeléeff worked out a complete system including all of the known elements and places for other elements to be discovered. He also stated the periodic law — the properties of the elements are periodic functions of their atomic weights. This system predicted the discovery of new elements, corrected the atomic weights of others, and correlated the properties of all of them.

More recently, Moseley, by means of X-ray spectra, arranged the elements in a row in nearly the same order as that determined by the periodic system but more definitely as to the relative positions of the doubtful elements and as to the number of elements yet to be discovered. The position of each element in this sequence is called its atomic number.

Fractional atomic weights were proved by Aston to be due to mixtures of isotopes — that is, atoms of the same element with different atomic weights. Every atom has an atomic weight expressible by an integer.

⇆

CHAPTER 21

ATOMIC STRUCTURE

Matter composed of electricity. It was brought out in Chapter 19 that the alpha rays emitted from radioactive substances are positively charged and that the beta rays are composed of negative electrons. This is in harmony with the conclusion expressed in Chapter 11 that all matter is electrical. In that chapter was described the charging of glass and rubber rods by rubbing, which pulled the charges apart so they no longer neutralized each other.

The electrical theory of matter also explains the action of the photo cell, in which an electric current is caused by light striking an alkali metal. The light simply jerks out some electrons already within the metal, throwing them across a gap, thereby causing them to return through the rest of the circuit. Moving electrons are an electric current. Ionization of electrolytes into equal numbers of negatively and positively charged particles also agrees with the idea of the electrical nature of matter.

Atomic models. The idea of a complex atom composed of heavy positive charges with light negative electrons rotating around them stimulated the imaginations of some deep thinkers, and they began to invent atomic models. The model that has succeeded in explaining the largest number of facts is that of Niels Bohr.

The value of the Bohr atom. It should be remembered that all atomic models are largely imaginative and lack the solid basis of fact of other chemical theories and conceptions. The theory of the structure of the atom is useful, however, for correlating a large number of facts and providing a mechanism for phenomena not explainable in any other way. Moreover, this theory has enabled chemists to predict and discover a few spectral lines not already known; it suggests the force that

Robert A. Millikan

1868–

As a youth Robert Millikan showed considerable athletic ability and partially worked his way through Oberlin College by acting as assistant in the gymnasium. Not being interested particularly in science, Millikan was taking a rather general course. One subject in which he did good work was Greek. One day his Greek professor called him in and said: "I have been put in charge of the introductory department, Millikan, and I want you to teach the elementary physics class."

In answer to Millikan's reply that he knew very little physics, the professor said: "You have done excellent work all year in my Greek class; I'll risk anyone who can do what you have done in that subject to teach physics."

Having to learn physics as he taught it caused Millikan to develop great interest in the subject. He put in as much time studying it as he could.

After graduation, he served as tutor for a while. Then getting a fellowship, he continued to study at Columbia University, where he got his Doctor's degree in 1895. Then followed some advanced study in Germany, after which he took a position at the University of Chicago.

Although primarily a physicist, Millikan did his greatest piece of work in measuring the charge of the electron, proving it to be a definite quantity and not a statistical average. This information is of as much value to the chemist as to the physicist, since electric charges play such an important role in chemistry. For this research he was given the Nobel Prize of $40,000.

One of Millikan's more recent studies has to do with the cosmic rays, very penetrating radiations from outer space. Millikan ranks as one of the outstanding American scientists and is recognized by scientists all over the world as a leader in his field.

holds the atoms together in chemical reaction; it gives a reason for the periodic law; it makes valence more real; and it accounts for the properties of an element.

The structure-of-the-atom theory. (1) All atoms are composed of protons and electrons in equal numbers, the number of both protons and electrons being equal to the number expressing the atomic weight.

(2) All of the protons are packed tightly into a central nucleus, cemented together by about half of the electrons. Another way of thinking of the nucleus is to consider the electrons as paired with protons to form neutrons. The nucleus, then, would consist of protons and neutrons.

(3) The rest of the electrons are revolving in one or more orbits about the nucleus.

(4) The orbits are fixed in position and no electron can rotate anywhere between orbits.

This picture of the atom is something like our solar system. The nucleus, in which is most of the mass, corresponds to the sun. The electrons represent the planets, differing in that there may be several electrons in one orbit in an atom while there is only one planet in each orbit in the solar system. Another difference between the proposed structure of the atoms and the solar system is that in the solar system the sun has a very large volume compared with the planets, while in the Bohr atom, even though the nucleus is supposed to contain most of the mass, its entire volume is supposed to be less than that of an electron.

Let us think of the orbit of Mercury (the innermost planet) as representing the innermost orbit of an atom; if other planets might drop into this orbit but be unable to get closer to the sun, we have a picture of the structure of the atom. In the case of the atom, never more than two electrons can occupy the innermost orbit at the same time, but as many as eight electrons can get into the next orbit — if there are that many electrons. The third orbit will hold up to eighteen electrons. The remaining orbits are not well worked out. The structures of the lighter atoms are outlined in the following scheme:

The hydrogen atom, the lightest of all, is thought to consist of only one proton as the nucleus and one electron rotating about it in the first orbit.

1 Hydrogen

Helium, the second lightest element, whose atomic number is two, has a nucleus of four protons, tied together with two electrons, which leaves two orbital electrons to revolve in the first orbit. These two electrons complete this orbit. The helium nucleus might also be thought of as having two protons and two neutrons.

2. Helium

In lithium, atomic number three and atomic weight seven, the atom consists of seven protons tied together into a compact nucleus with four electrons. Besides these four tying electrons there are three more, two completing the first orbit and one beginning a new orbit. The nucleus may also be thought of as four neutrons and three protons.

3. Lithium

Beryllium, of atomic number four, has its atomic weight of nine protons tied into the nucleus with five electrons. Of the other four electrons necessary to make a neutral atom, two are in the first completed orbit and two in the second orbit.

4 Beryllium

The fifth element, boron, in the atomic number row has an atomic weight of eleven. Its eleven protons require six electrons to tie them together as the nucleus. Its five other electrons fill the first orbit with two, and supply the next orbit with one more electron than occurs in the previous element, that is, three.

5. Boron

The carbon atom's twelve protons also need six electrons to tie them together. Since it has also twelve electrons, the second orbit can have one more electron than the atom of one lower atomic number, hence it contains four.

6. Carbon

We have discussed the first six elements in ascending atomic numbers in some detail. This is perhaps enough to show how the structure of an atom is related to that of the element immediately preceding it and to that of the one following it. *Each atom has one more electron outside of the nucleus than the atom of one less atomic number.* Several of the succeeding atomic models are now given in a group.

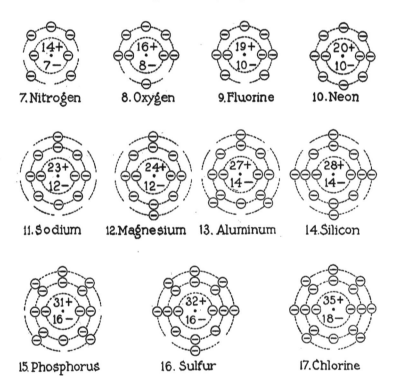

7. Nitrogen   8. Oxygen   9. Fluorine   10. Neon

11. Sodium   12. Magnesium   13. Aluminum   14. Silicon

15. Phosphorus   16. Sulfur   17. Chlorine

A close examination of the foregoing atomic models shows that they all agree with the four main statements of the structure theory of page 398. Another statement might be added to the four already given : *The number of electrons outside of the nucleus is always equal to the atomic number, and increases by one from each atom to the next.*

QUESTIONS OF FACT

1. What conclusion grew out of the attempts of the alchemists to change the cheaper metals to gold? (p. 378)

2. Approximately how long did Dalton's idea of the unchangeable atom hold sway? (p. 378)

3. What new conception of the atom was suggested by the study of radioactivity? (pp. 378–379)

4. What facts suggest that all matter may be composed of electricity?

5. Give a description of the Bohr atomic model.

6. What excuse is there for atomic-structure theories if they are partly imaginative?

7. Give the four suppositions in the structure-of-the-atom theory.

Which of the next five statements are true and which are false?

8. All the electrons in an atom are revolving about the nucleus.

9. The first orbit can never hold more than two electrons.

10. Hydrogen is the only atom that does not have the first orbit filled.

11. Sulfur has the first two orbits filled and the third over half filled.

12. Draw the atomic structures of each of the first eight elements of lowest atomic numbers. Do as much as you can from the rules without looking at the text.

QUESTIONS OF UNDERSTANDING

1. What element has the first orbit filled and the second one with four electrons?

2. Phosphorus has the first two orbits filled and how many electrons in the third?

3. Neon has how many cementing electrons in the nucleus?

4. Magnesium has how many electrons in the second orbit?

5. The way to determine the number of electrons in the nucleus is to subtract the sum of the electrons in the different orbits from the atomic weight. Is this statement true or false?

6. State the rule which tells the relation between the number of the revolving electrons in all the orbits and the atomic number of an element. Apply the rule to nitrogen.

7. Cesium should have how many electrons in the outer orbit?

8. Iodine should have how many electrons in the outer orbit?

9. The atomic number is the same as the sum of all the electrons in what part of the atom?

10. What is the only element whose first orbit is not filled?

11. Applying the rule given at the bottom of page 400, how many electrons are outside the nuclei of elements whose atomic numbers are: (a) 40; (b) 32; (c) 70; (d) 35; (e) 19.

12. Applying the same rule as in ex. 11 and looking up the atomic numbers in the back of the book, find how many electrons are outside the nucleus of the following elements: (a) radium; (b) iron; (c) chromiun; (d) iodine; (e) calcium; (f) bromine.

ADDITIONAL EXERCISE FOR SUPERIOR STUDENTS

1. With the periodic table before you, draw structural pictures of the following atoms, similar to those shown in the text: (a) argon, A; (b) potassium, K; and (c) calcium, Ca.

Atomic structural theory and whole number atomic weights. In 1815 Prout expressed a hypothesis that there should be no fractional atomic weights; hydrogen should be exactly one and all others exact whole numbers. This was little more than a guess on the part of Prout. He suggested that possibly hydrogen was the fundamental substance and the other atoms were complex structures built up from the hydrogen units. In the last chapter we learned that the discovery of isotopes has shown that their atomic weights *are* whole numbers. The rest of Prout's guess has not exactly been fulfilled. Recently electrical discharges of approximately a million volts have been used to speed up charged particles in attempts to break down the atoms of some of the elements. These high-speed bullets have succeeded in striking some of the nuclei of these elements and decomposing them into simpler elements. For instance, lithium has seemingly been decomposed into enough hydrogen and helium to show in the spectroscope. The reason so little decomposition results from these attempts is that the nucleus is such a small part of the atom that it is hard to hit. Thus Prout's guess has turned out to be almost right.

The atomic number. It is evident from the preceding discussion that the sum of all of the electrons outside of the nucleus equals the atomic number. Another way of expressing the same thing is to say the atomic number is equal to the excess of positive charges over negative charges in the nucleus of the atom.   When an atom becomes an ion, it is thought to have lost or gained an electron.   In this case, the first way of finding atomic number would no longer work, but the second method would.

Elements of higher atomic weight.   The greatest fault of the atomic structure theory is that no one has been able to work out an entirely satisfactory structure for the elements of higher atomic weights.   The third orbit is assumed to be filled with eight electrons, but for elements of higher atomic weights it is necessary to enlarge the third orbit to harmonize properties and structure.   This is not very satisfactory, as it is not evident when to crowd more electrons into inner orbits.   Ordinarily, in the atoms of smaller atomic weights each new electron is added to the outer orbit.   However, when we try to work out a structure for some of the elements of larger atomic weights we must think of electrons as having added to orbits other than the outermost.

Valence and atomic models.   In explaining valence, chemical affinity, and chemical reaction, some additional assumptions are necessary :

Assumption I.   The chemical properties of an element are determined by the number of electrons in the outer electron orbit.

If the outer orbit is entirely filled, the element is inactive. Helium has only the first orbit, which is completely filled with two electrons.   Since the outermost orbit is filled, helium does not react.   No helium compounds are known.   In the periodic system, a new group or family of elements (Group 0) has been added since the time of Mendeléeff.   It is made up of elements like helium which do not react or form compounds.

Neon, the second member of this group, has the second orbit filled.   Then come argon, krypton, and xenon with the third, fourth, and fifth orbits filled in turn.

Group I of the periodic classification of the elements contains lithium, sodium, potassium, rubidium, and cesium.   The elements were placed in the same family or group because their chemical reactions are practically the same. Atomic structure tells us why the chemical properties are the same. All have only one electron in the outer orbit. The inner completed orbits have little influence on chemical properties. Whether there is one completed orbit as in the case of lithium, two completed orbits as in sodium, or three completed as in potassium makes little chemical difference; the one electron in the last orbit gives them all the same chemical nature, that of the alkali metals.

Assumption II. The valence of an element is determined by the number of electrons in the outer orbit.

The elements of Group I all have one electron in the outer orbit, and all have a valence of one.

In Group II, beryllium, magnesium, calcium, strontium, and barium all have two electrons in the unfinished orbit. The valence of each element is determined by the electrons in this last orbit, as is also its chemical nature. Two electrons in the outer orbit mean a positive valence of two and a character less basic than the alkali metals, yet active enough to displace hydrogen from water.

The members of Group III as typified by aluminum have three electrons in the outer orbit. This gives them a positive valence of three. It also makes them less basic in their hydroxides. Boron hydroxide is a weak acid, and aluminum hydroxide is such a weak base that it is almost neutral.

The nonmetallic elements. In the periodic classification, each group is less metallic than the preceding group. According to the atomic-structure theory, members of each group have one more electron in the outer orbit than those of the preceding group. The most nonmetallic elements have the largest number of electrons in their unfinished orbits. The chlorine family of elements each has seven electrons in the unfilled orbit. The easiest way for chlorine or any other member of this family to get a completed outer orbit is to borrow one electron in chemical reaction; hence they are very reactive.

The oxygen family of Group VI each lacks two electrons to have a complete outer orbit. The nitrogen group needs to gain three electrons for completeness.   Thus we see that a negative valence of one means the element must obtain one electron to complete its outer orbit.   An element of valence minus two like sulfur must gain two electrons for a completed orbit.

We may conclude, then, that a metallic element has as many positive valences as it has electrons to lose, thereby leaving behind a completed orbit; and a nonmetallic element has as many negative valences as it needs to gain electrons to complete its outer orbit.

QUESTIONS OF FACT

1. State the assumption necessary when correlating structure of the atom theory with chemical properties.

2. What is the chemical nature of all elements whose atoms have completed outer orbits?

3. Name three elements whose atoms have completed outer orbits.

4. Are those elements (other than hydrogen) whose atoms have only one electron in the outer orbit metallic or nonmetallic?   Are their hydroxides bases or acids?

5. Give another assumption used when correlating atomic structure with valences.

6. Illustrate the assumption in ex. 5.

QUESTIONS OF UNDERSTANDING

1. Tell of Prout's hypothesis.

2. How nearly has it been proved?

3. How is atomic structure related to atomic number?

4. Draw the structure of the argon atom.   (The third orbit is filled with eight electrons.)

5. Draw the structure for the atom of potassium.

6. Give the valence of chlorine when it has gained one electron.

7. In $KClO_4$ chlorine is thought to have lost seven electrons. What is its valence?

8. If nitrogen gave away all its electrons in the outer orbit, what would be its positive valence?

9. Which situation probably exists for the nitrogen in NH_3?

10. What is the valence of nitrogen in nitric acid?

11. How many electrons are in the outer orbit of potassium?

12. How many electrons are in the outer orbit of iodine?

13. In terms of electrons, why are the alkali metals monovalent?

14. If an atom loses an electron, what is its net electrical charge?

15. What is the net electrical charge of an atom in each of the following:

(a) loses two electrons   (d) loses three electrons
(b) gains one electron   (e) gains four electrons
(c) gains two electrons   (f) loses four electrons

16. What has happened to the carbon atom in burning to CO_2?

17. According to this theory, what has happened to the carbon atom in the formation of CH_4?

18. Under what conditions have modern scientists changed the atoms artificially? In what quantities has this been done?

19. State the number and kind of valences of elements which have the following numbers of electrons in the outer orbits: (a) 6; (b) 2; (c) 3.

20. What is the greatest fault of the atomic-structure theory?

REFERENCE FOR SUPPLEMENTARY READING

1. Fisk, Dorothy M. *Modern Alchemy*. D. Appleton-Century Co., New York, 1936.

    (1) Tell how scientists know that there are neutrons and positrons, and how they are produced. (pp. 116–134, "New Particles")

    (2) Give a report on the artificial transmutation of the elements. (pp. 101–115, "Early Transmutations," and pp. 135–151, "Mechanical Transmutations")

The cause of chemical reaction. *Assumption III. Any element tends to react with other elements when the reaction results in leaving the atoms of that element with a completed outer orbit.*

A finished orbit seems to be the most stable condition for the atom. Reaction merely puts the atom into a more stable condition. This follows a general law of nature. Water tends to flow down hill to a more stable level. Electricity and heat also flow into more stable and uniform conditions. A generation ago chemists said elements reacted because of chemical affinity for each other. This was no explanation; it was simply another way of saying they do react.

Chemical force. According to this structure theory, the elements are held in chemical combination by electrical forces.

The quantitative law of electrical attraction is: *Two electrical charges of opposite kind attract each other with a force directly proportional to the product of the quantities of the charges and inversely as the square of the distance between them.*

The distance factor is more important than the quantity factor. If the distance is made one-half, the attraction is multiplied by four; if the distance between them is reduced to one-third, the attraction is nine times as great; and if the distance becomes one-tenth, the forces will increase a hundred times. The atoms of the elements are extremely small; hence the charges can get very close together, thus making the attraction great and the compound stable.

The structure theory applied to chemical reactions. Let us consider the elements sodium and chlorine, which react vigorously to form a very stable compound, table salt. Sodium, with only one electron in the outer orbit, readily gives up this electron, leaving it with the next inner orbit complete. Chlorine, with seven electrons in the outer orbit, fills this orbit by accepting the extra electron from the sodium atom. After the sodium atom has reacted with the chlorine atom and has given

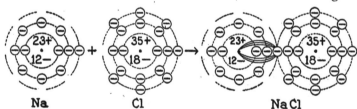

Na   Cl   NaCl

its electron to the chlorine atom, both have their outer orbits filled, which is a stable condition. When the sodium atom, which had an equal number of plus and minus charges, gives up a minus electron, it is left with an excess of one unit plus charge. The chlorine atom, which was also neutral with an equal number of protons and electrons, gained the electron and has an excess of minus one. The attraction of the plus charge on the sodium and the minus charge on the chlorine for each other holds the atoms in chemical combination. In the compound, sodium chloride, as shown in the figure, both the chlorine atom and the sodium atom have their outside orbits filled, which is a more stable condition than the atomic condition.

Hydrogen reacts with oxygen. In this case, the oxygen lacks two electrons of having a completed outer orbit. This is completed by capturing one electron from each of two hydrogen atoms, which, like the alkali metals, has only one electron.

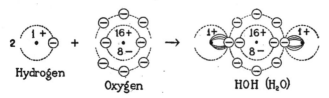

When the hydrogen loses its electron, it has one unneutralized positive charge; and when oxygen gains two electrons, it has two excess electrons. A valence of minus two, according to this theory, means that the atom must gain two electrons to complete its orbit. The metallic elements are those that attain their stable condition by giving up electrons, while the nonmetallic elements are those which have their outer orbits more than half filled with electrons and gain electrons to fill them.

Calcium unites with fluorine. Calcium has two electrons to give away. In other words, it has a valence of plus two.

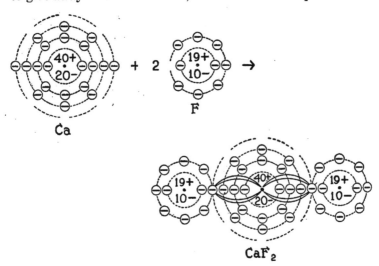

Here again we see why one atom of the metal is able to hold two atoms of the nonmetallic element. By the transference of electrons the atom of calcium and the two atoms of fluorine have completed outer orbits. Thus valence becomes something more than a name.

Ions and atomic structure. *Assumption IV. When a compound composed of two elements ionizes, each element becomes an*

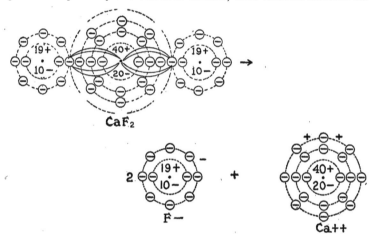

CaF₂

2 F−   +   Ca++

ion with a completed outside orbit.   In calcium fluoride, as shown, the fluorine would keep its captured electron to become a negative ion, while calcium has lost its two electrons and has an excess of two plus charges.

As another illustration of ionization, let us examine the reaction of NaCl below. When the parts separate, chlorine takes

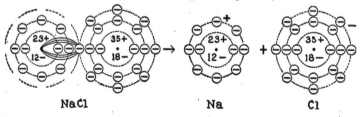

NaCl   Na   Cl

the extra electron which sodium originally had. This leaves sodium one electron short; hence the ion has one excess positive charge. The chloride ion, having gained this electron, has one excess negative charge. However, both ions have completed orbits. This tendency towards completed orbits may be considered the cause of the ionization.

Positive and negative valences for the same element. The atomic-structure theory does not solve all the mysteries of chemistry; far from it. However, those few points it makes more real are simplified to that extent. Some elements seem to have negative valences in some of their compounds and positive valences in others. The halogens — fluorine, chlorine, bromine, and iodine — are among this group. In most of their common compounds, their valence is minus one, as: $\overset{+}{\text{Na}}\overset{-}{\text{Cl}}$, $\overset{+}{\text{K}}\overset{-}{\text{Br}}$, $\overset{+}{\text{H}}\overset{-}{\text{I}}$. When we study their oxidized salts, however, we must consider them as having positive valences if oxygen is still negative; and it seems that oxygen is always negative in its valence. The following table will illustrate this condition.

VALENCES OF CHLORINE

COMPOUND	FORMULA	VALENCE
Sodium hypochlorite	$\overset{+}{\text{Na}}\overset{+}{\text{Cl}}\overset{=}{\text{O}}$	plus 1
Sodium chlorite	$\overset{+}{\text{Na}}\overset{+++}{\text{Cl}}\overset{=}{\text{O}_2}$	plus 3
Sodium chlorate	$\overset{++}{\underset{}{}}\overset{+}{\text{Na}}\overset{+++}{\text{Cl}}\overset{=}{\text{O}_3}$	plus 5
Sodium perchlorate	$\overset{+}{\underset{+++}{}}\overset{+}{\text{Na}}\overset{+++}{\text{Cl}}\overset{=}{\text{O}_4}$	plus 7

Interpreted by the structure theory, the chlorine has lost some of the electrons instead of gaining one. In $\overset{+}{\text{Na}}\overset{-}{\text{Cl}}$ the chlorine is minus one. When one oxygen is added, two electrons are taken from it with one excess plus charge. Each extra oxygen robs it of another pair of electrons until they are all removed from the unfinished ring. We see it stops with seven because there are no more electrons to remove. The inner completed orbit is too stable for chemical reaction.

Sulfur, also, may either gain or lose of its six electrons in the unfinished ring. In the sulfides such as $\overset{+}{\text{Na}}_2\overset{=}{\text{S}}$, the sulfur has gained two electrons to complete its ring. In the sulfites, $\overset{+}{N}a_2\overset{++=}{S}O_3$, the sulfur has lost four of its electrons, leaving it with an excess of four plus charges. In the sulfates, such as $\overset{+}{N}a_2\overset{+++=}{S}O_4$, the sulfur has lost all of the electrons in the unfinished orbit; and its valence is plus six.

Oxidation and reduction. We are now in a position to consider oxidation and reduction from another point of view. In Chapter 18 it was pointed out that oxidation might be considered as increasing or raising the valence of an element — that is, increasing a positive valence or decreasing the number of negative valences. If, as we have just learned, positive valences mean electrons given away, the only way to increase this valence would be to take away another electron. On the other hand, if negative valences mean electrons, the way to decrease this valence is to remove an acquired electron also. Oxidation then can be defined as a *loss of electrons*, and reduction, the reverse of oxidation, as a *gain of electrons*.

The periodic arrangement and atomic structure. When the periodic law was first formulated, many scientists thought the idea preposterous because they could see no reason for it. Investigation showed that it must undoubtedly be true, in spite of the fact that none could offer an explanation of why it should be true. Atomic-structure theory has given us the desired explanation. Let us recall the arrangement in terms of the atomic numbers:

Group	0	I	II	III	IV	V	VI	VII
Period 1	He (2)	Li (3)	Be (4)	B (5)	C (6)	N (7)	O (8)	F (9)
Period 2	Ne (10)	Na (11)	Mg (12)	Al (13)	Si (14)	P (15)	S (16)	Cl (17)

On examining the structures of these elements we find that in Period 1 lithium has one electron n the outer orbit. Each succeeding element has one more electron in its outer orbit than the preceding element, which would account for a gradual change from the most positive element, lithium, to the most negative element, fluorine. In atomic numbers two and ten, whose outer orbits are filled, there is found no tendency to react. In accordance with the theory, reaction is caused by the tendency of atoms having unfinished outer orbits to get into a condition where all orbits are completed. If the orbits are already complete, there is consequently no tendency towards a changed condition.

The reason for the periodic arrangement is evident from the structures of the atoms. After an orbit has been completed in one period, a similar series of elements will result from building up another orbit in the next period. All through this period each element will show the same properties as the corresponding element in the preceding series, since they are due to the same cause — namely, the number of electrons in the outer orbit.

The Groups 0–VII are explained on the basis of all atoms with completed outer orbits falling into Group 0, all with one outer electron into Group I, all with two outer electrons into Group II, and so on to the last element of the period, which lacks only one electron of completion.

Line spectra. The atomic models give us a mechanism for explaining the production of line spectra and even for predicting lines not previously discovered. However, this is too difficult to explain in a beginning course, as a considerable knowledge of physics is necessary to its understanding.

Weak points of the atomic-structure theory. The atomic-structure theory does not account for many facts of chemistry that it might be expected to explain. For example, if electrical attractions hold atoms together in compounds, why are not all compounds equally stable? As a matter of fact, some compounds are extremely stable, while others are very unstable. It does not explain how elements can have different positive valences, as gold with one and three, or mercury with one and two. The theory doesn't explain why chlorine, a negative element, can readily replace hydrogen, a positive element. The organic chemist has to assume that there are some bonds that are not polar — that is, not of an electrical nature.

QUESTIONS OF DEFINITION AND UNDERSTANDING

1. State the added assumption needed when applying atomic structure to the cause of chemical reaction.

2. What was the cause that was formerly given as the reason for the reaction of elements?

3. What was the objection to this explanation?

4. What forces are thought to hold atoms in chemical combination?

5. On the basis of this electrical attraction, how can the fact be accounted for that all compounds are not equally stable?

6. What holds the sodium and chlorine atoms together in the compound, NaCl?

7. Draw a picture equation, similar to the one in the text, for lithium uniting to chlorine.

8. State the quantitative law of attraction between electrical charges.

9. How many electrons has chlorine lost in each of the oxygen compounds of sodium and chlorine?

10. State two faults of the atomic-structure theory.

11. Draw a picture of the structure of the bromide ion.

12. How many electrons are needed by the chlorine atom for a completed outer orbit?

13. What does a negative valence of one mean?

14. In lithium chloride what is the condition of the outer orbits of both the lithium and the chlorine?

15. What is the resultant charge on the whole molecule?

16. Why do two elements react? Answer in terms of the atomic-structure theory.

17. In CaO how many electrons has calcium given away?

18. In the same substance how many electrons did oxygen gain to complete its outer orbit?

19. State the assumption used to explain ionization by atomic-structure theory.

20. In terms of atomic-structure theory, state what happens when KCl ionizes.

21. In terms of atomic-structure theory, how can an element sometimes show a positive valence and at other times a negative valence?

22. Explain oxidation and reduction in terms of electrons.

23. Explain the reason for the periodic classification of the elements in terms of atomic-structure theory.

24. Is this statement true? A family of related elements is composed of those elements with the same number of electrons in the outer orbit and differing in the number of completed orbits.

25. Is this true? A series in the periodic classification is composed of elements whose atomic structures differ only in the outer orbit.

SUMMARY

Based on the theory of the electrical nature of matter, atomic models have been invented. The Bohr model has met with most favor, since it best explains spectral lines, chemical forces, reasons for chemical reaction, the meaning of valence, and the periodic law. It embodies the following points:

(1) All atoms are composed of equal numbers of electrons and protons, the number of each in the case of each atom being equal to the atomic weight.

(2) The protons are packed into a small central nucleus with approximately half of the electrons.

(3) The rest of the electrons revolve about the nucleus in fixed orbits.

(4) No electron can take any position between the fixed orbits.

The number of electrons outside the nucleus is always equal to the atomic number. Since the atoms consist of protons of unit mass, there can be no fractional atomic weights.

The chemical properties of an element are determined largely by the number of electrons in its outer orbit. A metallic element is one with four or less electrons in its outer orbit. It gives these away in order to gain a completed orbit. Its valence is positive and equals the number of these available electrons.

A nonmetallic element is one whose atoms have four or more electrons in the outer orbit. It usually gains electrons to achieve a completed outer orbit. Its negative valence is the number of electrons it needs. Chemical reaction is caused by the tendency to gain or lose electrons. The attraction between the positive and negative charges of electricity holds atoms in chemical union.

During ionization the negative part of the compound retains its captured electron; hence it carries a negative charge. The positive part carries a positive charge because it has lost the electrons gained by the negative ion.

Oxidation is the loss of electrons, while reduction is the gain of electrons. The gradual change of properties within a period (of the periodic system) is explained by an increase of one electron in the outer orbit of each element, passing from left to right.

CHAPTER 22

←
→

SULFUR

Introduction. The yellow element sulfur has been known during all historical time. Under the name of brimstone, sulfur was associated with burning, demons, and the nether regions. Not until the invention of gunpowder, of which sulfur is a constituent, did this element take on any commercial

Courtesy of Texas Gulf Sulfur Co.

SULFUR BY THE WHOLESALE. This photograph gives some idea of the vast amounts of sulfur that are produced by the Frasch process.

importance. When the demand for sulfur arose, a sulfur production industry sprang up in the volcanic region of Sicily, which industry has continued until the present.

During his conquest of Mexico, Cortes once found his army short of gunpowder in a strange land. Knowing that sulfur is nearly always found around active volcanoes, Cortes sent a number of his men to climb the snow-capped volcano Popocatepetl in search of sulfur. After a hazardous trip to the top, their commander, Montano, found that they could not get

416

Herman *Frasch*

1851–1914

Coming to America from Germany in 1868, Herman Frasch took a position in the Philadelphia College of Pharmacy. After extensive study in chemistry, he moved to Cleveland and opened a chemical laboratory, at which he worked primarily at refining petroleum.

Soon he moved to London, Ontario, where he organized the Empire Oil Company. Previous to this, these Canadian oils could not be sold because of the skunklike odor of their sulfur compounds. Frasch succeeded in purifying these oils, making them salable on a par with any similar oils. His method was to remove the sulfur compounds with copper oxide. The oxide reacted with the sulfur to form copper sulfide, which effectively removed the sulfur, thereby destroying the odor.

This method had the very desirable feature of being cheap, in that the copper sulfide could be changed back into the oxide and used over again. In all he received twenty-one patents on the different phases of oil purification. The value of his oil-refining methods is illustrated by the fact that the price of certain oils was raised from 14 cents per barrel to $1 per barrel.

Great as his oil-refining inventions were, they were overshadowed by his method of bringing sulfur from underground by superheated water. Previous to Frasch's invention the United States produced only .5 per cent of the sulfur it consumed. Now, on the other hand, we export large quantities.

In developing his process for mining sulfur, Frasch experimented for many years at a cost of large sums of money. However, since the method was so successful, this money came back in profits many times over.

In 1912 Frasch received the Perkin gold medal in recognition of his great achievements in applying chemistry to industry.

down to the region where sulfur would be likely to be found. Realizing the desperate need for sulfur, Montano had himself lowered by rope 400 feet into the volcano, where he found plenty of the needed sulfur.

The United States has a great supply of sulfur in the underground deposits in Louisiana and Texas. Here enormous beds of nearly pure sulfur occur 500 feet underground. One company alone produces 5000 tons of sulfur daily.

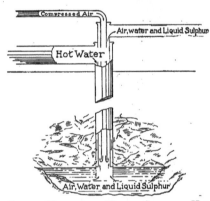

FRASCH METHOD OF MINING SULFUR. Hot water melts the sulfur, and compressed air pushes the mixture of water and sulfur to the surface.

Frasch method of mining sulfur. When sulfur was first discovered 500 feet underground in Texas, it was a great problem to mine it. It was too expensive to sink a shaft that deep, especially since it would have to pass through a layer of quicksand. An American chemist named Herman Frasch solved the problem in a very simple way and won both fame and fortune.

In the Frasch process a hole is drilled and piped down through the overlying deposits to the bottom of the sulfur bed. Inside the casing of this well are placed three concentric pipes, a 1-inch pipe inside a 3-inch pipe, which in turn is inside a 6-inch pipe. Steam under atmospheric pressure has a temperature of 100° C., but when under greater pressure its temperature goes above the melting point of sulfur, 117° C. Hot water and steam are forced down through the outside pipe to melt the sulfur. Hot compressed air forced down through the inside 1-inch pipe mixes with the molten sulfur and reduces the specific gravity of the liquid so that it can be raised through the intermediate pipe to the surface. The melted sulfur is run into large bins made of rough boards, where it soon cools into an enormous block of very pure (99 per cent) sulfur. The huge blocks, some of which weigh 100,000 tons, are broken up with dynamite and loaded into cars with steam shovels.

Purification of sulfur. The crude sulfur is now distilled from retorts and condensed in brick chambers. Part of the vapor sublimes on the walls as a very fine powder called *flowers of sulfur*. The rest becomes liquid and is cast into wooden molds 1½ inches in diameter and is called *roll sulfur*.

FLOWERS OF SULFUR. Flowers of sulfur is obtained by cooling sulfur vapor. The powder is used to dust plants that have mildew.

Allotropic forms of sulfur. No element shows a greater variety of allotropic forms than sulfur. It has two crystalline forms, one plastic form, three liquid forms, and two gaseous forms. If sulfur is crystallized from carbon disulfide, it forms clear yellow crystals like double pyramids (orthorhombic sulfur). When sulfur is melted and heated, it goes through some interesting and unusual changes. Just above its melting point it is a mobile liquid of a light amber color. As the temperature rises, the sulfur darkens rapidly

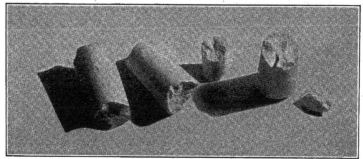

ROLL SULFUR. Molten sulfur is often cast into rolls and sold as roll sulfur.

ORTHORHOMBIC SULFUR. When sulfur is crystallized from carbon disulfide, it forms beautiful crystals called orthorhombic sulfur.

and thickens so that it can hardly be poured from the inverted test tube; on further heating, it again becomes less viscous and finally boils at 445° C., forming a pale orange vapor. When this vapor is heated still more, it becomes light yellow. If boiling sulfur is poured into cold water, it becomes a soft rubberlike plastic. This is an unstable form, because on standing a few days it changes to the common variety. Orthorhombic sulfur is the stable form under 96° C., since all other forms change to it below that temperature. Monoclinic sulfur is the stable solid form above 96° C. The explanation of the large number of allotropic forms is that different numbers of atoms may be

PLASTIC SULFUR. When sulfur is boiled and poured into cold water, it becomes somewhat like rubber. This is called plastic sulfur. It does not remain dark and elastic long but soon changes to ordinary yellow sulfur.

combined to form molecules of sulfur. This has been proved to be the case in the vapor state, where one form is S_2 and the other S_8.

Chemical conduct of sulfur. Sulfur, like oxygen, is very reactive at high temperatures and unites directly with many elements both metallic and nonmetallic. It burns in oxygen to form SO_2 and unites with metals to form sulfides. Most sulfides are insoluble in water, and many have characteristic colors. The blackening and tarnishing of metals is often due

MONOCLINIC SULFUR. This is another allotropic form of sulfur, which forms when liquid sulfur cools. The illustrations show the crystals individually and in clusters.

to the action of sulfur. Sulfur compounds pass into the air from the combustion of coal and illuminating gas. These compounds act upon silver, producing the black silver sulfide (tarnish) on silver of all kinds. The tarnishing of silverware by egg yolk or mustard is due to the sulfur compounds contained in these substances.

Uses of sulfur in the elementary form. Flowers of sulfur finds extensive use as a treatment for fungus diseases of plants, such as mildew of grapes and roses. The sulfur dust is blown on the plant. The oxygen of the air under the action of sunlight oxidizes the fine sulfur powder to sulfur dioxide, which attacks and kills the fungus. Sulfur is also burned in a closed room with cut fruit to bleach it. The SO_2 which results from the burning is what does the bleaching. Sulfur is boiled with lime to form a lime-sulfur dip and spray. Sheep are dipped into this solution to cure them of the scabies. Sulfur is also used in vulcanizing rubber. The chief use of sulfur is for the manufacture of its compounds, chief of which are carbon disulfide, sulfur dioxide, and sulfuric acid.

QUESTIONS OF FACT

1. Where does sulfur occur in the free state?
2. Describe the Frasch method of mining sulfur.
3. What is flowers of sulfur? roll sulfur?
4. Name the allotropic forms of sulfur.
5. How is sulfur related to the tarnish of metals?
6. Give four uses of sulfur.
7. What was the prehistoric name for sulfur?
8. Why did Cortes need sulfur while conquering Mexico?
9. Relate the story of Cortes's search for sulfur in Mexico.
10. How are the following produced: (a) orthorhombic sulfur? (b) monoclinic sulfur? (c) plastic sulfur?
11. How do sulfur compounds get into the air?
12. What evidence is there that sulfur occurs in mustard and eggs?

QUESTIONS OF UNDERSTANDING

1. In the Frasch method of mining underground sulfur:
(a) Would the method work if the melting point of sulfur were above the temperature of the steam?
(b) What is the reason for forcing hot air into the mixture?
(c) Suggest a reason why the sulfur mixture is forced out through the middle pipe.
2. Why is it better to produce dusting sulfur by condensing sulfur vapor rather than by crushing roll sulfur?
3. Give the evidence that there are different allotropic forms of liquid sulfur.
4. What is supposed to be the difference between the allotropic forms of sulfur?
5. What is meant by plastic sulfur being an unstable form?

ADDITIONAL EXERCISE FOR SUPERIOR STUDENTS

1. Write a 200-word history of sulfur. (See reference 3 in the following list.)

REFERENCES FOR SUPPLEMENTARY READING

1. Clarke, Beverly L. *Marvels of Modern Chemistry.* Harper & Brothers, New York, 1932. Chap. XIV, pp. 185–195, "Sulfur."
   (1) Give the use of sulfur in medicine. (p. 185)
   (2) Name several ores of important metals that are sulfides. (p. 186)
   (3) Discuss the chemical similarity of sulfur and oxygen. (p. 186)
   (4) Name two other elements chemically similar to sulfur. (p. 194)
2. Foster, William. *The Romance of Chemistry.* D. Appleton-Century Co., New York, 1936. Chap. XII, pp. 173–187, "Sulfur a Pillar of Industry."
   (1) Discuss the occurrence of sulfur. (pp. 174–175)
   (2) Compare the quantities that occur uncombined with those in which it is combined with metals. (p. 175)
   (3) Name some plant and animal products which contain sulfur. (p. 175)
   (4) What is the only form of sulfur which can be used by the human body? (p. 176)
   (5) Relate the results following a sulfur fire in Sicily. (p. 176)
3. Martin, Geoffrey. *Triumphs and Wonders of Modern Chemistry.* D. Van Nostrand Co., New York, 1922. Chap. XIII, pp. 297–313, "Sulfur and Its Compounds." (Very interesting.)

Sulfur dioxide. By far the easiest compound of sulfur to synthesize is sulfur dioxide. It is only necessary to burn sulfur or a metallic sulfide. The burning method is not entirely feasible in the laboratory because of the difficulty of collecting the gas free from air. Sulfur dioxide is therefore usually prepared by heating copper or some other difficultly soluble metal with concentrated sulfuric acid.

$$Cu + 2H_2SO_4 \longrightarrow CuSO_4 + SO_2 + 2H_2O$$

Sulfur dioxide is a colorless suffocating gas quite soluble in water. It is one of the easiest gases to liquefy and is kept in iron tanks as liquid. Sulfur dioxide is used chiefly in bleaching, as a preservative or disinfectant, and for making sulfuric acid. Straw, silk, wool, and materials that would be injured by chlorine are decolorized by it. A room to be disinfected with SO_2 should be tightly closed and the air kept moist, so that the disease germs may be killed by sulfurous acid formed by the action of SO_2 on water. Sulfur dioxide prevents the growth of molds in common beverages. The use of SO_2 for bleaching fruits has caused a great deal of controversy as to whether the presence of sulfurous acid in the fruit is harmful or not. However, it is still permitted in this country. Sulfur dioxide is used in one make of electric refrigerator. The most important use of sulfur dioxide is for making sulfuric acid.

Sulfur trioxide. Sulfur trioxide is another oxide of sulfur. It is a white solid, looking much like snow. When concen-

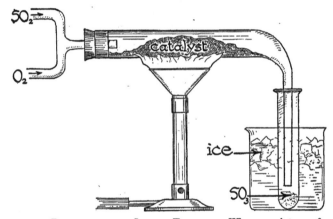

LABORATORY PREPARATION OF SULFUR TRIOXIDE. When a mixture of sulfur dioxide and oxygen is passed over a catalyst of platinum or vanadium dioxide, the sulfur dioxide and oxygen react to form sulfur trioxide.

trated sulfuric acid is heated to a very high temperature, white fumes are given off. These are due to sulfur trioxide. This oxide of sulfur is quite difficult to prepare. Its preparation from SO_2 and oxygen is one step in the preparation of sulfuric acid by the contact process.

Acid anhydrides. Since SO_2 will react with water to form sulfurous acid:

$$SO_2 + H_2O \longrightarrow H_2SO_3$$

and H_2SO_3 will yield SO_2 with heat:

$$H_2SO_3 \longrightarrow H_2O + SO_2$$

this oxide is said to be the *anhydride* of sulfurous acid. Anhydride means without water. We see that SO_2 differs from H_2SO_3 by 1 molecule of H_2O, so it is really H_2SO_3 minus water.

Sulfur trioxide has the same relation to sulfuric acid that sulfur dioxide has to sulfurous acid; that is, it is the anhydride of sulfuric acid.

$$SO_3 + H_2O \longrightarrow H_2SO_4 \text{ (dissolving } SO_3 \text{ in } H_2O)$$
$$H_2SO_4 \longrightarrow SO_3 + H_2O \text{ (vigorously heating } H_2SO_4)$$

The anhydride of an acid always forms the acid when it reacts with water. Most acid anhydrides show a readiness to react with water, but a few like silica (ordinary sand) show great reluctance to react.

Sulfurous acid. Sulfurous acid, as such, is not of wide usefulness because it is not very soluble and is quite unstable. When it is needed, its equivalent in the form of SO_2 or sodium sulfite is usually used. However, sulfurous acid is a strong reducing agent. Its reducing action is what bleaches the fruit when SO_2 is applied. This acid is used to remove dyes from cloth.

Sulfuric acid. Sulfuric acid is one of the most widely used chemicals. It has been said that the civilization of a nation could be measured by the amount of sulfuric acid it consumes. This is in a sense true, since there is scarcely a thing used today that does not require sulfuric acid directly or indirectly in its manufacture. A number of factors give sulfuric acid this prominence. In the first place it is by far the cheapest of all the acids. Where an acid is needed in any commercial process, it is natural that the cheapest acid should be used. Thus sulfuric acid finds extensive use in the manufacture of fertilizers, in refining petroleum, and in cleaning metals.

A cheap dehydrating agent. Concentrated sulfuric acid is a *dehydrating agent;* that is, it removes water from other substances, thus drying them. The vigor with which concentrated sulfuric acid reacts with water makes it dangerous to pour water into the acid. When it is necessary to dilute the acid, it must always be poured into the water to prevent sputtering which would endanger the eyes. Gases are often dried by bubbling them through concentrated sulfuric acid. Because of its great affinity for water, concentrated sulfuric acid destroys most organic material, pulling off the hydrogen and oxygen as water and leaving only carbon. A good way to test for carbon in a substance is to heat it with concentrated sulfuric acid and note the charring.

However, dehydration means more than violently reacting with water. If an anhydrous (free from water) substance contains H's and O's in its molecule, concentrated sulfuric acid may tear out these elements as if they were water. This dehydrating action produces a chemical change which is useful in many ways, particularly in combining two different molecules. An illustration is the synthesis of a fragrant substance called an ester from an alcohol and an organic acid.

$$CH_3CO_2H + HOC_2H_5 + H_2SO_4 \longrightarrow CH_3CO_2 \cdot C_2H_5 + H_2SO_4 \cdot H_2O$$
acetic acid   ethyl alcohol   ester

The sulfuric acid pulls off an H from the acid, leaving a loose end where it was attached. At the same time it pulls off an OH from the alcohol, leaving a loose end there also. The loose ends now join to make the ester. This reaction is typical of thousands of useful synthetic processes, including the manufacture of nitroglycerin, smokeless powder, and the dyes.

A cheap oxidizing acid. Hot concentrated sulfuric acid is an oxidizing agent. This gives it a third set of uses that help to make it the most widely used synthetic chemical substance. The further fact that sulfuric acid is not volatilized (made gaseous) at ordinary boiling temperatures adds to its usefulness. In the manufacture of the halogen acids (HCl, HBr, and HI), mentioned in a previous chapter, sulfuric acid is used. The removal of the halogen acid as a gas enables the reaction to proceed to completion.

QUESTIONS OF FACT

1. Give two methods of preparing sulfur dioxide.
2. Give three uses of SO_2.
3. How is most of the sulfuric acid produced in industry used?
4. Name four reasons why sulfuric acid is used in industry.

5. Give the formula for sulfurous acid and compare it with that of sulfuric acid.

6. How is sulfuric acid used as a dehydrating agent in the synthesis of new compounds?

7. What property of sulfuric acid is used in the following reaction:

$$NaCl + H_2SO_4 \longrightarrow NaHSO_4 + HCl$$

8. What effect does sulfurous acid have on the color of some dyes?

9. What property of sulfuric acid prevails when the acid is hot and concentrated?

QUESTIONS OF UNDERSTANDING

1. Explain the meaning of acid anhydride.

2. Of what acid is SO_2 the anhydride?

3. What is the anhydride of sulfuric acid?

4. What is formed when an acid anhydride reacts with water?

5. Upon what basis do some people object to eating fruit which has been bleached with SO_2?

6. Why is the amount of sulfuric acid used by a country said to be an index of its degree of civilization?

7. Outline the steps necessary in preparing sulfuric acid from sulfur. In which step is a catalyst needed?

8. Can you suggest any similarities between oxygen and sulfur?

9. Discuss a dehydrating agent:

(a) Does it react readily with water?

(b) Does it react with substances containing hydrogen and oxygen but which are anhydrous?

(c) In its action of dehydration, may the H come from one molecule and OH from another?

(d) Will the residues of the other molecules of (c) usually unite?

(e) May reactions such as (b) and (c) be useful?

(f) In forming an ester, which atom of the acid is replaced by a radical composed of carbon and hydrogen?

REFERENCE FOR SUPPLEMENTARY READING

1. Clarke, Beverly L. *Marvels of Modern Chemistry.* Harper & Brothers, New York, 1932. Chap. XIV, pp. 185–195, "Sulfur."

(1) Point out how sulfuric acid has been used in producing things in your home. (pp. 187–189)

(2) Explain the statement: "You may never see sulfuric acid yet you cannot get along without it." (p. 189)

(3) What great bank grew out of the manufacture of sulfuric acid? (p. 192)

Manufacturing sulfuric acid. The manufacture of sulfuric acid takes place essentially in three steps. The first is the formation of SO_2 by burning sulfur or sulfides. This step takes place by itself after the fire has been kindled.

$$S + O_2 \longrightarrow SO_2$$

or $$2\,CuS + 3\,O_2 \longrightarrow 2\,CuO + 2\,SO_2$$

The second step can be represented as:

$$2\,SO_2 + O_2 \longrightarrow 2\,SO_3$$

Heat alone will not cause this reaction to take place. A catalyst is needed. All recently constructed plants for the manufacture of sulfuric acid use what is called the *contact process*. A mixture of SO_2 and O_2 is passed through a heated tube containing the catalyst. Some plants use finely divided platinum as the catalyst, while others use vanadium oxide. SO_3 emerges from the end of the tube.

There is some difference of opinion among chemists as to the relative merits of the two catalysts. Two things are against the finely divided platinum: Its initial cost is great, and it is "poisoned" by arsenic and other substances. The vanadium oxide does not seem to be "poisoned" as the platinum is, but its efficiency is somewhat lower. However, by running the gaseous mixture over again, it is possible to complete the reaction in the presence of vanadium oxide.

The third step in the manufacture of sulfuric acid, that of causing the SO_3 to react with water, is quite simple and easily carried out by dissolving the SO_3 in dilute sulfuric acid to form the concentrated sulfuric acid.

Mineral springs of little value for curing disease. There are many so-called sulfur springs, whose water has a peculiar odor and taste. Such water is usually distasteful at first, but one can sometimes develop a liking for it. These springs are supposed to have health-giving value. This healthfulness of mineral waters is imaginary. In the first place, there is only a trace of mineral in the water; and in the second place, the mineral substances are valueless from a health standpoint. The only possible benefit from drinking mineral waters is in the increased intake of water they may encourage. It is the water that is beneficial and not the mineral. It would be far better to use fruit juices to make one drink more water.

Hydrogen sulfide. The odorous substance in sulfur springs is usually hydrogen sulfide (H_2S). Hydrogen sulfide odor suggests rotten eggs. This foul-smelling gas is quite poisonous. Students in the laboratory must be careful not to breathe too much of it. Hydrogen sulfide is used as a laboratory reagent to precipitate the metals as insoluble sulfides. This is especially useful in qualitative analysis. Hydrogen sulfide is made by the reaction of sulfuric acid on iron sulfide.

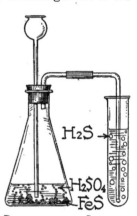

PREPARATION OF SULFIDES. In the generator at the left, hydrogen sulfide is produced by the reaction of sulfuric acid with iron sulfide. In the case of many metals, when hydrogen sulfide is passed into a metallic solution, the sulfide precipitates out.

$$FeS + H_2SO_4 \longrightarrow H_2S + FeSO_4$$

This reaction goes to an end because the H_2S is a gas and leaves the reacting solution. In solution H_2S acts as a weak acid, often called hydrosulfuric acid.

EXPERIMENT

Sulfides. Pass H_2S into solutions of the following, and note the color and insolubility of each:

(a) $Cu(NO_3)_2$ (b) $AgNO_3$ (c) $Pb(NO_3)_3$ (d) $Hg(NO_3)_2$ (e) $HgNO_3$
(f) $Bi(NO_3)_3$ (g) $Cd(NO_3)_2$ (h) $Sn(NO_3)_2$ (i) $As(NO_3)_3$ (j) $SbCl_5$

Sulfur compounds and bleaching. Sulfur compounds are used for bleaching, either in the open or in solution. Chlorine and SO_2 are the common gaseous bleaching materials. Chlorine reacts with water, liberating nascent oxygen, which oxidizes the colored substance to the colorless condition. As chlorine is injurious to silk, wool, or straw, SO_2 must be used for bleaching a straw hat, a silk gown, or wool coat. SO_2 unites with moisture to form sulfurous acid, which removes oxygen from the coloring matter and decolorizes it. For bleaching substances in a solution, such as wood paper pulp, a sulfite is used. Its similarity to sulfurous acid enables it to bleach by reduction. Dyers often use sulfite solutions to strip cloth of dye preparatory to dyeing it over.

BLEACHING WITH SULFUR DIOXIDE. Sulfur dioxide is used for bleaching fruit and straw hats.

Other useful sulfur compounds. Hypo is a sulfur compound that is used very much in photography. Its use depends upon the fact that it is the best solvent for silver salts. The chemical name for hypo is sodium thiosulfate, and its formula is $Na_2S_2O_3$.

Barium sulfate ($BaSO_4$) is sometimes used as a white pigment in paint.

QUESTIONS OF FACT

1. Write the three reactions necessary to prepare sulfuric acid from sulfur.
2. Which one requires a catalyst? What catalysts may be used?
3. What is this process of producing sulfuric acid called?
4. What substance gives the odor to sulfur springs?
5. Why is H_2S used in the laboratory?
6. Discuss sulfur compounds in relation to bleaching.
7. What substances are usually bleached with SO_2 gas?
8. What substance is the equivalent of SO_2 if used in solution?
9. What is the chemical name for "hypo" used by the photographer?
10. What sulfate is used as a white pigment?

QUESTIONS OF UNDERSTANDING

1. Sulfuric acid is a strong acid, while hydrogen sulfide is a weak acid. Explain the terms.

2. In what sense is sulfur dioxide the opposite kind of bleach from chlorine?

3. Which is the oxidizing bleach and which the reducing bleach?

4. Write the formulas of sodium sulfate and sodium thiosulfate near each other and answer the following:

(a) What are the differences in the formulas you have written?

(b) Is the thiosulfate the sulfate with one of its oxygen atoms replaced by sulfur?

(c) Is the fact that sulfur can thus replace oxygen in harmony with the ideas that these elements are chemically similar?

ADDITIONAL EXERCISES FOR SUPERIOR STUDENTS

1. Assuming oxygen to have a valence of two, calculate the positive valence of sulfur in each of the following: (a) H_2SO_3; (b) $CaSO_4$; (c) SO_2; (d) SO_3.

2. $Fe + S \longrightarrow FeS$. Which element is oxidized and which is reduced? Which substance is the oxidizing agent and which the reducing agent?

3. Give a reason why SO_2 is so severe on the lungs when breathed.

4. Calculate the molecular weight of "hypo" ($Na_2S_2O_3$) and the percentage of sulfur.

5. Suggest a reason why nitric acid would not be suitable for preparing H_2S from the sulfides.

SUMMARY

Sulfur is a common element, which occurs in metallic sulfides and as uncombined sulfur. Metallic sulfides are present in practically all mineralized regions. Uncombined sulfur occurs in volcanic regions and in vast underground deposits in Texas and Louisiana. These underground deposits are mined by the Frasch method of melting the sulfur with steam and forcing it to the surface with hot compressed air. There are two stable forms of sulfur — the diamond-shaped orthorhombic form below 96° C., and the needlelike monoclinic form between 96° C. and the melting point of sulfur. Rubberlike plastic sulfur is formed by boiling the sulfur vigorously and pouring it slowly into water.

Sulfur is used to dust grape vines, to make lime-sulfur spray, to vulcanize rubber, and to synthesize the various sulfur compounds. Sulfurous acid is useful in bleaching fruit and straw. Sulfur trioxide, the anhydride of sulfuric acid, is a white solid resembling snow in appearance.

Sulfuric acid, the most used of all synthetic substances, is the cheapest acid, a dehydrating agent, an oxidizing agent, and a nonvolatile acid. Each of these facts is the basis of several uses. The manufacture of sulfuric acid takes place in three steps: (1) burning the sulfur to sulfur dioxide; (2) oxidizing SO_2 to SO_3 by use of a contact catalyst; and (3) the addition of water to SO_3.

A small amount of hydrogen sulfide gives a strong odor to sulfur water but is of little medicinal value. The hypo used in photography is $Na_2S_2O_3$.

CHAPTER 23

CHEMICAL CALCULATIONS

Introduction. Some students like arithmetic. A problem to them is a challenge like a crossword puzzle or a riddle. Other students, however, find arithmetic difficult and uninteresting. This is unfortunate, for the person who uses chemistry will sooner or later meet some arithmetic problems that must be solved.

A chemist unable to work chemical problems would be in the same fix as a contractor who could not figure out what a building should cost or an automobile manufacturer who could not determine the cost of a car. Imagine the predicament of a mine owner who could not compute the weight of metal in a ton of ore or of a manufacturer of sulfuric acid who could not figure the amount of sulfur needed to make a ton of acid. The computation of weights in chemistry is particularly important. Farmers need to be able to handle chemical problems in connection with sprays, stock raisers in connection with dips for lice, laundrymen in connection with washing formulas, and pharmacists and doctors in connection with prescriptions.

The relation of symbols and atomic weights to problems. Most chemical weight problems depend upon chemical symbols, formulas, and atomic weights. At the time the determination of the atomic weights was being discussed in so much detail, the student probably wondered what they are good for anyway. Now we are going to use them. Inside the back cover of this book are the approximate atomic weights. These are close enough for our purposes. We think of each atom represented in a formula as weighing its atomic weight. Although atomic and molecular weights are expressed in atomic units (approximately the weight of the hydrogen atom), we may use them in working problems as though they were expressed in grams. Since molecular formulas are used mostly in fractions

433

and equations, any factor necessary to change from atomic units to grams will cancel out; and the results are the same whether the formula weights are expressed in grams or in atomic units. The same reasoning applies to the use of pounds as the unit of weight. Let us consider sulfuric acid:

H_2SO_4 stands for 2×1 g. of hydrogen, 32 g. of sulfur, and 4×16 g. of oxygen.

$Ca(NO_3)_2$ represents calcium 40, nitrogen 2×14, and oxygen 6×16 — all in grams or pounds or other weight units.

Weight problems from chemical formulas. Suppose you are asked the cost of 7 hats of a certain kind. Instinctively you realize that additional information must be forthcoming before you can give the answer. Your response may be something like this: Give me the cost of 1 hat — or of any number of hats — and I can tell you the cost of 7. If you are then told that 10 hats cost $30, you can easily figure the cost of 7 hats. Moreover, from this sample you can solve all problems involving this relation between hats and dollars.

One of the easiest ways of solving the foregoing problem is to set up the conditions in the form of an equality between fractions, thus:

$$\frac{\text{(dollars in sample) } 30}{\text{(hats in sample) } 10} = \frac{x \text{ (dollars in problem)}}{7 \text{ (hats in problem)}}$$

Then solve for x.

Now let us apply this method to the solution of a problem involving chemical weights:

How many pounds of sulfur are in 2000 lbs. sulfuric acid? The formula H_2SO_4 gives the sample relation.

$$H = 1 \qquad S = 32 \qquad O = 16$$
$$(2 \times 1) + (1 \times 32) + (4 \times 16) = 98$$

Thirty-two units represent the sulfur and 98 units the acid.

$$\frac{\text{(sulfur in problem) } x}{\text{(acid in problem) } 2000} = \frac{32 \text{ (sulfur in sample)}}{98 \text{ (acid in sample)}}$$

$$x = \frac{32 \times 2000}{98} = 653 \text{ lbs.}$$

The steps in the foregoing method may be summarized as follows:

(1) Set up one fraction from the weights (given and unknown) of the problem.

(2) Set up the other fraction from the corresponding weights as indicated by the formula.

(3) Set the two fractions equal to each other.

(4) Solve for the unknown.

PROBLEMS

1. Give four steps in solving a chemical weight problem.

2. What weight of oxygen could be obtained from 101 lbs. $HClO_3$?

3. How much carbon is needed to make 100 g. CO_2?

4. What weight of nitrogen would have to be fixed in order to make 1 ton HNO_3?

5. A copper ore consists of $CuCO_3$. How much copper is there in 1 ton of ore?

6. What weight of hydrogen is there in 100 lbs. H_2SO_4?

Weight problems from chemical equations. Sometimes the sample information is given by a chemical equation:

$$NaOH + HCl \longrightarrow NaCl + H_2O$$
$$(23 + 16 + 1) + (1 + 35.5) \longrightarrow (23 + 35.5) + (2 + 16)$$

A chemical equation always gives much more information than we need for any one kind of problem. Consider the problem: What weight sodium hydroxide is needed to neutralize 200 lbs. HCl?

$$\frac{\text{(acid in sample)}}{\text{(hydroxide in sample)}} \frac{36.5}{40} = \frac{200}{x} \frac{\text{(acid in problem)}}{\text{(hydroxide in problem)}}$$

$$x = \frac{200 \times 40}{36.5} = 219.2 \text{ lbs.}$$

The equation tells us what weights of sodium chloride and water result from the reaction. This information has nothing to do with this problem. Some other problem might require it.

Caution: In setting up your equation, always be sure to put the same substance in the numerator of each fraction. If sodium hydroxide is in the numerator of one fraction, it must also be in the numerator of the other fraction. The equation is easier to solve if the unknown (x) is always put in the numerator of its fraction. This means that it will be necessary to set up the problem fraction first, and match it in the other fraction.

In dealing with chemical equations that have more than one molecule of a kind, the weight of *all* molecules must be counted. Consider the equations below :

$$H_2SO_4 + 2KBr \longrightarrow 2 HBr + K_2SO_4$$
$$Ca(OH)_2 + 2 HNO_3 \longrightarrow Ca(NO_3)_2 + 2 H_2O$$

Suppose a problem deals with potassium bromide ; we must then use $2(39 + 80) = 2 \times 119 = 238$.

If nitric acid is involved in a problem, the weight value would be $2[1 + 14 + (3 \times 16)] = 2(1 + 14 + 48) = 2 \times 63 = 126$.

PROBLEMS

1. In the equation giving the sample relation, what is to be done when there is more than one molecule concerned?

2. What weight of oxygen is necessary to burn 10 lbs. of sulfur?

3. What weight of hydrogen could be obtained from 100 g. sulfuric acid? $Zn + H_2SO_4 \longrightarrow ZnSO_4 + H_2$

4. What weight of zinc sulfate would be formed in problem 3?

5. What weight of zinc was needed in problem 3?

6. What weight of nitric acid is needed to neutralize 100 g. potassium hydroxide? $HNO_3 + KOH \longrightarrow KNO_3 + H_2O$

Volume problems on gaseous reactions. Volume problems, where only gases are involved in a reaction, are simpler even than weight problems. Let us assume for the present that all volumes are reduced to standard conditions of temperature and pressure — that is, $0°$ C. and 76 cm. of mercury. As expressed by Gay-Lussac's law of combining volumes and interpreted by Avogadro's hypothesis, the volumes involved in a chemical reaction between gases are related in the same way as the numbers of molecules in the reaction. Let us illustrate with the equation for burning ethane :

$$2 C_2H_6 + 7 O_2 \longrightarrow 4 CO_2 + 6 H_2O$$

This equation tells us that 2 liters of ethane react with 7 liters of oxygen to produce 4 liters of carbon dioxide and 6 liters of water vapor. Thus we see that if one has the equation correctly written, he can work a large number of problems by inspection.

EXERCISES AND PROBLEMS

1. Why are volume problems simpler than weight problems?

2. When nothing is said about pressure and temperature, what pressure and what temperature are assumed?

3. State the relation between the numbers of molecules involved in a correctly balanced equation and the respective volumes concerned?

4. Must the equation be correctly balanced in order to represent the facts?

5. Answer the questions in the following:

$$(1) \quad 2 H_2S + 3 O_2 \longrightarrow 2 H_2O + 2 SO_2$$

(a) How many liters of oxygen are necessary to burn 4 liters of H_2S?

(b) How many liters of SO_2 result from burning 1 liter of H_2S?

(c) How many liters of water vapor are formed along with 8 liters of sulfur dioxide?

$$(2) \quad 4 NH_3 + 5 O_2 \longrightarrow 4 NO + 6 H_2O$$

(a) How many liters of oxygen are needed to burn 6 liters of NH_3?

(b) How many liters of NO will result when 2 liters of NH_3 burn?

(c) How many liters of NO are formed with 3 liters of water vapor?

Weight-volume problems. Somewhat more difficult are those weight problems in which the volume of a gaseous product is wanted instead of its weight. The weight must first be calculated and then changed into volume.

Sometimes the volume of a gaseous substance is known but its weight is desired, so it can be used to calculate what weight of some other substance reacts with the given volume.

In both types of problems, changing volume to weight or weight to volume, we can use the relation: *For any gaseous substance, 22.4 liters weigh as many grams as there are units in the molecular weight.* For instance, CO_2 with a molecular weight of 44 will have 44 grams in 22.4 liters. The same volume of H_2S, whose molecular weight is 34, will weigh 34 grams.

In both types of problem it is best to make the change from weight to volume or volume to weight by finding the number of grams in 1 liter of the gas. This is done by dividing the number expressing the molecular weight by 22.4. Let us now try a problem.

What volume of CO_2 would result from heating 400 g. of baking soda according to the equation.

$$2 \, NaHCO_3 \longrightarrow Na_2CO_3 + H_2O + CO_2$$

First step: Calculate the weight of CO_2 formed.
Looking up the atomic weights, we have:

$$\frac{x}{400} = \frac{44}{168} \qquad \therefore x = 104.7 \text{ g. } CO_2$$

Second step: Calculate the weight of 1 liter of CO_2.

$$1 \text{ liter } \frac{44}{22.4} = 1.96 \text{ g.}$$

Third step: In 104.7 g. there would be:

$$\frac{104.7}{1.96} = 53.4 \text{ liters}$$

EXERCISES AND PROBLEMS

1. What fact is used in order to go from weight to volume or from volume to weight?

2. What will 22.4 liters of the following gases weigh under standard conditions?

(a) N_2 (b) CO   (c) HCl   (d) H_2O (e) HBr
(f) NH_3 (g) S_2 (h) H_2 (i) O_3 (j) SO_2

3. Calculate the weight of 1 liter of each of the following gases measured under standard conditions.

(a) N_2O (b) C_2H_2 (c) N_2
(d) HCl   (e) SO_2 (f) H_2

4. What volume of ammonia will react with 60 g. HCl?

$$NH_3 + HCl \longrightarrow NH_4Cl$$

5. What volume of hydrogen results from the reaction of 5 g. of sodium with water:

$$2 \, \text{Na} + 2 \, \text{H}_2\text{O} \longrightarrow 2 \, \text{NaOH} + \text{H}_2$$

6. What volume of oxygen is needed to burn 8 g. of methane?

$$\text{CH}_4 + 3 \, \text{O}_2 \longrightarrow \text{CO}_2 + 2 \, \text{H}_2\text{O}$$

7. What volume of oxygen can be obtained from 50 g. of sodium chlorate?

$$2 \, \text{NaClO}_3 \longrightarrow 2 \, \text{NaCl} + 3 \, \text{O}_2$$

8. What volume of H_2S can be obtained by using 22 g. FeS?

$$\text{FeS} + \text{H}_2\text{SO}_4 \longrightarrow \text{FeSO}_4 + \text{H}_2\text{S}$$

ADDITIONAL PROBLEMS FOR SUPERIOR STUDENTS

1. A certain iron ore is 80 per cent Fe_2O_3. Calculate the weight of iron that might be obtained from a ton of ore.

2. Hydrogen sulfide gas is passed into a solution of $\text{Cu(NO}_3)_2$ and precipitates copper sulfide according to the following equation:

$$\text{H}_2\text{S} + \text{Cu(NO}_3)_2 \longrightarrow \text{CuS} + 2 \, \text{HNO}_3$$

What weight of precipitate can be obtained from 10 liters of H_2S under standard conditions?

3. Given the reaction $\text{FeS} + \text{H}_2\text{SO}_4 \longrightarrow \text{H}_2\text{S} + \text{FeSO}_4$. What volume of gaseous H_2S measured at 750 mm. pressure and 15° C. can be obtained from 500 g. of FeS?

SUMMARY

In working weight problems, it is first necessary to change symbols and formulas into weights by means of the atomic weights. Then:

(1) Set up one fraction from the problem with the unknown in the numerator.

(2) Set up another fraction in the same order from the correctly written formula or correctly balanced equation.

(3) Set the two fractions equal to each other.

(4) Solve for the unknown.

When a problem deals only with gaseous substances, the volumes involved are the same as the numbers of molecules in a correctly balanced equation.

When the volume of a substance is given instead of its we or when the volume is needed instead of the weight, volume be changed into weight or weight into volume by the rela that 22.4 liters of any gaseous substance weighs its molec weight in grams.

$$\xrightleftharpoons{}$$

CARBON AND ITS ELEMENTARY COMPOUNDS

Importance. Because of the great number of useful carbon compounds, the chemistry of carbon is thought of as a special branch of chemistry. Probably half of all chemical literature deals with carbon compounds. Over two hundred thousand compounds of this element have been studied, and hundreds of new ones are studied each month. The human body is largely made up of carbon compounds. Foods, naturally, are mostly compounds of carbon and the other elements found in the human body. Health and sickness depend to a considerable degree upon relationships between carbon compounds. Most animal and plant products are also carbonaceous. Included among them are wood and wood products, stock foods, clothing materials, leather, rope, practically all fuels, varnishes, and paper products — in fact, the major part of those things essential to life and well-being.

Organic chemistry. For a long time, chemists thought that carbon compounds could only be produced in the organs of living plants and animals. This gave the name "organic chemistry" to the study of the carbon compounds. In recent years the chemist has learned how to synthesize thousands of compounds of carbon from simpler carbon substances, including the carbon in coal. Although many compounds can be prepared apart from the organs of plants and animals, the old name of organic chemistry is still used.

Sources of carbon compounds. The carbon dioxide of the air is the ultimate source of practically all of the carbon compounds in plant products. Animals get their carbon from the plants. Nature has left us a vast heritage of carbonaceous material in coal and crude oil, two of our most abundant raw products. Certain rocks, including limestone, marble, chalk, and magnesite, are really carbon compounds, although

441

they are usually considered as part of inorganic chemistry, since in properties they resemble the other rocks, which are strictly inorganic.

Carbon as the element. The element carbon exists in three allotropic forms — charcoal, graphite, and the diamond. The properties of the grimy charcoal, the "greasy" graphite, and the sparkling diamond are very different, but these substances are proved to be the same by burning. Burn 1 milligram of diamond, 1 milligram of graphite, and 1 milligram of charcoal, separately, and in each case the product is $3\frac{1}{3}$ milligrams of carbon dioxide. The only possible conclusion is that they are different forms of the same element.

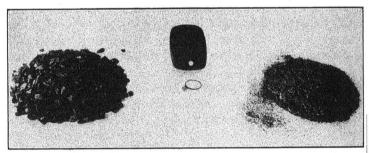

CHARCOAL, DIAMOND, AND GRAPHITE. The three allotropic forms of carbon are quite different, but all have their uses.

Why the forms of carbon are different. The difference between the three forms is due to the arrangement of the carbon atoms in the molecule, to the arrangement of the molecules themselves, or to both. The diamond is crystallized carbon, while charcoal, at least, is not. Crystals are due to a regular arrangement of the molecules.

Why the diamond is valuable. A good-quality diamond the size of a pea is worth several hundred dollars. The properties which make it valuable are hardness, high light-refracting power, insolubility, permanency, and rareness. It sparkles because it refracts the light and breaks a light beam into its colors. Discolored diamonds, although not highly prized as jewels, are used as the cutting tips of glass cutters and rock

drills.   Diamond is the hardest substance known, being 10 on
the mineralogist's scale of hardness.

Scale of hardness.   Minerals are grouped into 10 classes
according to hardness.   Those substances, such as talc, which
can be easily scratched with the fingernail are classed as having
hardness 1.   A substance with hardness 2 on this scale can be
scratched by the
fingernail only with
difficulty.   Miner-
als which cannot be
scratched with the
fingernail but which
can be scratched
with a brass pin
have a hardness of
3.   A knife blade
will scratch those
minerals with a
hardness of 5 but
not those of hard-
ness of 6.   Quartz
has a hardness of

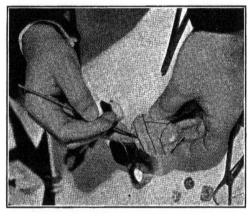

MARKING FOR CLEAVING.   Large rough diamonds are
usually split up into smaller stones, which in turn are
cut into regular patterns to bring out their brilliance.

7, topaz of 8, and corundum of 9.   The diamond's position at
the top of the scale of hardness indicates that it can scratch all
other minerals.

The hardness test may be used to tell the difference between
a real diamond and an imitation jewel of quartz or glass.
With a corner of the supposed diamond, rub the face of a
crystal of topaz.   If the jewel is diamond, it will scratch the
topaz; but if it is quartz or glass, it will not.

Artificial diamonds.   Chemists have always desired to make
diamonds artificially, but they have only partially succeeded.
The French chemist Moissan made some diamonds of microscopic size, but of no commercial value.   There are two common ways of obtaining crystals — a saturated solution may be
allowed to cool or evaporate, or a molten material allowed to
solidify.   Neither of these methods is feasible for crystallizing

carbon. Carbon is not appreciably soluble in any known solvent. Nor will it melt at any combination of temperature and pressure that man can produce. Nature produces diamonds by heating carbon to a very high temperature under enormous pressures underground. Moissan tried combining the two methods mentioned above. Carbon being more soluble in melted iron than in any other solvent, he saturated molten iron with carbon and then plunged the iron into cold water. Chilling the iron caused it to give up most of its carbon and to shrink with a tremendous internal pressure. This pressure crystallized the carbon into small black diamonds. Recently Karabacek of Vienna has invented a process by which he has been able to prepare good-quality diamonds ten times as large as those prepared by Moissan. This process has promise of commercial success. It uses the carbon in carbon monoxide, which is made to decompose under great pressure and heat.

Charcoal. Wood charcoal is made by heating wood in the absence of sufficient air to burn it. Gases and volatile liquids are driven off, leaving the charcoal. The gases are used for fuel, and the liquids are condensed and purified for the useful compounds they contain. The charcoal itself is used largely as a fuel.

Charcoal has several properties that make it useful — its resistance to chemical action, its color, and its ability to absorb large volumes of gases and colored substances. Experience has shown that charcoal made from peach and almond pits will absorb a greater volume of poisonous gases than will wood charcoal from other sources. For this reason, peach or almond charcoal is used in gas masks. Charcoal will not dissolve or tarnish or deteriorate from decay. This permanence, together with its blackness, makes it a choice black pigment, used especially in printer's ink.

Lampblack. Lampblack, or soot, is practically pure carbon. It is made by burning oils in a limited supply of air, while a cold object moves through the flame and cools the carbon below its kindling temperature. The unburned carbon is deposited as a soft, amorphous, slightly greasy powder. Lamp-black is used in the making of paint, shoe blacking, and printer's ink.

Boneblack. Boneblack, or animal charcoal, made by dry distillation of bones, consists of carbon mixed with a large percentage of minerals, chiefly calcium phosphate. All amorphous carbon has the property of absorbing large quantities of gases

CHARCOAL AS A DECOLORIZING AGENT. One of the principal uses of charcoal is for absorbing coloring substances. Notice how much lighter the liquid in the flask, which has been filtered through charcoal, is than the unfiltered liquid in the beaker.

and organic coloring matter. Boneblack has this property to a greater extent than any other form. This property makes it adapted for filtering processes. Boneblack is always used to decolorize the sirups in the refining of sugar. Charcoal lozenges are given to absorb distressing gases in certain stomach disorders.

EXPERIMENT

Decolorization with charcoal. Place a folded filter paper in a funnel. Fill it half full of boneblack. Pour through this some highly colored vinegar or water containing a little dye and note decolorization.

QUESTIONS OF FACT

1. What part of known chemistry deals with carbon compounds?
2. Name seven substances that are largely carbon compounds.
3. Why is the chemistry of the carbon compounds called organic chemistry?
4. Where do the animals and plants get their carbon?
5. Name some carbon compounds that are not ordinarily studied in organic chemistry.
6. Name the three allotropic forms of carbon.
7. How can it be proved that charcoal, graphite, and diamond are the same element?
8. How can there be allotropic forms of an element?
9. Recall two other elements that we have studied which exist in more than one form.
10. What are the difficulties in synthesizing a diamond?
11. What are some of the uses of charcoal?
12. Name some of the different varieties of charcoal and tell how they are made.
13. Discuss the permanence of charcoal.
14. How could one distinguish a real diamond from an imitation gem of glass?
15. How hard is a mineral of hardness 1?
16. What is the hardness of each of several minerals which respond as follows?

(a) Barely scratched with the fingernail.
(b) Barely scratched with a knife blade.
(c) Not scratched by the fingernail but is scratched by a brass pin.
(d) Scratched by quartz but not by a knife blade.

17. Describe two attempts to synthesize diamonds.

References for Supplementary Reading

1. Foster, William. *The Romance of Chemistry*. D. Appleton-Century Co., New York, 1936. Chap. XXI, pp. 341–359, "Carbon, Producer of Energy."
   (1) Discuss the occurrence of diamonds. (p. 342–343)
   (2) Does the diamond in a ring setting possess its natural crystal shape? (pp. 343–344)
   (3) Give the history of the largest diamond ever found. (pp. 344–345)
   (4) Where does graphite occur naturally? (pp. 345–346)

(5) How is graphite manufactured? (p. 346)

(6) Give some uses of graphite. (p. 347)

2. Martin, Geoffrey. *Triumphs and Wonders of Modern Chemistry.* D. Van Nostrand Co., New York, 1922. Chap. X, pp. 212–239, "The Element Carbon"; and Chap. XI, pp. 240–270, "Carbon Dioxide."

    (1) Discuss the importance of the element carbon. (pp. 212–213)

    (2) What weight of carbon is in a man weighing 154 lbs.? (p. 213)

    (3) Discuss the occurrence of carbon in outer space. (pp. 213–214)

    (4) What is unusual about the vaporizing temperature of carbon? (p. 215)

    (5) Under what conditions has it been estimated that carbon would liquefy? (p. 215)

    (6) In what sort of places do diamonds occur? (p. 218)

    (7) Relate the discovery of diamonds in South Africa. (pp. 219–221)

    (8) How was the nature of the diamond discovered? (p. 224)

    (9) Relate the story of one famous diamond. (pp. 226–228)

Courtesy of U. S. Bureau of Mines

MINING COAL. The value of the coal taken from the earth greatly exceeds that of any other mineral product.

Coal and coal products. Coal is the modified remains of prehistoric forests. This partially decomposed material contains a higher percentage of carbon than firewood. In every continent and in almost every country there exist seams of coal, varying in thickness from a fraction of an inch to scores of feet. Admiral Byrd reports finding coal high on the exposed mountain peaks of Antarctica. It has been estimated that in North America alone there are five trillion tons of coal. Coal is no doubt one of our most valuable heritages from the past. When our oil wells all run dry, we still will be able to synthesize motor fuels from coal.

Kinds of coal. Coal varies from anthracite, which is highest in carbon, through bituminous coal to peat, which is coal in the making. Anthracite coal is the highest in heat units, the lowest in smoke produced, the longest burning, and the freest from dust. It makes some of the best domestic fuel, although the anthracite with highest carbon content requires a heavy draught for burning. Anthracite coal is iron-black in color, does not soil the fingers, and burns with a short blue flame with little odor and no smoke.

Bituminous coal is a softer coal than anthracite. Its color varies from pitch-black to dark gray. This form of coal soils the fingers, burns with a long yellowish flame and a suffocating odor. There are many grades of bituminous coal. A very soft variety is cannel coal. The softer varieties of bituminous coal are valuable in the manufacture of fuel gas.

EXPERIMENT

1. Examine any sample of coal available and answer the following questions:
   (a) Does it soil the fingers?
   (b) Does it burn with a yellow or blue flame?
   (c) Does it suggest wood in texture?
   (d) Is it anthracite or bituminous coal?
2. If a sample of coal tar is available, note its odor and taste.
3. Recall any uses of coal tar.
4. Heat a piece of soft coal in a pyrex test tube, ignite the escaping gas, and note the tarry liquids formed.

Coke. Coke is an amorphous form of carbon obtained by the destructive distillation of bituminous coal and oil in gas manufacture. It is used in reducing iron and other metals from their ores. The iron ores are principally the oxides. When these are dumped into the hot furnace along with the coke, the carbon of the coke removes the oxygen from the metal and escapes with it from the furnace in the form of carbon monoxide. The United States alone uses over 50 million tons of coke annually.

BATTERY OF BY-PRODUCT COKE OVENS. In the modern manufacture of coke none of the valuable volatile constituents of the coal are lost.

In the early days of making coke no effort was made to save the volatile parts of the coal. However, in the modern by-product coking process, coal gas and a mixture of light liquids and heavy, tarry liquids are recovered. A complete discussion of the many substances occurring in the liquids and tar from coal would carry us half way through organic chemistry.

Graphite. Graphite mixed with clay is the so-called "lead" in lead pencils. Its value lies in its readiness to leave a mark.

The hardness is determined by the proportion of clay.   Graphite usually consists of flat scales which readily slip on one another.   This property makes it useful in lubricating greases, especially those used to lubricate hot surfaces.   At high temperatures, ordinary lubricating oils melt and run out.   Graphite, however, neither melts nor burns at the normal temperature

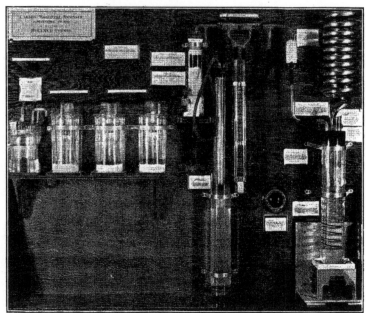

Courtesy of Port of New York Authority

CARBON MONOXIDE ANALYZER USED IN HOLLAND TUNNEL.   The air containing carbon monoxide enters at the left and is purified of acid, oil, moisture, etc. in the "train" of vessels along the shelf.   The long apparatus in the center keeps the air at a constant pressure.   That at the right "burns" the carbon monoxide to carbon dioxide with the aid of a catalyst.   The heat of combustion, which is proportional to the concentration of carbon monoxide, is measured by a delicate thermocouple.   When the carbon monoxide reaches a concentration of 3.5 parts in 10,000 parts of air, a signal light flashes on.

of any bearing; hence, it serves the purpose very well.   Because of its high resistance to heat, graphite is also used in crucibles and electric furnaces.

The oxides of carbon.   Carbon forms two oxides.   When carbon or a carbon compound partially burns in a limited supply of oxygen, carbon monoxide (CO) is formed. When the carbon is completely burned in plenty of oxygen, the resulting substance is carbon dioxide (CO_2).

Carbon monoxide. Carbon monoxide is a very dangerous substance. Occasionally we read of someone's being asphyxiated by this poisonous gas. Death has been known to result from burning a pan of charcoal in a closed room. Some fuel gas contains carbon monoxide as one of its constituents. Loss of life sometimes occurs from the accidental leakage of such gas. It is safest never to sleep in a tightly closed room with gas connections. Coal also is a possible source of danger from carbon monoxide if it is not burning under a good draft. For that matter, any carbonaceous substance may become a source of danger where burning is sluggish.

DEADLY CARBON MONOXIDE. Carbon monoxide strikes its victims without warning. This extremely poisonous gas is almost always present in the exhaust gases of a running automobile engine.

Waste gases from the automobile are a common source of carbon monoxide poisoning. The exhaust of any car contains more or less carbon monoxide. When the car is not working well, the exhaust gases are very rich in the monoxide. Although this gas is very poisonous, it strikes its victim down without warning. He detects no odor nor notices any peculiar sensation. He simply falls unconscious and, unless immediately removed to fresh air, soon dies. The gas destroys the red corpuscles of the blood. The recovery of a person saved from death following the inhalation of this gas is very slow.

NATURAL-GAS TANKS.   Billions of cubic feet of natural gas are burned every year.

Farmers sometimes kill ground squirrels by piping automobile exhaust gases from a too rich gasoline mixture into their burrows. The concentrated carbon monoxide makes short work of the pests.

Fuel gases. Nearly all fuels are either carbon or carbon compounds. Since carbon itself is always a solid, gaseous fuels must be mostly carbon compounds. The gaseous fuels used on a large scale by large cities are usually made from coal as the source of the carbon compounds.

Natural gas. In certain oil fields in the western

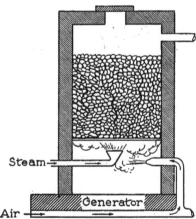

WATER GAS. A fuel gas is sometimes made by spraying steam into very hot coke. The products are hydrogen and carbon monoxide. Both of these gases are excellent fuels.

and midwestern parts of the United States, there are literally billions of cubic feet of natural gas under the ground. This gas is being piped hundreds of miles to large cities, where millions of cubic feet of it are used each year for cooking, for heating homes, and in the industries. The principal constituent in natural gas is methane (CH_4). Natural gas has a higher fuel value than most artificial gases and requires special burners to supply an extra amount of air.

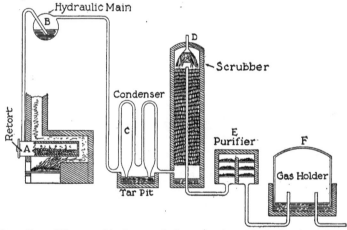

COAL GAS. Where coal is cheap, a fuel gas may be made by heating soft coal in the absence of air. The coal is heated in the retort A, liquids and gases being driven off by the heat. The trap B catches the high-boiling liquids. Most of the low-boiling liquids are condensed in C. Active gases, such as sulfur dioxide and the oxides of nitrogen, dissolve in water as the gas rises through the scrubber D. Any other undesirable gas will react with the chemicals in the purifier E. The cleaned gas collects in the tank F.

Water gas. In districts where hard coal is cheap, a gaseous fuel is made by spraying steam on white-hot coal:

$$C + H_2O \longrightarrow H_2 + CO$$

Since steam is gaseous water, it is natural that the product should be called water gas. The fuel content of water gas is 100 per cent, as both the hydrogen and the carbon monoxide are good fuels.

Coal gas. In communities where soft coal of good quality is available at a reasonable cost, the coal is heated in a closed vessel called a retort. The gas driven off is quite similar to natural gas; that is, it is composed of hydrocarbons — compounds of carbon and hydrogen. Coke remains in the retort. This gas is named coal gas.

Producer gas. Where low-grade coal is cheap, a limited supply of air is forced into the heated coal. Partial burning results

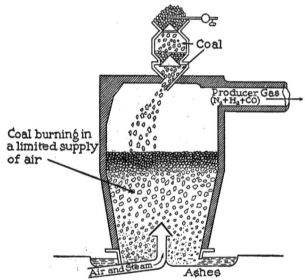

PRODUCER GAS. This fuel gas is made by partially burning coal with air. The carbon monoxide resulting from the partial combustion is an excellent fuel.

in carbon monoxide, mixed with large amounts of nitrogen and small amounts of hydrocarbons. This is called producer gas.

$$4 N_2 + O_2 + 2 C \longrightarrow 4 N_2 + 2 CO$$

If produced strictly according to the equation, only one-third (by volume) of the gas is combustible. In practice, however, soft coal is used instead of carbon, so small amounts of hydrocarbons enrich the gas and raise its fuel value to 35–40 per cent.

EXERCISES

1. Discuss the distribution of coal.
2. Describe anthracite coal. Do you use it at home?

3. Describe bituminous coal. Is it the kind usually used for domestic heaters?

What name is given to coal in the process of formation?

Is coke a natural substance or a man-made product?

How do we happen to need coke in large quantities?

What is the principal use of coke?

8. What is the meaning of the term by-product coking?

9. What is the substance referred to as "the lead" in lead pencils?

10. What property of graphite makes it a lubricant for bearings?

11. Under what conditions is graphite especially suitable for lubrication?

12. Under what circumstances is carbon monoxide likely to form?

13. Mention several situations where the formation of carbon monoxide is likely to become serious.

14. Discuss the relation of carbon to fuels.

15. Discuss natural gas as to: (a) supply; (b) principal constituent; (c) fuel value; (d) burners.

16. Discuss water gas as to: (a) constituents; (b) origin of the name; (c) how produced; (d) why very poisonous.

17. Discuss coal gas as to: (a) where use is feasible; (b) how made; (c) similar to what other gas; (d) by-product.

18. Discuss producer gas as to: (a) where use is feasible; (b) how made; (c) fuel constituent; (d) nonfuel constituent; (e) fuel value; (f) poisonous nature.

ADDITIONAL EXERCISE FOR SUPERIOR STUDENTS

1. Prepare an extensive report on the world's coal resources, using reference 3 in the following list.

REFERENCES FOR SUPPLEMENTARY READING

1. Glover, John George, and Cornell, William Bouck. *The Development of American Industries.* Prentice-Hall, Inc., 1933. Chap. XVI, pp. 331–356, "The Coal Industry."
   (1) When was the first authentic record of the use of coal? (p. 331)
   (2) What did Marco Polo tell about coal in China? (p. 331)
   (3) Mention early discoveries of coal in the United States. (p. 332)
   (4) What difficulty was experienced in marketing the early coal? (p. 332)
   (5) Give six classes of coal. (p. 333)
   (6) Discuss briefly the distribution of coal in the United States. (pp. 334–336)

(7) Describe the largest coal mine in the world. (p. 342)

(8) How many by-products of coal are there? (p. 343)

(9) In what connection are most of them obtained? (p. 343)

2. Howe, H. E. *Chemistry in Industry*. The Chemical Foundation, Inc., New York, 1924. Vol. I, Chap. IV, pp. 58–74, "Coal, Coke, and Their Products."

(1) Give one theory as to the origin of coal. (pp. 59–62)

(2) What improvements are needed in connection with the burning of coal? (p. 63)

(3) Of what importance is by-product coking in time of war? (pp. 66–68)

(4) List the weights of materials obtained from one ton of Pittsburgh coal. (p. 72)

(5) In making coke by the old process, what part of these substances is lost? (p. 72)

(6) What is the temperature of high-temperature carbonization? (p. 69) Of low-temperature carbonization? (p. 72)

(7) How are products of low-temperature carbonization different from those of high-temperature carbonization? (p. 73)

3. Moore, Elwood S. *Coal*. John Wiley and Sons, Inc., New York, 1922.

Carbon dioxide. We learned that carbon dioxide exists in the atmosphere to the extent of 3 parts out of every 10,000. Although this seems like a small percentage, it really represents an enormous weight, considering the entire atmosphere. This supply assumes great importance when we realize that all forms of life on the earth obtain their carbon from the carbon dioxide of the air as the ultimate source. Although animals cannot use the carbon dioxide directly, they get their carbon from plants, which take it from the air. The processes that restore carbon dioxide to the air just about balance its removal by plants, so that the concentration generally does not vary permanently. The breathing of animals, oxidation of carbonaceous materials, fermentation of sugarlike substances, bacterial decay in the soil, and volcanic heating of certain minerals all restore carbon dioxide to the air.

Carbon dioxide and human welfare. Since we exhale about 1.5 quarts of carbon dioxide every minute, air that contains much more than the normal amount of this gas should be regarded with suspicion. An excess of carbon dioxide is a good indication that air is not fresh. To the extent that this oxide has been formed, the oxygen of the air has been removed. Stale air is also objectionable for the reason that it contains certain more or less poisonous organic substances, which are breathed out by men and animals.   There is also a limit to the amount of carbon dioxide itself that we can stand.   If its concentration exceeds 2 per cent, the lungs have difficulty in eliminating it in the face of so much of the same substance trying to go in the reverse direction.

In a concentration of 8 per cent carbon dioxide, one would soon die.   Carbon dioxide, being heavier than air, often settles to the bottom of mines and wells.   If persons entering

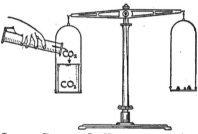

CARBON DIOXIDE IS HEAVIER THAN AIR. The experiment shows why carbon dioxide tends to collect in mines and deep wells.

such places are unaware of the danger, they may be suddenly overcome by the bad air and die of its effects.   No one should enter any closed place where there may be carbon dioxide in more than ordinary concentrations.   Recently the newspapers carried an account of men who had entered wine tanks for the purpose of cleaning them out and were killed by carbon dioxide formed in the fermentation of the wine.   Miners usually test the air before daring to enter a shaft which has not been ventilated recently.   Their method of testing the air is rather ingenious. A cage containing white mice or canaries is first lowered into the shaft, and the little creatures are watched for the effect of the air upon them.   Canaries and white mice are more quickly distressed by bad air than man.   A canary shows signs of distress as soon as it is put into bad air.   Its wings begin to droop and quiver.   White mice become uneasy and dart back and forth as if trying to escape the discomfort caused by the dangerous air.

There is a new possibility of danger from carbon dioxide in connection with the use of "dry ice," which is practically pure carbon dioxide.   As long as the ventilation is good, and large amounts are not evaporating, there is little danger; but should large amounts evaporate where the resulting heavy gas could settle into a deep basement, there might be danger. At this time it might be well to mention the possibility of freezing the hand by too long contact with dry ice. It must be remembered that the freezing point of carbon dioxide is about $-78°$ C., while that of water is only $0°$ C. The hand can be kept much longer in contact with ice without injury than with solid carbon dioxide.

In the face of so much danger from carbon dioxide, we may become unduly afraid of it. This substance is used to carbonate drinks. We like the effervescing "soda pop" with its tang better than the same drink without the carbon dioxide.

LABORATORY PREPARATION OF CARBON DIOXIDE. Any strong acid will react with any carbonate to release carbon dioxide.

No one has suggested that the small amount of carbon dioxide in carbonated drinks does us any harm. Stranger than the fact that we like carbon dioxide in our drinks, is the fact that a little carbon dioxide is needed in the lungs to stimulate breathing. It has been found that a person suffocated to the point that he has stopped breathing will respond more quickly to the pulmotor, a device for artificial breathing, when there is considerably more than ordinary concentration of carbon dioxide in the air administered.

The laboratory preparation of carbon dioxide. For the preparation of carbon dioxide in the laboratory, we use the action of an acid on a carbonate. Calcium carbonate is the cheapest carbonate, and hydrochloric acid is a suitable acid.

$$CaCO_3 + 2 HCl \longrightarrow CaCl_2 + H_2O + CO_2$$

Almost any combination of carbonate and acid can be used for the preparation of this oxide. The small boy often prepares himself a carbonated drink by pouring vinegar into a solution of baking soda. Although this drink cannot be generally recommended, it is a carbonated drink all right and has plenty of effervescence.

The test for carbon dioxide. To test a gas for the presence of carbon dioxide, it is passed through a freshly filtered solution of calcium hydroxide. The formation of a white precipitate of calcium carbonate serves as the test.

$$Ca(OH)_2 + CO_2 \longrightarrow CaCO_3 + H_2O$$

This reaction is typical of reactions which go to completion because of the insolubility of a product. The calcium carbonate, often called precipitated chalk, is chemically the same as marble or limestone. It is quite insoluble. This test for carbon dioxide is valuable also for detecting the presence of carbon in unknown substances, thus proving them to be organic. In this test the unknown substance is heated with copper oxide. If it contains carbon, this element is burned to carbon dioxide, which can be led into limewater and tested as above.

TEST FOR CARBON DIOXIDE. Carbon dioxide gas forms a white precipitate when passed into clean limewater. The drawing shows the test for carbon dioxide in the breath.

EXPERIMENT

Testing the carbon dioxide in the breath. Prepare some freshly filtered calcium hydroxide solution. Blow your breath through this limewater and note the formation of a white precipitate. This precipitate ($CaCO_3$) is a test for carbon dioxide. Continue to blow through the solution until the precipitate dissolves. The excess of carbon dioxide forms carbonic acid, which dissolves the carbonate.

Sources of carbon dioxide. The carbon dioxide of commerce is mostly obtained as a by-product of the fermentation industries, as it is always formed as one of the products of fermentation. Commercial alcohol for industrial use is largely obtained from the fermentation of molasses. Since carbon dioxide is a necessary by-product, it is natural that it should be marketed also. Other sources of carbon dioxide are limestone and magnesium carbonate, rocks of common occurrence. In the preparation of many useful compounds of calcium and magnesium,

A BLOCK OF DRY ICE EMERGING FROM THE PRESS. Ordinarily carbon dioxide is kept liquid only by keeping it under high pressure. When this pressure is released and the liquid allowed to flow into the chamber of a large press, part of the carbon dioxide vaporizes, and the rest is frozen by the loss of heat required to vaporize the first portion. The gaseous part is pumped to the compressor to be again liquefied, but the frozen part is pressed into a large cube by the piston of the press.

carbon dioxide is recovered as a by-product. Heat alone will drive off this carbon dioxide, leaving quicklime and magnesia.

$$CaCO_3 \longrightarrow CaO + CO_2$$
$$MgCO_3 \longrightarrow MgO + CO_2$$

The uses of carbon dioxide. A few years ago, when the carbonating of drinks was practically the only commercial use of carbon dioxide, more of this substance was produced by the alcohol manufacturers than could be marketed at a profit.

One of the large producers offered a prize for new uses of this oxide. Out of the resulting search for new uses came dry ice, the solid carbon dioxide used in refrigeration. One may wonder why the expression "dry ice." Since it is very cold, looks like ice, and serves some of the same purposes as ice, the term "ice" was natural and appropriate. The "dry" part of the name comes from the fact that solid carbon dioxide sublimes — that is, passes directly from solid to gas without going through the liquid state. In the case of ordinary ice, melting results in water. Provision must be made for the disposal of the water. If dry ice is used, no such provision is necessary.

EXPERIMENT

Examining dry ice. Examine dry ice (solid CO_2). Freeze water with it. Explain fuming in air. Use to extinguish a flame by allowing fumes to settle on it.

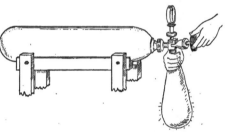

MAKING DRY ICE. When liquid carbon dioxide is allowed to escape from a container, part of it evaporates. The heat for vaporizing it comes from the carbon dioxide itself, thus the temperature of the remaining liquid is reduced to below the freezing point. The frozen carbon dioxide is the familiar dry ice.

The manufacture of dry ice makes use of the same principle as that employed in making liquid air. The steps are heating a gas by compression, removing the heat, and then allowing the gas to cool by expansion. Liquid carbon dioxide is sold in tanks for carbonating drinks. If the liquid is allowed to escape into a sack, part of it becomes solid at the expense of the heat necessary to evaporate the rest.

Solid or liquid carbon dioxide is an excellent weapon for fighting fire. Its low temperature and readiness to cool by evaporation tend to cool the burning substance below the kindling temperature, while the resulting heavy vapor settles around the fire and holds the oxygen back. However, the cost, as yet, limits this use of carbon dioxide.

Carbon dioxide in other processes.   The variety of processes in which carbon dioxide plays a part indicates that it is a very versatile substance.   It is responsible for the raising of bread. In yeast bread, carbon dioxide comes from the fermentation of sugar by the yeast plant.   Gas forms within the dough.   The resulting bubbles expand with the heat of baking, making the bread porous, or light.   In soda biscuits, the carbon dioxide results from the chemical reaction between baking soda and the

EXTINGUISHING A FIRE WITH CARBON DIOXIDE.   Because carbon dioxide refuses to take part in combustion, it is excellent for extinguishing fires.   Being heavier than air, it holds the oxygen away until burning ceases.

acid in the sour milk.   Baking powder contains both soda and acid constituents, which react upon the addition of moisture.

Carbonic acid.   When carbon dioxide is dissolved in water, some of it reacts with the water to form weak carbonic acid. (Carbonic acid must not be confused with the poisonous carbolic acid.)   One reason we like carbonated drinks is because of a little carbonic acid in them.   Carbonic acid never occurs in high concentrations, since it tends to decompose into carbon dioxide and water.

$$H_2CO_3 \longrightarrow CO_2 + H_2O$$

By holding the carbon dioxide in contact with the solution by means of pressure, the above reaction can be reversed to some extent, thereby increasing the concentration of the acid. There are two reasons why "soda pop" effervesces when the stopper is removed from the bottle; one is that carbon dioxide, like all other gases, is more soluble under high pressure than under lower pressure. When the pressure is reduced by uncorking the bottle, the excess gas begins to leave the solution. The other reason is that the carbonic acid in solution begins to break down as soon as the pressure is released and contributes more carbon dioxide to the rapidly forming bubbles. The foregoing reaction is a typical equilibrium reaction, which should be written:

$$H_2CO_3 \rightleftharpoons CO_2 + H_2O$$

Increasing the pressure makes the CO_2 more concentrated, so that mass action forces the equilibrium towards the left. Boiling a solution of carbon dioxide would have the reverse effect, and all the carbon dioxide would be expelled from solution.

QUESTIONS OF FACT

1. Name the sources of the carbon dioxide used in industry.
2. Under what conditions may carbon dioxide cause death?
3. Describe the test for carbon dioxide.
4. Explain the use of CO_2 as "dry ice."
5. What very poisonous gas is in the exhaust gases of an automobile?
6. What is carbonic acid?
7. What does an arrow pointing in both directions mean?
8. Explain why carbon dioxide is suitable for fire fighting.
9. Write the formulas for: (a) carbon monoxide; (b) methane (see page 453); (c) carbon dioxide; (d) carbonic acid.
10. Recall the percentage of carbon dioxide in the air.
11. Name five processes which add carbon dioxide to the atmosphere.
12. Where do animals get the carbon in their bodies?
13. How do miners test the safety of air in underground shafts?
14. State the part played by carbon dioxide in the resuscitation of asphyxiated persons.

15. Describe a test for carbon in organic substances.

16. Why does "soda pop" effervesce?

17. Why is the name "dry ice" especially suitable for solid carbon dioxide?

QUESTIONS OF UNDERSTANDING

1. How is solid carbon dioxide obtained?

2. Discuss briefly the relation of carbon dioxide to each of the following: (a) raising of bread; (b) hardening of mortar (p. 277); (c) hardening of whitewash. (p. 281)

3. Given the equilibrium reaction: $H_2CO_3 \rightleftharpoons CO_2 + H_2O$

(a) How can the equilibrium be shifted so there will be a greater concentration of carbonic acid in the solution?

(b) How can the reaction be made to go to completion towards the right?

4. Of what acid is carbon dioxide the anhydride?

5. Choose the correct word in each of the following sentences:

(a) While the "soda pop" is corked up in the bottle, the solution is (unsaturated, saturated, supersaturated) with CO_2.

(b) When the cork has been removed and effervescence has ceased, the solution is (unsaturated, saturated, supersaturated) with carbon dioxide.

(c) If some of the soda water should now be boiled, it would be (unsaturated, saturated, supersaturated) with CO_2.

6. What will 22.4 liters of each of the following weigh, if measured under standard conditions? (a) CO; (b) CO_2.

7. Balance the following equations:

(a) $BaCO_3 + HCl \longrightarrow BaCl_2 + CO_2 + H_2O$

(b) $CaCO_3 + H_3PO_4 \longrightarrow Ca_3(PO_4)_2 + CO_2 + H_2O$

8. Explain why solid carbon dioxide is more injurious to the hands than ordinary ice.

PROBLEMS

1. Given the equation:

$$CaCO_3 + 2 HCl \longrightarrow CaCl_2 + CO_2 + H_2O$$

What weight of carbon dioxide could be obtained from 2000 grams of $CaCO_3$?

2. What volume would the carbon dioxide from problem 1 occupy under standard conditions of temperature and pressure?

3. What weight and what volume of carbon dioxide could be obtained by completely burning 500 g. of graphite?

4. What weight of calcium chloride was formed in problem 1?

5. What volume would the water formed in problem 1 occupy if it were water vapor under 76 cm. of mercury and at 100° C.?

REFERENCE FOR SUPPLEMENTARY READING

1. Foster, William. *The Romance of Chemistry.* D. Appleton-Century Co., New York, 1936. Chap. XXI, pp. 341–359, "Carbon, Producer of Energy."

   (1) Relate a strange carbon dioxide story. (p. 355)

   (2) In what sense is carbon dioxide an essential food for plants? (pp. 356–357)

SUMMARY

Organic chemistry, the chemistry of the carbon compounds, is associated with health, foods, clothes, fuels, paper, etc. Coal, crude oil, and plant and animal products are the chief sources of organic compounds. As an element, carbon exists in the form of charcoal, graphite, and diamond. Diamond, the hardest substance known, is valuable because of its hardness and high refraction of light. Graphite is used in pencils and in lubricants. Charcoal is valuable as an absorbent for gases and coloring matter and as a black pigment.

Partially burned carbon, or carbon monoxide, is a very poisonous gas used mostly in gaseous fuels. The fuel gases are natural gas (a mixture of hydrocarbons), water gas (a mixture of hydrogen and carbon monoxide), coal gas (mostly hydrocarbons), and producer gas (mostly carbon monoxide).

Carbon dioxide is an important constituent of the atmosphere. Being emitted with the breath, its presence is a good index of the condition of ventilation; a concentration of 2 per cent is dangerous and of 8 per cent is fatal. However, the little CO_2 in carbonated drinks is safe. In the laboratory, carbon dioxide is formed by the action of acid on carbonates. The carbon dioxide of commerce comes mostly as a by-product in the fermentation of molasses. Carbon dioxide is important in bread making. To a certain extent carbon dioxide in solution is in the form of carbonic acid.

←
→

CHAPTER 25

SOME SIMPLE CARBON COMPOUNDS

Hydrocarbons. In crude oil (petroleum) we have a large number of compounds of carbon and hydrogen only. These we call hydrocarbons. Each year millions of barrels of oil are obtained from oil-bearing sandstones and conglomerate, called

AN OIL FIELD. An almost unbelievably large amount of oil is obtained from such a field as this. How such vast quantities of oil were formed is one of the mysteries of nature.

"oil sands." The oil is obtained by drilling wells through the overlying strata, sometimes to the depth of several thousand feet. Most of the petroleum is piped to some seaport to be refined. In one pipe line in the United States, oil is pumped from Oklahoma to the Atlantic coast, a distance of a thousand miles.

Nature of oil. Crude petroleum is a very complex mixture of hydrocarbons. These different hydrocarbons have boiling points ranging all the way from zero to several hundred

degrees centigrade. · For fuel purposes many crude oils can be used directly without any purification, or after the removal of a few impurities with sulfuric acid. The appearance and constituents of crude oil vary a great deal in the different localities and even from well to well. Some oils are thin and yellow, others greenish, and still others brown or black. There are all degrees of viscosity from the very light, thin oil up to the molasseslike asphalt oil, which in some cases is too thick to pump. Some oils are said to have an asphalt base as they contain large quantities of solid or semisolid asphaltum in solution. Others are said to have a paraffin base as they contain dissolved paraffin or "para-wax." In some wells the oils contain sulfur compounds and a few nitrogen compounds in solution.

Early history of petroleum. When petroleum was first discovered, it seemed to be of little use. A little of it was sold for medicinal purposes. This amount was not a "drop in the

Courtesy of Standard Oil Co of Calif

OIL REFINERY. The refining of crude oil into its various components is one of the major industries of the United States. Its methods are getting more complex each year as thousands of new products are needed. These products are being used as the raw material for the manufacture of synthetic resins, synthetic alcohols, and other synthetic products of many kinds. There appears to be no limit to the possibilities of synthesis from crude oil. The oil industries employ thousands of chemists, and the number will probably be increased from year to year.

bucket" as compared with the vast quantities of oil found. Soon it was learned that the oil would burn, and part of it began to be used in lamps under the name of kerosene. There was little sale for the lighter oils, so there was a tendency to put the lighter oils into the kerosene. This resulted in many serious explosions and fires. The government had to pass a law forbidding the presence of the low-boiling oils in kerosene.

After the internal combustion engine was invented, the demand for gasoline became much greater than that for kerosene.

Today we would not need a law to keep gasoline oils from being put into kerosene. Instead, oil chemists are busy inventing ways of getting more gasoline and less kerosene. At present one can throw a lighted match into a cup of kerosene without setting it on fire, because it contains only those high-boiling hydrocarbons that do not vaporize enough at room temperatures to take fire.

Distillation of hydrocarbons. A mixture of hydrocarbons can be separated into a large number of fractions by distilla-

Courtesy of Synthetic Hydrocarbon Co

OIL-CRACKING PLANT. The illustration shows the oil inlet tubes. When oils are heated to high temperature under great pressure, the quantity of gasoline is greatly increased.

tion. A few of these constituents are gases at ordinary temperatures and occur in petroleum only in solution. Here are some of the fractions, in the order of their boiling points: petroleum ether, used as a solvent; gasoline, fuel for the automobile; naphtha or distillate, for fuel; benzine, for dry cleaning; kerosene, for lamps and stoves; light, medium, and heavy lubricating oils; greases; vaseline; and paraffin.

Cracking of petroleum. Twenty years ago engineers were afraid the world supply of gasoline would soon be exhausted. A few men were even so bold as to predict that in twenty years there would not be enough gasoline to meet the demand and that the price would be prohibitive. The twenty years are now up, and their doleful prophecies have not come to pass. Not only have new oil fields been opened up, but also a new process in refining ("cracking") has been discovered, whereby the yield of gasoline has been increased in some cases as much as 50 per cent. Cracking consists of heating the oils to high temperatures under high pressures. The gasolines produced by cracking are better than the old uncracked gasolines. The future will see greater use of the cracking process.

In view of the new Bergius process of making gasoline out of coal and hydrogen (see page 79), engineers now have little fear for the future.

The saturated hydrocarbons. Most of the hydrocarbons in crude oil are very similar in their chemical reactions. In fact, they are quite resistant to ordinary chemical reaction. Only two reactions are common to them all: (a) They all burn; (b) they react with chlorine under the stimulus of sunlight. Because of their resistance to chemical reaction, these hydrocarbons are sometimes called saturated hydrocarbons.

A series of hydrocarbons. There is one hydrocarbon containing one carbon atom, one with two, one with three, and so on. All of these up to the one with twelve carbon atoms, and several others up to the one with sixty carbon atoms have been studied. Perhaps there are hydrocarbons in paraffin that contain more carbon atoms in the molecule than sixty, but because they are difficult to separate, no one has studied them.

METHANE SERIES (THE ALKANES) C_nH_{2n+2}

COMPOUND	FORMULA	MELTING POINT	BOILING POINT	SPECIFIC GRAVITY
Methane	CH_4	$-184°$	$-161.4°$	.415
Ethane	C_2H_6	$-172°$	$-88.3°$	.346
Propane	C_3H_8	$-169°$	$-44.5°$	.585
Butane	C_4H_{10}	$-135°$	$0.6°$	.650
Pentane	C_5H_{12}	$-131.5°$	$36.3°$	.627
Hexane	C_6H_{14}	$-95.3°$	$69°$	.66
Heptane	C_7H_{16}	$-90°$	$98.4°$	.683
Octane	C_8H_{18}	$-56.5°$	$124.6°$	.702
Nonane	C_9H_{20}	$-51°$	$150.6°$	.718
Decane	$C_{10}H_{22}$	$-32°$	$174°$	.747
Undecane . . .	$C_{11}H_{24}$	$-26.5°$	$197°$	.773

Each member of this series differs from the adjoining members by CH_2. This series is sometimes called the *methane* or *paraffin* series of hydrocarbons, or simply the *alkanes*. The formula of any member of this series can readily be written from the general formula C_nH_{2n+2}, where n is the number of atoms of carbon in the molecule. Thus to write decane, the one with 10 carbon atoms, we would write $C_{10}H_{(2 \times 10)+2}$, which is $C_{10}H_{22}$.

This is only part of the series, but it shows us how there can be so many compounds of carbon. It will be noticed that boiling point, melting point, and density increase with molecular weight.

QUESTIONS OF FACT

1. What are the principal substances occurring in crude petroleum?

2. How did the internal combustion engine change the demand for oil products?

3. How do the crude oils from different localities differ in physical properties and chemical composition?

4. How is the cracking of oils accomplished?

5. Name the two chemical reactions common to all the saturated hydrocarbons.

6. What is meant by a series of hydrocarbons?

7. What is meant by an asphalt-base oil?

8. What was the first use of mineral oil?

9. What is meant by a paraffin-base oil?

10. Give the uses for the following oil fractions: (a) benzine; (b) petroleum ether; (c) gasoline; (d) kerosene.

11. Why are certain hydrocarbons called saturated hydrocarbons?

QUESTIONS OF UNDERSTANDING

1. The invention of the internal-combustion engine rendered unnecessary the legal safety standards for kerosene. Explain.

2. How do the physical properties in a series of hydrocarbons vary with the molecular weights?

3. How can substances which are gases at ordinary temperatures be in crude oils? (Consider solution.)

4. What is the general formula for expressing the simple formula for any member of the methane hydrocarbons?

5. Give the formulas by applying this formula for the hydrocarbons with the following numbers of carbon atoms: (a) 4 carbon atoms; (b) 6 carbon atoms; (c) 8 carbon atoms; (d) 10 carbon atoms; (e) 12 carbon atoms; (f) 20 carbon atoms.

PROBLEM

1. Calculate the weight of CH_4 that reacts with 100 lbs. of chlorine according to the following equation: $CH_4 + Cl_2 \longrightarrow CH_3Cl + HCl$

EXERCISE FOR SUPERIOR STUDENTS

1. Prepare an extensive report on the petroleum industry. Use reference 2 in next list.

REFERENCES FOR SUPPLEMENTARY READING

1. Foster, William. *The Romance of Chemistry.* D. Appleton-Century Co., New York, 1936. Chap. XXVI, pp. 360–381, " Organic Chemistry."
   (1) When was the first organic substance synthesized in the laboratory? (p. 360)
   (2) What do we call two substances which have the same simplest formulas but different properties? (p. 361)
   (3) Discuss the distribution of petroleum fields. (p. 366)
   (4) Give the composition of mid-continent petroleum. (pp. 367–368)
   (5) Discuss the effect of cracking on gasoline yields. (pp. 368–369)
2. Egloff, Gustov. *Earth Oil.* The Williams and Wilkins Co., Baltimore, 1933.

Methane. Glancing at the boiling points of the hydrocarbons (p. 469), we see that the first four members of the alkanes are gases under average living temperatures. These occur in underground caverns and compose natural gas. Natural gas is the fuel of many of the cities of the United States. By heat-pressure cracking processes natural gas can be made to yield lampblack for pigment and hydrogen for the fixation of nitrogen and for other industrial processes. Methane (CH_4), the most abundant constituent of natural gas, is also a common gas in coal mines, where it has long been the cause of disastrous explosions.

In the early part of the nineteenth century, the coal mine explosions caused so many deaths in the British Isles that the government called in Sir Humphry Davy, one of the leading chemists of that time, to try to invent some way of avoiding these disasters, which were taking such a large toll of human life. The result of Davy's work was the Davy safety lamp, which greatly reduced the number of gas explosions in coal mines. The modern Davy safety lamp is a small kerosene lamp with a wire gauze chimney. This chimney is not only

Courtesy of U. S Bureau of Mines

EFFECTS OF A GAS EXPLOSION. Many years ago explosions in English coal mines were common and cost many lives. Davy's invention of the safety lamp greatly reduced the number of such explosions.

around the side like an ordinary glass lamp chimney, but is also over the top and under the bottom so that the flame is completely enclosed. A flame is a burning column of gas. Any gaseous explosion is the rapid burning of a large amount of combustible gas, causing a high temperature and great expansion of the products of combustion. In order to have a violent explosion of a gas, such as methane, the oxygen must be mixed with it. Any open flame or spark from metal striking rock can ignite such a mixture.   Once started, the flame travels rapidly throughout the entire mixture.   When the mixture in the coal mine reaches explosive concentrations, the flame in the safety

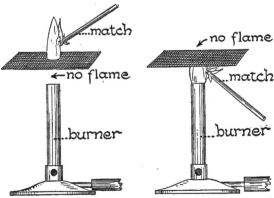

PRINCIPLE OF THE DAVY SAFETY LAMP. If the gas from a Bunsen burner is lighted above a wire gauze, it will not ignite the gas below the gauze, and vice versa. The gauze checks the spread of the flame by cooling the combustible mixture to below its kindling temperature.

lamp will be observed to jump out to the wire gauze chimney and stop; that is, when the mixture starts to burn, it cannot progress farther than the wire gauze.   The reason for its stopping there is that the wire gauze cools the gases in contact with it below the kindling temperature, which causes the burning to stop.

In ponds where vegetation is decaying at the bottom, bubbles of gas are often seen to rise to the surface.   These are methane.

Other constituents of the alkanes in petroleum. Propane and butane, the third and fourth members of the series, are liquefied by great pressure and sold in iron tanks, making possible fuel gas for homes in the country.   Under the name of petroleum ether, the lowest-boiling liquids are sold for priming tractors on cold mornings and for dry cleaning spots from clothes.   Another group of liquid hydrocarbons constitutes gasoline, the fraction most in demand

DAVY SAFETY LAMP

for motor fuel. Between the gasoline fraction and the kerosene fraction there is a group of hydrocarbons collectively called distillate.

Several years ago distillate constituted a considerable percentage of the volume of products from crude oil. With the cracking process, which changed part of the distillate to lower-boiling liquids, and the improvement in the ignition system of the automobile, which permitted still more of this fraction to be crowded into the gasoline fraction, the quantity of distillate has greatly decreased.

THE USES OF MOLECULES OF DIFFERENT SIZES

HYDROCARBON FRACTION	MOLECULAR RANGE	BOILING POINT RANGE	USES
Gases	CH_4 to C_3H_8	Below 0° C.	Fuels, source of lampblack and H_2 gas.
Liquid under pressure	C_3H_8 to C_4H_{10}	Below 1° C.	Fuels
Petroleum ether .	C_5H_{12} to C_6H_{14}	35° C. to 80° C.	Solvent, spotting clothes
Gasoline	C_5H_{12} to $C_{10}H_{22}$	40° C. to 225° C.	Automotive fuels and solvent
Kerosene	$C_{10}H_{22}$ to $C_{16}H_{34}$	225° C. to 300° C.	Illuminating and fuel
Fuel oil (stove distillate, Diesel oil)	Variable	275° C. up	Fuel; Diesel
Lubricating oils .	$C_{16}H_{34}$, etc.	Variable	Lubrication
Vaseline and greases	$C_{16}H_{34}$ to $C_{20}H_{42}$		Salves, ointments, greases
Paraffin	$C_{22}H_{46}$ to $C_{28}H_{58}$		Candles, waterproofing paper, acid bottles

Just above the kerosene fraction is obtained a higher-boiling liquid (stove oil) sold for use in furnaces and stoves. Another fraction heavier than stove oil has in the past been of little demand, and as a consequence has been relatively cheap. The Diesel engine is able to use this cheap high-boiling liquid for its fuel. Because it can use cheap oils and is very efficient, the Diesel engine should come into wide use in the future.

The heavier oils furnish a wide range of lubricating oils — very light oils for sewing machines, the light oils for cars, the medium oils, the heavy oils, and the semisolids, or greases.

The final fraction of these hydrocarbons is paraffin. Some paraffin melts so easily that a square cake may run out on the cupboard shelf on a hot summer day. There is also paraffin which requires a reasonably high temperature for melting. Hydrogen fluoride attacks glass, hence must be kept in paraffin bottles. Naturally, these bottles must have melting points

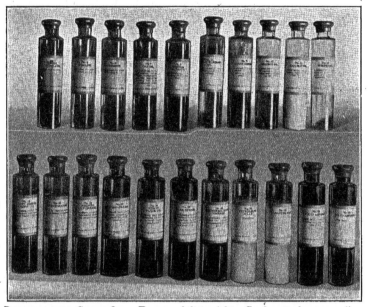

PRODUCTS FROM CRUDE OIL. Top row, left to right: Crude petroleum, gasoline, kerosene, gas oil, medium auto oil, steam turbine oil, liquid grease, soluble oil, soluble oil mixed with water, white oil. Lower row: Heavy Diesel engine oil, medicinal oil, medium cup grease, hard grease, heavy black oil, gear grease; universal joint grease, petroleum jelly, paraffin wax, liquid asphalt, solid asphalt.

far above any possible room temperature. Paraffin, like the other saturated hydrocarbons, is very resistant to reaction. This makes it useful for covering jelly and for protecting labels and different surfaces from active reagents.

In some petroleum oil fields, instead of the white paraffin as the base or ultimate product of distillation, there is a black substance called asphalt. Lubricating oils with a paraffin base are more valuable than those with an asphalt base. Asphalt is partially oxidized, which makes it more susceptible than paraffin to further oxidation. For this reason the asphalt base lubricating oils do not hold up so well as the others.

The alkyl radicals. When one of the hydrogen atoms of any member of the alkanes is replaced, the remainder is called an *alkyl* radical. For instance, CH_3 is methyl, C_2H_5 ethyl, C_3H_7 propyl, and C_4H_9 butyl. The name is obtained by replacing the *ane* ending of the name of the hydrocarbon with *yl*.

The unsaturated hydrocarbons. There is another series of hydrocarbons called the ethylene series, or alkenes. Some of these occur in certain kinds of oil. The chemical conduct of this series is the exact opposite to that of the saturated hydrocarbons of the methane, or paraffin, series. The members of the ethylene series are very reactive. They react vigorously with the chlorine family of elements, with hydrochloric acid, with sulfuric acid, and with all oxidizing agents. A test for the unsaturated hydrocarbons is the rapid decolorizing of potassium permanganate solution or bromine water. The general formula C_nH_{2n} makes it possible to write the formula for any member of the series. The lower members of the series are: ethylene or ethene (C_2H_4); propene (C_3H_6); butene (C_4H_8); pentene (C_5H_{10}); hexene (C_6H_{12}).

The physical properties of the members of this series vary in the same way as those of the methane series — from the lower gaseous members through low and high boiling liquids to solids. All members of this series are combustible, with a slightly more sooty flame than those of the methane series.

The peculiar unsaturated condition of the ethylene series of hydrocarbons leads to the assumption that they have a double bond between two carbon atoms as $H_2C{=}CH_2$. One reason for this assumption is that these hydrocarbons add two halogen atoms as:

$$H_2C{=}CH_2 + 2\,Cl \longrightarrow CH_2ClCH_2Cl$$

One chlorine atom always goes to each carbon atom. When the hydrocarbon adds HBr, the hydrogen goes to one carbon atom, and the bromine to the other.

$$H_2C{=}CH_2 + HBr \longrightarrow CH_3CH_2Br$$

When the hydrocarbon oxidizes, one hydroxyl always goes to each carbon on the two sides of the double bond.

$$H_2C=CH_2 + 2\,OH \longrightarrow CH_2OHCH_2OH$$

EXPERIMENT

Testing unsaturated hydrocarbons. (a) Shake a few drops of turpentine with some bromine water and note disappearance of color.

(b) Add a little sodium carbonate solution to some dilute potassium permanganate solution and shake with a little turpentine and note the disappearance of the purple color.

Uses of ethylene. Ethylene has two uses, as an anesthetic and as a catalyst for ripening fruit. As an anesthetic, ethylene promises largely to replace ether, as it leaves the patient less nauseated and recovery seems more rapid after its use.

When oranges are picked, some of them look green. Such oranges are placed in a room containing a little ethylene. In a day or so they have changed in color from green to golden yellow.

The discovery of the anesthetic properties of ethylene. During a flower show in Chicago a florist found that some of his flowers went to sleep and refused to open up the next day. Being greatly puzzled by this unusual occurrence, he called in a chemist. After running down every clue to explain the peculiar action of these flowers, the chemist found that it was caused by a trace of ethylene escaping from a gas leak.. This led to a study of the effect of ethylene on animals and the discovery of its anesthetic properties.

The discovery of the fruit-ripening properties of ethylene. As soon as navel oranges develop enough sugar to meet the legal requirements, they are picked regardless of color. Some of them are almost green in color. In one packing house the oranges were stored while being packed for shipment. It being freezing weather, a stove was kept going in the room with the oranges to keep them from being injured. It was noticed that the oranges improved rapidly in color, the green being changed to golden orange. It was supposed that the heat caused the change and the process was called "sweating."

One enterprising packer arranged a steam heating plant for sweating the oranges to improve their color, but the effort ended in failure. It was not heat that produced the change. The chemist was now called in. On analyzing the air in a packing house where the oranges did ripen, he discovered a trace of ethylene due to incomplete burning in the furnace. It was the ethylene that caused the ripening. Now ethylene is bought in cylinders, and ripening is caused by the addition of a small percentage of it to the air.

Ethylene has been especially useful in ripening pears. Pears must be picked practically hard to be shipped without bruising. It used to require a month or so fully to ripen them. Now they can be ripened in thirty-five hours with ethylene. Other fruits and vegetables also respond to the stimulating effect of ethylene.

Uses of the other members of the ethylene series. The higher members of the ethylene series are useful in synthetic work as starting points in the preparation of hundreds of useful carbon compounds. Their readiness to react is their best feature.

Acetylene. When lime is heated with coke in the electric furnace, calcium carbide is formed.

$$CaO + 3 C \longrightarrow CO + CaC_2$$

If this carbide is dropped into water, acetylene results:

$$CaC_2 + 2 H_2O \longrightarrow C_2H_2 + Ca(OH)_2$$

Bicycle lamps formerly burned acetylene, generated by allowing water to drip onto carbide.

This hydrocarbon is very reactive, or unsaturated. Hundreds of useful substances can be synthesized from coke by first making acetylene. During the World War when Germany was cut off from the ordinary supplies of carbon compounds, she was able to synthesize ten tons of rubber per day, with coal as the source of the carbon. The Du Pont Company has synthesized from acetylene a substance (duprene) which has most of the properties of rubber. Duprene is now successfully competing with the natural rubber for many purposes.

EXPERIMENT

Acetylene. Drop a piece of carbide into a test tube of water and ignite the resultant gas. Note the sooty flame.

Uses of acetylene. Acetylene mixed with pure oxygen makes a flame hotter than the oxyhydrogen flame. The oxyacetylene blow torch is used in almost every garage and machine shop for welding iron and brazing. It can even be used to cut through steel girders. Acetylene makes a very brilliant light when burned in a special type of burner with a gas mantle. Isolated farm houses often use an acetylene lighting system.

Acetylene series of unsaturated hydrocarbons (the alkynes). Acetylene is the first member of

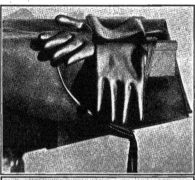

Courtesy of Du Pont Co.

SYNTHETIC "RUBBER" (DUPRENE). Duprene is made from acetylene and hydrochloric acid. It is able to compete with natural rubber at current prices.

another series of hydrocarbons with less hydrogen per molecule than the alkenes. The members of the series are very active chemically and participate in many reactions without catalyst or the need of heat. They react by (a) the addition of halogens, (b) oxidation, (c) the addition of hydrogen halides, such as HBr. These hydrocarbons are even more unsaturated than the alkenes. One molecule of an alkene can add two atoms of bromine, while the corresponding alkyne can add four atoms of bromine. In most of their reactions they can add twice as many elements or radicals as the members of the other series. This fact might be inferred from their general formula, C_nH_{2n-2}. The lower members of this series are:

acetylene or ethyne (C_2H_2)	hexyne (C_6H_{10})
propyne (C_3H_4)	heptyne (C_7H_{12})
butyne (C_4H_6)	octyne (C_8H_{14})
pentyne (C_5H_8)	nonyne (C_9H_{16})

Most of the members of this series may be distinguished from the alkenes by the precipitate formed with silver oxide, which is explosive when dry. The alkenes do not react with silver oxide.

QUESTIONS OF FACT

1. How do the unsaturated hydrocarbons differ from the saturated ones?
2. Name two uses of ethylene.
3. How can acetylene be prepared from calcium carbide?
4. Give two uses of acetylene.
5. Relate the history of the discovery that ethylene has anesthetic properties.
6. Relate how the fruit-ripening properties of ethylene were discovered.
7. How is calcium carbide prepared?
8. Name a valuable substance made from acetylene.
9. What part does acetylene play in the metals industry?
10. Are the alkynes more like the alkanes or the alkenes?
11. Compare the ability to react of C_2H_2 and C_2H_4.
12. Name an explosive substance that can be made of alkynes but not of alkenes.

QUESTIONS OF UNDERSTANDING

1. Why is an asphalt-base lubricating oil undesirable?

2. Why is the future of the Diesel engine very promising?

3. What is an alkyl radical?

4. What are the following radicals? (a) propyl; (b) methyl; (c) butyl; (d) ethyl.

5. What is an unsaturated hydrocarbon?

6. Name four kinds of substances which react with unsaturated hydrocarbons.

7. Using the general formula, give the formulas for the alkenes with the following numbers of carbon atoms: (a) 2 carbon atoms; (b) 7 carbon atoms; (c) 5 carbon atoms; (d) 10 carbon atoms.

8. How do physical properties within a series of hydrocarbons change with the number of carbon atoms in the molecule?

9. Do the unsaturated hydrocarbons react by addition of other elements or by replacing hydrogen?

10. Do entering radicals always enter in pairs?

11. How are these reactions explained by the conception of a double bond between carbon atoms?

12. Why isn't there an alkene with only one carbon atom?

13. How does the triple bond idea explain the fact that an alkyne can react with four bromine atoms, while an alkene can react with only two?

Other series of hydrocarbons. Still other series of hydrocarbons are found in coal tar. The first member of one series is benzene, used as a solvent by the painter and as a starting point for many synthetic dyes and medicines.

Naphthalene, the material of ordinary moth balls, is representative of another series. These different series of hydrocarbons and their derivatives when oxygen, nitrogen, and halogens are attached to them include hundreds of thousands of useful compounds. Organic chemistry is a study of these substances. A student might study for a lifetime and then know only a part of organic chemistry.

The halogen derivatives of the hydrocarbons. When hydrocarbons in the gaseous condition are mixed with chlorine and exposed to sunlight to catalyze the reaction, the chlorine replaces the hydrogen atoms one after another, as the following will illustrate :

$$CH_4 + Cl_2 \longrightarrow CH_3Cl + HCl$$
$$CH_3Cl + Cl_2 \longrightarrow CH_2Cl_2 + HCl$$
$$CH_2Cl_2 + Cl_2 \longrightarrow CHCl_3 + HCl$$
$$CHCl_3 + Cl_2 \longrightarrow CCl_4 + HCl$$

The first product CH_3Cl, methyl chloride, is used in electric refrigerators; $CHCl_3$ is chloroform, a familiar anesthetic; CCl_4, carbon tetrachloride, is used as a dry-cleaning liquid and in small fire extinguishers.

Carbon tetrachloride is used in large quantities in dry-cleaning establishments. It is a good solvent for greases. Because of its safety it is widely used in dry cleaning. Gasoline was formerly used as a dry-cleaning solvent. The constant fire hazard has caused many states to forbid its use. Since in carbon tetrachloride the carbon is attached by all of its bonds to a very negative element, chlorine, it will not burn. Burning, or uniting with oxygen, is primarily a process of the positive elements uniting with the negative element, oxygen. Since, however, chlorine is even more negative in its nature than oxygen, the latter cannot get the carbon away from it.

This same property of incombustibility makes this substance a good filler for fire extinguishers. When sprayed on a small fire, the heavy vapor of carbon tetrachloride settles over the burning substance and holds the oxygen away until burning stops. The evaporating liquid also has considerable cooling action.

This shows us that for every hydrocarbon there are several chlorine derivatives, each one probably useful for some special purpose. There will also be an equal number of bromine derivatives, and another set of iodine derivatives. In general, these derivatives take part in dozens of useful reactions. In many cases where we do not as yet know any other use for them, they serve as starting points for other useful substances.

Hydroxyl compounds, or alcohols. If methyl chloride is treated with moist silver hydroxide, the chlorine is replaced by the hydroxyl radical, making methyl alcohol (wood alcohol).

$$CH_3Cl + AgOH \longrightarrow CH_3OH + AgCl$$

There is a whole series of alcohols corresponding to the halogen compounds. Methyl alcohol, or wood alcohol, is a colorless liquid with a characteristic alcoholic odor and a biting taste. It is very poisonous, causing blindness. It is a good fuel, burning with a hot nonluminous flame. The principal uses of wood alcohol are as a solvent and in antifreeze radiator mixtures.

Grain alcohol. Grain alcohol, or ethyl alcohol (C_2H_5OH), made by fermenting the sugar from grains, fruits, and vegetables, is the next member of the series of alcohols and perhaps the most important. It is valuable as a fuel and as a solvent for varnishes. Ethyl alcohol is a good antiseptic and preservative, but not a good beverage. It is said "alcohol will preserve a dead man, but will kill a live one." Beer and wine are made from the fermentation of malt and fruit juices, respectively. They may contain from 3 to 30 per cent alcohol. Whisky, brandy, rum, and gin are all distilled liquors and contain very high percentages of alcohol. The greatest danger from the moderate use of alcohol as a beverage is its tendency to create a craving for more. In certain cases, alcohol so weakens the judgment and will of the individual that he loses the ability to discern right from wrong, and the power to follow the right.

Some people claim that light wines and beers are not intoxicating and are therefore harmless. However, research has shown that all physiological responses, such as recognizing a sound or writing on the typewriter or thinking, are delayed and are more or less erratic when the person has drunk the normal amount of light wines or beer. This slowness of response averages about 25 per cent. Let us suppose a person who has partaken of some wine or beer is approaching a crossing at high speed in an automobile. Suppose other machines are also approaching, and a critical situation develops. If this person is 25 per cent late in sensing the danger, then 25 per cent late in deciding what to do, then 25 per cent late in executing the decision, a collision may result, where had all of these responses been normal, the accident would have been averted. As faster machines are produced from year to year, the drinking of alcoholic beverages becomes increasingly dangerous.

Industrial alcohol. Scores of industries require alcohol for industrial purposes. Governments, in their zeal to obtain revenue from the sale of alcohol as a beverage, often impose restrictions on its use in the industries. This should not be. The industrial alcohol is usually "denatured" or made undrinkable. Wood alcohol was formerly used for denaturing, but it resulted in many deaths of people who took a chance and drank it. More recently alcohol is denatured with an oil derivative that has such a disagreeable taste that it is practically undrinkable.

QUESTIONS OF FACT

1. Name the importance and uses of benzene.
2. How are the halogen derivatives prepared from the hydrocarbons?
3. Name the following substances and give one use of each: (a) CH_3Cl; (b) $CHCl_3$; (c) CCl_4.
4. What are the alcohols?
5. Give the chemical names and formulas for (a) wood alcohol, (b) grain alcohol.
6. What is denatured alcohol?
7. Discuss the importance of grain alcohol to industry.
8. What is the name of the constituent of moth balls?
9. What is necessary to cause the halogens to replace one or more hydrogen atoms of an alkane?
10. Discuss the physiological effects of beverages of low alcohol content.

QUESTIONS OF UNDERSTANDING

1. Why does CCl_4 not burn?
2. Explain how the carbon tetrachloride fire extinguisher works.
3. What is meant by distilled liquors?
4. Is there any limit to the alcohol they contain?
5. Why is alcohol denatured?
6. Discuss a denaturing substance from the point of view of: (a) poisonous nature; (b) nauseating effect; (c) boiling point near that of ethyl alcohol.

7. Explain how there can be at least as many alcohols as there are hydrocarbons.

8. The scientific names of the alcohols are obtained by dropping the *e* from the name of the hydrocarbon and replacing it with *ol*. Name the following alcohols :

(a) CH_3OH from methane   (b) C_2H_5OH from ethane

(c) C_3H_7OH from propane   (d) C_4H_9OH from butane

REFERENCES FOR SUPPLEMENTARY READING

1. Howe, W. E. *Chemistry in Industry*, Vol. I. The Chemical Foundation, Inc., New York, 1924. Chap. III, pp. 34–57, "Alcohol and Some Other Solvents."

   (1) What is meant by denaturing alcohol? (p. 40)

   (2) Tell how ethyl alcohol is used in connection with :

      (a) polishes in the home (p. 40)

      (b) heating and lighting (p. 40–41)

      (c) wood lacquers (p. 41)

      (d) making pyroxylin articles (pp. 41–42)

      (e) jellies and food dyes (p. 42)

      (f) medicines (p. 43)

      (g) articles around the home (pp. 43–44)

      (h) antifreeze liquids (p. 44)

      (i) artificial leather (p. 44)

      (j) automobile finishes (pp. 44–45)

      (k) motor fuels (pp. 45–46)

      (l) war (pp. 46–48)

      (m) surgery (p. 48)

      (n) dyes (pp. 48–49)

      (o) artificial silk (p. 49)

      (p) other articles (pp. 49–50)

   (3) What is absolute alcohol? (p. 51)

2. Emerson, Haven. *Alcohol and Man*. The Macmillan Company, New York, 1932.

   (1) Give a report on the poisonous nature of alcohol. ("The Human Toxicology of Alcohol," pp. 126–151)

   (2) Prepare a short essay on the physiological action of alcohol. ("The Physiological Action of Alcohol," pp. 1–23)

   (3) Prepare a written report on alcohol and the mind. ("The Prevalence of Mental Disease Due to Alcoholism," pp. 344–371)

Ethers or oxides. When most metallic hydroxides are heated to high temperatures, the oxides are formed by the loss of water. Likewise when the alcohols are made to lose water by sulfuric acid and heat, an organic or carbon compound oxide results. These oxides are called ethers:

$$2 \, AgOH \longrightarrow Ag_2O + H_2O$$
$$H_2SO_4 + 2 \, C_2H_5OH \longrightarrow (C_2H_5)_2O + H_2O \cdot H_2SO_4$$

These ethers are colorless liquids with an ethereal odor. Ethyl ether (ordinary ether), whose formula is given in the above equation, is used as the common anesthetic. There may be a vast number of other ethers, but they are of little importance as yet.

Aldehydes. In carbon compounds certain radicals seem to give similar properties to all of the compounds in which they occur. For instance, all alcohols look alike and conduct themselves alike. The same is true for compounds containing any other radical. If any typical alcohol is gently oxidized, the aldehyde radical results. Compounds containing this radical are called aldehydes. Ethyl alcohol oxidizes to *acetaldehyde:*

$$CH_3CH_2OH + O \longrightarrow CH_3CHO + H_2O$$

Methyl alcohol oxidizes to formaldehyde (HCHO). The aldehyde group is CHO. It gives certain properties to compounds. It is a very reactive group. It reduces silver nitrate to metallic silver. It adds HCN and NH_3. These are tests for aldehydes. A 40 per cent solution of formaldehyde — the most important aldehyde — is used as an antiseptic and disinfectant and as a preservative. Aldehydes find extensive use in the manufacture of bakelite.

Organic acids. If aldehydes are oxidized or alcohols are strongly oxidized, there result compounds containing the radical CO_2H, all of which yield hydrogen ions in solution. They are all acids. Acetic acid (CH_3CO_2H) is produced by the fermentation of cider or the destructive distillation of wood. More recently, it has been synthesized from acetylene. It occurs in vinegar up to 5 per cent and gives it its most important characteristics. Butyric acid gives the odor to rancid butter or Limburger cheese. Tartaric acid from grapes is used in baking powder. Citric acid occurs in lemons and oranges.

Fatty acids from soap. To a solution of some soap add a little sulfuric acid. Note the fatty acid that separates out. The relative values of two samples of liquid soap can be compared in terms of the depths of the fatty acids in equal-sized test tubes.

Amines. Another important class of carbon compounds contains the amines. These result when one or more of the hydrogen atoms of ammonia are replaced with radicals composed of carbon and hydrogen. The amines occur in decayed meat, especially fish. It used to be thought that the amines cause ptomaine poisoning. Recently, however, it has been shown that other products of bacterial decay are the extremely poisonous products, the amines being only moderately poisonous. They are important in that they form the basis of most of the synthetic dyes. Dyes of all shades of color and degrees of stability can be made from the amines. The most important amine is aniline, basis of the aniline dyes. Aniline occurs in bone oil and coal tar but is mostly made from the hydrocarbon benzene and nitric acid with a reducing agent.

As different compounds of carbon formed in plant and animal products may contain two or more of these radicals, some parts of organic chemistry are very complex and are still fruitful fields for study.

Esters. Another class of carbon compounds is the esters, which parallel the salts. An inorganic acid and base react to form a salt:

$$HCl + NaOH \longrightarrow NaCl + H_2O$$

So likewise will an organic acid and hydroxyl body if heated with a little sulfuric acid to remove the water as fast as it is formed.

$$CH_3CO_2H + CH_3OH \longrightarrow CH_3CO_2CH_3 + H_2O$$

The esters are important because of their pleasant taste and fragrant odor. The flavor of apples, pears, bananas, and most perfumes is due to esters. A familiar ester is oil of wintergreen. All fats are mixtures of the higher esters — that is, those of 12 or more carbon atoms and the corresponding amounts of

ORANGE FLAVOR. Most fruit flavors are esters. This one is called octyl acetate.

BANANA FLAVOR. The ester that flavors bananas is named amyl acetate.

PINEAPPLE FLAVOR. A compound called methyl butrate is responsible for the distinctive flavor of pineapple.

hydrogen. Lard is softer than tallow because its fats have lower melting points. Heating fats with concentrated alkalis or with steam under pressure breaks them up (hydrolyzes them) into the corresponding acid and alcohol. The alcohol entering into the formation of all fats is an alcohol with 3 hydroxyl radicals called glycerin. Glycerin is a thick oil with a sweetish taste. The alkali with which the fat is heated neutralizes the acid, forming the sodium or potassium salt, which is a soap. The potassium salts form soft soap. The sodium compounds are solids and compose ordinary hard soap.

EXPERIMENT

Formation of esters. Mix a few cubic centimeters of any anhydrous organic acid and alcohol and add a few cc. of concentrated sulfuric acid. Warm almost to boiling and set aside corked for a few minutes. Note the odor of the ester formed. If phthalic anhydride and methanol are used, the very fragrant oil of wintergreen is formed.

Soaps. Soap may be a mixture of the salts of any of the solid acids, such as palmitic acid from palm oil, oleic acid from lard, or stearic acid from tallow. Any cheap fat, such as animal fats, coconut oil, olive oil, or cottonseed oil, may be used. The fat is boiled for about two days with sodium hydroxide in huge kettles. The soap is then "salted out" of solution by adding salt to the solution, as soap is not appreciably soluble in brine. Cheap laundry soaps are made from rancid and impure fats. The sodium hydroxide has not been removed; hence they are too caustic for toilet purposes and for washing silks and woolens. Castile soap is a pure form which has been freed from the excess of caustic soda. Many adulterants are added to soaps. An undesirable one is sodium silicate

SALTING OUT SOAP. The addition of salt to a soap solution causes the soap to separate out and rise to the surface of the liquid.

(water glass), as it injures certain fabrics. Rosin makes laundry soaps yellow. It is not harmful and is of some advantage in that it increases the lathering quality of the soap. Some soaps are 25 per cent water, which fact should guide us in comparing their cost. Naphtha soap has naphtha dissolved in it, which helps to dissolve dirt. Scouring soaps contain some sort of abrasive, such as sand or pumice.

QUESTIONS OF FACT

1. Give the uses of aldehydes.
2. What radical is common to all organic acids?
3. For what is ethyl ether used?

4. Which acid comes from grapes? Which one from lemons?
5. What acid gives Limburger cheese its disagreeable odor?
6. Name the organic compound that corresponds to the inorganic salt.
7. Where do the esters occur naturally?
8. Fats are esters of what alcohols?
9. What is hard soap? soft soap?
10. What is an ether?
11. Might ethyl ether be called diethyl oxide?
12. How is an aldehyde prepared from an alcohol?
13. What is the acid of vinegar?
14. What is the acid of rancid butter?
15. Name two reagents that react with aldehydes.
16. Where do amines occur?
17. Name one amine that is of great industrial importance.
18. What alcohol is obtained by the forced hydrolysis of fats?
19. Give the compositions of several kinds of soaps.

QUESTIONS OF UNDERSTANDING

1. What radical is common to all aldehydes?
2. What aldehyde results from the oxidation of CH_3OH?
3. What do aldehydes oxidize to?
4. How is glycerin made?
5. Do all sour fruits contain acid?
6. If one or more of the hydrogen atoms of NH_3 are replaced by alkyl groups, what are the resultant substances called?
7. How are esters related to acids and alcohols?
8. When esters were mentioned previously, what part did sulfuric acid play in the reaction? (p. 426)
9. In what way are the amines related to ammonia?
10. Fruit contains a characteristic ester and traces of others; synthetic flavors, only the one. Which is the better tasting?
11. Explain the terms "soap" and "fatty acid."
12. Identify each formula as one of the following: hydrocarbon, alkyl halide, ether, alcohol, aldehyde, acid, amine, or ester:

(a) C_2H_5OH
(b) CH_3Br
(c) CH_3NH_2
(d) CH_3CHO
(e) CH_3CO_2H

(f) C_2H_5CHO
(g) $(C_2H_5)_2O$
(h) $(C_2H_5)_2NH$
(i) CHI_3

(j) $CH_3CO_2C_2H_5$
(k) $CH_3CO_2CH_3$
(l) HCO_2H
(m) C_5H_{10}

(n) $(CH_3)_2O$
(o) $C_5H_{11}OH$
(p) $(CH_3)_3N$
(q) C_2H_5I

Graphic formulas. The chemistry of the carbon compounds could never have developed as far as it has, had it not been for the use of graphic formulas. Graphic formulas picture the relationship between the atoms within the molecule in much the same way as a blueprint pictures the relationship between the rooms of a house. To illustrate the meaning and value of graphic formulas, let us consider the two compounds, ethyl alcohol and methyl ether. The common formulas, often called the empirical formulas, are identical for these two substances (C_2H_6O). Here are two compounds with most of their properties different, which show no difference in their empirical formulas. Their graphic formulas, however, are different.

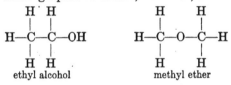

ethyl alcohol   methyl ether

From the formula for ethyl alcohol it can be seen at once that the carbons are connected together, while the oxygen is in a hydroxyl group. From the other formula it appears that the carbons of methyl ether are connected not directly but through the oxygen atom. It seems reasonable that the properties of the two substances should be quite different when it is considered that the atoms of carbon, oxygen, and hydrogen are arranged so differently in the two compounds. Since it is rather troublesome to draw a completely graphical formula, chemists usually use formulas that are only partially graphic yet near enough to reveal the important relationships. Ordinarily the two formulas just illustrated would be written CH_3—CH_2—OH and CH_3—O—CH_3.

Aldehydes, organic acids, and esters are better understood from their graphic formulas than from their empirical formulas. The table on page 492 illustrates the first few compounds in each of the three classes of substances just mentioned.

Benzene. The value of graphic formulas in organic chemistry is greatest for explaining differences between substances, predicting new substances, and correlating the properties of

ALDEHYDES, ACIDS, AND ESTERS

NAME	COMMON FORMULA	GRAPHIC FORMULA
Aldehydes		
Methanal	HCHO	$\underset{\displaystyle H}{\overset{\displaystyle H}{\big|}}$ H—C=O
Ethanal	CH_3CHO	H—C—C=O (with H above and below first C, H above second C)
Propanal	C_2H_5CHO	H—C—C—C=O (H's attached)
Butanal	C_3H_7CHO	H—C—C—C—C=O (H's attached)
Acids		
Formic	HCO_2H	OH above, H—C=O
Acetic , ...	CH_3CO_2H	H OH above, H—C—C=O, H below
Propionic	$C_2H_5CO_2H$	H H OH above, H—C—C—C=O, H H below
Esters		
Methyl formate	HCO_2CH_3	H above, H—C—O—C—H, O below first C (double bond), H below second
Methyl acetate	$CH_3CO_2CH_3$	H H above, H—C—C—O—C—H, H O H below
Ethyl formate	$HCO_2C_2H_5$	H H above, H—C—O—C—C—H, O H H below

new and old substances in a large group of compounds called the aromatic compounds, because many of them have aromatic odors. Most of the aromatic compounds can be thought of as derivatives of benzene. Of the many interesting substances obtained as by-products in the gas industry, benzene is one of the most important. From the heating of one ton of soft coal, there is obtained three gallons or more of a light oil, 60 per cent of which is benzene.

Benzene, which has the empirical formula C_6H_6, was one of the conundrums of chemistry. In light of the fact that carbon usually has a valence of four, it seemed strange that there could be a substance with only six hydrogen atoms for six carbon atoms. Moreover, the problem was made more perplexing by the fact that benzene is not reactive, which indicates that all of its valences must be occupied. There are a few carbon compounds in which the carbon is known to have only three of its valences occupied by other radicals. These substances, however, are extremely reactive — entirely different from benzene.

There is a story that the German chemist Kekulé, while trying to understand benzene, went to sleep and dreamed that six carbon atoms tied together came into the room in the form of a snake. When this carbon snake arrived in the middle of the room, it turned and swallowed its tail to form a ring. At this Kekulé awoke with a start, saying, "This is the answer to the problem; benzene has a ring structure." Kekulé's formula is accepted today as best describing the structure of benzene. Instead of an exact ring, Kekulé considered benzene as a hexagon with one carbon atom and one hydrogen atom at each vertex as :

This formula still seems inappropriate, since one would expect a compound with double bonds to be unsaturated like ethylene. We know that benzene does not show the usual reactions of unsaturated hydrocarbons containing double and triple bonds. Its stability is explained by assuming that the double bonds alternate between pairs of carbons and thereby prevent the ordinary reactions of unsaturated bodies. In organic chemistry these theories of valence are extended to explain the properties of benzene and its derivatives. In a beginning course, however, it is not advisable to attempt further explanation.

Other compounds related to benzene. . Since the six hydrogen atoms in the benzene molecule can all be replaced by other radicals in various combinations, millions of compounds related to benzene are possible. Among the simplest of the benzene derivatives are :

phenol   aniline   toluene

Phenol, the active constituent of carbolic acid, is used in large quantities in the manufacture of bakelite.

Aniline, which is benzene with one hydrogen replaced by an NH_2 radical, is the starting point for many of the aniline dyes.

Toluene, as will be seen from its formula, is benzene with one of its hydrogen atoms replaced by a CH_3 radical. Trinitrotoluene, the powerful explosive used so extensively in war, is made by nitrating toluene.

Another compound related to benzene is nitrobenzene, which can be prepared by heating benzene with a mixture of concentrated nitric acid and concentrated sulfuric acid. It is a yellow oily liquid with a fragrant odor, often noticed in shoe blacking.

EXPERIMENT

Nitrobenzene. In a test tube mix a little concentrated nitric acid with some concentrated sulfuric. Add a few drops of benzene and heat to boiling. Pour into some water and note the formation of yellow oil with the odor of shoe blacking. This is nitrobenzene, $C_6H_5NO_2$.

Naphthalene. Ordinary moth balls are composed of naphthalene, whose peculiar odor keeps the moths away. The graphic formula of naphthalene is a double benzene ring,

naphthalene

Thousands, or perhaps millions, of derivatives of naphthalene are also possible. Other carbon rings even more complex than naphthalene are known.

REVIEW EXERCISES

1. What does a graphical formula tell that an empirical formula does not? Illustrate with the formulas of ethyl alcohol and methyl ether.

2. Relate the dream of Kekulé which gave the idea of ring compounds.

3. How are the following related to benzene? (a) aniline; (b) toluene; (c) phenol.

4. What sort of ring structure does naphthalene have?

5. On a piece of blank paper draw graphical formulas of the following:

(a) CH_3CHO (b) HCO_2CH_3 (c) $(CH_3)_2O$ (d) $C_2H_5CH_2OH$
(e) $CHCl_3$ (f) CH_3CO_2H (g) C_2H_4 (h) $HCHO$ (i) CH_3OH

PROBLEM

1. Calculate the weight of bromine needed to react with 100 grams of methane. $CH_4 + Br_2 \longrightarrow CH_3Br + HBr$

Summary

Hydrocarbons (compounds containing only hydrogen and carbon) occur extensively in natural gas, crude oil, oil shale, and bituminous coal. Crude oil is fractionally distilled in order to separate it into its many useful fractions — naphtha, gasoline, kerosene, fuel oil, and various lubricating oils. The gasoline fraction can be increased at the expense of the heavy oils by cracking — that is, heating to high temperatures under pressure.

Some hydrocarbons take part in few chemical reactions and then only by replacing the hydrogen atoms. These are called saturated hydrocarbons. The unsaturated hydrocarbons take part in many reactions by direct addition. Hydrocarbons of like properties can be arranged into a series according to the number of carbon atoms per molecule.

Methane (CH_4) is the principal constituent of natural gas and of the explosive gas in coal mines. It is the first member of the saturated series of hydrocarbons. Ethylene, the first member of an unsaturated series called the alkenes, is useful as an anesthetic and in ripening fruit. Acetylene, the first member of the alkynes, another unsaturated series, is used for fuel, in welding, and in synthetic reactions.

Of the halogen derivatives, chloroform ($CHCl_3$) and carbon tetrachloride (CCl_4) are important, especially the latter, which is used as a fire extinguisher and dry-cleaning solvent.

Of the alcohols, wood alcohol (CH_3OH) and grain alcohol (C_2H_5OH) are very important as solvents.

The aldehydes are the compounds containing the —CHO radical. Formaldehyde (HCHO) is used in preserving pathological specimens and in synthesizing bakelite and similar substances.

Characterized by the radical —COOH, the organic acids occur in fruits. Acetic acid (CH_3CO_2H), the acid in vinegar, is one of the most important.

When one or more of the hydrogen atoms of ammonia are replaced by carbon radicals, we get the amines, such as methyl amine (CH_3NH_2). The amines are very important in dyes.

Fats are esters. The characteristic flavors of fruits are due to esters. Esters can be prepared synthetically by removing the acid H from an acid and the OH from an alcohol by means of concentrated sulfuric acid.

The soaps are the metallic salts of the acids with large molecules. The solid soaps are the sodium salts, while the liquid soaps are the potassium salts.

Graphic formulas tell much more about an organic substance than the usual empirical formulas. Benzene and related compounds especially require graphic formulas to explain their properties. In this case a ring structure gives the most satisfactory description of the compound. The naphthalene of moth balls requires a double ring to explain its properties.

⇄

CHAPTER 26

THE MORE COMPLEX COMPOUNDS OF CARBON

There are thousands of very important carbon compounds. In a beginning course it is possible to study only a few of those compounds that we meet in our daily lives.

The sugars. Two great food industries deal with the production of cane sugar and corn sirup. The latter is known chemically as glucose or dextrose; it is also called grape sugar because it was first obtained from grapes. Both cane sugar and grape sugar are characterized by a sweet taste. There are dozens of similar sugars found in nature, although they are not used in such large amounts as the two just named.

Uses of the sugars as food. The sugars are an important item in our diet. In the United States each person uses close to one hundred pounds of cane sugar per year. This amount is consumed as table sugar, and as a constituent of candies, ice cream, cakes, cookies, jellies, etc. A large part of our sirup is glucose, which is made from corn. Some of our candies, such as taffy, are often partly made from glucose. The bakers use another sugar called malt sugar. It is not so sweet as cane sugar and glucose, but it makes dough hold together better than the other sugars.

The chief value of the sugars, aside from their sweetening qualities, is the readiness with which they can be used by the body. Most food has to go through a process of digestion taking more or less time before it can be assimilated by the body. Glucose apparently is already in a form that can be assimilated. Cane sugar needs the saliva in the mouth to digest it, but by the time it reaches the stomach it is ready for use. Soldiers on the march are sometimes given sugar for the immediate relief of fatigue. Cane sugar, called sucrose by the chemist, has one thing against it; .it tends to dissolve calcium compounds. Since the teeth are largely calcium compounds, there is danger that the sugar may remove some of the calcium and thereby open up cavities, especially where the surface enamel has already been broken. Hard candies like "all-day suckers" should be avoided. There is always danger of injuring the teeth in trying to chew hard candies. Besides, the long time

Courtesy of Pan-Pacific Press Bureau

OAHU SUGAR PLANTATION. A cane field in blossom is a beautiful sight. From 11 to 18 per cent of the stalk of the sugar cane is sugar.

that they are held in the mouth favors the reaction between sugar and calcium that forms the soluble *calcium sucrate*.

Occurrence of sucrose. Sucrose occurs in the roots, grasses, stems, and fruit of many plants. It is possible to obtain sugar in considerable quantities from sugar cane, sugar beets, sorghum, maple trees, corn stalks, carrots, and sweet potatoes.

About two-thirds of our sugar of commerce comes from the beet, and one-third from the sugar cane. From 11 to 18 per cent of the stalks of the cane, 12 to 16 per cent of the sugar beet, 15 per cent of sorghum, and 2 per cent of maple sap are cane sugar.

Preparation of sugar from sugar cane and beets. In preparing sugar from sugar cane, the stalk is ground and the juice pressed out. The acids in the liquid are neutralized with milk of lime to prevent the hydrolysis of the cane sugar into simpler

A Trainload of Sugar Beets. Every day during the beet harvest trainloads of sugar beets pull into the factories. Americans use more sugar per capita than do any other people. A large part of this sugar comes from the white sugar beet which thrives in a cooler climate than the sugar cane.

sugars. The juice is then heated to coagulate the soluble albumen and to cause the lime to unite with other proteins and precipitate them out. The juice is now filtered through bone black to decolorize it by removing the vegetable coloring matter. It is next evaporated in vacuum pans so the water can be mostly removed without the heat injuring the sugar. The crystals are separated from the sirup residue by centrifuging. Beet sugar, when purified, is equal to cane sugar in every respect, since they are the same chemical compound.

The hydrolysis of cane sugar. When cane sugar is boiled in a solution containing acid or alkali, it reacts with water and

Ira Remsen

1846-1927

After finishing the lower schools of New York City, Ira Remsen entered the College of the City of New York, specializing in Latin, Greek, and mathematics. However, before he had time to complete the course, his father had him apprenticed to a physician. He later entered the College of Physicians and Surgeons, where he could better prepare in this field. After graduating, however, Remsen found that he disliked the practice of medicine. His real interest was chemistry.

Since chemistry was not well developed in the United States at that time, he went to Germany to study. Here he studied under Liebig and Wöhler and after five years of study and research got his Doctor's degree.

Returning to the United States, Remsen accepted a professorship at Williams College. In 1876 Johns Hopkins University was founded as a school of research and advanced study. Remsen was chosen as head of the chemistry department. At this institution he did research work on benzene compounds, organic acids, and carbon monoxide. Remsen founded the *American Chemical Journal*, which in 1914 was replaced by the *Journal of the American Chemical Society*. For eleven years after 1901 Remsen served as president of Johns Hopkins University. One of the outstanding discoveries of Remsen is saccharin, a substance 550 times as sweet as sugar. The story is told that coming home to lunch hurriedly after synthesizing some new sulfur compounds, Remsen neglected to wash his hands. While eating, he noticed a very sweet taste. Rushing back to the laboratory, he found that one of the new substances which he had prepared was responsible for the sweet taste. This substance, related to benzene, sulfuric acid, and ammonia, because of its extremely sweet taste was named saccharin, after the sugars which are sometimes called saccharides.

forms two sugars — glucose and a similar sugar called fructose or fruit sugar.

$$C_{12}H_{22}O_{11} + H_2O \longrightarrow C_6H_{12}O_6 + C_6H_{12}O_6 \text{ (boil in acid)}$$

cane sugar   -   glucose   fructose

This same reaction takes place during digestion. It is catalyzed by a substance in the saliva called ptyalin. There are sixteen sugars with the same formula, $C_6H_{12}O_6$. They are very similar, differing only in some minor property or chemical reaction. These sugars are called the hexoses, *hex* meaning six and *ose* meaning sugar, the name indicating that they contain six carbon atoms.

EXPERIMENT

Making taffy. Mix 2.5 cups of sugar, .5 cup of butter, 4 tablespoons molasses or sirup, 3 tablespoons of water and two tablespoons of vinegar. Boil 20 minutes. Add 1 teaspoonful of vanilla, cool and pull.

EXPERIMENT

Circus lemonade sweetened with saccharin. Into a pitcher of water slice a lemon and add tartaric acid to the right tartness. Divide into two parts. Sweeten one half to taste with sugar and the other half with saccharin.

Disaccharides. Cane sugar is composed of two 6-carbon units joined together chemically by the elimination of one molecule of water from the two molecules of 6-carbon sugar. Nature has made this synthesis in many cases, but the chemist has prepared only sucrose from its constituent hexoses. Even this synthesis is not a commercial success. There are several other 12-carbon sugars that also hydrolyze into the 6-carbon sugars. The 6-carbon unit is sometimes called a monosaccharide and the 12-carbon sugars, or double units, are called disaccharides. *Mono* means one, *di* means two, and *saccharide* means sugar. *Monosaccharide* means a sugar composed of one unit, and *disaccharide* means a compound of two sugar units.

Sweetness of the sugars. Cane sugar is quite sweet. Glucose is about .6 as sweet as sucrose. Some of the other sugars have very little sweetness. There have been some attempts to pass laws declaring the cheaper glucose an adulterant or substitute and forbidding its use for sweetening foods. As a matter of fact, glucose is as good a food as, if not better than, cane sugar. However, the law might justly require the label to state that a food is sweetened with glucose. Even this has been opposed on the grounds that unfavorable publicity has prejudiced some people against glucose.

Saccharin. Professor Remsen discovered a substance (saccharin) that is 550 times as sweet as cane sugar. However, saccharin is not a food. If only $\frac{1}{550}$ as much is used as sugar, it tastes sweet with a little suggestion of bitterness. Saccharin is poisonous if eaten in the same quantities as sugar. Two college students once made some lemonade and put a spoonful of saccharin in each glass. It made them very ill. Had they used just a pinch of the substance, there would have been no ill effects.

Maple products. Maple sirup is 62 per cent cane sugar; hence it is natural that it should be more expensive than corn sirup, which is glucose. Maple sugar is 83 per cent cane sugar. However, the value of maple sugar and maple sirup depends more on the maple flavor than on the sucrose content.

Molasses. Molasses, or the residue of the sirup from which the sugar crystals come, contains some uncrystallizable sugar, some cane sugar, some gum coloring matters, and some mineral salts. As the process of sugar extraction is perfected more and more, less and less cane sugar stays in the molasses, and it becomes stronger in taste. Twenty years ago molasses was often used in place of sirup. Today, however, real molasses is too bitter for such use. Industrially, molasses is the raw product from which alcohol is made by fermentation.

Lactose. Lactose, or milk-sugar ($C_{12}H_{22}O_{11}$), is a disaccharide occurring in cow's milk to the extent of 4.5–5 per cent.

It is not quite so sweet as cane sugar. Lactose hydrolyzes to glucose when it is heated with acid. Considerable quantities of lactose are found as a by-product of the dairy-products industry.

Maltose. Maltose ($C_{12}H_{22}O_{11}$) is the sugar produced from malting barley and other grains. It will hydrolyze to glucose and galactose. Maltose can be prepared by the hydrolysis of starch. Maltose finds use in the fermented liquor and baking industries. It seems to help the dough hold the carbon dioxide in the bubbles until the bread has thoroughly risen.

Caramel. When cane sugar or glucose is heated above 160° C., there is an effervescence of water, and the sugar turns dark red. This is due to the formation of a group of substances collectively called caramel. When separated, they have a dark brown color and a bitter taste. They reduce blue cupric salts to red cuprous oxide and are not acted upon by ferments. Caramel gives the flavor to taffy and the caramel candies.

QUESTIONS OF FACT

1. Name the two sugars most used in our food industries.
2. What does sugar do for the body?
3. Compare cane sugar and glucose as to (a) sweetness; (b) ease of assimilation by the body.
4. What danger is there in the "all-day sucker"?
   Outline the preparation of sugar from cane.
   What products are formed from the hydrolysis of cane sugar?
   Explain "monosaccharide" and "disaccharide."
   Is saccharin a good substitute for sugar?
9. What are (a) lactose? (b) maltose? (c) caramel?
10. What are two other names for grape sugar?
11. What sugar is used by bakers for another reason than that of sweetening?
12. Name a list of plants containing sucrose.
13. What catalyst hydrolyzes sucrose in mouth digestion?
14. What is the meaning of the term hexose?
15. How is caramel made?
16. What is the greatest use of molasses?

17. Select the disaccharides and the monosaccharides from the following:

(a) Glucose, $C_6H_{12}O_6$ (d) Sucrose, $C_{12}H_{22}O_{11}$

(b) Maltose, $C_{12}H_{22}O_{11}$ (e) Galactose, $C_6H_{12}O_6$

(c) Fructose, $C_6H_{12}O_6$ (f) Lactose, $C_{12}H_{22}O_{11}$

Carbohydrates. The sugars are all included, along with several substances which are not sweet but can be hydrolyzed into sugars, under the general term *carbohydrates*. The non-sugar carbohydrates are principally starch, dextrin, cellulose, pectin, glycogen, and certain gums.

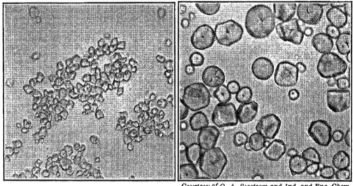

Courtesy of O A Sjostrom and Ind and Eng Chem.

STARCH GRANULES (Magnified 10 times). Granules from different plants vary in size. The rice starch granules (left) are quite small as compared with the corn starch granules (right).

Starch. Starch is the digestible carbohydrate in grains, roots and potatoes, and all vegetables. All cereals are more than half starch. Potatoes are three-fourths starch. Rice and tapioca are almost pure starch. Starch may be prepared from wheat, maize, rice, potatoes, arrowroot, tapioca, and sago. The granules from different sources vary in size, and each variety can be identified under the microscope. Starch turns iodine blue; this reaction is the common test for it. Starch will hydrolyze to dextrin and finally to glucose. In the stomach of the cow or horse, starch hydrolyzes readily at the body temperature. In the laboratory, starch needs heat and acid to promote hydrolysis. Corn sirup is made by heating corn starch with dilute acid, which hydrolyzes it to glucose. The acid is now neutralized with a mild alkali, and the sirup purified as in preparing cane sugar.

Dextrin. Dextrin may be prepared by either heating or hydrolyzing starch; hence it occurs in toast and bread crust. Pure dextrin is a white substance, soluble in water. It is used in the manufacture of mucilage, label gums, and "sizes" for giving a glazed finish to textiles, cardboard, and paper. The purpose of toasting bread is to change the starch into the more digestible dextrin.

Glycogen. Glycogen is similar to dextrin. It is the reserve carbohydrate in animals and has an important role in the oxidation processes in the body.

Pectin. Pectin is a carbohydrate occurring in varying amounts in all fruits that are not too ripe. Pectin is the most important substance in jelly making. Fruit juices with correct amounts of pectin, fruit acids, and sugar will always jell. If any of the three substances are below a minimum concentration, the juice will not jell. The pectin content is too low for jelling in overripe fruit and in certain fruits, such as strawberries. In these cases, it is necessary to add commercial pectin in making jelly. Most failures in jelly making are due to too little pectin.

Cellulose. Cellulose, the principal constituent of the vegetable textiles, is also the woody fiber in all wood. Cotton and filter paper are practically pure cellulose. Cellulose is composed of carbon, hydrogen, and oxygen and belongs to the same class of compounds (carbohydrates) as the sugars. The reason cellulose is classed with the sugars is that when it is boiled a long time with acids, it hydrolyzes to form grape sugar. Cellulose is a comparatively inactive substance chemically, as only a few reagents attack it. However, it dissolves readily in cold concentrated sulfuric acid and in strong solutions of certain salts in hydrochloric acid; mercuric chloride and zinc chloride are among these salts.

Cellulose, being a carbohydrate, has the characteristics of an alcohol with many —OH groups. These —OH groups are capable of reacting with an acid to form several esters. Some of the nitrates of cellulose, such as guncotton, are very explosive, while the lower nitrates form collodion and artificial silk. Cellulose is nitrated by treatment with a mixture of nitric and concentrated sulfuric acids. Cellulose nitrates are commonly called nitrocelluloses.

Occurrence of cellulose. Cellulose occurs in the stems, roots, and leaves of all plants. About 30–40 per cent of hay is cellulose, and almost all of cotton and flax. Cellulose from different sources has a varying texture; that in hemp is tenacious and flexible; that in wood, hard; that in pith, elastic; and that in seeds, spongy. Cellulose is used in making rayon, guncotton, paper, and automobile lacquers.

The chemistry of wood. Wood will probably always be our most abundant cellulose material. In removing her forests, America has wasted enough cellulose to meet her needs for an almost indefinite time. Other chemicals besides cellulose are obtainable from wood. The recovery of chemicals from wood began with the dry distillation of the hard woods, such as oak, maple, and birch. The dry distillation of wood is the process of heating it in the absence of air. This treatment drives off a large volume of combustible gases, a considerable volume of liquids, and some tar, and leaves a residue of charcoal. The latter was formerly the principal product, since it was needed in the metallurgy of the metals. About 40 per cent of the original wood forms charcoal, which is still used as domestic fuel, in making black powder, in gas masks, and in poultry foods.

From the liquids distilled from wood are obtained several useful chemicals, including acetic acid, wood alcohol, and acetone.

Dry Distillation of resinous woods. The wood distillation industry was later extended to include the soft woods. On cut-over pine lands was an almost inexhaustible supply of stumps, practically worthless for any other purpose. From these old stumps are distilled tar, pitch, rosin, turpentine, and pine oil. Because of their use in connection with shipping, these products are known as naval stores. In time, the demand for charcoal decreased, but the demand for naval stores increased. This led to the development of a new solvent extraction process, which increases the naval stores but produces no charcoal. There are at least fifty uses for turpentine, among the most important of which are its use as a paint thinner, a lacquer solvent, and a constituent of stains, waxes, and greases. Of the eighty uses for rosin, calking ships, making lathering soaps, sizing for paper, printer's inks, waxes, and polishes are only a few. Pitch is used to make fly paper, and tar to preserve rope.

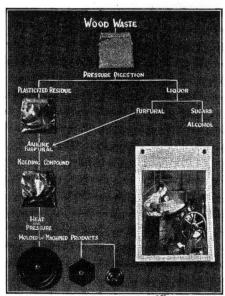

Courtesy of U S Forest Service

USEFUL PRODUCTS FROM WOOD WASTE.

Wood as a possible source of food. Wood, being largely cellulose, is a possible future raw product in the production of foods for both animals and man. In Germany the chemist Frederick Bergius has succeeded in changing over 60 per cent of dry wood into a mixture of sugars, equal to barley as stock food. These sugars might be further purified as food for man, or fermented into alcohol for commerce and industry. One ton of dry wood yielded 85–90 gallons of high-grade alcohol. Only 80 per cent of the sugars obtained by the hydrolysis of wood with hydrochloric acid are fermentable into alcohol. The unfermentable 20 per cent is composed of a mixture of xylose (a 5-carbon sugar) and galactose (a 6-carbon sugar), and can be used in other ways.

Less than half of the weight of a fallen tree is marketable as lumber. In the future, more and more of the wasted half will be converted into useful substances by the chemist.

REVIEW QUESTIONS

1. Define carbohydrate.
2. Why is starch classed as a carbohydrate?
3. Name several carbohydrates that are not sugars.
4. What is the purpose of toasting bread?
5. What carbohydrate is responsible for fruit juices jelling?
6. What carbohydrate is cotton?
7. What is the difference between starches from different sources?
8. What is the ultimate product of starch hydrolysis?
9. How are the dextrins prepared?
10. Give some of the uses of the dextrins.
11. What is the reserve carbohydrate stored in the animal body?
12. Is cellulose chemically active or inactive?
13. How is cellulose nitrated?
14. Discuss the occurrence of cellulose.
15. What is called "naval stores"?
16. Name several uses of turpentine.
17. Discuss the possibility of making food from wood.
18. Name the substances obtained from the distillation of wood.

PROBLEMS

1. The formula of sucrose is $C_{12}H_{22}O_{11}$. Calculate the molecular weight.

2. When sucrose hydrolyzes the equation is

$$C_{12}H_{22}O_{11} + H_2O \longrightarrow C_6H_{12}O_6 + C_6H_{12}O_6$$

(a) What weight of water is necessary to react with 10 lbs. of sucrose?

(b) If the first $C_6H_{12}O_6$ is glucose and the second is fructose, calculate the weight of fructose that can be obtained from 100 lbs. of sucrose.

ADDITIONAL EXERCISES FOR SUPERIOR STUDENTS

1. Review in Chapter 21 the number of electrons in the outer orbit of carbon. (a) In the compound CCl_4, has carbon lost these four electrons or has it gained other electrons? (b) Explain how

electrical forces may be holding the atoms together in the compound. (c) Has the carbon been oxidized or reduced, compared to its condition in charcoal?

2. (a) In the substance methane (CH_4) if each hydrogen has given up an electron, what is the condition of the outer orbit of carbon? (b) Explain the forces which hold the atoms together in the molecule. (c) Has the carbon been oxidized or reduced from its condition in charcoal?

REFERENCE FOR SUPPLEMENTARY READING

1. Hawley, L. F. *Wood Distillation.* The Chemical Catalogue Company, Inc., New York, 1923.

Write a report on the distillation of wood, emphasizing the following topics:

"Importance of Wood Distillation Products" (pp. 31–34)

"Products of Wood Distillation" (pp. 64–71)

"Steam Distillation and Extraction Process" (pp. 91–94)

"Uses for Resinous Wood Distillation Products" (pp. 96–102)

Natural cellulose fibers.   Cotton and linen are practically pure cellulose. These fibers have been used in making cloth throughout most of historic time. The raising of cotton and flax have long been important industries.

Of all the fiber-producing plants, cotton is the most important. Cotton clothes are worn to some extent in every country on the globe. This important fiber comes from the flower of the cotton plant. These little hollow fibers, extending from the outer end of the pod to the individual seeds, convey the pollen to the seeds for fertilization. When the pod ripens and breaks open, the cotton remains as a tuft of white. When picked, the seeds still cling to the cotton. In the ginning process the fiber and seeds are combed apart. The seeds are then sent to the cottonseed-oil plant. The cotton is manufactured into cloth, twine, rope, and other products. The first ginning process leaves a few short ravelings hanging to the seeds. A second ginning with finer combs removes these short ends — the cotton linters used in filling cheap mattresses and in making rayon and other cellulose products.

Flax, the oldest fiber known, is used in making linen. Chemically linen is similar to cotton, both being cellulose.

These fibers, however, are quite different physically. Linen fibers come from the stem of the plant and run its whole length between the outer skin and the woody center. Linen thread is stronger than cotton thread largely because it is made by twisting together very long fibers, while cotton thread is made by twisting fibers ranging from one-half inch in upland cotton to one and one-half inches in sea-island cotton.

In preparing flax, the gummy substance attaching the fibers to the outer skin is either rotted away or removed chemically. Next the outer shell is crushed and rubbed free from the desired fibers, which are finally combed out, bleached, and twisted into thread.

CELLULOSE NITRATE (PYROXYLIN). When cotton is treated with concentrated nitric acid in the presence of sulfuric acid, from one to three nitrate groups per 6-carbon unit may enter the cellulose molecule. The entrance of a single nitrate group greatly changes the properties of cotton. The resulting pyroxylin still looks like cotton, but it is soluble in many liquids that do not dissolve pure cellulose.

Cellulose nitrates. Cellulose is a substance with a very large molecule. Because of its insolubility, its molecular weight has not been determined. Since it will hydrolyze with continued boiling with strong acid only into glucose ($C_6H_{12}O_6$) and since it adds water during hydrolysis, its formula is usually written as $(C_6H_{10}O_5)x$. This means that an unknown number of glucose units are joined together to form the cellulose molecule. When heated with a mixture of concentrated nitric acid and concentrated sulfuric acid, the cellulose can receive nitrate radicals up to three NO_3 groups per glucose unit of six carbons. The fully nitrated cellulose is known as guncotton, which constitutes smokeless powder. It is smokeless because there is enough oxygen in the nitrate part to make gases out of all the carbon in the substance on explosion.

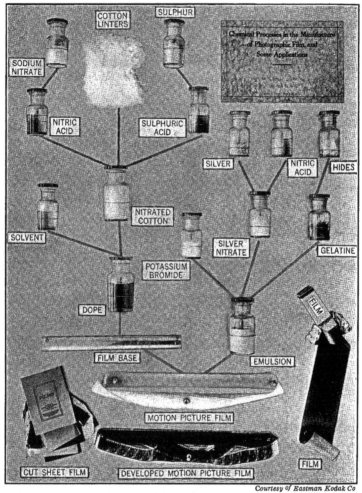

Courtesy of Eastman Kodak Co

MATERIALS USED IN MAKING PHOTOGRAPHIC FILM.   The development of photographic film has popularized still photography and has made the motion picture possible.

With fewer nitrate radicals per molecule the substance is known as pyroxylin, which is the basic substance in most of our lacquers or quick-drying paints.   The pyroxylin is dissolved in a volatile liquid.   When sprayed or spread on the surface to be coated, the solvent soon evaporates, leaving the pyroxylin as a tough protective coating. Pyroxylin or guncotton dissolved in a mixture of ether and alcohol is sold as collodion for covering cuts and abrasions on the skin. Cellulose nitrate dissolved in camphor makes celluloid, from which hairpins, combs, etc., are shaped. Kodak film is mostly cellulose nitrate. The moving picture became possible only because of the invention of cellulose nitrate.

EXPERIMENT

Nitrating cellulose. Mix 10 cc. each of concentrated nitric acid and concentrated sulfuric acid. While hot add about one-half gram of cotton. and allow it to remain 3 minutes. Remove the partially nitrated cotton, wash thoroughly, and dry. Treat some of the material with a mixture of equal volumes of alcohol and ether. Allow some of the liquid to evaporate from a watch glass. Examine the film and test its inflammability.

Nitrate rayon. Rayon is a name given to all synthetic cloth. Synthetic cloth has become so beautiful that a large part of our clothing is made from it. The first rayon was obtained from cellulose nitrate. The nitrate is dissolved in enough solvent to form a sirup. This sirup is forced through a number of small holes into a solution of NaHS,

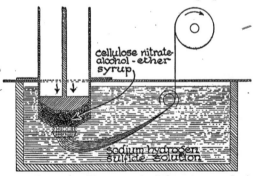

RAYON PROCESS. The paste of cellulose nitrate and solvent is forced through many fine holes into the regenerating solution. The regenerated cellulose filaments are wound up as fast as they are formed.

which removes the nitrate groups and changes the substance back into cellulose. After "spinning" — forcing of the substance through small holes — and regenerating, we have a continuous fiber of cellulose. As the original source of cellulose, cotton linters were used. Although the linters are too short to twist into a cotton thread, the resultant rayon fiber can be made as long as desired. Rayon threads are weaker than cotton threads for some reason we do not fully understand. In order to increase the tensile strength of rayon thread, it is made by twisting together a large number of fine filaments. Engineers in bridge building have learned that the steel cable made by combining many small wires is much stronger than a single rod of the same weight of material. The rayon manufacturers make use of the same principle.

Cellophane. Recently there has come upon the market a cellulose product that has taken the country by storm. This substance is called cellophane. Because it is made in tough, thin transparent sheets, it is used to wrap practically everything that can be soiled or contaminated. Cellophane is made by squeezing the cellulose nitrate sirup (see page 513) through a narrow slit into the regenerating NaHS solution, from which it emerges as a continuous sheet of cellulose.

Other cellulose rayon processes. There are two other rayon processes which start with cellulose and return to cellulose in the finished thread. One of these processes starts with cotton linters as did the nitrate process. The cellulose is dissolved in ammonia and copper hydroxide. After being "spun," the material is regenerated by passing through an acid bath.

The other process, the most recent of the four common rayon processes, uses the cellulose of wood as raw product. The shredded wood is cooked with NaOH, and carbon disulfide passed into it. The resulting sirupy product is "spun" as in the other processes and regenerated by an acid bath. The sirupy material is known as viscose; and the process, the viscose process. Because wood cellulose is so abundant, this process produces more rayon than any of the others.

Acetate rayon. One rayon process does not regenerate the filament into cellulose but leaves it as cellulose acetate. If cellulose is heated with acetic anhydride (a product made by splitting water from two molecules of acetic acid), the acetate results. For spinning, the acetate is dissolved in enough organic solvent to form the sirup. The volatile solvent evaporates from the small filament, leaving it cellulose acetate. Some of the most beautiful rayon cloth is the acetate. This rayon has properties somewhat different from the others. It may melt under a very hot iron or dissolve in certain spotting

Exhibit of Viscose-Process Rayon. Rayon makes beautiful cloth, which is growing in popularity from year to year. The bottles (left to right) contain wood chips; shredded, bleached, and compressed wood; sodium hydroxide; alkali cellulose; carbon disulfide; cellulose xanthate; viscose; and rayon thread.

solvents, such as acetone. For this reason, more care must be used in handling it than the others.

Safety motion-picture film. Formerly all motion-picture film was made of the very combustible cellulose nitrate. This nitrate film sometimes catches fire and causes dangerous conflagrations. Now cellulose acetate is much less inflammable than cellulose nitrate. For this reason, film made of cellulose acetate is called safety film. At the present time acetate film is replacing much of the more combustible product.

QUESTIONS OF FACT

1. What two vegetable fibers are almost pure cellulose?
2. Explain the origin of cotton.
3. Explain the origin and meaning of cotton linters.
4. What part of the flax furnishes the linen fiber?
5. Describe the process of preparing linen.
6. What is the provisional formula for cellulose?
7. Why do we not know the exact formula of cellulose?
8. What is pyroxylin?
9. How does pyroxylin differ from guncotton?
·10· What is celluloid?
11. In making nitrate rayon: (a) How is the cellulose made soluble? (b) What is the cellulose nitrate dissolved in? (c) How is it given the shape of a filament? (d) What substance regenerates the cellulose? (e) What substance is the final rayon?
12. What principle is used to make rayon threads as strong as possible?
13. What is cellophane?
14. Explain in detail how cellophane is made.
15. Explain the acetate rayon process.

QUESTIONS OF UNDERSTANDING

1. Pick out the three substances from the following that are cellulose nitrate:

(a) cellophane	(e) collodion
(b) wool	(f) guncotton
(c) pyroxylin	(g) cotton
(d) acetate rayon	(h) palmitin

2. Pick the substance in the second column that explains the word in the first column.

(1) sucrose	(a) cellulose
(2) nitrate rayon	(b) safety film
(3) cellulose acetate	(c) cellulose nitrate
(4) photographic film	(d) cane sugar

3. Explain the viscose rayon process including: (a) raw product; (b) treatment of the wood; (c) making of viscose; (d) spinning; (e) regeneration.

4. Why is more rayon made by the viscose process than all others?

5. Which rayon is not regenerated to cellulose? ·
6. What are photographic film and safety film?
7. Explain why smokeless powder is smokeless.

REFERENCES FOR SUPPLEMENTARY READING

Allen, Nellie B. *Cotton and Other Useful Fibers.* Ginn and Company, Boston, 1929.
(1) Prepare a report on cotton from planting the seed to the finished product.
  "The Importance of the Cotton Plant" (pp. 69–75)
  "Growing Cotton" (pp. 8–13, 76–114)
  "Marketing Cotton" (pp. 14–22, 130–136)
  "Manufacturing Cotton Products" (pp. 50–68, 137–176)
  "Useful Products from Cotton Seeds" (pp. 177–186)
Avram, Mois H. *The Rayon Industry.* D. Van Nostrand Co., Inc., New York, 1929.
Prepare a report on the production of rayon, with emphasis on the following points:
  "Historical" (pp. 1–15)
  "Importance and Growth" (pp. 16–44)
  "Products Other than Rayon Obtained from Cellulose Solutions" (pp. 120–132)
  "Fundamental Principles in Rayon Production" (pp. 319–352)
  "Principal Raw Materials" (pp. 414–435)
  "Viscose Process" (pp. 436–522)
  "Cupra-Ammonium Process" (pp. 523–543)
  "The Nitro or Chardonnet Process" (pp. 544–558)
  "The Acetate Process" (pp. 559–580)
Glover, John George, and Cornell, William Bouch. *The Development of American Industries.* Prentice-Hall, Inc., New York, 1933.
Prepare a report on one of the following topics:
  "The Pulp and Paper Industry" (pp. 95–130)
  "The Book Publishing Industry" (pp. 131–140)
  "The Newspaper Industry" (pp. 141–155)
  "The Textile Industry" (pp. 157–220)
  "The Cotton Growing Industry" (pp. 221–226)
Howe, H. E. *Chemistry in Industry,* Vol. I. The Chemical Foundation, New York, 1925. Pp. 75–84, "Cotton and Cotton Products"; pp. 357–372, "Chemistry in the Textile Industry."
(1) In what sense can it be said that cotton won the World War? (p. 75)
(2) Name the products that are made from cotton products. (pp. 75–76)
(3) How many principal commercial products are made from cotton seeds? (p. 77)

(4) Make a list of products made from the cotton seed. (p. 78)

(5) What colors predominated in the natural dyes? (p. 358)

(6) About how much money was spent in producing the first pound of indigo dye? (p. 361)

The fats. The fats as well as the carbohydrates are a food source of heat and muscular energy. Just as coal burned under the boiler of a steam engine supplies heat and does work, so the fats and carbohydrates oxidized in the animal body furnish the body with heat and make possible muscular effort. Pound for pound, the fats yield about twice as much energy as the carbohydrates; but they are hard to digest. The fats are the basis of the soaps, the most important cleaning materials. Nature has supplied us with three fats in large amounts:

stearin $(C_3H_5)(C_{17}H_{35}CO_2)_3$, from beef and mutton tallow;

palmitin $(C_3H_5)(C_{15}H_{31}CO_2)_3$, palm oil and fruit fats;

olein $(C_3H_5)(C_{17}H_{33}CO_2)_3$, fish oil, olive oil, lard.

When these fats are heated with alkali or superheated steam, they react with water to form three molecules of acid and one molecule of glycerin for each molecule of the fats.

$$(C_3H_5)(C_{17}H_{35}CO_2)_3 + 3\ H_2O \longrightarrow 3\ C_{17}H_{35}CO_2H + C_3H_5(OH)_3$$
stearin   stearic acid   glycerin

$$(C_3H_5)(C_{15}H_{31}CO_2)_3 + 3\ H_2O \longrightarrow 3\ C_{15}H_{31}CO_2H + C_3H_5(OH)_3$$
palmitin   palmitic acid   glycerin

$$(C_3H_5)(C_{17}H_{33}CO_2)_3 + 3\ H_2O \longrightarrow 3\ C_{17}H_{33}CO_2H + C_3H_5(OH)_3$$
olein   oleic acid   glycerin

When these acids are neutralized with sodium hydroxide, hard soap is formed. Its composition is $C_{17}H_{35}CO_2Na$, $C_{15}H_{31}CO_2Na$, or $C_{17}H_{33}CO_2Na$. If potassium hydroxide is used to neutralize the acids, the result is soft or liquid soap with the composition $C_{17}H_{35}CO_2K$, $C_{15}H_{31}CO_2K$, or $C_{17}H_{33}CO_2K$.

The student is not expected to memorize these formulas, but he should know that the fats are made of two parts, one coming from the acid and one from the alcohol (glycerin). He should remember that a fat will react with water under the stimulus of acid, base, or superheated steam to form acid and glycerin. The acids are used for making soap, and the glycerin for making dynamite.

The proteins. While we are studying some of the principal classes of compounds used as food, we must include the proteins. The proteins are a group of very complex carbon compounds containing also hydrogen, oxygen, and nitrogen. Some of them also contain sulfur and phosphorus. The proteins are the building materials of the lean meat of the animal. Proteins are found in vegetable substances also, especially in beans, nuts, and grains. The animal kingdom as a whole gets its protein from the vegetable kingdom. Some proteins are soluble, as the white of an egg; others are soft, as lean meat; and others are tough and insoluble, as the skin, horns, and hoofs of animals. A growing animal needs much more protein than one that has finished growing. A work horse needs much carbohydrate and little protein, but the growing colt needs more protein and less carbohydrate.

EXPERIMENT

Test for protein. With a stirring rod touch a drop of concentrated nitric acid to the top of the finger nail. After a few seconds neutralize this with sodium hydroxide solution and note the orange yellow color. This test is called the xanthoproteic test for proteins.

The protein fibers. Two of the common fibers used in cloth making, silk and wool, are proteins. Rayon was once called artificial silk, since it had a shininess like silk. However, it is not silk in any true sense. Silk, being a protein, contains nitrogen, as does wool. According to its weight the silk fiber is the strongest substance known. Rayon is a much weaker fiber. In general, the animal fibers are protein, but the fibers of vegetable origin are cellulose. Attempts have been made to produce a fiber resembling wool by attaching nitrogen to the cellulose molecule. Although the addition of nitrogen to the cellulose has increased the tensile strength somewhat, it has not given any of the other desirable characteristics of the animal fibers.

Burning serves as a qualitative test for distinguishing between the protein fibers and the cellulose fibers. Burning protein fibers smell like burned feathers, but the cellulose fibers smell like burning paper or wood.

For the quantitative determination of the percentages of wool and cotton in a piece of cloth, boil the weighed sample of the cloth in 5 per cent sodium hydroxide solution. The wool will dissolve, leaving the cotton, which can be washed, dried,

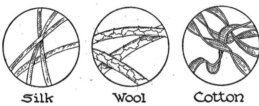

Silk Wool Cotton

SILK, WOOL, AND COTTON FIBERS. The drawings show the appearance of the natural fibers under a low-power microscope.

and weighed. From the weight of the dried residue, the percentages of both cotton and wool can be calculated.

Dyes. Of all the complex compounds of carbon, the dyes illustrate the possibilities of synthetic chemistry as well as any other class of substances. As late as three decades ago, the colors used in dyeing clothing were few in number. All stockings were dyed black, not because our grandparents liked black stockings any better than we do, but because the only dye for cotton that was both cheap and permanent was black. Today we can have our stockings and other garments practically any color, or any tint or shade of the color, that the heart desires.

Most of the dyes known before the last century were of plant origin. Among the best of these were indigo from the indigo plant and alizarin from the root of the madder plant. Indigo is the blue used for jeans and overalls, but alizarin is a red dye known as turkey red. There were only two dyes of animal origin, royal purple and cochineal. The former was obtained

William Henry Perkin

1838–1907

When only eighteen years old, young Perkin was given by his college professor the difficult problem of trying to synthesize quinine.

From coal tar, which at that time was a worthless by-product of the gas industry, they were able to separate aniline. The teacher and pupil reasoned that since aniline is an amine, and quinine is an amine which also contains oxygen, it might be possible to oxidize aniline to quinine. The reaction failed as far as quinine was concerned, but as a result of this attempted synthesis, a new industry sprang up, that of synthesizing dyes.

From the mixture resulting from the oxidation of aniline sulfate with potassium dichromate, Perkin separated a purple dye, which is now called mauve. Recognizing the commercial possibilities of his discovery, eighteen-year-old Perkin patented his process and with the aid of his father and brother built a factory for synthesizing the new dyestuff.

Perkin's patent is interesting. It read:

"The nature of my invention consists in producing a new coloring matter for dyeing with a lilac or purple color stuffs of silk, cotton, wool, and other materials in the following manner:

"I take a cold solution of sulfate of aniline, or a cold solution of sulfate of toluidine, or a cold solution of sulfate of xylidine, or a mixture of any one of such solutions with any others or other of them, and as much of a cold solution of a soluble bichromate as contains base enough to convert the sulfuric acid in any of the above-mentioned solutions into a neutral sulfate. I then mix the solutions and allow them to stand for ten to twelve hours, when the mixture will consist of a black powder and a solution of neutral sulfate. I then dry the substance at 100° C. and digest it repeatedly with coal-tar naphtha until it is free from a brown substance which is extracted by the naphtha."

from a mollusk in Asia Minor, and the latter from a plant louse in Mexico.

Synthesis of phenolphthalein. Place a pinch of phenol and a pinch of phthalic anhydride in a test tube. Add four drops of concentrated sulfuric acid. Stir with a thermometer and heat to 160° C. for two minutes. Pour the hot melt into 10 cc. of water. Make part of this solution alkaline and note the color of phenolphthalein.

The royal purple was too expensive for common use. The bare speck of the substance in each mollusk limited its use to royalty, hence the name. However, when compared to the synthetic dyes today, it is really not very stable towards light. It is not used now because we can make so many better purple dyes.

Cochineal, the powdered remains of a Mexican aphis, or plant louse, was formerly used to give candy a pink color. It served all right as long as the person eating the candy did not know its origin. However, we are glad that candy today is colored with synthetic dyes.

The first synthetic dye was discovered accidentally when the English chemist William Perkin tried to synthesize quinine. He failed in his purpose, but he showed that dyes could be synthesized and started dozens of other chemists on the quest for the chemical rainbow, a quest which has succeeded far beyond the dreams of the early workers. We know now that all the possibilities of synthesizing dyes will never be exhausted. The 50,000 dyes already synthesized are only the beginning of what can be done.

REVIEW QUESTIONS

1. Compare the energy value of fats and carbohydrates.
2. Name three common fats and give two sources of each.
3. Name two classes of substances obtained from fats.
4. How is soap made from fats?
5. What parts of the animal body are proteins?

Name four elements in all proteins.

Name two other elements in some proteins.

8. Where could you find a soluble protein? an insoluble protein?

9. How should the diet of a growing colt differ from that of a work horse?

10. Name the protein fibers.

11. What is a dye?

12. What was the fortunate result of Perkin's failure to synthesize quinine?

13. Why did our grandparents wear black stockings?

14. Discuss the present situation in regard to synthetic dyes.

15. Give a practical test to distinguish between cotton and wool.

16. How could the percentage of wool in a sample of cloth be determined?

17. Name some of the naturally occurring dyes: (a) of animal origin; (b) of plant origin.

PROBLEMS

1. The fat, stearin, has the formula $(C_{17}H_{35}CO_2)_3C_3H_5$

(a) Calculate the molecular weight.

Stearin hydrolyzes as follows in acid solution:

$(C_{17}H_{35}CO_2)_3C_3H_5 + 3\ H_2O \rightleftharpoons 3\ C_{17}H_{35}CO_2H + C_3H_5(OH)_3$ (glycerol)

(b) Calculate the weight of glycerol obtainable from 100 g. of stearin.

2. If in sodium hydroxide solution the stearic acid of problem 1 is made into soap as follows:

$$C_{17}H_{35}CO_2H + NaOH \longrightarrow C_{17}H_{35}CO_2Na + H_2O$$

calculate the weight of soap obtainable from 500 g. of stearic acid.

ADDITIONAL EXERCISE FOR SUPERIOR STUDENTS

Review the structural representations of hydrogen, sodium, and sulfur in Chapter 21. Draw a picture of the compound $\overset{+}{N}a\overset{+}{H}\overset{=}{S}$.

REFERENCES FOR SUPPLEMENTARY READING

1. Allen, Nellie B. *Cotton and Other Useful Fibers.* Ginn and Company, Boston, 1929.
   Prepare a report on wool (pp. 292–351); on silk (pp. 258–281).
2. Howe, H. E. *Chemistry in Industry*, Vol. II. The Chemical Foundation, New York, 1926. Pp. 360–377, "Soap-Cleanliness through Chemistry."

(1) Describe the process of making home-made soap. (pp. 360–361)
(2) Describe one process of making soap. (pp. 363–369)
(3) Explain how glycol is recovered as a by-product of soap making. (p. 366)
(4) In the lime saponification method how is the soap recovered from the insoluble calcium soap? (pp. 368–369)
(5) Describe the making of toilet soap. (pp. 370–371)

Summary

In the juices of certain plants occur substances of sweet taste called sugars. Sucrose, the common sugar of food, is obtained from both sugar cane and sugar beets by neutralizing organic acids, precipitating albuminous substances with heat, decolorization with bone black, and crystallization in vacuum pans. Sucrose hydrolyzes into equal molecular parts of glucose and fructose (fruit sugar). Because it hydrolyzes into two sugar units, sucrose is called a disaccharide. Other disaccharides of industry are lactose, the sugar in milk, and maltose, the sugar in malt barley. Sugars and those substances which hydrolyze to sugars only are called carbohydrates, many of which are important in our foods as energy producers.

Starch, a very important carbohydrate, is the body of such foods as potatoes, rice, and tapioca. Heating changes starch to dextrin. Glycogen, an animal carbohydrate, is prepared in the livers of animals. Pectin is the carbohydrate which causes jelly to solidify.

Cellulose, the fiber of wood and cotton, is also a carbohydrate. The higher nitrates of cellulose form guncotton (smokeless powder). The lower nitrates make pyroxylin (common photographic film), which, dissolved in alcohol or ether forms collodion, in camphor forms celluloid, in volatile solvents forms quick-drying lacquers.

Cellulose nitrate is made into rayon by squeezing it through small holes into a NaHS regenerating solution. Cellophane is made in exactly the same way except that the nitrate is forced through a narrow slit. Two other rayon processes begin with cellulose and finally regenerate the cellulose. The acetate process, however, leaves the substance as cellulose acetate.

Fats, another class of energy foods, are the glyceryl esters of the higher fatty acids.

Proteins, the chief constituents of lean meat, are the cell-building materials of our foods. Silk and wool are protein fibers.

Dyes are the colored organic amines and acidic compounds, thousands of which have been synthesized.

FRACTIONAL DISTILLATION. An important operation in organic analysis.

⇄

CHAPTER 27

CHEMISTRY AND HEALTH

Health a question of chemistry. The human body is a chemical laboratory in which thousands of chemical reactions are taking place. If these reactions are all going on as they should, we are happy and robust and healthy. We are full of hope and like to work. Hard times and hard work cannot make us discouraged. If, on the other hand, some reaction goes wrong, we are more or less ill. According to one theory, cancer is caused by glucose oxidizing the wrong way; in other words, the chemical reaction is not going the way it should. We do not as yet know why the glucose does not oxidize in the normal way when a person has cancer. When we are able to know the reason for this abnormality and correct it, we will probably be able to cure cancer. If the glucose in the blood does not oxidize, instead of serving as a food for the body, it poisons the system, and the person suffers from diabetes. Here sickness results because the reaction does not take place. The reason this reaction does not take place is that insulin, a powerful catalyst, is not being produced by the pancreas, one of the organs of our bodies.

Is disease inevitable? Why should so many people be sick? Some are ill all the time, and most of the others are ill part of the time. The normal and reasonable condition is health. Sickness is the abnormal or unnatural condition. In fact, we ought not to be sick.

Animals are seldom sick. The horse and cow are far more likely to live their allotted span of life than their owner. Animals as a rule live about 5 times as long as it takes for them to reach full maturity. A horse requires about 4 years to become full grown. His average life is about 20 years. The cow matures in 3 years and lives about 15 years. The human is not fully grown until he is 25 years old. On the same basis the average human life ought to be 125 years. The Biblical three score years and ten (70 years) is far short of this span. The life insurance companies tell us the average length of life is now about 58 years. Twenty years ago the average length of the human life was hardly 40 years. This increase has been brought about by saving the lives of the babies. With better care at birth and more intelligent care during the first year, infant mortality has been greatly lessened. Although this has greatly increased the average life span, it has not increased the chances of those who live to be 40 years old. In other words, those who reach 40 years of age are now no better able to reach 80 than they were before.

One reason the babies are doing so much better than they did twenty-five years ago is that we have learned what they should eat, and we feed it to them. Older people, on the other hand, eat almost anything without regard to its effect on their health.

Kinds of disease. If disease means that the chemical reactions of the body have gone wrong, what makes them do that? We sometimes classify the diseases according to the nature and cause of the trouble into deficiency diseases, infectious diseases, and degenerative diseases.

The deficiency diseases, such as scurvy, pellagra, beriberi, and rickets, are known to be due to something lacking in the diet, either some necessary food or some catalyst that nature has placed in the food to aid in its digestion and assimilation. These diseases are caused primarily by the lack of certain essential substances known as vitamins.

An infectious disease results from some disease germ which gets into the body. The germ multiplies rapidly and secretes poisons, which cause the chemical balance of health to be upset. Among the infectious diseases are tuberculosis, influenza, diphtheria, common colds, smallpox, typhoid fever, blood poisoning, pneumonia, and lockjaw. The treatment for these diseases takes a double form, that of killing the germs and of neutralizing their poisons.

The causes of degenerative diseases are not so well known. In this group are heart trouble, kidney trouble, nervous troubles, and circulatory troubles. Overwork, worry, faulty habits, and poor diet may each contribute its part. The diseases chiefly responsible for deaths are in order: (1) heart disease, (2) circulatory diseases, (3) nephritis, (4) cancer, (5) tuberculosis. This arrangement shows a degenerative

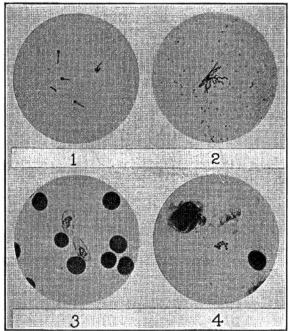

Courtesy of U. S. Marine Hospital Service

PHOTOMICROGRAPHS OF DISEASE BACTERIA. (1) Lockjaw bacilli, with spores; (2) typhoid bacillus with flagella; (3) bacilli of *Plasmodium malariae* in human blood; (4) staphylococci in pus.

disease at the head of the list. We should know something about the causes of these diseases in order to avoid them.

Nature's healing agencies and body defenders. Nature has provided certain agencies that enable the body to overcome invading germs and correct its chemistry. These agencies are sunlight, exercise, fresh air, rest, water, and good food. Our great-grandparents did not pay a great deal of attention to diet, but they lived mostly in the fresh air and sunlight, had

plenty of exercise, which in turn caused them to drink plenty
of water, and had a reasonable amount of rest.   Their food
was simple and largely as nature gave it, which means that
the vitamins and minerals were still in it.   Preventive medicine was little known, yet the chances of living to eighty years
of age were better for a person of forty then than now.

Photo by Dwight Bentel

EXERCISING IN THE SUNLIGHT.   Reasonable exercise is necessary for good health.
The ultraviolet rays of the sun are very effective in killing disease germs and in
stimulating the defense processes of the body.

Sunlight.   Sunlight serves the body in two ways; its ultraviolet rays kill disease germs and stimulate the normal chemical reactions in the body.

Exercise.   Exercise assists in the elimination of waste products, which interfere with the healthy chemical reactions in
the human mechanism.

Fresh air.   Although fresh air is desirable because of its
freedom from germs, it is important primarily as a source of
oxygen needed to maintain the oxidation processes in the body,
which keep up the body heat and furnish the energy for all
forms of activity.

Rest. Rest gives the body a chance to get caught up on its work of repairing worn-out cells and removing waste products and other poisons. If we give the body a chance to clean up all of today's troubles before tomorrow, disease germs will have little chance of gaining the upper hand.

Water. In the chapter on water, it was shown that a great deal of water is needed daily. All of the chemical reactions in the body take place in solution. Water must be available for the solutions. Water treatment, or hydrotherapy, was also discussed in that chapter.

Good food. Food is the building material for the body. If it does not include everything that is needed, trouble is sure to follow. Even though a deficiency disease does not materialize, the body cannot work to its best advantage if it is not properly supplied with food, and some germ is likely to get a foothold and cause disease.

Elements in the human body. The human body contains about eighteen elements. However, only thirteen occur in quantities large enough to concern us. These are shown in the accompanying table.

ELEMENTS IN THE BODY

ELEMENT	PER CENT	ELEMENT	PER CENT
Oxygen	65	Sulfur	0.25
Carbon	18	Sodium	0.15
Hydrogen	10	Chlorine	0.15
Nitrogen	3	Magnesium	0.05
Calcium	2	Iron	0.004
Phosphorus	1	Iodine	trace
Potassium	0.35		

The absence of a sufficient quantity of any of these elements from the diet will result in some form of ill health. The first principle of dietetics, the science of correct diet, is that *the diet must contain sufficient amounts of all of the thirteen elements required by the body.* Even if iodine, which is required in an immeasurably small amount, is missing from the diet, the disease of goiter results.

Oxygen. If we drink enough water and have all the food we care to eat, we need have no concern about our oxygen except what we need for breathing purposes. Water is eight-ninths oxygen, and practically all foods contain high percentages of this element. The three great food groups, fats, carbohydrates, and proteins, all contain large percentages of oxygen. However, if we do not get plenty of good fresh air, we suffer for lack of oxygen for purifying our blood.

Carbon. The next element in abundance, carbon, also is present in large percentages in the three food classes. If we eat the normal quantity of food, we will get enough carbon whether the food be considered good, bad, or indifferent.

Hydrogen. Hydrogen, the third element in abundance in the animal body occurs in water and the three food classes. No ailment is known to be due directly to a lack of hydrogen so long as the person is not starving or suffering from thirst.

REVIEW QUESTIONS

1. What have the chemical reactions in the body to do with sickness?
2. What reaction is thought to be wrong in cancer?
3. What reaction is failing in diabetes?
4. Upon what basis do we think we ought to live 125 years?
5. How has infant mortality been lessened in the last quarter century?
6. What are the deficiency diseases?
7. What is meant by a degenerative disease?
8. Name some infectious diseases and give their cause.
9. Name some of the probable causes of degenerative diseases.
10. Name nature's six healing agencies.
11. Name two ways sunlight helps to preserve health.
12. What does exercise do to preserve health?
13. What is the relation of fresh air to health?
14. Explain how plenty of rest helps to keep the body in good health.
15. Why is water necessary in our food?
16. Name the thirteen elements necessary for perfect health.
17. State the first law of dietetics.

18. Why do we not have to concern ourselves about oxygen if we have plenty of fresh air to breathe?

19. Need we concern ourselves about carbon as an element in our food, if we have all we wish to eat?

20. Is hydrogen likely to be deficient in our food?

REFERENCE FOR SUPPLEMENTARY READING

1. Sadtler, Samuel Schmucker. *Chemistry of Familiar Things.* J. B. Lippincott, Philadelphia, 1924.
   (1) What enzyme is in the blood? (p. 248)
   (2) How is carbon dioxide carried to the lungs? (p. 249)
   (3) What is the acid-resistant substance in the teeth? (p. 249)
   (4) What causes tooth decay? (p. 250)
   (5) What does exercise do to the chemical reactions of the body? (p. 252)

Nitrogen. Nitrogen does not occur in all of the food classes; only the proteins contain this element. However, the proteins are quite common. Poorly fed people may be thin from lack of protein; but where people are not starving, they are more likely to suffer from too much nitrogen rather than too little. The proteins are the building materials of the body. For the growing child a great deal of protein is needed, and the danger of too much protein is slight. However, the grown person, who needs only to repair worn-out cells, may easily eat too much protein. The protein that is in excess of the building needs of the body is used for fuel. The trouble with this is that the body has difficulty in getting rid of the waste products of the proteins that have been burned up. The nitrogen residues are substances that are hard to eliminate and cause various kinds of trouble. The fats and carbohydrates are the normal energy foods. When they are burned, the resulting carbon dioxide is easily eliminated in the out-going breath. The little water formed by the oxidation of the hydrogen is not noticeable in the presence of the water normally in the body liquids.

Calcium. It is a strange fact that the elements that occur in our bodies in small amounts are the ones most likely to cause trouble by their lack. Although calcium forms only 2 per cent of our bodies, many people lose their teeth early in life because they did not get sufficient calcium in their food while their teeth were growing. Mothers often lose their teeth because they do not get enough calcium before their children are born. Nature often sacrifices the mother for the child. The calcium is literally taken from her teeth to make bone for the child. Our bones and teeth are largely calcium phosphate. The growing child must have a sufficient supply of calcium, or his bones will not develop as they should. Rickets and weak

FOODS RICH IN ESSENTIAL MINERALS. Mineral foods are especially important for the building of strong bones and teeth.

joints frequently result from a diet deficient in calcium. In this connection the question naturally arises: Which foods contain calcium? The foods richest in calcium are milk, cheese, turnips, celery, spinach, tomatoes, lentils, and rice. We will see as we go along that milk is an almost perfect food. It is the perfect food for the baby and the young child. It is also an excellent food for the grown-up. It should be remembered that all foods containing milk, such as cheese, cottage cheese, puddings, custards, and creamed foods, are also sources of calcium.

When we consider these calcium foods, we see that most of them are vegetables. Unfortunately many children and older people do not like vegetables. Says one, "I do not like turnips." Another says, "I cannot eat spinach." If you happen to be such a one, remember that if you do not get enough calcium while you are growing, your teeth will be poor. Toothaches and dentists will worry you most of your life. The reason some of us do not like spinach, turnips, and other vegetables is that parents used to think children should not have anything but milk and milk products until after they were a year old. The result of this unfortunate theory has been that many of us did not acquire a taste for vegetables at the time when our food habits were forming, and now do not like

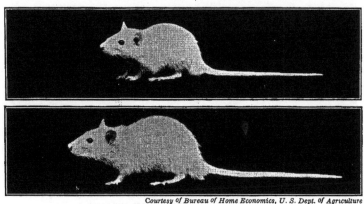

A LITTLE PHOSPHORUS MAKES A LOT OF DIFFERENCE. These rats are from the same litter and received the same food, except that the one above received too little phosphorus. The one that received the normal amount weighs almost twice as much as his brother.

many vegetables. However, good health requires that we get plenty of the minerals and vitamins that are mostly found in vegetables. Fortunately, one can learn to like most vegetables if he makes an honest effort to do so.

Phosphorus. Phosphorus comes next in amount (one per cent) in the human body. A large part of it is in the calcium phosphate of the bones. It is also an essential part of the nucleus of the cells and is especially concentrated in the brain and nerve cells.

We get our phosphorus largely as we get our nitrogen, in the proteins of lean meats, milk, eggs, peas, beans, grains, and nuts. A meat-eating people, who can afford what they like, are apt to get too much protein rather than too little; and since all lean meat and fish are rich in complete protein — that is, proteins containing phosphorus and sulfur — there need be no concern about protein phosphorus. Moreover, we probably do not get all of our phosphorus as protein. Some of it is taken as phosphate in vegetables. Spinach is very rich in phosphorus. Although we do not know exactly in what form phosphorus is used by the body, it is a reasonable guess that since the bones are largely calcium phosphate, they can use the vegetable phosphates readily, while the phosphorus in the brain and nerves, being of a protein nature, require the protein phosphorus of milk, eggs, meat, and nuts.

Potassium. Potassium (0.35 per cent in the body) is found in the soft tissues. It is obtained primarily from vegetable foods, as are the other mineral elements. The way we cook vegetables determines whether we will get all the minerals or not. Recently there has been developed a method of cooking vegetables with practically no addition of water; they are cooked in their own juices, and none of the juice is thrown out. In this way none of the mineral salts are lost. The juice in which vegetables or potatoes are cooked should never be thrown away. The minerals are needed. Nature put them in the foods for us, so we should not throw them away. If there is no other use for the water drained off potatoes or vegetables, it should be saved for soups and gravies.

We do not know much about the ailments that are caused by a deficiency of potassium. However, we may rest assured that if we eat plenty of vegetables, we will avoid these ills.

Sulfur. Sulfur constitutes 0.25 per cent of the human body. It occurs in the protein part of the nucleus of the cell. If we eat enough complete protein, we will get enough sulfur. There are no ailments known to be due to a deficiency of sulfur in our food. The natural conclusion is that we get a sufficient amount without giving it any thought.

Sodium and chlorine. We consider sodium and chlorine together, as they occur in chemical union as salt, or sodium chloride. Chlorine also occurs in the stomach as hydrochloric acid, but the body manufactures hydrochloric acid from salt. The blood and other body liquids contain salt. A shortage of salt is soon felt by the body. When an animal is deprived of salt, an insistent craving drives it to extremes, such as traveling long distances to a salt lick or salt spring.

Since we add salt to food to flavor it or — as in the case of salt meat and salt fish — to preserve it, there is practically no danger of our not getting enough salt. There is more danger of getting too much. Some physicians say we eat far too much salt. There is a danger along this line. We are very much creatures of habit; we like what we are used to. If the person who prepared our food while we were young salted it freely, we will always want our food salted heavily. Sometimes the habit for some reason creeps up on us. We should always keep in mind the possibility that we may eat too much salt. It is possible that nature has put nearly enough salt in our foods already, especially in meat and eggs. Salting eggs is a matter of habit. They taste very well without salt. The water in some alkaline districts already contains considerable salt. It stands to reason that one should not add so much salt to food cooked in such water.

Magnesium. Our present habits of eating probably provide us enough magnesium. We may feel reasonably safe if we eat vegetables, which are the chief source of the body minerals.

REVIEW QUESTIONS

1. What class of food contains nitrogen?

2. Why does a growing boy need more nitrogen food than a grown man?

3. If more protein is eaten than is needed for growth and repair, what becomes of the excess?

4. Is there any objection to too much protein?

5. Where is too little calcium in a person's diet likely to be noticed?

6. What calcium salt forms a large percentage of the bones?

7. Name some foods rich in calcium.

8. Name the body elements that we get in sufficient quantities with our present diet.

9. What is meant by a complete protein?

10. Why should vegetable juices not be thrown away?

11. Where does sulfur occur in the body?

12. What sodium compound is needed in the body?

13. Name two places where chlorine occurs in the human body.

14. Do we need to worry about our magnesium food requirements?

15. Name two kinds of body tissue rich in phosphorus.

16. Can our bodies use uncombined phosphorus?

17. Do you think proteins containing phosphorus are an essential part of our diet?

18. What foods contain sodium and chlorine in the form of sodium chloride?

PROBLEM

Calculate the weight of the following elements in a person weighing 130 lbs.: (a) nitrogen; (b) calcium; (c) chlorine; (d) phosphorus; (e) magnesium.

FOODS CONTAINING IRON. One whose diet is deficient in iron is likely to be pale and anemic.

Iron. Although iron occurs only to the extent of 0.004 per cent in the body, lack of it in the food causes anemia. This trouble may also come from the inability of the body to use iron that is present in the food. Iron occurs in the hemoglobin of the red corpuscles. It is absolutely necessary.

Sometimes physicians prescribe medicines containing iron compounds, but the safest and surest way of getting iron is to eat those foods that contain iron prepared in nature's way. Leafy vegetables, tomatoes, and meat are rich in iron. Green and red foods contain iron. Beef liver and kidney are rich in iron. Raisins and prunes are advertised for their iron content. They do contain iron and it is good, but the same weight of spinach contains ten times as much iron. This fact again emphasizes the value of the vegetables. Whether we like them or not, the fact still remains that they are the best source of most of our useful minerals.

Iodine. The last element known to be essential to our well-being is iodine. The amount needed is so small that it would seem to be negligible. However, if we do not get it, we begin to develop thyroid trouble, or goiter. Ordinarily our foods do not contain it. We get our supply from the drinking water. Wherever the water has run considerable distances through the soil, enough iodine has been picked up. In countries where the water is obtained in a pure form directly from melted snows, iodine is deficient and goiter is common. By putting iodine into the drinking water or in the table salt, goiter can be prevented. In communities where goiter was formerly very common, it has practically disappeared due to the use of iodized salt.

Vitamins. The science of dietetics is not so simple as one might gather from the study so far. Even if we had a diet varied enough to contain all of the necessary elements in sufficient amounts, we might still be troubled with some of the deficiency diseases. Certain foods require special catalysts before the body can assimilate them. Until about a hundred years ago sailors were often afflicted with a disease called scurvy. It had been observed that the absence of fresh vegetables seemed to cause this disease. When this was thoroughly established, the French Government offered a prize for a method of preserving fruit and vegetables in such a form that their antiscorbutic properties would not be destroyed. The inventor of a successful process of canning fruit and vegetables won the prize. This process of saving food has grown to be one of the world's greatest industries. We have become so accustomed to the use of canned foods that it is said that a bride today does not have to know how to cook ; all she needs is a can opener. The catalyst in green-colored vegetables that prevents scurvy is called a *vitamin*.

The antiberiberi vitamin. At the time of the Russian-Japanese War, the Japanese found many of their soldiers becoming seriously ill with a disease they called beriberi. Since this disease seemed to be peculiar to rice eaters, the doctors came to the conclusion that the trouble must be in their diet. The diet was largely polished rice and fish. It was then learned that if the soldiers ate the unpolished rice — that is, rice with the hulls still on — they did not get beriberi.

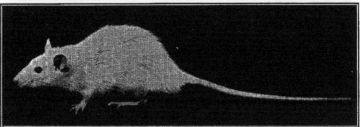

Courtesy of Bureau of Home Economics, U S Dept of Agriculture

VITAMINS AND HEALTH.   The pictures of two rats from the same litter emphasize the importance of the vitamins to health.   The diet of the rat in the upper picture was deficient in a single vitamin — vitamin G.

Nature had put a vitamin into the hulls of the rice, which was necessary for the proper assimilation of this food.   This change in diet was so effective in eradicating beriberi that a large hospital, which formerly had taken only beriberi patients, had to close its doors.

Search for vitamins. The success of the Japanese in determining the cause of this deficiency disease started investigators all over the world in a search for vitamins.   Their combined attack has put dietetics on a solid footing.   Several diseases, including rickets and pellagra, were shown to be deficiency

Foods Rich in Vitamin A. This vitamin is necessary for growth and for preventing eye diseases and infections of the breathing passages.

diseases due to lack of one or more vitamins. It was found that young animals were stunted and often died before reaching maturity if one of the vitamins was left out of their diet. It was observed that if a cow lacked a certain vitamin in her diet, her calf would be born dead. This knowledge of the essential vitamins has been of the greatest help to babies. Thousands of infants now live that would have died had we not learned about vitamins. Thousands of others are strong and healthy where they would have been sickly had we not known how to give them the vitamins.

What are the vitamins? The vitamins are complex organic compounds that are stored in parts of our food and are necessary for the proper assimilation of this food. They are needed in such minute traces that the chemist has had the utmost difficulty in separating them from the foods. They are wonderful catalysts.

Names of the vitamins. The vitamins are named in the order of their discovery by using the letters of the alphabet. The letter is frequently accompanied by a statement regarding the solubility of the vitamin.

Vitamin A. Vitamin A is often called "fat-soluble A" because it occurs in the fats, especially in milk and cod-liver oil. It is the growth-producing vitamin. Young rats do not grow if it is kept out of their food. Growing children need it. They should have plenty of milk and butter, which contain vitamin A.

In many respects oleomargarine is a satisfactory substitute for dairy butter, but growing children need the vitamin A found in butter. Since cod-liver oil contains this vitamin in a very concentrated solution, it is often given to children who do not seem to be doing well. Promotion of growth is only one of several uses of this vitamin. Grown people as well as children suffer many ailments when vitamin A is missing from the diet. Vitamin A is found in cod-liver oil, butter fat, egg yolk, fat, liver, leafy vegetables, and yellow pigmented vegetables. This vitamin is an anti-infective, which helps our bodies overcome invading germs.

Vitamin B. Vitamin B is sometimes called "water-soluble B." Its absence from the food causes beriberi among the rice eaters and pellagra among the corn eaters. It also is necessary for growth in young animals. Vitamin B occurs especially in oranges and other fruits, the cereal germ, spinach and other vegetables, and cream and milk.

Vitamin C. Vitamin C both prevents and cures scurvy. It is present in fresh vegetables and fruits. Tomatoes and oranges are particularly rich in vitamin C, so we see another reason why babies are given orange juice.

Vitamin D. Rickets is a disease of malnutrition caused by an insufficiency of calcium in the bones. Calcium alone in the diet cannot prevent rickets and other ailments caused by

FOODS RICH IN VITAMIN B. An infant who does not get enough vitamin B does not grow properly; an adult who does not get enough is likely to develop neuritis.

VITAMIN C FOODS. This vitamin prevents scurvy. Vitamin C is quickly destroyed by the heating, salting, canning, drying, and aging of foods.

calcium deficiency. Vitamin D must be present also before the body is able to use the calcium in the food. Rickets is prevented by vitamin D. Children need no longer have rickets. The ultraviolet light in sunlight is just as effective in curing the difficulty as is food containing the vitamin. Vitamin D is also soluble and occurs in fats, milk, eggs, butter, cod-liver oil, and the leafy vegetables. Sunshine synthesizes vitamin D in the skin.

Vitamin E. Vitamin E is fat soluble. It is necessary for the development of the young animal before birth. It is found in wheat and other raw cereal germs, seeds, green leaves, and egg yolk. It is rarely lacking in the normal diet.

Vitamin G. Vitamin G is water soluble. It is sometimes called B_2. Pellagra is prevented by this vitamin. Pellagra

FOODS RICH IN VITAMIN D. One result of a deficiency of vitamin D is rickets, a faulty bone development, often occurring in children.

shows itself by mental depression and nervous disorders. This vitamin is found in eggs, liver, lean meat, and vegetables.

Refined foods. Many of our foods are changed a great deal from their natural state. For example, in the manufacture of white flour, the hull of the grain is discarded in the bran. With the bran goes a very essential vitamin. The bran itself supplies us with roughage, which helps in the elimination of waste products.

If a pigeon is fed on white bread alone, its wings droop and it soon gets too weak to stand up. If its diet is not changed, it will die. If, however, even after it is unable to stand, a little water in which the bran of the grain has been soaked is poured down its throat, it soon revives and becomes normal. When we reject the bran, we lose this vitamin.

Part of the protein of the grain is also discarded in making white flour, as the flour looks whiter and keeps better without it. However, the phosphorus and the iron go with the protein. We are now turning back to the whole-wheat bread, which is much more wholesome. We are beginning to learn that most of our attempts to refine the foods nature has given us have really made them less valuable as foods.

REVIEW EXERCISES

1. Where does iron occur in our bodies?
2. Name foods rich in iron.
3. What disease is due to iodine deficiency?
4. How do we get our food requirements of iodine?
5. What is a vitamin?
6. Where did nature put the vitamin necessary for the proper digestion of rice?
7. What has our knowledge of vitamins meant to the present generation of babies?
8. What is the chemical nature of a vitamin?
9. Name some foods rich in vitamin A.
10. Name two diseases caused by vitamin B deficiency in the food.
11. Name the foods rich in vitamin B.
12. Which vitamin prevents and cures scurvy?
13. Where do we get vitamin C?

14. Name a disease caused by lack of vitamin D.
15. Name foods containing it.
16. State the relation of sunshine to vitamin D.
17. What is the danger in too highly refined foods?
18. Describe the experiment with the pigeon on a white-bread diet.
19. Compare white bread and whole-wheat bread.
20. What is the likelihood of vitamin E deficiency in a diet?
21. Relate the story of diet as a cause of scurvy.
22. Tell how people learned to can foods.
23. Relate the story of beriberi.
24. Which vitamin prevents rickets?

REFERENCE FOR SUPPLEMENTARY READING

1. Stieglitz, Julius. *Chemistry in Medicine.* The Chemical Foundation, New York, 1928.
    (1) Write a history of the discovery of the vitamins. ("The Story of the Discovery of the Vitamins," pp. 112–144)
    (2) Prepare a report on one of the following:
        (a) Conquest of Rickets. (pp. 145–164)
        (b) The Disappearance of Scurvy. (pp. 165–172)
        (c) The Advance against Pellagra. (pp. 173–179)
        (d) The Needless Sacrifice to Beriberi. (pp. 180–190)

Acid-base balance in the body. The gastric juice in the stomach has an acid reaction, while the saliva, the bile, and the blood are alkaline. If the stomach should turn alkaline, trouble would result, and if the blood should become acid, we would die. The acid or alkaline condition must stay quite constant, or we become ill. Sometimes, as a result of improper diet, our blood becomes less alkaline than it should be. The urine begins to give an acid test with blue litmus paper. The doctor says we have "acidosis." Acidosis may manifest itself in different ways, some of which are lumbago and other forms of rheumatism.

The acid or basic condition of the body depends upon the ultimate decomposition products of the food we eat, rather than upon its immediate reaction. For instance, one may increase the alkalinity of the blood by taking lemon juice. Lemon juice, as we all know, is in itself decidedly acid. However, it is largely composed of organic acids and calcium salts.

The acids are oxidized in the body to carbon dioxide and water. The carbon dioxide is expelled in the breath, and the water is neutral. The ultimate oxidation product of calcium salts is calcium oxide, which is alkaline in water. This explains how we can increase the alkalinity of the body by eating acid fruits. We sometimes describe food in terms of acid ash or alkaline ash, in which ash refers to the final products of oxidation. Ordinarily there is no danger of the body becoming too alkaline. We need to know the acid foods only to be careful not to eat too much of them, and the alkalinizing foods for their alkalinity. The alkaline-ash foods are the fruits and vegetables, but the acid-ash foods are the meats and eggs and grains.

Food requirements of the body. One of the food requirements of the body is enough protein to build new cells and repair old ones. This must include enough complete protein to furnish protein iron and phosphorus. There must be enough of the energy foods — fats and carbohydrates — to supply body heat and furnish energy for the work a person does. About one-fourth of the total heat units should be in fats. There should be enough minerals, enough water, considerable bulk (or roughage), plenty of vitamins, and plenty of alkaline-ash foods. The quantity of protein depends on size and growth, while the quantity of energy foods depends largely upon the amount of exercise taken.

How the energy content of foods is measured. The unit of measuring heat energy is the calorie. We can tell how much energy is in the different foods by burning them in a vessel called a calorimeter and measuring the heat. It is assumed that the body gets an equal amount of heat from the food when it is eaten and oxidized. This is approximately true, as was proved by an elaborate experiment in which a man was put into a large-room calorimeter and allowed to do a large amount of work on a machine. The work he did turned out to be approximately equivalent to the energy in his food. We know that the stomach and other digestive organs do not digest and use the energy in the roughage — shreds and husks. Burning the food in a calorimeter would measure the energy contained in husks and all.

Energy needed by the average person. The number of calories needed by a person depends upon size and activity, as the following table shows.

CALORIES ACCORDING TO WORK

CONDITION	CALORIES PER DAY PER POUND OF NORMAL WEIGHT
Resting in bed	11–14
No exercise, but out of bed	13–16
Small amount of work, bookkeeping	15–18
Moderate work, walking	17–23
Heavy work, running, football	22–30

Use the next table to find your normal weight. The weights should be slightly different for older people. These figures are for persons of the high-school age.

NORMAL WEIGHTS

Boys

Height (ft. : in.)	5 : 0	5 : 1	5 : 2	5 : 3	5 : 4	5 : 5	5 : 6	5 : 7	5 : 8
Normal weight (lbs.)	107	109	112	115	118	122	126	130	134

Height (ft. : in.)	5 : 9	5 : 10	5 : 11	6 : 0	6 : 1	6 : 2	6 : 3	6 : 4
Normal weight (lbs.)	138	142	147	152	157	162	167	172

Girls

Height (ft. : in.)	4 : 8	4 : 9	4 : 10	4 : 11	5 : 0	5 : 1	5 : 2	5 : 3	5 : 4
Normal weight (lbs.)	101	103	105	106	107	109	112	115	118

Height (ft. : in.)	5 : 5	5 : 6	5 : 7	5 : 8	5 : 9	5 : 10	5 : 11	6 : 0
Normal weight (lbs.)	122	126	130	134	138	142	147	152

A reducing diet. Many people, especially women, fear getting too fat. It is a thing to be feared, for statistics show that fat people are more susceptible to many diseases than people of normal weight. However, thousands of people reduce unwisely and do themselves great injury. In fact, many have practically wrecked their health by starving themselves for periods of time. Being too fat is sometimes an hereditary characteristic, sometimes due to overeating of the energy foods, and sometimes a disease. In reducing weight by restricting the diet, all the necessary requirements for good health except the energy foods must be supplied, or ill health results. These include protein, complete protein, calcium, phosphorus, iron, and all the vitamins; the basic balance must at least be kept. The body may get its energy largely from stored fat, but the food would naturally add some energy foods. Fats, however, should be eliminated as far as possible, and most of the carbohydrates.

Things to avoid in a reducing diet. In a reducing diet one should avoid the fats primarily, and then the carbohydrates. At the same time, one should not overeat of the proteins, since any excess above the requirements for cell repair and growth is oxidized in the body, leaving products hard to eliminate. The fat foods to be avoided are butter, oleomargarine, mayonnaise, and all fat meats. All pork products — bacon, ham, pork chops, etc. — contain far too much fat to warrant their inclusion in a reducing diet. All good meat tends to contain fat; hence only small amounts of lean beefsteak and chicken should be eaten, and then only a little at a time. In general, fried foods should be avoided because of the grease. The carbohydrates to avoid are sugar and sugar products, as jams, jellies, and sirups, and the starchy foods, as macaroni, potatoes, rice, and white bread.

Permissible foods in a reducing diet. Since the vegetables contain plenty of vitamins, minerals, and base-forming elements, one meal each day should contain all the vegetables the person cares to eat. It must be kept in mind that mayonnaise and butter should not be used if the person has much reducing to do. Some vegetables are in season any time of the year, as the following partial list will show: spinach, asparagus, string beans, cabbage, beets, cauliflower, carrots, celery, lettuce, cucumbers, tomatoes, green onions, green peas, and egg plant.

In choosing the protein for repairing the cells, and for growth if the person is in his teens, one should select those foods that do not have too much fat or carbohydrate also. Some of these are brown beans, lima beans, cottage cheese, dried peas, eggs, whole-wheat bread, and a little lean meat.

Milk is such a good food that it should be used moderately except in extreme cases.

Although fruits contain more or less carbohydrates, they are needed to supply certain vitamins. Since the body must keep up its heat and exercise, we cannot remove all carbohydrate from our diet for long. The body may use up its store of fat for a while, but this cannot last indefinitely. If we eat our fruit as nature gives it to us, we need have little fear. This means that canned fruits ought to be used unsweetened. Since fruit digests faster than vegetables, they should not be eaten in the same meal. In selecting our cereal for breakfast we may select some with less carbohydrate than others. Bran, puffed wheat, the roman meal are low in carbohydrate.

Desserts are hard to plan for the person who is trying to reduce. Fruits, both fresh and canned, are good. Agar gelatin is also a good dessert. Although it is a carbohydrate, it does not digest. It serves the purpose of roughage in food.

Water. Water is one of the most essential parts of our diet. It has been estimated that nearly half of the people in the United States are not in perfect health because of not drinking enough water. The doctors tell us a grown person needs from $1\frac{1}{2}$ to 2 quarts of water daily. Possibly a third of this amount is taken in our food, and therefore we must drink at least a quart more. If it is a warm summer day, one will naturally have a thirst. The hard-working man will develop a thirst even in winter. One employed indoors is unlikely to have a natural thirst for sufficient drinking water. Any person being constipated or feeling tired should check up on the quantity of water he drinks. If you do not like water in the winter, add a few drops of lemon juice.

Conclusion. Health is the normal condition in the body. Sickness is abnormal. If conditions are right, the chemical reactions in our bodies will be normal, and we will be in good health. Good health is the factor most needed for our happiness. We should learn all we can about the chemical reactions in our bodies. Only by making possible correct reactions can we enjoy good health.

REVIEW EXERCISES

1. Explain how we can increase the alkalinity of the blood by eating acid fruits.

2. Is the blood ever of acid reaction in a living person?

3. How are the calories in a food measured?

4. Why does this method give too high results?

5. What two things determine the energy requirements of a person?

6. What five necessary parts of our food must not be left out of a weight-reducing diet?

7. What kinds of food should be omitted in a weight-reducing diet?

8. How much water does a grown person need daily?

REFERENCE FOR SUPPLEMENTARY READING

1. Sadtler, Samuel Schmucker. *Chemistry of Familiar Things.* J. B. Lippincott Co., Philadelphia, 1924.
   (1) Which foods are rich in the following: (*a*) protein and fat? (*b*) protein and carbohydrate? (p. 186)
   (2) Which foods are lacking in fats? (p. 196)
   (3) Discuss eggs as a food. (p. 198)
   (4) Compare the composition of human and cow's milk. (p. 199)

SUMMARY

There are three kinds of diseases — degenerative diseases, infectious diseases, and deficiency diseases. The deficiency diseases are caused by the lack of something in our diet. The real healing agencies are sunlight, exercise, fresh air, rest, water, and good food. Eighteen elements are necessary for good health, but only part of them need be occasion for worry. With normal food and good air, oxygen, carbon, hydrogen, and nitrogen need not worry us.

Calcium is needed for the teeth and the bones. Foods rich in calcium are milk, cheese, turnips, celery, spinach, tomatoes, lentils, and rice.

Phosphorus, the next element in abundance in the human body, occurs in the bones. It is also an important constituent of brain and nerve cells. We probably get most of our phosphorus from proteins in meat products, milk, nuts, and soy beans, although some of it may come from vegetable phosphates.

We seem to get enough potassium, sulfur, sodium, chlorine, and magnesium in any normal diet.

Lack of iron in the blood is associated with anemia. This may be due to either a deficiency in the food or the inability to assimilate iron. All green vegetables, tomatoes, and meat are rich in iron.

Although only a trace of iodine is found in the body, it is nevertheless essential. Its absence produces goiter. Iodine usually comes dissolved in the drinking water, but it may be taken in salt.

Vitamins are substances occurring in foods in small amounts, which are essential for complete health. Several deficiency diseases, such as rickets, beriberi, scurvy, and pellagra are caused by vitamin deficiency in the food.

Fat-soluble A, the growth-producing vitamin, occurs in cod-liver oil, butter fat, egg yolk, leafy vegetables, and carrots.

Water-soluble B prevents pellagra and beriberi and is necessary for growth. It occurs in oranges, fruit, the cereal germ, spinach, cream, and milk.

Vitamin C in fresh vegetables prevents and cures scurvy. Vitamin D is necessary for bone development and prevents rickets. It is found in fats, milk, eggs, butter, fish oils, and leafy vegetables.

Vitamin E, the vitamin necessary for embryo development, occurs in cereal germs, seeds, green leaves, and egg yolk.

The energy requirement of the body depends on work, size, age, etc., and comes mostly from fats and carbohydrates. Water, too, is a necessary element in our diet.

CHAPTER 28

⇄

HOW CHEMISTRY HELPS THE DOCTOR

Introduction. Very few people die of old age. Among uncivilized peoples, such as the natives of Central Africa, a large percentage of the children die before they are one year old. We now know that there are hundreds of diseases, each one of which requires a different treatment. Before the nature of disease was understood, men hoped to find a cure-all — a medicine that would cure all ailments. Today we know that no one substance could possibly cure all diseases.

Patent medicines. We know that some remedies are "specifics" for one disease; for instance, quinine is able to cure malaria. By a specific, we mean it cures the one disease only. Quinine taken in reasonable doses will kill billions of malaria germs in the blood without injuring the person who takes it. In spite of the knowledge that no one remedy can hope to cure a large variety of diseases, the American people spend millions of dollars for "patent medicines," which, according to the printed claims and testimonials, are supposed to cure almost every ailment from toothache to Bright's disease. Most of the claims are absolutely ridiculous, and the medicines practically worthless for them all. Many of these patent medicines are positively harmful. Pain killers were extensively sold a few decades ago which contained habit-forming opiates. Many a person became a drug addict from their use. The idea of a pain killer is fundamentally wrong. In the first place, pain is nature's way of telling us something is wrong. The sensible thing to do is to remove the trouble — not merely punish the nerves for telling us and allow the trouble to continue.

The Pure Food and Drugs Act made the manufacturers of patent medicines print the names of the ingredients on the label, but this does not always protect the sick person, since

Paracelsus

1483–1541

Although modern medicine is sometimes said to begin with the works of Hippocrates and Galen, other persons also had much to do with the scientific beginnings of this practice. One of these was a lecturer at the University of Basel, Switzerland. He had the awkward name Philippus Aurelius Theophrastus Bombastus von Hohenheim but is now commonly called Paracelsus, a name which he adopted after the Roman physician Celsius. Paracelsus stood at the crossroads which separated the old alchemists, whose main goal was the preparation of gold from other substances, and the iatrochemists, whose chief concern was to cure disease.

Living during a time of great stress, that of the Reformation, Paracelsus lived an eventful life. In one of the first episodes we find him burning the scrolls of the alchemists at the university and calling the students to drop the old theories that were obstructing progress and to step out with him to a new knowledge. With the zeal of a religious fanatic he shattered all of the treasured beliefs of his day.

Having denounced the local physicians of his town for bleeding and otherwise torturing their patients, and the pharmacists for excessive charges for their medicines, he was driven out of Basel and began wandering over the continent studying at different centers of learning.

After being driven from Basel, Paracelsus wandered over Europe doctoring the poor and waging war against the false medical theories of the time. Among these theories were the following: An excess of phlegm makes one phlegmatic; too much blood makes one sanguine; and an abundance of bile makes one choleric. Another absurd practice was that of bandaging to a wound the weapon which had caused it.

he may not know how harmful a substance is. The government bureau responsible for enforcing the Pure Food and Drugs Act examined a supposed remedy for diabetes. It sold for $12 per pint, consisted of a water extract of a weed, and was of no value in curing the disease. Although in itself this patent medicine was harmless, it encouraged the patient to go

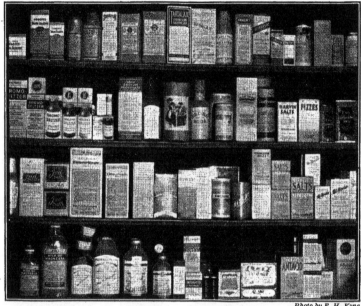

Photo by F H. King

PATENT MEDICINES. The people of the United States pay millions of dollars annually for patent medicines, most of which are of doubtful value, while the rest are common remedies parading under assumed names and frequently sold for extravagant prices.

without the attention of a physician until too late ; whereas he might have lived had he called a doctor at once. There are a few patent medicines that are legitimate, but they should be recommended by a physician.

One should not depend too much on testimonials, as they are easy to get. Dr. Grenfell, the famous doctor for the natives in Labrador, tells of how he arrived at a port with all his medical supplies exhausted. The natives demanded medicine

Courtesy of Merck & Co

PRIVATE LABORATORY OF A RESEARCH CHEMIST. Chemical research each year
provides the physician with many new remedies.

and could not understand his refusal to grant their requests. Finally in desperation he put some red pepper into water and gave it to them. The next time he called at this station, he found numerous persons ready to vouch for the wonderful cures effected by the medicine.

Pure drugs. One of the services the chemist renders to the medical profession is the preparation of pure drugs. In a few cases, harmful impurities have resulted in death, especially impurities in ether used as a general anesthetic. Practically all of the inorganic remedies are prepared by the chemist. There are hundreds of them.

Laxatives. Perhaps the commonest remedy is the laxative. There are few persons in the United States who have not at some time taken a laxative. These may be of plant origin, such as castor oil or cascara, or they may be inorganic compounds, such as Epsom salts ($MgSO_4 \cdot 7\ H_2O$), Glauber's salts ($Na_2SO_4 \cdot 10\ H_2O$), milk of magnesia ($Mg(OH)_2$), or the purgatives, calomel ($HgCl$) and magnesium citrate, a more complex substance. The chemist prepares them all in pure form.

The chemist improves plant remedies. Many plants have been found to possess valuable properties. The chemist has improved them by separating the effective ingredient from the other constituents of the plant. When this is obtained in pure form, the doses will be uniform in action. Quinine has been purified for malaria, chaulmoogric acid has been separated for leprosy, strychnine has been obtained pure for a heart stimulant, atropine has been isolated for dilating the pupil of the eye, and pure grain alcohol has been distilled for preparing tinctures, etc.

General anesthetics. Before 1832 surgical operations were performed without an anesthetic. The patient simply had to endure the pain. Operations were horrible beyond expression. The terrible agony often killed the patient. One of the great tragedies of all time is the fact that ether was known five centuries before it was used as an anesthetic in 1846. Chloroform was used as a general anesthetic fourteen years before ether, but ether soon replaced it in the United States as a safer substance. Recently there has appeared a new general anesthetic named cyclopropane, the formula of which is:

$$H_2C\underline{\quad\quad}CH_2$$
$$\diagdown\diagup$$
$$CH_2$$

Ethylene has already been mentioned in a previous chapter as being a general anesthetic. It has been used successfully in over 50,000 operations on people ranging in age from infants to the aged. It is one of the best general anesthetics developed so far.

Local anesthetics. The natives of the Andes region of South America were found by the early explorers chewing coca leaves to relieve fatigue and pain. The chemist was able to separate a substance from this plant that would relieve pain locally when given hypodermically.. Cocaine, as this substance is called, became very useful for tooth extractions and other minor operations. Cocaine is quite poisonous, has some undesirable after effects, and has occasionally even caused death.

Cocaine has been studied by many organic chemists. Its structure is fairly well established as:

$$CH_2\underline{\quad}CH\underline{\quad}C\underline{\quad}H \quad CO\underline{\quad}O\underline{\quad}CH_3$$
$$N\underline{\quad}CH_3\underline{\quad}CH\underline{\quad}O\underline{\quad}CO\underline{\quad}C_6H_5$$
$$CH_2\underline{\quad}CH\underline{\quad}CH_2$$

The beginning student cannot hope to understand what all this chemical symbolism means. However, it is plain that the dotted line divides the molecule into two different-looking parts. On the left there is a seven-carbon ring with a nitrogen bridge across it. On the right are two straight chains containing two oxygen atoms each. Investigation showed that the poisonous properties of cocaine are caused by the ring bridged with nitrogen, and the local anesthetic properties are caused by the oxygen groups, called in organic chemistry "ester groups." The organic chemist next tried to synthesize a substance that would have the anesthetic properties without the poisonous properties of cocaine. After a number of compounds were prepared, one was obtained that is satisfactory. It is sold now under the name of procaine (formerly novocaine). This substance has practically replaced cocaine in minor surgery, especially by the dentist. The formula for procaine is also too complicated for the beginner to understand :

$$H_2NC_6H_4CO—O—CH_2CH_2N(C_2H_5)_2 \cdot HCl$$

He can see, however, the ester group with its two oxygen atoms — here indicated between the dotted lines.

Atropine is another substance obtained from plants. It is a local anesthetic especially useful to the oculist. A solution can be sprayed into the eye, where it will dilate the pupil for certain tests. Again the chemist has improved on nature and made a substance that serves the same purpose but whose effects wear off in a small fraction of the time needed by atropine. These two illustrations are only samples of what the chemist has already done. The future will see the organic chemist improving other natural remedies and synthesizing new ones that are much better.

REVIEW QUESTIONS

1. Name two dangers from patent medicines.
2. Give the names and formulas for three laxatives.
3. What is a general anesthetic?
4. How long was ether known before its anesthetic properties were known?
5. Tell how the two different kinds of properties of cocaine depend upon a different part of the molecule.
6. What new anesthetic resulted from the study of cocaine?
7. Discuss quinine as a specific for malaria.
8. Give two reasons why a patent medicine is not likely to cure one when ill.
9. Discuss the danger of pain killers.
10. Why are patent medicines dangerous even though the ingredients are harmless?
11. How has the chemist improved natural remedies?
12. What is the structural formula for cyclopropane?
13. What is cyclopropane used for?
14. To what extent has the chemist improved atropine?

PROBLEMS

1. If a druggist heated 100 g. $MgSO_4 \cdot 7 H_2O$ until all of the water of hydration had been driven off, what weight would he have left?

2. What per cent of crystallized sodium sulfate, $Na_2SO_4 \cdot 10 H_2O$, is water?

3. $MgSO_4 + 2 NaOH \longrightarrow Mg(OH)_2 + Na_2SO_4$. What weight of $MgSO_4$ is needed to prepare 200 g. $Mg(OH)_2$?

REFERENCES FOR SUPPLEMENTARY READING

1. Ehrenfeld, Louis. *The Story of Common Things*. Minton, Balch and Company, New York, 1932. Chapter VI, pp. 99–109, "Chemistry and Health."
   - (1) When did Glauber introduce sodium sulfate as a medicine? (p. 103)
   - (2) When did Paracelsus live? (p. 100)
   - (3) Who discovered ether? About what date? (p. 103)
   - (4) When was ether found to be an anesthetic? (p. 104)
2. Foster, William. *The Romance of Chemistry*. D. Appleton-Century Company, New York, 1936. Chap. XXV, pp. 420–435, "The Physician's Dependence on Chemistry."
   - (1) When did Hippocrates, who is called the "father of medicine," live? (p. 421)
   - (2) How many drugs did Hippocrates mention? (p. 421)
   - (3) Upon what did Hippocrates depend primarily for curing disease? (p. 422)
   - (4) Explain the origin of the barber's pole. (p. 422)

Antiseptics for external use. An antiseptic is a chemical substance that kills germs. Several different substances have been used as antiseptics since Pasteur and Lister proposed the germ theory of disease. Among these are corrosive sublimate ($HgCl_2$), phenol (better known in its impure solution as carbolic acid), hydrogen peroxide, and alcohol. Corrosive sublimate and phenol are very poisonous and can only be used externally. Although alcohol is often drunk in quite concentrated solutions as gin or whisky, it is a strong poison. Iodine is used in alcoholic solution called tincture of iodine. Iodoform (CHI_3) is also used as an external antiseptic. Recently there have been developed some new antiseptics, which are less poisonous to man, but more poisonous toward germs. Two of the better known of these are mercurochrome and hexylresorcinol. Mercurochrome is the mercury salt of a red dye. Its formula is very complicated. Hexylresorcinol is described as seventy times as poisonous as phenol towards disease germs but is almost nonpoisonous to human beings, even when taken internally.

Internal antiseptics. Physicians have long wished for an antiseptic which could be taken by mouth and would kill germs in the kidneys, bladder, and blood. It has been a difficult task to develop a substance harmless to human beings yet poisonous enough to kill germs as it passes through any part of the body, often in a highly diluted state. Hexylresorcinol has partly fulfilled this hope as far as the bladder is concerned. Certain organic dyes with complex formulas have been found to help in internal infections. Among these are acriflavine and gentian violet. It is remarkable how the chemist has learned to modify chemical compounds, increasing their poisonous properties towards disease germs and at the same time reducing their poisonous action towards the human body. This work has involved years of careful study and the preparation of thousands of compounds. After all the possibilities along a certain line have been tried, one or two of the best products are given to the medical profession. Hexylresorcinol illustrates this point. Resorcinol has the structural formula illustrated. This structure is known as the benzene ring. When other groups are added to this ring, the properties are changed. After trying groups of various sizes and numbers of carbon atoms, chemists found that the hexyl group of six carbons gave the most desirable properties.

```
        C—OH                      C—OH
      /      \\                  /      \\
   HC         CH           HC         CH
    ||         |            ||         |
   HC        C—OH          HC        C—OH
      \      //                \      //
        CH                       C—CH₂—CH₂—CH₂—CH₂—CH₂—CH₃
     resorcinol              hexylresorcinol
```

The outstanding achievement of synthetic chemistry. The outstanding achievement in synthetic chemistry in behalf of medicine was accomplished by the German chemist Paul Ehrlich and his helpers. Ehrlich started out to find a specific for the disease syphilis. Syphilis has ever been one of the major scourges of mankind. About one-fourth of the hopelessly insane are made that way by syphilis. Many babies are born dead or otherwise defective because of syphilis in one of the parents. Syphilis is caused by a small spirallike germ that belongs to the animal kingdom.

Ehrlich began his study with the organic compounds of arsenic. All ordinary arsenic compounds are highly poisonous. Ehrlich worked on the theory that, since all arsenic compounds are not equally poisonous, possibly a substance could be synthesized that would not be poisonous enough to kill the human host but still poisonous enough towards the germs to kill them. After having prepared and tried over six hundred arsenic compounds, Ehrlich produced one, "606," which could be injected into the veins of the patient without fatal results, yet produced a cure in a large percentage of cases.

It would take a considerable knowledge of organic chemistry to get much meaning out of the structural formula of Ehrlich's arsphenamine, as he called his 606th compound of arsenic. However, the formula is given to satisfy the curiosity of the thoughtful student.

$$As \equiv\!\equiv\!\equiv\!\equiv As$$

From the formula it will be seen that the molecule has two arsenic atoms, two benzene rings, two OH groups, and two amine groupings (NH_2).

Medical research advances. Recently there has been synthesized a new organic chemical which is a specific for those diseases caused by the type of bacteria known as the cocci. One strain of cocci known as the hemolytic streptococcus was particularly deadly. Meningitis caused by this streptococcus was practically 100 per cent fatal. In a series of 45 cases of this disease treated by the new remedy, 36 recovered. In another series of 19 cases, 14 survived. The new remedy has the chemical name of para-amino benzene sulfonamide and the structuré:

$$
\begin{array}{ccc}
& \overset{\text{H}\quad\text{H}}{\underset{}{\text{C}=\text{C}}} & \\
\text{SO}_2\text{NH}_2-\text{C} & & \text{C}-\text{NH}_2 \\
& \underset{\text{H}\quad\text{H}}{\overset{}{\text{C}-\text{C}}} &
\end{array}
$$

The effectiveness of the compound has been shown to depend upon the part to the left of the benzene ring.

Other ailments cured by this specific are erysipelas, childbed fever, gonorrhea, and other infections of the urinary tract. We little realize what a discovery such as this means to the many people suffering and facing almost certain death. Neither do we realize the great thrill that must have come to Gerhard Domagk, the German scientist, when he learned the curative properties of his new medicine.

Photo by F H King

HABIT-FORMING DRUGS. One of the worst misfortunes that can happen to a person is to acquire the drug habit.

Hypnotics. There are times when the doctor must give something to quiet a patient's nerves and produce sleep. That which soothes the nerves is called a sedative, and that which produces sleep a hypnotic. The two are more or less related. Bromides, such as KBr, were the first sedatives. Morphine is a more effective sedative, but since it is the worst habit-forming drug known, its use is dangerous. The old-fashioned "pain killer" was a sedative and often habit forming. Perhaps the worst misfortune that can befall a person is to become a drug addict. There is no form of slavery comparable to slavery to a habit-forming drug. The end of the drug addict is always premature death.

Recently the chemist has made a series of hypnotics, such as nembutal and pentabarbitol, that are not habit forming.

Hormones. There are several ductless glands in the human body that secrete certain body regulators and add them to the blood. These complex substances are known as *hormones.* The pancreas secretes a hormone that regulates the oxidation of sugars in the blood. If this hormone is absent, the body is really poisoned with glucose sugar. This diseased condition is called diabetes. Dr. Banting and his associates came to the rescue of the diabetics by separating this hormone (insulin) from the pancreases of animals. Insulin has saved the lives of thousands of people.

Another hormone, called thyroxine, is secreted by the thyroid gland in the neck. This hormone controls the heat-producing processes of the body. If the gland is unable to produce enough thyroxine, it becomes enlarged. This diseased condition of the gland is known as goiter. The chemist, in this case, not only has separated and purified thyroxine but has actually synthesized it.

Another hormone, adrenaline, is secreted by little glands on top of the kidneys. This hormone is a heart stimulant. It contracts the capillaries and gives relief from asthma. Adrenaline has been purified and synthesized, another great chemical achievement. These achievements are samples of what the chemist may do in other cases.

To mention all of the remedial compounds the chemist has made for the physician would require more space than we can spare. These few outstanding illustrations will give the beginning student some idea of the importance of the chemist to medicine. How great are the possibilities for relieving human suffering and misery! If the people of the world would spend half as much money each year for medical research as they now spend for armies and navies, wonderful advances in medicine could be made. Perhaps those dreaded diseases — cancer, tuberculosis, influenza, and pneumonia — could be cured.

EXERCISES

1. Name four antiseptics for external use.
2. Name two new synthetic antiseptics.
3. Name two internal antiseptics.
4. Tell the story of Ehrlich's work.
5. Name the hormone that cures diabetes.
6. Name two hormones that have been synthesized.
7. How is resorcinol related to benzene?
8. How is resorcinol related to phenol?
9. How many benzene rings are in arsphenamine?

PROBLEMS

1. Hexylresorcinol has the following formula: $C_6H_4(OH)_2C_6H_{13}$. Calculate its molecular weight.
2. Calculate the percentage of oxygen in hexylresorcinol.

ADDITIONAL EXERCISES FOR SUPERIOR STUDENTS

1. Write a 500-word essay on hormones.
2. If a person took 100 g. of anhydrous magnesium sulfate and crystallized it from a water solution, what would the crystals weigh?
3. Calculate the molecular weight of cocaine. (See structural formula on page 556.)
4. What is the simplest substance with a ring structure similar to that of hexylresorcinol?
5. Give the number of each kind of atom in arsphenamine, whose formula is shown on page 560.

REFERENCES FOR SUPPLEMENTARY READING

1. Darrow, Floyd L. *The Story of Chemistry*. Blue Ribbon Books, Inc., New York, 1930. Chap. VII, pp. 253–292, "Chemistry and Disease."
   (1) Discuss the use of dyes in medicine. (pp. 265–266)
   (2) Tell about Bayer 205. (p. 268)
   (3) Who first isolated an enzyme? (p. 285)
   (4) How many drugs are known? (p. 291)

2. Holmes, Harry N. *Out of the Test Tube.* Ray Long and Richard R. Smith, New York, 1934. Chap. XXIII, pp. 267–278. "Shall We Have Medicines and Anesthetics to Order?"
   (1) What did Lister discover in 1867? (p. 269)
   (2) What chemical is given for hookworm? (p. 272)
   (3) Which hormones have been synthesized? (p. 271)
3. Stieglitz, Julius. *Chemistry in Medicine.* The Chemical Foundation, Inc., New York, 1928.
   (1) Write an essay on one of the following topics:
       Internal Secretions. (pp. 191–204)
       The Hormones of the Suprarenal. (pp. 205–231)
       The Story of Thyroxine. (pp. 232–255)
       Iodine in the Prevention and Treatment of Goiter. (pp. 272–296)
       Insulin to the Rescue of the Diabetic. (pp. 297–314)
       The Internal Secretion of the Parathyroid Glands. (pp. 315–321)

SUMMARY

Some of the earliest medicinal substances prepared by the chemist for use by the doctor were the laxatives: Epsom salts, magnesium sulfate; Glauber's salts, sodium sulfate; calomel, mercurous chloride; and magnesium citrate. Plant chemicals separated by the chemist include quinine for malaria, chaulmoogric acid for leprosy, strychnine for heart stimulation, and atropine for dilating the pupil of the eye.

All the general anesthetics — ether, chloroform, ethylene, and cyclopropane — are synthetic chemicals. Chemistry produced procaine, the best of our local anesthetics.

Most of the antiseptics are likewise chemical products. Among these are iodoform, mercurochrome, and hexylresorcinol.

One of the outstanding achievements of chemotherapy is Ehrlich's arsphenamine, used to kill the organism causing syphilis.

The chemist has also produced sedatives and hypnotics in large numbers.

Perhaps most important of all is his separation of the hormones, the body regulators produced by the ductless glands. Among those which have been synthesized are thyroxine and adrenaline.

←
→

CHAPTER 29

THE ALLOYS

What alloys are. Any metallic substance containing two or more metals is called an alloy. By melting metals together, it is possible to extend their properties over very wide ranges. Gold, for example, although one of the most beautiful of the metals, is too soft to wear well. A ring of pure gold would wear away quite rapidly and would be readily scratched and dented by bumping against iron or any other hard object. However, after being alloyed with varying amounts of copper, it retains all of its beauty, becomes hard enough to wear for decades, and does not scratch or dent easily. Pure iron has several properties that make it unsuitable for certain uses. Where resistance to rust is desired, chromium and nickel are alloyed with the iron to overcome this fault of the pure metal. Where a material of low expansion with heat is desired for pendulums and measuring tapes, nickel is alloyed with the iron to

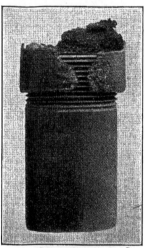

Courtesy of Republic Steel Corp

CORROSION RESISTANT IRON. The resistant-iron pipe shows little corrosion, while the sleeve of ordinary iron is practically rusted away.

form invar, whose expansion for ordinary temperature changes is almost negligible. If we had to depend upon the unmodified properties of the different metals, their uses in industry would be much restricted. By alloying the different metals, it is possible to get a substance with almost any combination of properties desired. Every year new alloys are being made. There seems to be no limit to what can be done with alloys

Courtesy of German Railways

FRAME OF A DIRIGIBLE. Light alloys have made the airship possible.

Nature of alloys. The making of an alloy by melting two metals together and allowing them to cool, is not so simple as it might seem. Some metals, for example, lead and antimony, mix in all proportions, as do the liquids, alcohol and water. Other pairs of metals dissolve each other only to a limited extent, as do water and carbolic acid. Finally there appears a dividing line between the two elements. On the bottom is a saturated solution of the lighter metal in the heavier; the upper layer consists of a saturated solution of the heavier metal in the lighter.

Other metals unite in definite proportions, which shows that they form definite chemical compounds. Still other metals are insoluble in each other, as oil is insoluble in water. When the melted metals cool, a complicated mixture often results, consisting of crystals of the pure metals, solid solutions of one metal in the other, or crystals of the compound formed between the metals.

Low-melting alloys. It is a general principle that small amounts of one substance lower the melting point of the other. This principle applies to the mixture of metals in alloys. If lead is added little by little to a fixed quantity of antimony, the melting point drops from 630° C., the melting point of pure antimony, until it finally gets down to 247° C. On the other hand, the gradual addition of antimony to lead brings its melting point down from 327° to 247°. This mixture, which has the lowest possible melting point obtainable by alloying antimony and lead, is called the *eutectic* mixture of the two metals. Many of the commercial alloys are eutectic mixtures, because these mixtures have a fine-grained structure, which gives them strength and firmness.

If a low-melting alloy is desired for solder, for electric fuses, or for plugs in automatic fire-fighting equipment, two things are needed: (1) low-melting metals and (2) the eutectic mixture of the metals chosen.

Solder. Solder as used by the plumber or tinsmith is an alloy of tin (melting point 231.8° C.) and lead (melting point 327.5° C.). An alloy of two-thirds lead and one-third tin melts at 228°, which is lower than the melting point of either tin or lead. Half-and-half solder, composed of equal parts of these metals, melts at 188° C., and a solder of two-thirds tin melts at 121° C. Hard solders used in "brazing" contain nearly equal amounts of copper and tin with less than 4 per cent zinc. Brazing, however, is done at much higher temperatures than ordinary soldering.

EXPERIMENT

Preparing solder. In a large porcelain crucible mix 7 grams of powdered tin with 3 grams of powdered lead. Cover the crucible and heat until the tin is melted. With the tongs shake the crucible until the lead dissolves in the melted tin. Pour the melted solder into water to cool. Examine the product.

Electric fuses. All electric currents heat the wires more or less. The heating increases rapidly as the current increases.

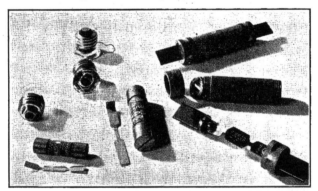

ELECTRIC FUSES. An electric fuse is essentially a strip of low-melting alloy, which will carry a current only up to a certain definite strength. Beyond this point, the fuse melts and breaks the circuit, thus preventing fires and protecting equipment from overloading.

A fuse consists of a short section of the circuit that will melt at a definite low temperature and shut off the electricity when the current increases beyond a certain amount because of a short circuit or an overload. By varying the constituents of the low-melting alloy, it is possible to shut the current off when it reaches any desirable strength. Two of these alloys are given definite names as follows.

LOW-MELTING ALLOYS

ALLOY	M. P.	PB	SN	BI	CD
Rose's metal	93.8	28	22	50	
Wood's metal	60.5	25	12.5	50	12.5

It seems strange that Wood's metal can melt at 60.5° C. when the lowest melting point of any constituent is 231.8° C. This illustrates the peculiarity of eutectic mixtures. A practical joke sometimes played is to give a person a teaspoon made of

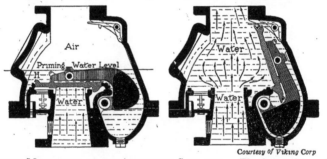

Courtesy of Viking Corp

RELEASE MECHANISM OF AN AUTOMATIC SPRINKLER. The low-melting alloy hook *H* holds the plug down tight over the top of the water pipe. In case of fire the hook melts, releasing the water into the spraying pipes.

Wood's metal. Imagine his surprise and embarrassment when the spoon melts in his coffee and leaves him holding a part of the handle.

It is possible to get alloys that melt at lower temperatures than Wood's metal by using sodium or potassium, but their great tendency to react with water makes them impractical. Mercury, the lowest melting metal, is used in the special class of alloys called *amalgams*. If the amalgam contains much mercury, it is often too soft for most purposes. The dentist adds considerable mercury to his silver fillings.

Automatic sprinklers. Many buildings are equipped with sprinklers plugged with Wood's metal. In case of fire, the alloy melts and turns on the water, which usually puts out the fire.

High-melting alloys. High-melting alloys depend for their efficiency upon the high melting points of the constituent metals together with their resistance to burning. Nickel alloyed with 10–20 per cent chromium has high electrical resistance and high resistance to oxidation and is, therefore, an alloy suitable for heating units in electric toasters, stoves, etc. Platinum-

MONEL METAL. This alloy is both beautiful and resistant to corrosion. It is composed of 67 per cent nickel, 28 per cent copper, and 5 per cent iron or manganese.

iridium alloy is used in resistance thermometers for measuring very high temperatures.

Structural alloys. Structural alloys are of two kinds: (1) those in which weight-compression resistance is the all-important property, and (2) those in which lightness is the chief concern and structural strength is secondary.

Alloys of the former class are primarily the alloy steels in which the iron is alloyed with nickel. In alloys of the latter class, one of the metals must contribute lightness, while the other or others contribute strength. One of the principal alloys of this class is duralumin. Duralumin contains aluminum and magnesium as the light metals, with copper to give strength. This alloy is used in airplane and dirigible balloon parts. It is 95 per cent aluminum, yet it is remarkably strong as compared with pure aluminum.

Magnalium, a silver-white alloy used for castings and mirrors, contains aluminum and magnesium. Both metals are light, hence the alloy is light, yet it is strong enough to use in dirigible parts.

Corrosion-resistant alloys. One of the common corrosion-resistant alloys is monel metal (67 per cent nickel, 28 per cent copper, 5 per cent iron or manganese), which is a tough metal of high tensile strength and very resistant to corrosion. Monel metal is used for propeller blades, kitchen sinks, and washing machines for laundries. The nickel-chromium steels are very resistant to oxidizing acids, which severely corrode most metals. These steels may contain so much chromium that they are no longer magnetic.

Courtesy of Republic Steel Corp

STAINLESS-STEEL TUBES. These tubes, which have been in use for six years, can be used again, although the walls of the plant must be rebuilt.

Duriron is an alloy steel containing a high percentage of silicon. The silicon makes the steel hard and brittle but very resistant to acid corrosion.

Alloy coins. Most modern coins are alloys. Both gold and silver coins contain 10 per cent copper. In these coins the copper is added to increase hardness and reduce wearing. The nickel is 75 per cent copper, and the penny is composed of

95 per cent copper, 3 per cent tin, and 2 per cent zinc. The penny is really a bronze, similar to some of the earliest prehistoric metallic objects. Many old ornaments and statues are of bronze.

Alloys and the automobile. Without the various alloys the automobile could never be as cheap and as durable as it is. Let us consider a few alloys used in its manufacture. Starting at the front and working backwards, we find a radiator of brass, an alloy of copper and zinc. Then come the front springs, made of alloy steel containing (besides the iron) chromium,

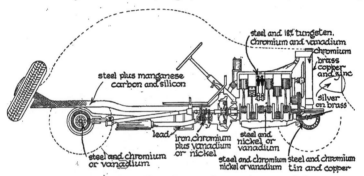

ALLOYS IN THE AUTOMOBILE. This diagram shows only part of the numerous alloys that go into the manufacture of a modern motor car.

nickel, or vanadium. The pistons above are aluminum hardened by alloying with manganese, and the crankshaft is nickel or vanadium steel. The vanadium contributes the property of great shock resistance. This is what automobile springs and many other parts need. The engine valves get their toughness and resistance to pitting from 18 per cent tungsten, plus some chromium for corrosion resistance and vanadium for shock resistance. The crankshaft bearings require an alloy easy to cast and soft enough not to wear the crankshaft itself. The bearing metal used here is an alloy of antimony, tin, and copper. It wears better than ordinary Babbitt metal, which was the metal used in bearings for several decades. Babbitt is an alloy of antimony, lead, and tin. The frame of the automobile is a steel containing manganese, carbon, and silicon. The manganese contributes strength, the carbon hardness, and the silicon resistance to rust. The gears are composed of iron, chromium, and vanadium or nickel — the chromium for toughness and the vanadium for shock resistance. The axles of an automobile, being subject to great shock and twisting forces, are made of chromium and vanadium steels to best resist these

forces. It is said that a good axle can be twisted through a complete circle and then bent double while cold without breaking. This is truly remarkable in view of the fact that twisting of a wrought iron bolt through two-thirds of a circle or less will break it off. One more alloy used on the automobile is found in the "points" of the spark plug, which are made of nickel, manganese, and silicon for resistance to wear by the electric spark.

A BRONZE STATUE. Bronze is an alloy of copper and tin, with occasionally small amounts of phosphorus, zinc, aluminum, or lead added for special purposes.

Other alloys. Our study of alloys will not be complete until we have included a few more with special properties. Aluminum bronze is an alloy of copper and aluminum. If the percentage of copper is quite high, the bronze resembles gold. Its principal use is in ornamentation, as imitation gold. If the alchemists in their attempt to change other metals to gold had had access to aluminum, they would probably have thought their quest successful when they heated the metals together and got this bronze.

One of the oldest alloys in point of time is bell metal. The ancients used to cast their bells with much ado and with secret and magical formulas. However, the pleasing sound of the bells depends upon the alloy of copper and tin.

German silver is a substance quite resistant to corrosion, which takes a high polish. It is misnamed, however, in that it contains no silver but only has a silverlike appearance. The real constituents are copper, zinc, and nickel — all metals of considerable resistance to corrosion.

Type metal possesses the rather exceptional property of expanding slightly when setting. This makes the type fit the form snugly and have a sharp, well-defined edge. The metal is an alloy of lead, antimony, and tin.

REVIEW QUESTIONS

1. How is an alloy made?
2. State four different ways that metals may act towards each other in alloys.
3. State the effect of adding small amounts of one metal to another.
4. Explain the meaning of an eutectic mixture.
5. What is solder? How does the melting point of solder compare with that of the constituent metals?
6. What is Wood's metal?
7. Explain the mechanism of an automatic sprinkler.
8. Name two high-melting alloys.
9. What alloy is used in airplane construction?
10. Name the metals in magnalium.
11. Give the elements in Monel metal.
12. What are the corrosion-resistant steels?
13. What is duriron?
14. Why are coins alloyed? Give the composition of gold, silver, and nickel coins.
15. Tell what properties the following metals contribute to steel: (a) nickel; (b) vanadium; (c) silicon; (d) tungsten.
16. What are the constituents of the brass in the radiator?
17. What alloy is used for light automobile pistons?
18. What is the composition of the bearing metal?
19. Give the composition of (a) bronze, (b) bell metal, (c) German silver, (d) type metal.

Summary

Alloys, made by melting two or more metals together, can have almost any combination of metallic properties. By fusing low-melting metals, still lower-melting alloys can be obtained, which are useful as solder, fuse plugs for water sprinklers, and electric fuses.

By fusing metals of high melting point, alloys of high melting point also can be obtained. These serve as heating elements in electrical devices, such as toasters, irons, and stoves.

Structural alloys which are useful because of lightness come from the light metals. Machines, pipes, tanks, etc. which must resist corrosion usually contain one or more of the metals (copper, nickel, and chromium) which contribute this property to the alloy. A high percentage of chromium alloyed with iron will resist corrosion by many of the strong corroding substances, such as nitric acid.

The precious metals are alloyed with harder metals to keep them from wearing too rapidly.

Alloys have made the automobile what it is today. Scarcely any of its metallic parts are a single metal except the copper wire in the electrical wiring.

Among our most important industrial alloys are the alloy steels, in which different properties, such as retaining hardness while hot, resistance to wear, and resistance to penetration, have been accentuated.

CHAPTER 30

PROTECTING AND CLEANING SURFACES

Magnitude of corrosion. It is estimated that each year over two billion dollars' worth of useful things are lost by corrosion. Most of this loss is of iron products. Gasoline is one cent per gallon higher in price because of the corrosion of pipe lines, oil-refining machinery, and other iron products used in obtaining

RUSTING MACHINERY. The rusting of machinery and equipment costs the farmers of America millions of dollars annually.

and refining the oil. Cities pay more for fuel gas because of the corrosion of the underground pipe lines. Freight rates are higher because of the rusting away of rails and rolling stock. Even the farmer loses more than he realizes from the rusting of his machinery and implements.

Nature of the metal. Whether or not a metal will corrode depends primarily upon its fundamental nature as indicated by its position in the replacement series. Those metals near the bottom of the series tend to resist chemical reaction. Other things being equal, the lower a metal is in the series, the less it tends to rust, or oxidize, or react with the atmosphere. Platinum and gold are near the bottom of the series, and mercury and silver are close by. As we would naturally expect, these metals are all resistant to corrosion. However, there are some metals quite high in this series that seem to be permanent in the atmosphere and appear as exceptions to this arrangement. These apparent exceptions are due to the fact that when corrosion starts, there forms a protective coating that stops further corrosion. Sometimes this coating is the oxide, and sometimes another resistant compound of the metal. Often the layer is so thin that one has to use the microscope to see that it is really there. Copper, chromium, nickel, and tin fall into this class. Unfortunately, the metals at the bottom of the series are relatively scarce, which limits their use.

COPPER-BOTTOMED BOILER. The copper bottom of a discarded wash boiler will be fairly well preserved after the tinned iron upper part has rusted entirely away.

Protective coatings. The following table gives the common metals that form protective coatings and the nature of the protecting compound.

PROTECTIVE COATINGS

METAL	PROTECTIVE COATING	APPEARANCE OF COATING
Al, Zn, Sn, Cr	oxide of the metal	dulled luster
Cu, Au, Ag, Pb	sulfide	black
Pb	sulfate (in contact with water)	white
Cu, Ni	basic carbonate	greenish

The protective coating of a metallic oxide is very thin and is light in color. The housewife does not realize that her aluminum teakettle has lost its bright luster because of oxide formation. The other protective oxide coatings also are hardly noticeable. The sulfide coatings, however, are very conspicuous. The silver teapot that has been on the shelf for a few years may be entirely black and look nothing like silver. The tarnish on copper, lead, and nickel is plainly visible. In any of these cases the metal, although covered up, is all there.

Protecting iron with other metals. Iron, our cheapest metal, is very susceptible to rusting. It is common practice to use

GALVANIZED IRON. Its surface is covered with a protective coating of zinc.

iron for its strength and to protect it with a coating of a more resistant and more costly metal. Galvanized iron is such a product, in which the iron is dipped or sprayed with a protective coating of zinc. Tin plate is iron dipped in molten tin for the protective coating.

A more general way of applying protective coatings is by electroplating. Chromium plating, copper plating, silver plating, and gold plating are common. It happens that silver or gold will not plate on iron. In silver-plating iron, it is necessary to plate it first with copper and then to put a layer of silver on top of the copper.

Corroding materials and conditions. The commonest type of corrosion is atmospheric corrosion. The action of the atmosphere depends upon several things, principally the humidity and the content of SO_2. In very dry climates rust, decay, and all similar chemical reactions are very slow. In the old cave dwellings of New Mexico are found many objects which would have been soon destroyed in a humid climate, but which have kept in good shape for centuries in the dry desert air.

Coal contains more or less sulfur. In large cities and industrial centers where large quantities of coal are burned, considerable sulfur dioxide is liberated. With moisture this oxide forms sulfurous acid, a very effective corrosive acid in starting and promoting rust. Ninety-five per cent of all corrosion is acid corrosion.

Temperature of corroding materials. The speed of most chemical reactions increases rapidly with the temperature. Ordinarily we do not think of iron as reacting with water, yet steam reacts with iron in a heated tube. Oxidizing acids attack metals slowly when cold but rapidly when hot. In the tropics, where the average temperature is high, iron rusts faster than it does in cooler climates.

Length of time of contact of corroding substance. Many chemical reactions take place so slowly that a short test would lead to the conclusion that there is no reaction, yet on long standing the metal corrodes away. If we should dip aluminum in a strong solution of table salt, we might conclude that it did not react. However, a woman once put a brine solution in an aluminum kettle to preserve some olives. After a few months' time, the kettle had holes in it. Most tarnishing reactions take a long time, either because the reaction is slow or because the amount of the attacking chemical is small.

Purity of the metal. The resistance to corrosion of metals increases with the purity of the metal. Zinc is considered fairly soluble in dilute sulfuric acid. However, if chemically pure zinc is used, it is almost insoluble in dilute sulfuric. If a piece of copper is dropped in so as to touch the pure zinc, the latter begins to dissolve in the acid. This shows that the dis-

solving is an electrochemical action.   When there are impurities in the zinc, they form a cell action within the zinc, causing
it to dissolve.

Condition of surface.   The condition of the surface of a metal
determines to some extent its tendency to corrode.   The sportsman knows that if he can keep the inside of the gun barrel
smooth and bright, it is not likely to rust.   A rough surface collects moisture, which dissolves carbon dioxide to form carbonic-
acid solution, which in turn catalyzes rusting and tarnishing.

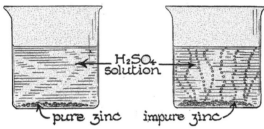

H_2SO_4 solution

pure zinc   impure zinc

PURE ZINC DOES NOT DISSOLVE.   Impure zinc readily dissolves in dilute acid as
a result of electrochemical action.

When the surface of iron is treated with very concentrated
nitric acid, the surface becomes more resistant to rust.   The
iron is said to assume the passive state, which is a very resistant
condition.   Iron treated with a phosphate becomes resistant
also.   Perhaps in these cases the surface becomes protected by
a thin layer of insoluble oxide or phosphate.

EXPERIMENT

Passive iron.   Clean a piece of sheet iron.   Completely immerse
the iron in fuming nitric acid.   As the iron enters the acid, a vigorous
evolution of oxides of nitrogen occurs for a few seconds.   When the
reaction ceases, pour off the acid and rinse with water.

Compare the action of ordinary concentrated nitric acid on this
and also on an untreated piece of iron.

Dip this piece of iron and also another which has not been treated
into copper sulfate solution.   Note that the passive iron does not
become coated with copper as does the other.

Prevention of corrosion. Two methods of preventing corrosion have already been suggested — that of polishing the surface and that of producing a protective coating. Frequently some simple modification of conditions may greatly reduce corrosion. Recently a restaurant was troubled by expensive aluminum kettles developing holes. It was suggested that the holes were caused by electrolysis. A college student happened to be working in this restaurant. He borrowed some electrical-measuring instruments and found that there was over a volt difference in electrical driving force between the aluminum kettle and a stainless steel spoon used in the kettle for dipping out the food.

ENAMEL COOKING UTENSILS. Enameled ware is prepared by coating the iron surface with a glass for protection against rust.

His current-measuring instrument had a range of only one-half ampere, but the needle swung clear past the scale, which showed that it did not measure all the current. In this case the substitution of a wooden spoon eliminated the difficulty.

When the corroding solution is not food, it is sometimes possible to add to it a negative catalyst that inhibits or prevents the formation of tarnish.

Protective coatings and resistant materials. Recently the industries have developed ways of applying protective coatings to metals, such as glass linings for tanks and rubber coverings for wires. Resistant alloys are also being developed rapidly. Resistant organic plastic materials, which can be molded into any shape and set, or hardened, by heat and pressure, are of growing importance. Bakelite was the first of these products. It has become a standard material for electrical appliances. Dozens of similar products are now synthesized for buttons, dishes, fountain pens, and even furniture.

Enamels. One method of protecting iron surfaces from tarnish and chemical action is to coat it with a thin coating of glass. This type of utensil is called enamel ware. Enamel cooking utensils, sinks, bath tubs, and automobile number plates are quite common. Since glass can be colored practically any color, enamel ware is very pretty. It is resistant to most chemicals, too, until the enamel becomes chipped off.

REVIEW QUESTIONS

1. Discuss the magnitude of the corrosion problem.
2. What does the replacement series tell us about the tendency of metals to corrode?
3. Name the metals that form protective coatings of : (a) oxide; (b) sulfide; (c) sulfate; and (d) basic carbonate.
4. Name two metals that are used to protect iron.
5. What is the process of plating iron with silver?
6. Name the two most important factors in atmospheric corrosion.
7. Describe the effect of temperature, length of time of contact, and purity of metal on corrosion.
8. How does the condition of the surface affect corrosion?
9. Name two methods of preventing corrosion of metals.
10. How are buttons of plastic material superior to iron ones?
11. What is meant by the passive state of iron?
12. How is iron made passive?
13. Relate how electrochemical action injured serving kettles.
14. Name other ways of protecting iron surfaces.
15. What is enamel ware?

Methods of removing tarnish and dirt. *Physical methods.* Physical methods of removing tarnish consist of polishing and scouring. Scouring includes everything from removing the entire surface with a scratch brush to rubbing with a little fine powder. Scouring, to give the best results, should leave the surface smooth. Polishing places the emphasis more on smoothing the surface than on the removal of material, but it removes thin layers of tarnish. An excellent metal polish is made by mixing very finely powdered talc with oleic acid and a little ammonia. Physical methods always remove more or less of the metal.

Chemical methods. Chemical methods of removing tarnish dissolve the tarnishing substance. Most of the metallic oxides are soluble in acids. All metals whose oxides are soluble in acid can be cleaned by "pickling." Pickling consists of soaking the metallic article in acid. Before plating iron, it must be cleaned by pickling.

The housewife often cleans her silverware by an electro-chemical method. She takes an aluminum pan, puts her silverware into it, covers it with water, adds a teaspoonful of soda and a couple of spoonfuls of table salt, and heats it almost to boiling. The silver is soon cleaned. It should not be heated in the solution too long, or there may be some pitting. The explanation of this method depends upon the replacement series of the metals. The stain on the silverware is black silver sulfide. Aluminum is above silver in the series, hence can replace the latter in its compounds according to the following equation:

$$3 \, Ag_2S + 2 \, Al \longrightarrow Al_2S_3 + 6 \, Ag$$

The good feature of this process besides its simplicity is that it does not remove any of the silver. Of course a little aluminum dissolves but only a little.

Cleaning metal and other surfaces. The problem of keeping metallic and other surfaces in good shape is usually that of removing grease and other forms of dirt. Immediately we think of soap, the king of detergents. Next to soap comes some mild alkali like trisodium phosphate. This pair of cleansers will solve most of our problems in cleaning surfaces.

Laundering clothes. The laundering of clothes by the public laundry has become quite a scientific process. In Germany, many of the large laundries have a full-time chemist.

Soap, the most important detergent by far, is the main ingredient of all washing solutions. Soap, alone, does not do well at the high temperatures of the steam-heated washing

584 PROTECTING AND CLEANING SURFACES

suds. The heat causes hydrolysis to the free fatty acids. The fatty acids are greasy and sticky in nature. If the soap is allowed to hydrolyze, the cloth takes on a dingy gray appearance due to a multitude of transparent greasy specks.

To prevent the hydrolysis of the soap, an excess of alkali is added to the suds. The excess alkali tends by mass action (see p. 363) to reverse the reaction.

$$\underset{\text{soap}}{C_{17}H_{35}CO_2Na} + \underset{\text{water}}{H_2O} \rightleftharpoons \underset{\text{fatty acid}}{C_{17}H_{35}CO_2H} + \underset{\substack{\text{sodium} \\ \text{hydroxide}}}{NaOH}$$

Like other equilibrium reactions, the effect of adding more of one of the products is to use up most of the companion substance; that is, adding more $NaOH$ uses up the $C_{17}H_{35}CO_2H$ and keeps it out of the clothes.

The alkali serves other purposes besides preventing the hydrolysis of the soap. A large percentage of the dirt on clothes is acidic. The alkali neutralizes this acid and increases the activity of the soap.

Although much sodium hydroxide was once used by laundries, it is not a safe alkali unless a chemist has studied the situation so that too much is not used. Excess sodium hydroxide attacks the cloth, dissolving the surface of the cellulose, leaving it rough and unable to stand much wear. Sodium carbonate, sometimes called washing soda, is a far better alkali. Since this salt supplies alkali only as fast as it hydrolyzes, it never has a high concentration of $\overline{OH}$ ions at any one time but has a large reserve in the unhydrolyzed part. If the acid dirt uses up the sodium hydroxide, more forms to replace that used up. Sodium hydroxide, itself, is not so good as the soda, for the reason that sodium hydroxide is all ionized; that is, it is all active with no reserve. If enough sodium hydroxide is put into the suds to guarantee the neutralization of acid substances on the cloth, it may cause trouble in case the given batch of clothes should not contain the acid expected.

Washing formula for cotton clothes. The laundry usually uses three suds solutions: (1) a lukewarm suds of high alkali to neutralize the acid and remove any stains that might be set by heat; (2) a hot suds; (3) another hot suds, to which the bleach is added. Not all stains can be removed by washing alone; some bleaching is needed. The housewife sometimes depends entirely upon the sunlight to bleach her clothes, but as a rule, her work does not equal that of the laundry.

Following the third suds, there are often as many as five rinses. Some mild acid and bluing are put into the last rinse. The acid neutralizes any alkali that may not have been

HARD WATER AND SOAP. The two flasks in the center each contain 100 cc. of hard water. Two test tubes were filled with hard-water soap (left) and ordinary soap (right). Enough soap was shaken out of each to make a lather of the hard water. Notice how much more of the ordinary soap was required than of the sulfonated soap.

rinsed out of the cloth. If alkali is left in the fabric, the cotton will be weakened under the heat of the iron. The bluing is necessary to offset a certain yellowish tint in old cloth.

Washing wool and silk. When washing the protein fabrics, the washing formula must be different from that used with cotton. Because these fibers tend to shrink, the water must not be hot — merely warm. Since alkali dissolves wool, no base or base-forming substance should be used. Only the purest soap should be used to wash silk and wool.

Hard-water soaps. Ordinary soap will not lather in hard water until all of the dissolved calcium and magnesium are precipitated out as calcium and magnesium soap. In other words, the soap will not lather until the water has been softened with soap. This is an expensive way to soften water. Laundries cannot afford it. They find it worth-while to install water softeners costing $8000 to $10,000.

Recently there have been invented soaps that lather in hard water. These are still too costly for general use, but in time we may be able to find cheaper ways of producing them. These hard-water soaps are quite similar to other soaps in structure except that they have a molecule of sulfuric acid attached to each soap molecule.

$$C_{12}H_{25}CO_2Na \qquad\qquad C_{12}H_{25}CH_2SO_4Na$$

ordinary soap hard-water soap

From the formulas, it will be seen that the last carbon atom in the molecule must have its oxygen replaced by hydrogen before the sulfate part can be added. This is where the cost comes in — the reduction of the fatty acid before combining it with the sulfate part.

EXERCISES

1. Name (a) one physical method of removing tarnish, (b) one chemical process, and (c) one electrochemical process.

2. Write hydrolysis reactions for the following soaps:

> (a) $C_{17}H_{35}CO_2K$ (c) $C_{15}H_{31}CO_2Na$
> (b) $C_{17}H_{33}CO_2Na$ (d) $C_{12}H_{25}CO_2Na$

3. Why is sodium carbonate better than sodium hydroxide in soap suds?

4. Why is a weak acid put into the last rinse? why bluing?

5. What does the fact that hard-water soap will lather in hard water tell us about the solubility of the calcium and magnesium salts?

6. Why is the "break" usually kept at a moderate temperature?

7. Why is an alkali needed in the suds?

8. Why, in washing silk and wool: (a) Is not the temperature allowed to get high? (b) Should no alkali be in with the soap?

9. How does the formula for hard-water soap differ from that of ordinary soap?

ADDITIONAL EXERCISES FOR SUPERIOR STUDENTS

1. If some hard water contained 100 lbs. of $Ca(HCO_3)_2$, (a) what weight of NaOH would precipitate it as $CaCO_3$? (b) what weight of $CaCO_3$ would be formed?

2. If chlorine can react as follows:

$$Na_2CO_3 + Cl_2 \longrightarrow NaClO + NaCl + CO_2$$

(a) What volume of Cl_2 gas would be necessary to react with 100 lbs. of Na_2CO_3?

(b) What weight of NaClO would result?

3. If 2 lbs. of Na_2CO_3 remained in a batch of laundry, what weight of $HC_2H_3O_2$ would be necessary to change it to $NaHCO_3$?

$$Na_2CO_3 + HC_2H_3O_2 \longrightarrow NaHCO_3 + NaC_2H_3O_2?$$

REFERENCE FOR SUPPLEMENTARY READING

1. Howe, H. E. *Chemistry in Industry*, Vol. II. The Chemical Foundation, Inc., New York, 1926, Chap. XX, pp. 346–359, "Rust-resisting Metals."

(1) What causes the most commonly observed examples of corrosion? (pp. 346–347)

(2) By what time had iron implements superseded bronze? (p. 347)

(3) List the factors influencing corrosion. (p. 349)

(4) Discuss rustless iron. (pp. 356–358)

SUMMARY

The metals at the bottom of the replacement series are least susceptible to tarnish. Aluminum, zinc, tin, and chromium form protective coats of oxide. Copper, gold, silver, and lead cease to react when their surfaces are covered with sulfide. Nickel and copper form a basic-carbonate protective coating. Corrosive metals, such as iron, are often protected by coating their surfaces with more resistant metals. Tin plate and galvanized iron are examples of this type of protection.

The extent of metallic corrosion increases with rise of temperature, length of exposure, humidity of the atmosphere, and impurity of the metal.

Metallic surfaces may be protected with glass (forming enamels), rubber, and synthetic resins.

Tarnish is removed by scouring, polishing, pickling, and electrochemical methods.

In laundering, alkali is added to the suds to prevent the soap from hydrolyzing to fatty acids. Commercial washing formulas usually require three suds operations, a bleach, and five rinses, the last one being the bluing and sour. Since silk and wool are proteins, suds for washing them must not contain alkali. Hard-water soaps are formed by reducing the fatty acids to alcohols and then forming sulfates, which are then neutralized to form sodium salts.

⇄

CHAPTER 31

PAINTS AND VARNISHES

Introduction. The most extensive method of protecting exposed surfaces is through painting or varnishing. Keeping a building well painted adds greatly to the length of its usefulness. If more people appreciated this fact, there would be less neglect of painting. Machinery of any kind can be preserved from rust by a good coat of paint.

What is paint? Paint consists essentially of a colored pigment, a drying oil, and a paint drier. The pigment serves the double purpose of giving a beautiful color and helping preserve the surface.

UNPAINTED HOUSE. Without paint wood deteriorates very rapidly.

The drying oil is an unsaturated oil that is quickly oxidized into a tough film, which serves the double purpose of protecting the surface and holding the pigment in place. The paint drier is an efficient catalyst to oxidation, which shortens the time it takes the air to toughen the oil into a film to about one-fourth of what it would be without the drier.

Oil paints compared to water colors. In everyday language painting includes the application of color in other ways than in oil. For instance, water colors are used in art and for coloring photographs. These are simply aniline dyes that are soluble in water. The water colors, although available in almost unlimited variety of colors in their various tints and shades, are not very permanent towards strong light. Because of their

589

solubility, water colors cannot be used where they will get wet. The pigments used in the oil paints, especially for outdoors, are insoluble inorganic substances as far as possible. Where it is not possible to find insoluble inorganic substances that will give the desired colors, we have to use the insoluble metallic salts of the acid dyes — the *lakes*. There are a few lakes that are lasting enough for use indoors, but none are very good for the rigors of strong sunlight and the extremes of weather. The pigments for oil paints must be ground to extreme fineness, or they will not stay suspended in the oil while they are being applied to the surface.

The names, the colors, the formulas, and the characteristics of some well-known commercial pigments are shown in the accompanying table.

PIGMENTS

TRADE NAME	COLOR	FORMULA	REMARKS
White lead	White	$Pb_2(OH)_2CO_3$	Darkens with age, black PbS, good covering power
Zinc white	White	ZnO	Permanent color; tends to crack
Titanox	White	TiO_2	Very great covering power
Cadmium yellow . .	Yellow	CdS	Very permanent
Chrome yellow . .	Yellow	$PbCrO_4$	
Lemon yellow . . .	Yellow	$BaCrO_4$	
Ocher	Yellow	Hydrated iron oxide	
Sienna	Yellow	Hydrated iron oxide	
Strontian	Yellow	$SrCrO_4$	
Viridian	Green	Cr_2O_3 (hydrated)	
Chrome green . . .	Green	$PbCrO_4 + Fe_4(FeC_6N_6)_3$ (yellow + blue)	
Prussian blue . . .	Blue	$Fe_4(FeC_6N_6)_3$	
Cobalt	Blue	$Co_3(AlO_3)_2$	
Ultramarine . . .	Blue	Complex clay	
Cobalt violet . . .	Purple	$Co_3(PO_4)_2$	
Indian red	Red	Fe_2O_3	
Red lead	Red	Pb_3O_4	
Light red	Red	Roasted yellow ocher	
Burnt sienna . . .	Brown		
Burnt umber . . .	Brown		
Lampblack . . .	Black	C	Very resistant and insoluble

Permanency of color is the chief point desired in a paint pigment. Next comes covering power. Some paints will cover more surface with the same brilliancy of color; this is due to their greater opaqueness or reflecting power. White lead is outstanding in this respect. Some paints are useful because of certain properties; thus poisonous paints tend to keep barnacles from attaching themselves to the bottoms of ships. Red lead, because of its ability to retard the rusting of iron, is almost always used as the first coat for steel structures, such as bridges and steel buildings.

Chemistry and color. In the colored organic compounds (the dyes), chemists have discovered just what groups of atoms are necessary to produce color. This subject, however, is too complicated for an elementary course. In inorganic chemistry, which deals largely with the metals, color is not well understood. We can only say that color seems to be a property of certain metals. For instance, such metals as sodium, potassium, and lithium never give colors other than white to their compounds, but copper, cobalt, iron, chromium, lead, sulfur, and manganese usually have colored compounds. Color is partially a matter of particle size in solids. Mercuric oxide exists in two colors, red and yellow. If the red substance is powdered in a mortar, its color becomes lighter and finally approaches that of the yellow oxide. It is quite generally true that smaller particles favor a lighter color. Some of the naturally occurring pigments, such as ferric oxide, exist in a variety of colors varying through red, orange, and yellow. It is hard to match the naturally occurring pigments by synthesis, since we do not get the particles the same size as the pigment we are trying to match; hence the synthetic product, although chemically the same substance, does not have the same shade or even the same color.

Drying oils. Drying oils, the binders of the pigments in paints, are not numerous. Linseed oil, from the seed of the flax, and tung oil, from the seed of a certain Chinese tree, are the best of the natural drying oils. Poppy-seed oil finds a limited use for artists' colors. Certain fish oils are drying oils, but partly because of the smell and partly because of the slowness of oxidation they are inferior to the two oils of vegetable origin. However, at the high temperatures to which boiler paints are subjected, the fish-oil paints stand up the best. Recently a synthetic drying oil is being made from acetylene, which dries in 60–90 minutes into a substance that is not dissolved by any solvent, and is attacked only by strong oxidizing agents.

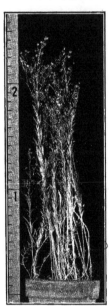

PUNJAB FLAX. Linseed oil, one of the principal drying oils used in the paint industry, is obtained from flax seed.

Linseed oil is used mostly in outdoor paints. It tends to yellow with age, but the bleaching action of the sunlight keeps it clear. The tung oil, or Chinese wood oil, is used for flat indoor paints and in varnishes of the spar type. It does not have so glossy a surface as linseed oil. The Chinese practice of adulterating their tung oil has made it very unsatisfactory for many uses. From plantings of the Chinese tree in the southeastern part of the United States, it is now possible to obtain a uniformly pure product of this drying oil. Linseed oil and tung oil differ in another respect; linseed oil begins to harden on the surface and to get tough farther in, but tung oil hardens uniformly throughout.

Paint driers. Without a drier to catalyze the oxidation of the oil, the paint remains sticky too long. Besides the danger of having it touched or marred by something, it tends to catch gnats and gather dust. The driers are usually certain lead, manganese, and cobalt compounds, such as the oxides, acetates, and soaps. Red lead, being both a pigment and a drier, serves a double purpose.

Varnishes. A plant louse living on the twigs of the trees of India and Burma protects itself by secreting a resinous substance called lac. The natives of those countries gather this secretion. It is then mixed with a little rosin, heated in boiling water, strained, and spread to dry in thin sheets. The resulting shell-like material is called shellac. When a thin layer of an alcoholic solution of shellac is spread over a surface, the alcohol evaporates, leaving a thin layer of the shellac as a shiny surface coating. This is called a varnish.

Other varnishes. As shellac varnish stains readily with water, other more resistant varnishes are preferred for many purposes. It has been found that the most resistant resins are the gums of certain trees — including fossil gums from trees long dead. Zanzibar copal is such a fossil gum. Gums from living trees include dammar from Malaysia and copal from Manila and the Congo. The substitution of other gums for shellac has made necessary the substitution of other solvents for alcohol. Among these solvents are benzene, amyl

Photo by Calvin Coover

TUNG TREE. The tung tree is a native of China but is now being raised in the United States.

acetate, and ether. Eventually tung oil came to be used as a solvent, and in many respects it is the best of them all. It oxidizes into a coating much more resistant to many reagents than the resins alone. It is insoluble in most solvents and resistant to hot water and oil. Perhaps you have seen advertisements showing the housewife pouring hot water from the teakettle onto the varnished table. This can be done safely only with the spar varnishes, which use tung oil.

Enamels. An enamel is a varnish colored with a pigment. The pigment also increases the opacity of the varnish. The first enamel was black. It came from the juice of the sumac shrub in Japan and was called Japan lacquer. It was found that the appearance and durability of enamels were greatly improved by baking. For quite a while, baked enamels were used on automobile bodies. Today, however, the pyroxylin (cellulose-nitrate) enamels have replaced the baked enamels almost entirely. These can be applied with a spray gun. They dry in 30–60 minutes, and are much more resistant than

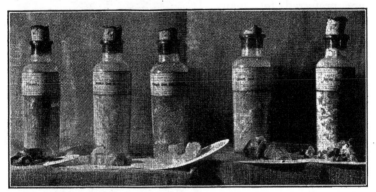

RESINS USED IN VARNISHES. Left to right: amberol, demar, congo, kauri, and ester.

the former baked enamels. Formerly it took anywhere from a week to a month to manufacture an automobile ; the introduction of pyroxylin enamels reduced this time to two days or less. The pyroxylin enamels have come to be known in everyday language as "lacquers," although that word came originally from *lac,* and the dictionary uses lacquer in the same sense as varnish.

The manufacture of pyroxylin. Pyroxylin is synthesized by heating cotton with a mixture of concentrated nitric and concentrated sulfuric acids. The resulting cellulose nitrate looks somewhat like cotton, but it is very different in solubility. It dissolves readily in many organic liquids. This solution is spread on the article to be varnished with a brush or sprayed on with compressed air. The liquid in which it is dissolved evaporates and leaves the cellulose nitrate as a smooth coating.

QUESTIONS OF FACT

1. Name the three essential constituents of paint.
2. Tell what each constituent does.
3. What is a lake used as a pigment?
4. Compare white lead, zinc white, and titanium oxide as pigments for white paint.
5. What pigment is used in the first coat of steel structures?
6. Name four elements that usually have colored compounds.
7. Name two drying oils of vegetable origin.
8. What is the advantage of fish-oil paints?
9. What is the special advantage of the synthetic drying oil?
10. Name some paint driers.
11. Tell how shellac is prepared.
12. What is the objection to shellac varnish?
13. What are the constituents of a varnish that resists hot water?
14. What is an enamel?
15. How is pyroxylin synthesized?

QUESTIONS OF UNDERSTANDING

1. In the following methods of preparing pigments, the pigment is underlined. Tell why the reaction goes to completion:

(a) $Na_2CrO_4 + Pb(NO_3)_2 \longrightarrow \underline{PbCrO_4} + 2\ NaNO_3$
(b) $Zn(OH)_2 \longrightarrow \underline{ZnO} + H_2O$ (calcined)
(c) $CdCl_2 + H_2S \longrightarrow \underline{CdS} + 2\ HCl$
(d) $3\ CoCl_2 + 2\ Na_3PO_4 \longrightarrow \underline{Co_3(PO_4)_2} + 6\ NaCl$

2. In question 1, part (a), is the solution from which the lead chromate is obtained, unsaturated, saturated, or supersaturated?

3. Calculate the molecular weights of the following:

(a) $BaCrO_4$ (b) $Co_3(AlO_3)_2$

4. Match the names of pigments in the first column with formulas from the second column:

(a) Chrome yellow   (1) Cr_2O_3
(b) Zinc white   (2) C
(c) Cadmium yellow   (3) $Pb_2(OH)_2CO_3$
(d) White lead   (4) ZnO
(e) Lampblack   (5) TiO_2

(f) Red lead	(6) CdS
(g) Cobalt violet	(7) $SrCrO_4$
(h) Prussian blue	(8) $PbCrO_4$
(i) Strontian	(9) $Fe_4(FeC_6N_6)_3$
(j) Viridian	(10) $Co_3(PO_4)_2$
(k) Titanox	(11) Pb_3O_4

ADDITIONAL EXERCISES FOR SUPERIOR STUDENTS

1. The ochers are hydrated iron oxide. What does this mean?
2. Calculate the weight of cobalt in 20 lbs. of $Co_3(AlO_3)_2$.
3. Calculate the molecular weight of Prussian blue.
4. How many grams of lead chromate make a gram molecular weight?
5. One lake has the formula:

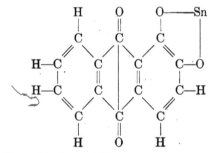

Calculate its molecular weight.

REFERENCES FOR SUPPLEMENTARY READING

1. Glover, John George, and Cornell, William Bouck. *The Development of American Industries*. Prentice-Hall Inc., New York, 1933. Chap. XXV, pp. 479–493, "The Paint, Varnish, and Lacquer Industry."
   (1) What chemical pigments did the American Indians use? (p. 479)
   (2) What paint pigments were mentioned in 1772 as imported from England? (p. 480)
   (3) When was the first paint factory established in the United States? (p. 480)
   (4) When were pigments first ground in oil? (p. 480)
   (5) What three natural pigments are iron oxide? (p. 482)
   (6) Explain the making of varnishes. (p. 485)
2. Howe, H. E. *Chemistry in Industry*, Vol. II. The Chemical Foundation, New York, 1926. Chap. XIV, pp. 232–259, "Paints, Varnishes and Colors."

(1) In a sentence describe paint making. (p. 232)
(2) In a sentence state what varnish making is. (p. 232)
(3) Who was the first chemist to apply scientific knowledge to paint
    problems? (p. 234)
(4) When was the manufacture of white lead started? (p. 234)
(5) When was zinc oxide pigment discovered and by whom? (p. 235)
(6) What is the composition of lithopone? (p. 235)
(7) What is the value of paint to farm buildings? (pp. 237–238)
(8) What pigments inhibit corrosion? (p. 238)
(9) Tell of the problems of paints for ships. (pp. 240–241)
(10) Describe tests applied to paints. (pp. 255–256)
(11) How did spraying paint come to be invented? (pp. 258–259)

Summary

Paint is a mixture of finely ground pigment, drying oil, and
paint drier. When spread on the surface, the oil oxidizes into
a tough film, which holds the pigment and protects the surface.
For outdoor use, the pigments must be inorganic substances
of great stability. Color in the pigments, although primarily
dependent on the nature of the substance, tends to be lighter
in tint with increased fineness of size.

The common drying oils are linseed oil, tung oil, and fish oils.
Of these tung oil is the best for general purposes, although
fish oils withstand heat the best.

Paint driers, which shorten the time of hardening of paint
oils, include lead, manganese, and cobalt compounds.

Varnishes are solutions of resins in alcohol and other solvents.
Enamels are the varnishes that require baking. The pyroxylin
varnishes have become known as lacquers.

CHAPTER 32

SILICON AND ITS COMPOUNDS

Introduction. Next to oxygen, silicon is the most abundant element in the earth's crust. Although quite similar to carbon in its chemical conduct, it is not found free in nature. Most of the common rocks and sands and clays are compounds of silicon. Silicon forms possibly as many compounds as carbon, but since most of them are relatively insoluble and of minor importance, chemists have studied them but little. Few uses have been found for silicon metal. It is difficult to separate from its compounds.

Courtesy of Amer. Museum of Nat. Hist.
QUARTZ CRYSTALS.

Silicon dioxide. Silicon dioxide (SiO_2) is known also as *silica* and *quartz*. Pure quartz frequently occurs in six-sided transparent crystals. White sea sand is often pure quartz. Common sand is an impure quartz. Quartz is extremely hard — it scratches glass — and is difficult to fuse. Pure quartz is especially suitable for laboratory glassware for the reason that it has a low coefficient of expansion — that is, does not expand or contract much with changes in temperature. A quartz dish may be heated to redness and dropped into water without breaking. Quartz dishes look like ordinary glassware. They are so difficult to make that they are unlikely to come into extensive use. Quartz must be melted in the electric furnace, hence the high cost of quartz dishes.

When quartz sand is heated with certain basic compounds, such as those of sodium and calcium, the mixture readily fuses to form metallic silicates, which are familiar as common glass.

Silicon dioxide, either crystallized or uncrystallized, may be colored by traces of other substances to form various semi-precious stones. Smoky quartz (cairngorm stone), amethyst, chalcedony, agate, carnelian, onyx, sardonyx, chrysoprase, jasper, flint, wood opal, and opal are all examples of such stones.

SAND DUNES.   Photo by Dwight Bentel

Occasionally trees and animals petrify. The process by which the change is accomplished is interesting. The cell structure is gradually replaced by silica from solution, which reproduces the form of the material in wood opal (hydrated silica) of various colors.

Silicates. Most of the common silicate minerals are the normal or acid salts of complex silicic acids. The simplest is sodium silicate (Na_2SiO_3), or water glass. It has a glassy appearance and is practically the only silicate soluble in water. It is used for protecting labels on bottles and for preserving eggs.

OPAL, WOOD OPAL, AND MOONSTONE. These are all impure forms of silicon dioxide.

Other common silicates are the micas, talc, serpentine, feldspar, beryl, garnet, tourmaline, and topaz.

EXPERIMENT

A chemical garden. Fill a large jar with sodium silicate solution of specific gravity 1.1. Into different parts of the solution drop a crystal of each of the following : cobalt nitrate, nickel sulfate, uranium nitrate, manganese sulfate, ferrous sulfate, copper sulfate, potassium ferrocyanide, and potassium chromate. Under these different crystals, silicates will begin to form. As they form, they will spread out into treelike structures. This will require several hours, during which time the jar should not be disturbed. Moss agates are supposed to have been formed similarly to these.

Clay. Clay, or kaolin, is a hydrated aluminum silicate, $H_2Al_2(SiO_4)_2 \cdot H_2O$. In its purest form it is almost white. Sometimes clay is colored with oxides of iron and manganese. From such colored clays are made the well-known pigments — ocher, umber, and sienna. Because clay is plastic when moist and gets hard without melting when heated, it is used in making china and earthenware. Flower pots, roof tile, and brick are colored red by the iron oxides in the clay. For white chinaware the clay must be very pure.

All earthenware vessels must be glazed before they will hold water. This is done by throwing salt into the furnace where they are being baked. The salt causes the surface to melt and run together, making it smooth and impervious to water.

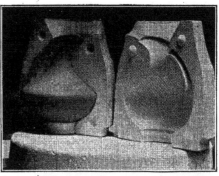

CASTING A CLAY PITCHER. The mold is of plaster of Paris.

Glass. Glass is a mixture of complex silicates of calcium, sodium, aluminum, and (for special glasses) certain other elements. The ingredients dumped into the glass furnace include sand (SiO_2), sodium carbonate, lime, and a little feldspar to furnish the 1–3 per cent aluminum oxide. Each constituent contributes certain properties. The SiO_2, which constitutes from 72–76 per cent of the glass, furnishes the transparency, low coefficient of expansion, and resistance to cracking with heat changes. The sodium gives the glass softness and ease of melting. The calcium increases the strength and durability. The aluminum makes the glass harder, more durable, and more brilliant. The addition of borax in pyrex glass makes the

Courtesy of Ill. Glass Co.

A GLASS HOUSE.

finished glass more brilliant, harder, more durable, and much more resistant to heat changes. Dishes of pyrex glass can be used in the oven as baking dishes. Pyrex glassware has reduced breakage in the chemical laboratory to about one-twentieth what it used to be with soft glass. It is quite common to see a student set a flask of water on an iron ring and heat it with the direct flame of a Bunsen burner. In the days of soft glass, the flask would have been sure to break.

Courtesy of Corning Glass Works
GLASS BLOWING.

Combustion glass. Glass made by replacing sodium with a double amount of potassium has a very high melting point. Such glass is used for combustion tubes.

Cut glass. If about 43 per cent lead oxide is substituted for silica, the finished product has great brilliance and sparkle as a result of its high refractive power — that is, the property of bending light rays. This glass, because of its high lead content, is quite heavy. When used in vases and dishes this type of glass is known as "cut glass"; in lenses and prisms, as "flint glass."

Making of glassware. The making of glassware is both an art and a science. Glass blowing requires a great deal of skill. Almost any kind of article can be made by a skillful glass blower.

In making window glass the workman first takes a mass of the molten glass on the end of a long iron blowpipe and blows it into a large bubble. This is drawn out into a cylinder by swinging it and rolling it on a plate. The ends of the cylinder are cut off, a cut is made lengthwise, and the glass is spread out flat. Plate glass is poured in the molten condition onto a bronze table and rolled out with a hot iron cylinder. Bottles are blown in a mold.

Annealing. Annealing glassware consists in cooling it so slowly that no part of it is left in a strained condition. If it were cooled suddenly, it would break at any slight jar or change in temperature. Annealing is usually accomplished by passing the glass slowly through a tunnellike oven, which is much hotter at one end than at the other.

Coloring glass. Glass is colored by dissolving various substances in the molten mass. The greenish color of cheap bottles is from iron in the sand. Chromium gives a rich green. Compounds of gold give a ruby red; copper and cobalt, a blue; manganese, pink to violet; silver, yellow; manganese and iron, yellow to brown; and calcium fluoride, a white translucent appearance.

QUESTIONS AND EXERCISES

1. Give two common names for SiO_2.
2. Describe quartz glass.
   Name the uses of quartz glass.
   Tell of the difficulties of manufacturing quartz glassware.
   Which silicate is soluble? What is its common name?
   Give the uses of sodium silicate.
   Name ten semiprecious stones that are essentially SiO_2.
8. What is clay? Of what use is it?
9. Account for the red color in tile and brick.
10. How is earthenware glazed?
11. What is glass?
12. What properties of glass are due to SiO_2?
13. What properties of glass are due to calcium?
14. What properties of glass are due to aluminum?
15. What element gives the heat resistance to pyrex glass?
16. What are the desirable properties of cut glass?
17. What element is essential to cut glass?
18. Name the elements that color glass (a) green, (b) blue, (c) pink, (d) yellow.
19. What element makes combustion tubing difficult to melt?

20. Describe the making of window glass and plate glass.

21. Why is glass annealed? What is annealing?

ADDITIONAL EXERCISES FOR SUPERIOR STUDENTS

1. Prepare a talk on some of the precious stones that are silicates. Use the encyclopedia.

2. Prepare a paper or a talk on glass; the different kinds, methods of coloring, cutting, and blowing.

REFERENCES FOR SUPPLEMENTARY READING

1. Ehrenfeld, Louis. *The Story of Common Things.* Minton, Balch and Company, New York, 1932. Chap. III, pp. 51–66, "Glass"; Chap. XI, pp. 175–191, "Ceramics."
    (1) What was the first industrial enterprise in the United States? (p. 55)
    (2) What lead compound is used in making cut glass? (p. 57)
    (3) Describe glass blowing. (pp. 59–60)
    (4) Describe the making of plate glass. (p. 61)
    (5) How small are the gold particles in ruby glass? (p. 63)
    (6) What color does copper give glass? (p. 63)
    (7) Why are so many clay articles round? (p. 180)
    (8) Describe the kiln for firing clay. (p. 183)
    (9) What is the glaze? (p. 184)
    (10) Describe brick making. (pp. 186–187)
    (11) Describe porcelain. (p. 189)
2. Foster, William. *The Romance of Chemistry.* D. Appleton-Century Co., New York, 1936. Chap. XIX, pp. 308–322, "The A B C of Pottery and Glass."
    (1) What is meant by the term *ceramics?* (p. 308)
    (2) How is clay formed? (p. 308)
    (3) Mention early uses of brick. (pp. 309–310)
    (4) What is the melting point of pure clay? (p. 311)
    (5) Explain how ceramics are colored. (p. 313)
    (6) Prepare a paper or a talk on glass — the different kinds; methods of coloring, cutting, and blowing.
3. Glover, John George, and Cornell, William Bouck. *The Development of American Industries.* Prentice-Hall, Inc., New York, 1933. Chap. XXI, pp. 419–435, "The Glass."
    (1) How early were glass blowers pictured on Egyptian tombs? (p. 419)
    (2) Tell about the Portland Vase. (p. 420)
    (3) Name several glass articles made by the Romans. (p. 420)
    (4) Discuss the American glass industry today. (pp. 424–426)
    (5) How is colored glass made? (pp. 429–430)

4. Howe, H. E. *Chemistry in Industry*, Vol. I. The Chemical Foundation, New York, 1924. Chap. X, pp. 130–146, "Glass: One of Man's Blessings."

    (1) To what extent are we dependent upon glass? (pp. 131–132)

    (2) What happens to glass that has not been heated enough? (p. 132)

    (3) Describe the making of window glass. (pp. 134–135)

    (4) Describe bulletproof glass. (p. 136)

    (5) What is glass wool used for? (p. 139)

    (6) How many light globes, water tumblers, and bottles are made in the United States? (p. 144)

5. Martin, Geoffrey. *Triumphs and Wonders of Modern Chemistry*. D. Van Nostrand Co., New York, 1922. Chap. XII, pp. 271–296, "Silicon and Its Compounds."

    (1) What chemical substance is flint? (pp. 276–277)

    (2) What are some colors of opals? (pp. 277–278)

    (3) Describe some valuable jewels that are opals. (p. 278)

    (4) Tell about the search for rock crystals. (pp. 279–281)

    (5) Describe some geysers and mineral springs. (pp. 282–284)

    (6) What does silica do for plants? (p. 285)

    (7) Describe the microscopic structure of rock. (pp. 288–289)

    (8) Describe the bricks in the temple of Belus. (p. 294)

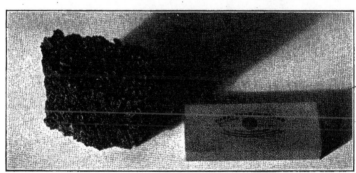

CARBORUNDUM.

Carborundum. The artificial carbide of silicon is extremely hard, being next to the diamond in hardness. It is valuable as an abrasive, and has largely replaced emery.

Silicon fluoride. Silicon fluoride is the only known gaseous compound of silicon. The fact that there is a gaseous compound of silicon is important in qualitative analysis. Most naturally occurring minerals contain more or less silica or silicious materials. To remove the silicon, the substance to be analyzed is treated with HF. This changes the silicon in whatever form it may be to the gaseous SiF_4 which is driven off with heat.

Portland cement. Portland cement is made by calcining (dry heating to very high temperatures) limestone and clay, and finally adding some calcium sulfate to keep it from setting prematurely. It has been defined as "a finely pulverized material consisting of certain definite compounds of lime, alumina, and

ROTARY CEMENT KILN.　　*Courtesy of Portland Cement Assn*

silica, which, when mixed with water, has the property of combining slowly with the water to form a hard, solid mass."

More thorough studies have shown that the mixture of the best product is not quite so simple as the above definition would imply. Small amounts of $CaSO_4$ and MgO are essential. Usually the mixture contains some iron, although it is not known to be necessary. The average dry cement mixture is approximately that shown in the table on the next page.

The formulas used here are mostly of the type used by the mineralogist instead of regular chemical formulas. The reason the mineralogist is limited to simple formulas is that, dealing with insoluble solids, he had no way of knowing the exact nature of the substance therein. This type of formula gives some idea of the composition, but says nothing about the structure of the molecule.

CEMENT

SUBSTANCE	MINERALOGICAL FORMULA	PER CENT
Tricalcium silicate	$3\ CaO \cdot SiO_2$	50
Dicalcium silicate	$2\ CaO \cdot SiO_2$	25
Tricalcium aluminate	$3\ CaO \cdot Al_2O_3$	12
Calcium ferro aluminate	$4\ CaO \cdot Al_2O_3Fe_2O_3$	8
Calcium sulfate	$CaSO_4$	2.9
Magnesia	MgO	2.4

Origin of Portland cement. Portland cement was invented by Joseph Aspdin, a bricklayer of Leeds, England. It was named Portland cement because the finished rock resembled a building stone obtained from the Isle of Portland. The first large project using cement was the Thames River Tunnel, in 1828.

The setting of cement. The setting of cement is a very complicated group of reactions that are not completely known. Cement undergoes an initial set, which happens in a short time, and then a continuous hardening that extends over months.

The initial set is largely composed of reactions of hydration, or adding water to form crystalline substances, such as $3\ CaO \cdot Al_2O_3 \cdot 10\ H_2O$ and $3\ CaO \cdot Al_2O_3 \cdot 3\ CaSO_4 \cdot 31\ H_2O$. The strength of the product depends upon the size of the interlocking crystals.

The final setting is largely brought about by a set of hydrolysis reactions by which finely divided substances are formed. These fill in between the crystals and harden. This holds the interlocking crystals in a firm compact mass.

Concrete. Cement is too expensive to use alone. When a mixture is made of about two-thirds gravel and one-third cement, it is called concrete. Concrete can be made very strong by imbedding steel rods within it. This type of construction is used in bridges and tall buildings. It is called "reinforced concrete." Cement and concrete have almost completely replaced building stone and to a large extent brick.

A stone or brick building can be shaken down in an earthquake, with great danger to life. A well-built reinforced concrete building is quite safe in such times.

Concrete buildings are safe, cool, and durable. The future will use cement more and more in all kinds of construction. It is said that concrete houses are now being poured in practically one piece. This artificial stone has many advantages

BRIDGE OF REINFORCED CONCRETE.

over natural stone; there is no hewing, or cutting, yet it can be cast into any shape.

The cement in Boulder Dam. Boulder Dam, which will eventually back up the Colorado River for over one hundred miles, and make a lake averaging eight miles wide, is the largest single piece of concrete ever built. The dam is 220 yards thick at the base, 725 feet high, and a quarter of a mile long across the canyon.

Heat of reaction of the setting cement was a problem in building the dam. Had the research staff not developed a special low heat cement and the builders not used 300 miles of pipe through which cold brine circulated, the cement would have required one hundred years to cool. Perhaps then the cement would have become cracked and brittle. The follow-

BOULDER DAM.

Courtesy of United Air Lines

ing table shows how the low-heat cement compares in composition with ordinary cement.

COMPARISON OF CEMENTS

COMPOUND	ORDINARY CEMENT	BOULDER DAM CEMENT
$4\,CaO \cdot Al_2O_3 \cdot Fe_2O_3$	8 per cent	14 per cent
$3\,CaO \cdot Al_2O_3$	12 per cent	6 per cent
$2\,CaO \cdot SiO_2$	25 per cent	54 per cent
$3\,CaO \cdot SiO_2$	50 per cent	18 per cent

REVIEW EXERCISES

1. Give the constituents and uses of carborundum.
2. What is the only gaseous compound of silicon?
3. What part does silicon fluoride play in chemical analysis?
4. Name five substances in cement.
5. Who invented Portland.cement?
6. What kind of reactions occur in the initial setting of cement?
7. What type of reactions occur in the final hardening of cement?
8. What is reinforced concrete?
9. Compare the composition of the cement used in Boulder Dam with that of ordinary cement.
10. Why did the cement used in Boulder Dam have to be different from ordinary cement?
11. Why did special means of cooling the setting cement in Boulder Dam have to be used whereas ordinarily no attempt is made to cool setting concrete?

ADDITIONAL EXERCISE FOR SUPERIOR STUDENTS

1. Prepare a paper or talk on the history and use of cement.

REFERENCES FOR SUPPLEMENTARY READING

1. Clarke, Beverly L. *Marvels of Modern Chemistry.* Harper and Brothers, New York, 1932. Chap. XVII, pp. 214–224, "Sand and Clay."
   (1) What is silica? What percentage of granite is silica? (p. 214)
   (2) What substances are heated in the electric arc to produce carborundum? (pp. 214–215)
   (3) How long since glass was first manufactured? (p. 215)
   (4) What is the composition of feldspar? (p. 222)

2. Howe, H. E. *Chemistry in Industry*, Vol. II. The Chemical Foundation, New York, 1926. Chap. XV, pp. 261–278, "Portland Cement."
   (1) Who discovered gypsum? (p. 262)
   (2) What led to the invention of cement? (p. 263)
   (3) Why is high temperature fusion in cement necessary? (p. 264)
   (4) What are the raw materials in cement manufacture? (p. 267)
   (5) What happens in the setting of cement? (pp. 274–275)

SUMMARY

Silicon, the second element in abundance, is hard to obtain from its compounds and does not have many uses in the free state. Silicon dioxide as quartz and sand finds several uses in making quartz glass dishes and in the manufacture of ordinary glass. Other substances which are mostly silicon dioxide are amethyst, chalcedony, agate, carnelian, onyx, sardonyx, chrysoprase, jasper, flint, opal, and petrified wood.

Water glass (Na_2SiO_3) is the only soluble silicon compound. It is used to protect labels and other surfaces from atmospheric attack. Another naturally occurring silicon compound, commonly called clay but scientifically known as kaolin, is a hydrated aluminum silicate. It is the necessary material for making all kinds of earthenware, flower pots, roof tiling, bricks, and dishes.

Ordinary glass is made by mixing sand, sodium carbonate, lime, feldspar, and aluminum oxide in the furnace. By partly replacing the sodium carbonate with potassium carbonate, the melting point of the glass is greatly raised. Increasing the percentage of calcium increases the strength and durability of the glass. Pyrex glass, which has greatly reduced breakage in the laboratory, is made by adding borax, which reduces shrinkage with cooling. The addition of lead to glass increases its sparkle in cut glass.

Freshly blown glassware must be annealed or cooled very slowly to avoid internal strains, which make the glass easily broken.

Carborundum, or silicon carbide, the second hardest substance known, is used in cutting points for lathes and in abrasives.

Because silicon fluoride is gaseous, silicon is removed from mineral samples in quantitative analysis by treatment with hydrofluoric acid.

One of the most important silicious products is the artificial rock called Portland cement. This substance is made by calcining lime, alumina, silica, and iron oxide. When the resultant powder is mixed with water, crystallized hydrated substances form into a mass of interlocking crystals, similar to ordinary rock. Some hydrolysis results in finely divided substances, which fill in between the larger crystals, making the substance firm.

Concrete is a mixture of cement with rock, gravel, and sand. With steel rods within, the mass forms reinforced concrete. The properties of cement can be varied through a wide range by varying the proportions of the constituents.

CHAPTER 33

THE CHEMISTRY OF COOKING

Introduction. Cooking is a chemical process in most of its phases. The part that is not necessarily chemical is the killing of germs. Although the full chemistry of cooking is not known, the general process is understood. A better understanding of the chemistry of cooking will result in better cooking.

Frying. Frying is the process of heating food in grease at a rather high temperature. In the case of meats and eggs, the heat coagulates the albumins and changes the proteins. There is some reaction between the grease and the protein, making a combination difficult to digest. In hot cakes and fried potatoes the heat partially hydrolyzes the starches and changes the vegetable proteins.

Boiling. Boiling is a very important process for making food more digestible and tasty. The principal result of boiling, aside from exploding the cell structure, is hydrolysis. The starches are partially hydrolyzed to simpler carbohydrates, especially sugars. Proteins are also hydrolyzed to simpler structures. Cooking vegetables slowly in their own juices improves the taste and nutrient value.

Poaching of eggs. Heat coagulates albumins. When the proteins of the egg are coagulated by the heat, their digestibility depends on the temperature at which they are cooked. The lower the temperature, the more digestible the food. In poaching eggs by cooking in water or milk or soft boiling, it is better to cook for a longer time at a temperature considerably below boiling than a shorter time in boiling liquids.

Cooking meats. In baking or roasting meats, the flavor is retained best by quickly searing the outside and then cooking slowly. The quick searing hardens the outside which keeps in the juices which contain the flavor. Gradual prolonged

613

heating, especially in the case of boiling, cooks out all the juices and leaves the meat tasteless.

Toasting. Toasting makes bread more digestible by breaking the starch into simpler substances called dextrins. Zwieback (a German word meaning twice baked) and other hard breads are really much better for us than the soft white bread

so commonly eaten. The chewing of hard bread scours the teeth and rubs the gums, both of which actions are good for the mouth. The dextrins, products of heating starch, are more easily reached by the digestive juices than sticky bread, which tends to form lumps. Sometimes dextrin is white and sometimes brown, which indicates that it is not a single substance.

TOASTING. Toasting changes starch to dextrins.

Yeast bread. Bread making is usually thought of as an art, yet there is a certain amount of chemistry underlying it. The rising of the bread is caused by the expansion of carbon dioxide which is generated within the dough. The formation of carbon dioxide within the dough depends upon two chemical reactions. The first reaction beginning with cane sugar and resulting in a mixture of glucose and fructose was shown on page 502. The next reaction begins with these substances and ends in alcohol and carbon dioxide.

$$C_6H_{12}O_6 \longrightarrow 2\,CH_3CH_2OH + 2\,CO_2$$

Without the yeast plant neither the hydrolysis of cane sugar nor the formation of alcohol and carbon dioxide can take place. While growing and multiplying, the yeast plant produces within its cells two remarkable substances called *enzymes*. An enzyme is a complex organic substance that catalyzes one specific chemical reaction. This it usually does rapidly and at low temperatures. The first enzyme, called invertase, brings about the hydrolysis of cane sugar into the simpler sugars. Zymase, the other enzyme, catalyzes the second reaction. The baking kills the yeast, drives out the alcohol, and expands the dough. Gluten in the flour keeps the dough around the bubbles of gas from breaking and allowing the bread to fall. Certain kinds of wheat contain more of this protein than others. The manufacturers of flour have to see that their flour has the right mixture of starch and gluten. In order to obtain flour that will make the best bread, the chemist with the flour company must analyze the grains for gluten and then add enough of the high-gluten wheat to the low-gluten wheat to bring the gluten up to the optimum percentage.

The starch is often washed out of high-gluten flour, and the remaining gluten is boiled and eventually fried. When properly flavored, this cooked gluten resembles meat in taste and possesses several advantages over meat. Some animal did not have to die to produce it; it is free from the bacteria often found in meat; and it is free from possible decomposition products of meat. Vegetarians use gluten largely in place of meat.

Enzymes. Mention has already been made of the fact that enzymes are specific catalysts for only one reaction each. One enzyme causes sugar to ferment to alcohol, but another enzyme causes it to produce some other substance. Molds, yeasts, and bacteria all produce enzymes. Besides these, enzymes are produced in the animal body. In the saliva the enzyme ptyalin produces the hydrolysis of starch in the food. In the stomach the enzyme pepsin hydrolyzes the proteins into the acids, which are assimilated by the blood. In the intestines the fats are digested by other enzymes. It has been said that if we had the right enzymes for successive treatments of alfalfa hay, it would be possible to produce milk without the aid of a cow.

Sour-milk-and-soda biscuits. In sour-milk-and-soda biscuits the carbon dioxide is generated by the action of the lactic acid in the milk upon the soda. The housewife has no accurate way of knowing how much acid is in the sour milk. If there is too much soda for the acid, the results are not satisfactory. The extra soda changes into washing soda — alkaline and bitter of taste.

$$2 \, NaHCO_3 \longrightarrow Na_2CO_3 + H_2O + CO_2$$

If the acid is in excess, the results are not so bad. A little lactic acid will not be noticed.

QUESTIONS OF FACT

1. What is an objection to fried foods?
2. Name two things accomplished by boiling.
3. What chemical change is accomplished by toasting?
4. What gas causes yeast bread to rise?
5. What part does the yeast plant play in yeast bread?
6. What substance must be there for the yeast to work on?
7. In soda biscuits what happens if there is too much soda for the sour milk?
8. Why is it not so serious if the lactic acid is in excess?
9. What part do enzymes play in yeast bread?
10. What part do enzymes play in digestion?
11. What protein from flour is used as a meat substitute?

QUESTIONS OF UNDERSTANDING

1. How could dextrin be prepared?
2. What is meant by saying an enzyme catalyzes a specific reaction?
3. How is the digestibility of a cooked egg related to the temperature of cooking?
4. What is the purpose of searing a roast before slowly roasting?
5. Why would a poached egg be more advisable for a person recovering from illness than a fried egg?
6. If a batch of sour milk biscuits tastes bitter, what conclusion can one make?

Baking powders. Baking powders play a large part in modern cooking. The active ingredients of all baking powders are baking soda and some solid substance which is either an acid or a substance that hydrolyzes to form acid when it is dissolved in water.   The reaction between the carbonate and the acid forms the CO_2 which causes the bread to rise.

There has been much controversy over the relative merits of the different baking powders.   Attempts were made to have the cheaper baking powders declared unhealthful and condemned by the pure-food administrators.   The housewife is asked four times as much for one baking powder as for another.   Naturally she wonders if the price is in any sense a measure of baking quality and healthfulness.   The squabble over baking powders has resulted in requiring that the ingredients be listed so the user can at least know which kind he is using.

Bread-raising efficiency of baking powders.   If baking powders consisted only of the soda and the acidic substance, a spoonful of one kind would give a quite different amount of carbon dioxide from a spoonful of another kind.   This difference in the yield of carbon dioxide is largely equalized by the addition of starch.   Starch is added to protect the mixed soda and acidic substance from the moisture of the air.   To the extent they absorb moisture, the ingredients will react and deteriorate.   Egg albumen is also added to some baking powders to strengthen the containing walls of the carbon-dioxide bubbles.   Too much or too little carbon dioxide is not desirable.   If the powder does not yield enough of this gas, the biscuits will not rise enough.   On the other hand, if there is too much carbon dioxide, the bubbles will run together, making large empty spaces in the dough and often allowing the bread or cake to fall.

Tartrate baking powders.   All baking powders use baking soda as the source of the carbon dioxide.   No other carbonate is satisfactory in so many respects as soda.   It is cheap, mild tasting, and soluble.   The difference between baking powders is primarily in the acid constituent.   The baking powders can be divided into three classes, although some of the powders contain mixtures of two classes.   The tartrate powders use cream of tartar ($KHC_4H_4O_6$) as the acid part.   Sometimes a little tartaric acid is added to speed up the action.   Cream of

TARTRATE BAKING POWDER.

tartar is obtained as a by-product of the wine industry. Large quantities of this acid tartrate are deposited in the wine tanks. Recently tartaric acid has been made by the electrolysis of grape sugar. The reaction by which the carbon dioxide is formed from the soda and cream of tartar is as follows:

$$KHC_4H_4O_6 + NaHCO_3 \longrightarrow KNaC_4H_4O_6 + H_2O + CO_2$$

$$\underset{\text{cream of}}{} \qquad \underset{\text{soda}}{} \qquad \underset{\text{Rochelle salts}}{} \qquad \underset{\text{water}}{} \quad \underset{\text{carbon}}{}$$
$$\underset{\text{tartar}}{} \qquad\qquad\qquad\qquad\qquad\qquad\qquad\qquad \underset{\text{dioxide}}{}$$

The tartrate baking powders are usually the most expensive and for that reason have been assumed by manufacturers to be the best. However, as has been shown by the studies of the United States Pure Food Administration, price is in no sense a measure of the quality of baking powders.

The phosphate baking powders. Another class of baking powders uses calcium acid phosphate as the acid constituent. The carbon dioxide is produced by the following reaction:

$$2\,NaHCO_3 + Ca(H_2PO_4)_2 \longrightarrow Na_2HPO_4 + CaHPO_4 + 2\,H_2O + 2\,CO_2$$

PHOSPHATE BAKING POWDER.

It might be said in favor of the phosphate powders that the body does need phosphate for bone building. The acid phosphate is a pure form of the superphosphate made from phosphate rock by the action of sulfuric acid.

The sulfate baking powders. The sulfate baking powders use sodium' aluminum sulfate or simply aluminum sulfate as the acid part. Recalling that acids are hydrogen compounds that form hydrogen ions in solution, we might wonder how these salts can take the place of an acid. The explanation is

SULFATE BAKING POWDER.

this — they hydrolyze or react with water to form sulfuric acid, which in turn reacts with the soda.

$$Al_2(SO_4)_3 + 6 H_2O \rightleftharpoons 2 Al(OH)_3 + 3 H_2SO_4$$
$$2 NaHCO_3 + H_2SO_4 \longrightarrow Na_2SO_4 + 2 CO_2 + 2 H_2O$$

The sulfates are quite cheap, which makes the sulfate powder the cheapest. The fact that these powders could undersell others led certain interested persons to try to have them declared unhealthful, on the ground that the aluminum sulfate is so nearly like alum, which is admittedly not good for food.

Healthfulness of baking powders. Baking powders can only be condemned because of the residues they leave in the bread. When the baking powder dispute came up, Dr. H. W. Wiley, who administered the pure food laws, appointed a referee board of physicians to study the question of the sulfate powders. These doctors fed college students bread made with sulfate baking powders for over a year and studied their health in every way. Their conclusion was that the body did not assimilate any aluminum from the baking powders and that this type of baking powder, therefore, was no more harmful than either of the others.

All three classes of powders leave a laxative residue in the bread. Rochelle salts. ($KNaC_4H_4O_6$), disodium phosphate (Na_2HPO_4), and Glauber's salts (Na_2SO_4) are all sold at the drug stores as laxatives. However, it is not believed that the small amounts left in the bread from the baking powders do any harm. Moreover, aluminum hydroxide, one of the products formed from the sulfate powders, is both insoluble and inactive. The sulfate powders leave less than half as much active residue as the tartrate baking powders, and but a little more than the phosphate powders. On general principles, the phosphate powders have a slight advantage over the others; they leave the least amount of laxative residue, and this residue may serve as a source of phosphate.

QUESTIONS OF FACT

1. What kinds of substances are put in baking powder? State the purpose of each.
2. Why is there little difference in the amount of CO_2 given by the different kinds of baking powders?
3. What is cream of tartar?
4. Where is it obtained?
5. Why do all baking powders use the same carbonate?
6. What residue is left in the bread by the tartrate powders?

QUESTIONS OF UNDERSTANDING

1. Write the reaction between baking soda and $Ca(H_2PO_4)_2$.
2. Name one thing in favor of the phosphate powders.
3. What are the acid constituents used in sulfate baking powders?
4. In what sense is $Al_2(SO_4)_3$ acidic?
5. Write the reaction for the hydrolysis of aluminum sulfate.

PROBLEMS

1. Calculate the weight of $KNaC_4H_4O_6$ that would result from 10 g. $KHC_4H_4O_6$.
2. Calculate the weight of Na_2SO_4 that could be obtained from 10 g. of $Al_2(SO_4)_3$.

3. Calculate the volume of carbon dioxide that comes from 10 g. of $NaHCO_3$ in baking powder.

4. Calculate the volume of carbon dioxide obtainable by the fermentation of 10 g. of glucose.

5. Calculate the volume of carbon dioxide obtainable from 100 g. of $Ca(H_2PO_4)_2$.

6. If 10 cc. of carbon dioxide forms at a temperature of 40° C., what volume will it have at 200° C., the pressure remaining practically the same?

REFERENCE FOR ADDITIONAL READING

1. Darrah, Juanita E. *Modern Baking Powder.* The Commonwealth Press, Inc., Chicago, 1927.

Write a 500-word paper on baking powders.

SUMMARY

Frying coagulates and changes proteins and hydrolyzes starches. The same changes occur in boiling except that the hydrolysis of starch can take place to better advantage in solution. Frying tends to form greasy combinations that are difficult to digest. Boiling avoids these. Toasting changes starch to dextrin.

The digestibility of poached eggs is increased by cooking at temperatures below boiling.

In cooking meats the flavor is best retained by first searing the outside of the meat by a short intense surface heating before the regular cooking.

In sour-milk biscuits the carbon dioxide which causes the rising comes from the reaction between the soda and the acid in the milk. Baking powders are mixtures of baking soda and some acid constituent or constituents which may include aluminum sulfate, calcium acid phosphate, cream of tartar, and tartaric acid. Some starch is added to the mixture to keep the moisture of the air from causing the reaction to take place prematurely. Egg albumen is often added also to help confine the carbon dioxide in bubbles in the bread. In healthfulness and efficiency all baking powders are about the same.

⇌

CHAPTER 34

THE COLLOIDAL STATE

Introduction. In our study so far we have been mostly interested in the *chemical* properties of substances. In most of the reactions and processes studied, it has been assumed that the individual molecules are the reacting units. In this chapter, on the other hand, characteristics that are primarily dependent upon the size of the reacting particles will be considered. The reacting particles in every case are clusters of molecules varying through a definite range of diameters. Here we are interested

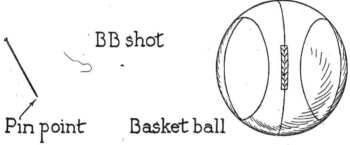

COMPARATIVE SIZES OF HYDROGEN MOLECULE AND COLLOIDAL PARTICLES. The magnification that would enlarge a hydrogen molecule to the size of a pin point would enlarge the smallest colloidal particle to the size of a BB shot and the largest one to the size of a basket ball.

in a state of matter (the colloidal state) rather than in a class of compounds. Substances of any kind may exist in the colloidal state, although certain classes of substances are more stable in this state than others. Any substance in the colloidal state is said to be a *colloid*.

What is the colloidal state? One way of defining the colloidal state is to say that when the units into which the substance has been divided vary in diameter from one ten-thousandth of a millimeter (.0001 mm.) to one-millionth of a millimeter (.000,001 mm.) the substance is colloidal. Another way of saying the same thing is that if the particles of a substance are anywhere in size from just too small to be visible in the highest power compound microscope down to a size of 60 times the diameter of a hydrogen molecule, it is a colloid. Within this range of particle size, substances possess properties different from those of the same substances in either the molecular or the larger particle states.

The importance of colloids. The colloidal state is of the utmost importance in all phases of chemistry. A knowledge of colloidal chemistry is essential to the proper understanding of cement, brick, pottery, glass, enamels, oils, greases, soaps, candles, ·glue, starch, adhesives, .paints, varnishes, lacquers, rubber, celluloid, leather, paper, textiles, filaments, casts, pencils, crayons, inks, asphalt, graphite, cream, butter, cheese, and milk. Likewise, such processes as cooking, washing, dyeing, printing, ore flotation, water purification, sewage disposal, smoke prevention, photography, pharmacy, and physiology require a knowledge of colloids. Life itself is a process of colloidal chemistry. Death by electric shock or poisoning is explainable in terms of colloidal changes. The physician, above all others, needs a thorough knowledge of colloidal chemistry. Even a limited knowledge of colloids will enable one to live a richer life than he would without such knowledge.

Preparing colloids. Perhaps the most important colloids are those prepared for us by nature. Among these are most of the substances in plants and animals, crude oil, clays, adobe soil, textile fibers, greases, rubber, and milk. However, many of the substances which are partly valuable for their being in the colloidal state are prepared by man. There are several methods of preparing colloids.

By grinding. In making paint, the pigments are reduced to colloidal dimensions by grinding in ball mills. If the pigment particles were not colloidal in size, they would not remain suspended until the paint was spread on the surface. This would result in a lack of uniformity in appearance as the tint would change as the particles settled out.

By the electric arc. An electric arc between silver electrodes under water forms a colloidal silver solution. *Argyrol,* the mild antiseptic used in the nose and throat, is a solution of colloidal silver, plus a protein protective colloid. Argyrol is, however, prepared by a chemical process rather than an electrical one.

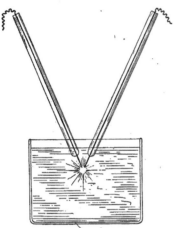

APPARATUS FOR PREPARING COLLOIDAL SILVER. An electric arc between silver electrodes under water forms colloidal silver, which settles out slowly if at all.

By stirring or beating. The whipping of cream, the beating of an egg, and the stirring of oil, vinegar, and egg white into mayonnaise bring about the colloidal state by stirring.

By the clustering of particles. When small particles of moisture separate from the air, they are often attracted to electric charges on molecules of air. On reaching colloidal dimensions, these small droplets form a fog or a cloud. Following many chemical reactions, particles collect into aggregates of colloidal size. In qualitative analysis many of the precipitates are not easily separated by filtration because their particles are of colloidal dimensions and the smaller ones pass through the filter paper.

Electric charges on colloidal particles. It can be proved that colloidal particles carry electric charges. If the electrodes from a high-voltage battery are dipped into a colloidal solution of liquid or solid particles floating in a liquid, the particles begin to drift to one or the other electrode. Some colloids, such as arsenious sulfide, wander to the positive terminal, which means the particles carry negative charges. On the other hand, colloidal ferric hydroxide moves towards the negative electrode, which proves it carries positive charges. In many colloidal solutions the repulsion between the charges keeps the particles from collecting into larger aggregates.

Advantage is taken of the fact that colloidal particles are charged in a very useful process of manufacturing rubber articles. Rubber latex is a colloidal solution resembling milk. Its whiteness comes from colloidal particles of rubber, which carry negative charges. If the positive electrode is shaped as the rubber article is going to be, the rubber is deposited upon

Photo by Vida Fisher Meyer

SMOKE A COLLOID. Smoke is a colloidal solution in the air. Most of the particles of smoke are too small to be seen individually.

the form to produce such articles as bathing caps and hot-water bottles.

Colloids in colloidal solutions can be made to coagulate and settle out by neutralizing the charge by an electrolyte. If an electrolyte is added to colloidal arsenious sulfide, the change on the latter charge attracts the positive ion. After being neutralized, the arsenious sulfide coagulates into a curdy precipitate, which is now easily filtered. Deltas form at the mouths of muddy rivers such as the Mississippi and the Ganges. When the colloidal soil particles reach the salt water, their charges are neutralized by ions in the salt water, after which coagulation results and the sediment settles out.

In dyeing, cloth in acid solution becomes positively charged. In this condition it attracts the negatively charged acid dyes. On the other hand, in alkaline solution the cloth is negatively charged. It now attracts and holds the positively charged basic dyes.

QUESTIONS OF FACT

1. Define the colloidal state.
2. Must there always be at least two substances to make up a colloidal solution?
3. Name five important substances that are colloidal at some time.
4. Name four processes that are colloidal in nature.
5. Name four methods of getting particles of colloidal size, and name one colloidal substance prepared by each method.
6. Give one use of (a) colloidal silver in water, (b) a colloidal pigment in oil, and (c) colloidal olive oil in vinegar.
7. Explain the formation of deltas.
8. What are the limiting sizes of colloidal particles? Define in two ways.
9. Name several colloids prepared by nature.

QUESTIONS OF UNDERSTANDING

1. Discuss the importance of colloids to the doctor.
2. What would be the consequences if the colloids in the body should coagulate?
3. What kind of process is coagulation of the blood?
4. Are chemical precipitates sometimes of colloidal dimensions? What makes you think as you do?
5. How does the smallest colloidal particle compare in diameter with that of a hydrogen molecule?
6. How are paint particles made of colloidal size?
7. Why should paint particles be of colloidal size?
8. Name two processes used in the preparation of foods which have to do with colloids.
9. How do the colloidal particles of fog form?
10. What shows some colloidal particles carry electric charges?
11. How do the electric charges prevent coagulation?
12. How can electrified colloids be made to coagulate?

13. Describe how rubber articles are prepared by electrolytic deposition.

14. How do colloidal phenomena aid in dyeing cloth?

REFERENCE FOR SUPPLEMENTARY READING

1. Clarke, Beverly L. *Marvels of Modern Chemistry*. Harper and Brothers, New York, 1932. Chap. VIII, pp. 110–127, "Colloids."

(1) Describe the observations of Thomas Graham which led to the study of colloids. (p. 110)

(2) What did the term "colloid" originally mean? (p. 110)

(3) Why is the division of substances into colloids and crystalloids not a valid distinction? (p. 110)

(4) Explain the difference between sugar and sand in water. (p. 111)

(5) Explain the term "zone of maximum colloidality." (p. 112)

(6) What do we mean by saying gelatin is a colloid? (p. 113)

(7) Illustrate the sizes of colloidal particles. (pp. 114–115)

(8) What sets the lower limit of particles that can be seen in the compound microscope? (p. 115)

(9) Describe the behavior of a jelly. (p. 119)

(10) What is a liquid-liquid colloidal solution called? (p. 120)

(11) Describe a method of diagnosing one type of insanity by testing the spinal fluid. (pp. 120–121).

Relation of colloid to dispersion medium. Some colloids, such as colloidal gold, do not absorb any of the dispersing medium. These keep themselves from coagulating by their motion and the repulsion of their electric charges. This class of colloids is sometimes called the *suspensoid* colloids.

The other class, which absorbs more or less of the dispersion medium, is called the *emulsoid* colloids. Because of the absorption of the liquid the effects of the electric charges are not so pronounced as in the other class. The greater the absorption of the liquid, the more stable the colloid. Such emulsoid colloids as gelatin and albumen are not precipitated by electrolytes until they become very concentrated.

Protective colloids. After gelatin has been added to colloidal gold, it requires a much greater concentration of electrolyte to precipitate the latter. Gelatin in this capacity is called a protective colloid, because it protects the other colloid from coagulation. Gelatin or albumen is added to ice cream as a protective colloid to prevent graininess. The protective colloid lactalbumin in human milk prevents curdling. Babies fed on cow's milk, which contains less lactalbumin than mother's milk, are often troubled by the milk's curdling. Barley water is then added to protect this milk from curdling. The egg added to mayonnaise as a protective colloid increases its stability.

Other important colloid problems. Photography is dependent upon colloids in several respects, most important of which is the sensitive emulsion. Strictly speaking, the so-called emulsion on a dry film is not an emulsion, which is defined as

EMULSIONS. All emulsions are colloidal in nature. Shown here are mayonnaise, oil-water emulsion, rosin-water emulsion, petroleum-agar emulsion, and milk.

a colloidal solution of a liquid dispersed in a liquid. On the film the sensitive layer consists of finely divided silver bromide in gelatin, which, being a protective colloid, keeps the silver particles from gathering into chunks. Being in the colloidal state, the silver salt can be quickly reached by the light.

Printing depends upon colloids to the extent that printer's ink is a colloidal substance. Rubber cement is colloidal rubber dispersed in benzene. The paste inside of the so-called dry cell is a colloidal mixture. In the manufacture of soap, the product exists largely in colloidal solution. To recover the soap, the solution is saturated with salt, which, being an electrolyte, neutralizes the charges on the colloidal soap particles, causing them to gather into solid soap. In washing, soap in its capacity as a colloid gathers on the surface of the rough particles of dirt, making them so smooth they can no longer cling to the fiber but wash away with the water.

The liquid gel. When a colloidal solution has set into a jelly, the nature of this gel is dependent somewhat upon how much of the dispersing liquid is present. Glue, library paste, or adobe soil shrinks as it dries out and swells again as the water is reabsorbed. The coagulation or gelling is caused by

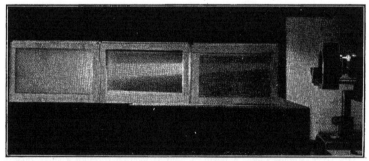

BEAM OF LIGHT IN A COLLOIDAL SOLUTION. Left to right, the three glass tanks contain a suspension of silver chloride in water, a colloidal solution of rosin in water, and clear water. The clear water does not intercept the beam of light from the lantern; the beam is plainly marked in the colloidal solution; the silver chloride precipitate scatters and absorbs the light. The tank of clear water remains dark because the water particles are too small to scatter the light.

the collection of the smaller units into larger units. If the process is reversed and the particles made to separate into smaller units, the process is called peptization. The larger units in the jellies are more or less stringy or threadlike. The interlocking of these shreds is what holds the jelly together. One per cent of agar-agar in hot water can cause it to set into a jelly, which shows that the colloid does not have to represent a large part of the final gel.

The ultramicroscope. Although we defined the colloidal state as consisting of particles too small to see in the highest powered microscope, they can be watched as points of light by means of an optical device called the *ultramicroscope*.

We can illustrate the principle of the ultramicroscope by allowing a beam of sunlight to shine into a darkened room. If we look across the beam from the side, we can see a large number of particles of dust dancing back and forth in the beam of light. With daylight in the room, the particles of dust cannot be seen, which shows they are too small to be seen with the unaided eye. In the beam of sunlight in the darkened room, however, they can be seen moving about. In fact, we do not see them there, but we see the splashes of light that hit them and are reflected in various directions. If we view a distant

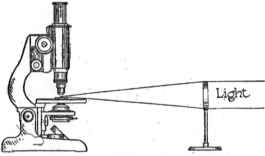

THE ULTRAMICROSCOPE. A beam of light is focused through a small quantity of a colloidal solution, which is viewed at right angles to the beam with a compound microscope. Particles too small to be seen directly are visible as patches of reflected light.

street lamp at night we see, not the lamp itself, but the sphere of light radiating from it. A parallel situation exists in the case of the dust particles, which are turning the sunlight in several directions. Although the particles are so small as to be invisible, the scattered spot of light is large enough to see.

In the ultramicroscope, also, a beam of light passes through the substance being examined. Instead of the unaided eye, a compound microscope is used in examining the side of the beam. Here again particles too small to be seen in the compound microscope are visible as spots of radiating light in the illuminating beam. The little spots of light representing the colloidal particles are seen to be in perpetual motion. This motion is called the Brownian movement after its discoverer, a botanist named Brown.

In the Brownian movement, the particles are perpetually in motion, like gnats dancing in the sunlight. They seem to move a short distance in a straight line; then they suddenly change direction and travel in a different straight line until there is another abrupt change. These sudden changes are supposed to be caused by head-on collisions between the visible particle and the invisible molecules of the suspending fluid.

THE BROWNIAN MOVEMENT. Colloidal particles viewed under the ultramicroscope are seen to be in perpetual motion due to collisions with the invisible molecules.

The Brownian movement is direct evidence of the existence and motion of molecules.

The extent of colloidal phenomena. Colloidal phenomena cover a very wide range of subjects. A solid may be dispersed as a colloid in a liquid, as clay soil floating in water, rosin suspended in water, colloidal silver, gold in water, milk of magnesia, and milk of lime. Liquids colloidally suspended in liquids (the emulsions) are quite common; among them are milk, oil sprays, mayonnaise, and many medicinal preparations. Fog and clouds are liquids dispersed in gases. Smoke and dusts are solids of colloidal size suspended in gases. Liquids dispersed in solids are not

ADOBE SOIL. Adobe soil is colloidal and shrinks as it dries out.

so common as other colloids, but moist clay and damp adobe soil are examples. Froths and foams are examples of colloidal gases in liquids. Solid particles dispersed in solids are represented by colored glasses and metallic alloys. The remaining possibility of gases dispersed in solids would seem to be the least probable in occurrence. However, soap that will float is really an illustration of this situation. Even white hair is thought to be white not because it contains a white pigment but because it has small bubbles of air dispersed throughout it.

QUESTIONS OF FACT

1. Define a protective colloid. Why are protective colloids added to ice cream?

2. Give one reason why human milk is superior to dairy milk for infants.

3. What is rubber cement?

4. Why is the so-called emulsion on a dry photographic film not really an emulsion?

5. Discuss water in relation to a gel.

6. What is the Brownian movement?

7. Give an illustration of each of the following colloidal particles in a dispersing medium: (a) solid in liquid; (b) liquid in liquid; (c) gas in liquid; (d) liquid in solid; (e) solid in solid; (f) gas in solid; (g) solid in gas; (h) liquid in gas.

8. Define the suspensoid colloids; the emulsoid colloids.

9. In which class of colloids is the effect of the electric charges less pronounced?

10. Name colloids that have to be very concentrated before they can be precipitated by electrolytes.

11. Is printer's ink colloidal?

12. What is peptization?

13. What percentage of agar can gel?

QUESTIONS OF UNDERSTANDING

1. Harmonize the statements that colloidal particles are invisible yet we can watch them in an ultramicroscope.

2. Give two illustrations of the use of a protective colloid.

3. Do you consider the constant movement of colloidal particles discovered by Brown sufficient to prove the existence of molecules?

4. Explain the working of the ultramicroscope.

5. Why would soap be less effective as a detergent if it were not colloidal?

REFERENCE FOR SUPPLEMENTARY READING

1. Findlay, Alexander. *Chemistry in the Service of Man.* Longmans, Green and Co., New York, 1931. Chap. XIII, pp. 264–282, "The Colloidal State."
    (1) State the analogy between a dissolved substance and a gas. (pp. 264–265)
    (2) Describe the Tyndall phenomenon. (pp. 267–268)
    (3) Describe the preparation of colloidal arsenious sulfide. (p. 270)
    (4) What is adsorption? (p. 270)
    (5) Name two substances of high adsorptive capacity. (pp. 270–271)
    (6) How do some colloidal particles acquire electric charges? (p. 271)
    (7) Explain the meaning of cataphoresis. (p. 272)
    (8) Give two ways of neutralizing the electric charges on colloidal particles. (p. 273)
    (9) How does filtration through soil purify sewage water? (p. 275)
    (10) How long does the Brownian movement of colloidal particles continue? (p. 280)
    (11) How did Z. Sigmondy describe the motion of colloidal gold particles? (p. 281)

SUMMARY

Any kind of substance may exist in the colloidal state — that is, be made up of particles too small to be seen directly in a compound microscope but still fairly large as compared with molecular dimensions. Many peculiar properties are dependent only upon the colloidal size of their constituent particles. A large percentage of the substances met in daily life and a large percentage of the industrial processes are best understood as colloids.

Colloids can be prepared by grinding, by the electric arc, by stirring, and by chemical reaction. Colloidal particles are electrically charged, which tends to keep them from collecting, as like charges repel. Coagulation of these colloidal particles can be fostered by neutralizing the charges with an electrolyte.

Some colloidal particles absorb some of the dispersing medium, which tends to modify the effect of the charges on the particles. The addition of small amounts of one colloid may make another more stable in its colloidal solution. Such a stabilizing substance is called a protective colloid. Protective colloids in milk tend to prevent curdling, and protective colloids in mayonnaise keep it stable. As the dispersing liquid evaporates, some colloids form a gel which will dry out, accompanied by shrinking. When the liquid is added again, it is absorbed, accompanied by swelling back to the original gel.

Colloidal particles in liquid can be watched under the ultramicroscope and be seen to be in perpetual motion due to collisions with the invisible molecules of the liquid. Substances in any phase — solid, liquid, or gas — may be dispersed as colloids in other media of almost all of the same three phases.

CHAPTER 35

⇄

CHEMISTRY AND TRANSPORTATION

Introduction. We live in an age of transportation and travel. People and goods are moved about at speeds beyond the imagination of people in the days when Jules Verne wrote his fantastic stories. Reality today has gone, in most cases,

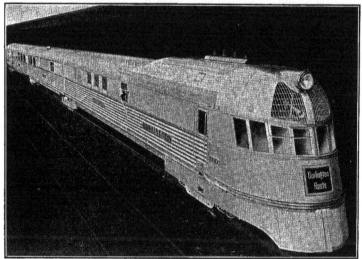

Courtesy of Burlington Railroad

STREAMLINE TRAIN. Chemistry has contributed much to the efficiency and economy of modern transportation.

beyond Jules Verne's imagination. In the early days of railroad transportation some person figured out theoretically that if a person were to travel at the rate of 40 miles per hour he would not be able to breathe. To anyone who has driven in a modern automobile or ridden in an airplane, these old ideas are a huge joke. Our modern streamline trains can go at 115 miles per hour, our racing automobiles up above 200 miles per hour on a straightaway, and airplanes have reached a speed at least 100 per cent greater than the fastest automobile. All this speed would have been impossible without many discoveries in chemistry.

The wagon. In the days of the wagon, chemistry was not called in directly to help improve it. It is true chemistry was used in producing the wrought iron and steel used in its metallic parts, but no study was given to make this iron especially durable or suited to this use. Likewise, the paint used for protecting the wood required the use of chemical processes in its manufacture. However, paint manufacture of the wagon days was largely carried on by rule-of-thumb methods without the guidance of the research chemist. The preparation of leather, too, is really a chemical process, yet it is only recently that chemists have been called in to study and direct the process of tanning leather, which has resulted in simpler processes and better leather. A little chemistry, also, was involved in producing the axle grease so necessary to wagon transportation. Although the problems of lubricating modern high-speed machinery are many times more complex than those involved in greasing a wagon, they are much more intelligently handled, because dozens of research chemists are studying the relative lubricating values of the different constituents of oils and greases, and methods of separating the most effective constituents from the complex mixtures in which they occur. This has resulted in lubrication specifically adapted to a given purpose. In the wagon days, on the other hand, available greases were used without much attempt to find better ones.

Compared to the mileage now obtained in the life of an automobile, that obtained from the wagon was insignificant. Those people old enough to remember the horse-and-buggy days will recall a collection of rattles and squeaks, very unlike the marvelous machine which glides swiftly and smoothly over our modern highways.

The horseless carriage. The early automobiles were called horseless carriages, which is almost what they were, since they had high wheels like carriages and otherwise resembled them. In their dependability, however, they were not equal to the horse and buggy, which usually got to the destination. Likely as not, something would go wrong with the automobile, which would require the services of a mechanic before it could proceed. Axles would break, bearings would burn out, wheels would get loose, rims would slip off, radiators would leak, gears would strip, valves would corrode, and frames would crack.

These are just a few of scores of things that went wrong with the early automobiles. Men of vision, however, were not discouraged by the difficulty of the task. They saw the great possibilities of the new type of transportation, which, unlike the horse, did not require feed and attention when not at work,

HORSELESS CARRIAGE. This is the way the automobile looked before the scientist took a hand in its manufacture.

which did not have to be hitched up and unhitched, which did not require a large storage space for hay, and which was free from barnyard muss and smell. It was seen that difficulties due to the greater mechanical complexity of the automobile must be overcome. This gave rise to the development of large research staffs, including chemists, who are continually trying to develop materials more enduring and more suitable to the purpose in mind.

The modern automobile. As compared with early automobiles and wagons, the modern automobile is a marvel of beauty, endurance, and efficiency. Twenty years ago a car was worn out when it had gone 15,000 miles. Today it does not even need the valves ground or new piston rings until it has gone 35,000 miles. It is not unreasonable to expect a car to be still in running order after it has covered 100,000 miles. These improvements have been the result of research, a large part of which has been chemical.

Alloys and the automobile. The life of the automobile is largely dependent upon the growth of our knowledge of alloys.

Charles

Goodyear

1800–1860

As a young man Charles Goodyear worked for his father in his hardware store and factory. Due to a too liberal policy of extending credit the Goodyears became bankrupt.

Noticing a rubber life preserver in a show window, Charles went in to examine it. Seeing that the valve of this article was rather crude, young Goodyear designed a better one and submitted it to the proprietor. The latter said it would be worth much more if he could perfect a process of curing rubber in such a way as to prevent it from sticking, melting, and decomposing in summer.

Goodyear now went home and began to experiment with rubber. He began by kneading all kinds of materials into the rubber. For a year he kept up these experiments without success.

Finally he patented a process for overcoming the stickiness of rubber by a surface application of metals.

Although destitute, he got the use of the machinery of a factory which had gone bankrupt trying to produce rubber articles. A man by the name of Hayward had also invented a method of overcoming the stickiness of rubber by mixing turpentine, sulfur, and oil with it. Hayward joined Goodyear, and they combined their processes.

One day while arguing with some men, Goodyear accidentally dropped some of the mixture of sulfur and rubber on a red-hot stove. This resulted in a nonsticky product which had all of the desired properties. He had discovered vulcanization.

After several years of experimentation he perfected the process but found himself $50,000 in debt. Because of his obligations, he was forced to sell licenses far below their worth; and although he created one of the great American industries, one which has made the automobile possible, he never was able to get out of debt.

Where well-known metals were inadequate for the work in hand, the lack was filled by new alloys. In fact the automobile is largely an assemblage of special kinds of alloys. Without them, it would still be the rattling junk heap it was thirty years ago.

The present cheap price of automobiles depends to a large extent on the development of rapid-working machinery. In the early days parts were shaped and threads were cut by ordinary steel. If the cutting were speeded up, the cutting tool lost its hardness and ceased to cut. At best, the life of the tool was short. Research developed tungsten-and-manganese steel,

Photo by Frederick King

A RUBBER TIRE WITH A RECORD. This tire has gone thirty thousand miles.

which can be used until it becomes red hot without loss of hardness. This means that dozens of automobile parts can be made now in the time formerly required to make one. Moreover, the life of the cutting tool is many times what it was formerly.

The rubber tires. Rubber tires were first used on carriages, but the pneumatic tire came primarily with the automobile. Less than two decades ago, one company, a little more daring than the others, guaranteed their automobile tires for 5000 miles. Today, if we do not get 15,000 miles out of our tires, we feel we are entitled to an adjustment. Twenty-five years ago an unused inner tube became cracked and worthless in three months' time.   Tires left hanging in the garage also soon became useless from oxidation.   For some unknown reason the tires in actual use on the car did not suffer so much from oxidation as did those that were idle.

Rubber is a most useful substance that has had an interesting history.   In one of his trips to the West Indies, Columbus noticed the natives playing with small elastic balls.   Columbus took some of the balls to Europe.   Later Priestley accidentally learned that the substance would erase pencil marks, so it was called rubber.   The natives obtained this rubber from the sap of a tropical tree.   Pizarro found the natives of South America putting rubber on shoes and cloth to make them water proof.   This substance was sticky when warm, which was a decided disadvantage.   Mackintosh placed the rubber between two pieces of cloth to lessen the stickiness.   This was the beginning of the mackintosh rain coat.

Courtesy of Goodyear Rubber Co

MAKING A RUBBER TIRE.   The illustration shows vulcanization and inspection.

Many different ways were tried to cure rubber of its stickiness when hot and to prevent its rapid deterioration.   Charles Goodyear, in 1839, discovered accidentally that the addition of sulfur and the application of heat change the properties of rubber and make it permanently elastic, and keep it from being sticky when hot.   This process (vulcanization) solved many problems in connection with the manufacture of rubber.

The improvements in rubber tires since 1915 have come in two ways.   In the first place it was learned that certain compounds, such as sulfur chloride, will vulcanize the rubber at lower temperatures than will sulfur, and make a better product.

The discovery of various accelerators of vulcanization still further improved the product. Oxidation has been mentioned as a defect of early rubber articles. The introduction of antioxidants — negative catalysts, which prevent oxidation — was the second great step forward in the improvement of tires. The most effective accelerators and antioxidants are usually complex organic compounds. They have tripled the life of nearly all rubber products. The wearing quality of automobile tires is really remarkable. A rubber tire on an automobile will go farther in its life time than an iron tire three-fourths inch thick would go on a wagon wheel.

REVIEW EXERCISES

1. In the horse-and-buggy days, was chemical research employed to make metals and paints especially suitable for the wagon or did they just use the best products available at that time?

2. How has the situation suggested in ex. 1 changed with reference to improving the automobile?

3. Could the automobile be the marvel of efficiency it is without the aid of hundreds of research men, chemists, and engineers?

4. Name a few things that often went wrong with the first automobiles.

5. Compare the mileage obtained in the lifetime of the modern automobile with that obtained from the automobile of twenty years ago.

6. What part of the automobile uses the following: (a) brass; (b) iron, tungsten, and chromium; (c) iron and nickel; (d) aluminum and manganese; (e) iron, chromium, and vanadium; (f) iron, manganese, and carbon; (g) nickel, manganese, and silicon? (p. 572)

7. Account for the name "rubber."

8. What are the two principal objections to raw rubber?

9. How was rubber used by the natives of South America?

10. How did Mackintosh avoid the stickiness of natural rubber in rainproof clothing?

11. What discovery by Goodyear overcame entirely the stickiness of rubber?

12. What does vulcanizing consist of?

13. What was the expected mileage of an automobile tire in 1915?

14. What is it now?

15. What two discoveries have produced the change?

16. What is meant by an accelerator of vulcanization?

17. What causes old rubber to crack and get brittle?

18. How is this now greatly eliminated?

19. An antioxidant is said to be a negative catalyst. What does this mean?

REFERENCE FOR SUPPLEMENTARY READING

1. Howe, H. E. *Chemistry in Industry*, Vol. II. The Chemical Foundation, Inc., New York, 1926. Chap. XII, pp. 204–220, "Lubricants."

(1) Describe the appearance of smooth surfaces under the microscope. (p. 204)

(2) What does a lubricant do when applied to surfaces? (pp. 204–205)

(3) What happens if the lubricant is too thin? (p. 205)

(4) What is the objection to too thick a lubricant? (p. 205)

(5) Discuss intermittent lubrication. (pp. 208–209)

(6) Give five different methods of lubrication and discuss them. (pp. 210–212)

(7) Name some laboratory tests on petroleum lubricants and give reasons for the tests. (pp. 212–214)

(8) Discuss the problems of automobile lubrication. (pp. 217–218)

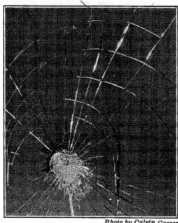

Photo by Calvin Coover

SHATTERPROOF GLASS. This is what happens when a shatterproof windshield is struck with a rock.

Shatterproof glass. Another improvement of the automobile that means much to safety is shatterproof glass. Two pieces of glass are cemented together by a transparent synthetic organic substance. When the glass is struck, it cracks but does not fly, since the plastic substance holds it. In the near future windshields are likely to be made of transparent synthetic materials, much more elastic and safer than glass.

Automobile lacquers. The early automobiles were given an enamel finish. Enamel is a baked varnish. The baking of the enamel required both time and large ovens. Present-day cars are finished with cellulose-nitrate lacquers. The cellulose nitrate is dissolved in a volatile organic solvent along with a dye. The mixture is applied to the car with a spray gun. The volatile solvent evaporates, leaving the dye and the cellulose nitrate as a continuous tough film, which is several times more durable than the old enamels. The time of finishing has been reduced from days to hours, and the baking ovens have been abandoned.

Automotive fuels. With the coming of the internal combustion engine, the demand for gasoline changed the entire nature of oil refining. Before this time the demand was principally

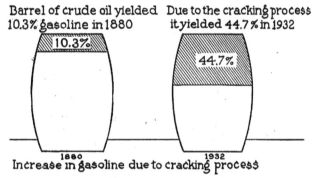

Barrel of crude oil yielded 10.3% gasoline in 1880 Due to the cracking process it yielded 44.7% in 1932

Increase in gasoline due to cracking process

CRACKING INCREASES THE YIELD OF GASOLINE. When heated under high pressure, the large oil molecules are broken up, and the percentage of gasoline is greatly increased.

for kerosene for lighting lamps and stoves. The demand for gasoline was so slight that gasoline was said to have been wasted because there was no market for it. The automobile engine made the demand for gasoline greater than that for kerosene. It looked for a while as if the demand would increase to the point where it would be impossible to get enough gasoline.

At this point the chemist came to the rescue by inventing the process of cracking of oils in refining petroleum. Cracking consists in heating the oil at great pressures. The larger molecules break up into smaller molecules, and much larger quantities of gasoline are obtained from the crude oil. This increased the supply about 50 per cent.

Cracking has also improved the quality of gasoline. Its antiknock properties have been greatly increased, so more power and mileage can be obtained from it.

Antiknock gasoline. The antiknock properties of gasoline have been increased by the addition of tetraethyl lead. Knocking jars the walls of the automobile cylinders by the explosion instead of steadily pushing the piston back. Knocking represents the wasting of a large percentage of the energy. The wasting of energy is not the worst thing connected with knocking; it injures the car. Engineers calculated that they could design automobile engines with much greater power and pickup if they could avoid knocking. They planned to get more power by increasing the compression ratio; in other words, they would compress the mixture of gasoline vapor and air more before exploding it. At the time of the development of tetraethyl lead, the compression ratio commonly used was 4.5 to 1; that is, when the piston was entirely back, the volume of the gases was 4.5 times the volume when it was fired by the spark. With the coming of tetraethyl lead, the compression ratios of practically all automobile engines were increased to 5.5 to 1 and greater.

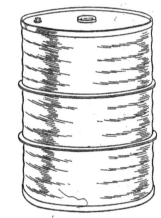

RATIO OF TETRAETHYL LEAD TO GASOLINE. A little goes a long way.

It was discovered that iodine would prevent knocking. Iodine, however, is very expensive and corrodes the cylinder walls. These factors make it unsuitable for knock prevention. Having found one substance that prevents knocking, investigators began a search for other substances which would do the same. Of all the things tried for this purpose, tetraethyl lead, $Pb(C_2H_5)_4$, was the most effective. However, at the time the antiknock property of tetraethyl lead was discovered, this substance was just a laboratory curiosity and very expensive. Intensive research soon developed cheaper ways of preparing the new knock preventer, and it became a commercial product. Now tetraethyl lead is added to many of the better grades of gasoline.

The antiknock property of gasoline is rated in terms of the hydrocarbon isooctane, which possesses high antiknock qualities. A gasoline of rating 80 is a gasoline with antiknock properties equal to a mixture of 80 per cent isooctane. The power obtained from gasoline increases rapidly with the octane rating. There is from 20–30 per cent increase in power of gasoline of octane number 100 over that of octane number 87.

The chemist has invented ways of refining lubricating oil. Nowadays oil is selected to fit the conditions under which it is to be used. The problem of lubrication is much more difficult for an engine driving a car at 60 miles per hour than for one driving a car at 30 miles per hour. At the latter speed almost any oil will do, but at the former speed, only the correct oil must be used.

Railway transportation. All along in the development of railway transportation, the need of better metals, paints, lacquers, and lubricants has paralleled that of the automobile. In addition to these, there was the water problem, which took two forms: (1) that of keeping boiler scale out of the boiler; and (2) that of supplying good drinking water. With one-eighth inch of boiler scale, 15 per cent more heat is required to run the engine. Every pound of scale kept out of the engine saves the company 13 cents a day. This is a big factor when it is realized that an engine pulling a passenger train evaporates 50,000 gallons of water in going 600 miles. The chemist has succeeded in softening the water and in removing the dissolved oxygen which tends to corrode the iron in the boiler. The drinking water supplied in stations and trains has been filtered and chlorinated.

Air transportation. The airplane demands some alloys not emphasized by the automobile. Among these are the light alloys magnalium and duralumin.

One of the principal problems in connection with the airplane engine has been to decrease the ratio of weight to horsepower. The first airplane engine weighed 117 pounds per horsepower. Recently, the weight has been brought down to 1.75 pounds per horsepower.

The dirigible balloon needs strong fabrics and a noninflammable gas in addition to an efficient engine and a strong framework. The strongest gas bag to date is a three-ply cotton cloth cemented with rubber and vulcanized. The development of noninflammable helium has been mentioned elsewhere.

The radio. When we think of radio in connection with transportation, we think of riding leisurely down the highway in our automobile listening to our favorite program. Enjoyable as this is, it is not the most important service of radio to transportation. The S. O. S. bringing help to the disabled ship at sea, and the radio beam guiding the airplane through a fog are far more important. The chemist has supplied the pure copper for radio circuits, the pure metal for the elements of the radio tubes, and the filaments that evaporate such a rich supply of electrons. The storage battery, another chemical achievement, fills an important niche in the development of the automobile and the radio.

REVIEW EXERCISES

1. Describe shatterproof glass.

2. What were the automobile enamels?

3. What are the lacquers and how are they applied?

4. To what extent have quick-drying lacquers shortened the time necessary to produce an automobile?

5. Give the formula of tetraethyl lead.

6. Explain the meaning of compression ratio.

7. Increase of compression ratio has what effect on the efficiency of an automobile engine?

8. What difficulty arose as the compression ratio was increased?

9. What substance did research produce for the purpose of preventing knock?

10. In terms of what hydrocarbon are the antiknock properties of gasoline rated?

11. Name some of the chemical problems in connection with railway transportation.

12. What is the daily cost of one pound of boiler scale in a locomotive engine?

13. Name some of the problems peculiar to air transportation.

14. What has chemistry done for radio?

PROBLEMS

1. Calculate the molecular weight of tetraethyl lead.

2. Calculate the percentage of each element in tetraethyl lead.

3. What weight of bromine is needed to form 100 g. of 1–2 dibromoethane according to the following reaction?

$$C_2H_4 + Br_2 \longrightarrow C_2H_4Br_2$$

4. What volume of ethylene measured under standard conditions was needed in problem 3?

5. What volume would the ethylene obtained in problem 4 have if measured at 27° C. and 780 mm. of mercury?

References for Supplementary Reading

1. Howe, H. E. *Chemistry in Industry.* Vol. II. The Chemical Foundation, Inc., New York, 1926. Chap. XVII, pp. 292–304, "Railroad Chemistry."

    (1) For what purpose did the Pennsylvania Railroad establish chemical testing laboratories in 1875? (p. 293)

    (2) Name the kinds of metals which go into a locomotive. (p. 294)

    (3) What tests are applied to paints to be purchased for painting locomotives? (pp. 294–295)

    (4) What is the largest single problem of a railroad? (p. 295)

    (5) Name the impurities in natural waters. (p. 295)

    (6) How do boiler compounds prevent boiler scale? (p. 297)

    (7) Name two problems connected with increasing the life of locomotive engines. (pp. 297–299)

    (8) Discuss the problem of lubrication as applied to railroading. (p. 300)

    (9) What is the most expensive single item in running a train? (p. 301)

    (10) What part does the chemist have in promoting railroad safety? (pp. 302–304)

2. Howe, H. E. *Chemistry in Industry,* Vol. II. The Chemical Foundation, Inc., New York, 1926. Chap. II, pp. 15–35, "The Chemist's Contributions to Aviation."

    (1) When did the Wright brothers make the first airplane flight of any distance? (p. 15)

(2) What is the cost of hydrogen gas per thousand cubic feet? (p. 17)

(3) What advantage does cellulose acetate dope have over cellulose nitrate dope for airplane use? (pp. 21–22)

(4) In what connection do airplanes need luminous paints? (pp. 27–29)

(5) Discuss glues in connection with the airplane. (pp. 29–30)

(6) What is the cost of helium per thousand cubic feet? (p. 34)

Summary

The introduction of new alloys has made the modern automobile possible. Its strength, its durability, and its beauty are directly traceable to these combinations of metals. Moreover, the use of alloys in high-speed cutting tools has accelerated automobile manufacture and reduced the cost. As a result of scientific research, the modern automobile is a vast improvement over the early cars. Before a modern car is finally worn out, it will have run at least 100,000 miles as against 15,000 miles for the first models. There has been a corresponding decrease in the expense for repairs and the replacement of parts. Better vulcanization and the use of accelerators and antioxidants have greatly improved rubber, so that tires now wear much longer than formerly. Shatterproof glass, synthetic lacquers, and antiknock fuel are scientific contributions that have helped the automobile industry. The discovery that the addition of tetraethyl lead to gasoline will reduce engine knock, and striking improvements in the refining of lubricating oils are achievements of the industrial chemist. For the future, he has promised us a satisfactory synthetic gasoline when our oil wells give out. Among the chemist's other gifts to transportation are strong, tight fabrics and helium, a noninflammable gas, for dirigibles, and tough, light alloys for airplanes. Finally, his contributions to radio include batteries, wires, and efficient vacuum tubes.

CHAPTER 36

CHEMISTRY AND ENERGY

What is energy? Energy includes a variety of phenomena. Some of its varieties are mechanical motion, heat, light, electricity, sound, chemical energy, energy of position, and energy of state. We are well aware that a moving object may do a variety of things all included in the mechanical definition of work. Work in the mechanical sense is the displacement of a force through a distance. The moving drive shaft of an engine, a moving automobile, or a moving football player can all displace a force that may oppose them. Heat also by causing a gas or a solid to expand can do mechanical work, as is seen in the moving piston of an engine.

RADIOMETER. The radiometer is a windmill-like device, with veins silvered on one side and black on the other. Light (or other forms of radiant energy) is reflected from the silvery surface but absorbed by the black one. Thus the black side warms up and heats the adjacent air, which expands and causes the radiometer to rotate.

It is not quite so easy to see how light in itself can do work, but even this is possible by means of the little light wheel called a radiometer. Indirectly light may do work by heating up some gas, which can expand and do the work.

Gilbert Newton Lewis

1875–

The advanced study of the energy relations of chemistry is known as thermodynamics. This study has supplied industry with one of its most valuable tools in that it enables chemists to predict whether a reaction will or will not take place under a given set of conditions and to determine the conditions under which it will take place most efficiently. Professor Lewis began studying thermodynamics at a time when it was partly worked out, but still was so indefinite in other parts that scientists could not fully use what was known. By adding the third law of thermodynamics and introducing the idea of "free energy," Professor Lewis made the knowledge clear-cut and workable. He is co-author of a textbook on the subject, a book widely used in the universities of America. Having shown how thermodynamic data about chemical substances is of value to industry, his pupils and others have done a great deal of valuable research in determining this data.

Professor Lewis is equally well known because of his atomic-structure theory, a modification of the Bohr atom.

Some of the other fields of Professor Lewis' research are as follows: equilibrium in numerous reactions; electrode potentials of the common elements; properties of solutions and activities of ions; and rates of chemical reactions.

At the present time Dr. Lewis is professor of chemistry and dean of the College of Chemistry at the University of California. The following honors have been given Professor Lewis: the Davy medal of the Royal Society of Great Britain in 1929; the Willard Gibbs medal of the Chicago section of the American Chemical Society; the Distinguished Service medal for war service; and Chevalier, Legion d'honneur. In the World War he was a major, lieutenant colonel, and chief of the defense division of the gas service.

Electricity by means of the electric motor is a very common way of performing useful work.

Sound ordinarily is not used to produce work. However, sound can easily be used to put a string or a diaphragm into motion. This happens every time we talk into the telephone. Putting the diaphragm of the telephone transmitter into motion is work as truly as the moving of larger objects.

Chemical energy means the energy that may be obtained from a chemical substance by causing it to undergo some chemical change. For instance, we can get a great deal of work by burning gasoline.

Before the gasoline is burned, its energy is perfectly quiescent, yet it is real, just waiting for the right mechanism to release it.

Another form of inactive energy is the energy of position. A lake of water on the top of a mountain can be led into a power plant and be made to do a great deal of work. A lake of equal size at sea level could not be made to do a similar amount of work.

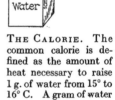

THE CALORIE. The common calorie is defined as the amount of heat necessary to raise 1 g. of water from 15° to 16° C. A gram of water has a volume of 1 cc.

The lake on the mountain had energy of position given to it when the winds carried the water up there from the ocean. A rock on the hillside has energy that a similar rock at the bottom does not have. This energy does not show itself in any way until the rock is loosened. Then it does work against the things it strikes while rolling down hill. This energy of position is called potential energy. Another form of inactive energy is the excess energy of the liquid state of a substance over the solid state, or the much greater excess energy of the gaseous state over the liquid state. As we learned in Chapter 3, if 1 g. of ice at 0° C. is melted, it requires 80 units of heat, yet the thermometer still reads zero. This 80 units represents the heat that the liquid state of water has in excess of the solid state at the same temperature. It takes 540 units of heat to vaporize 1 g. of water at 100° C. The gaseous state of water,

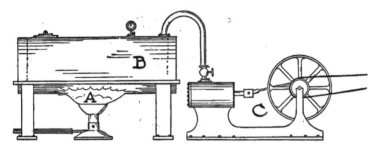

TRANSFORMATION OF ENERGY.   Any form of energy may be transformed into any
other form.   Combustion changes the chemical energy of the gas (A) into heat
energy in the boiler (B).   The heat energy becomes energy of motion in the

then, represents a much larger amount of energy than the liquid
state.   However, until the water passes from the gaseous state
to the liquid state, this energy is hidden or inactive.

The only definition for energy that will include all of these
various forms, quiescent, active, and potential, is the following :
Energy is that which can do work.

The law of conservation of energy.   One of the fundamental
laws of nature is : *Energy can neither be created nor destroyed.*
This law is called the *law of conservation of energy.*   Matter
and energy were long thought to be the two fundamental
entities of the universe.   Matter also was thought to be indestructible and uncreatable, hence the *law of conservation of
matter.*   As was mentioned in a previous chapter, it is now
believed that under extreme conditions matter may be changed
into energy, and somewhere in the universe the reverse change
of energy into matter may be taking place, but ordinarily we
do not have to consider these two possibilities.

One form of energy may be changed into other forms.
Chemical energy is changed into heat energy by burning the
fuel.   Heat energy is changed into mechanical energy by means
of the steam engine.   If the engine turns a dynamo, the
mechanical energy is largely changed into electrical energy.
The electrical energy can now be changed into sound energy

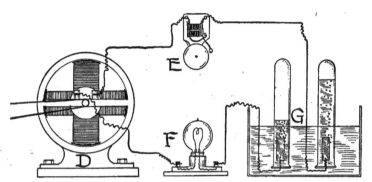

cylinder of the engine (C). The dynamo (D) transforms this energy into electrical energy, which becomes sound energy in the bell (E), light energy in the bulb (F), and again chemical energy in the electrolytic cell (G).

by means of an electric bell, into light energy in the light globe, into heat energy in the toaster, into chemical energy by charging a battery, and into mechanical energy by the motor. Some of the energy is frittered away in friction, heating the wires, and other losses; but none is destroyed.

Importance of energy. Energy is just as important as matter. Each generation is spending more than the previous one for energy. We buy chemical energy in fuels to heat our homes, to run our automobiles, and to cook with. We buy electrical energy to furnish light, heat, sound, and to cool our food in electrical refrigerators. We buy energy foods to supply our body heat and enable us to work. In the study of chemistry, heat and electrical energy are the most important forms. We may make firecrackers for the noise or colored fires for light, but heat and electricity are the factors that must be considered in connection with all chemical reactions.

Exothermic and endothermic reactions. Every chemical reaction falls into one of two classes. Either it will turn loose a given amount of heat when the reaction takes place, or it must be furnished a certain amount of heat before it will take place. If the reaction gives out heat, it is said to be an *exothermic reaction*, but if it must be given heat, it is called an *endothermic reaction*.

QUESTIONS OF FACT

1. What is the mechanical definition of work?

2. Name eight different forms of energy and point out how each can do work.

3. Name three forms of inactive energy, and tell how they may be made active so they can do work.

4. Define energy so it will include all of the phenomena included under the name.

5. Compare the energy content of a substance in the liquid state with that of the same substance in the solid state. In the case of one gram of water, how many calories' difference is there?

6. Compare the energy content of a substance in the gaseous state with that of the same substance in the liquid state. In the case of water and steam at 100° C. how many calories' difference is there per gram?

7. State the law of conservation of energy.

8. What new extension may be necessary to the laws of conservation of matter and energy?

9. Tell how the following energy transformations can be brought about: (a) chemical energy into heat energy; (b) mechanical energy into electrical energy; (c) electrical energy into sound energy; (d) electrical energy into light energy; (e) electrical energy into heat energy; (f) electrical energy into chemical energy; (g) electrical energy into energy of motion.

10. Discuss the importance of energy.

11. Compare exothermic and endothermic reactions.

REFERENCE FOR SUPPLEMENTARY READING

1. Howe, Harrison E. *Chemistry in the World's Work.* D. Van Nostrand Company, New York, 1926. Chap. XI, pp. 162–178, "Power."

   (1) What gas will not appear in the flue gases of a perfect burner? (p. 163)

   (2) How can it be determined when a burner is working efficiently? (p. 163)

   (3) How is the percentage of carbon dioxide in a gas measured? (p. 163)

   (4) What can the chemist learn about coal? (p. 163)

   (5) What percentage of the heat value of coal is utilized on the open grate? (p. 164)

   (6) Describe the invention of fuel gas. (p. 165)

   (7) What has made anthracite coal popular? (p. 167)

   (8) How is the ash content of coal reduced? (p. 169)

Heat of formation. The unit of heat usually used is the *calorie*, which was defined in Chapter 3 as that quantity of heat necessary to raise the temperature of 1 g. of water 1° C.

The *heat of formation* of a compound is the number of calories liberated or absorbed when one *mole* of the compound is formed from its elements. The mole is sometimes called the *gram molecular weight*, which, as we learned in a previous chapter, means one gram for each unit in its molecular weight.

$$C + O_2 \longrightarrow CO_2 + 96,900 \text{ cal.}$$
$$12 \text{ g.} + 32 \text{ g.} \longrightarrow 44 \text{ g.}$$

Hence 96,900 cal. is the heat of formation of CO_2.

We see now why some reactions are endothermic. Looking at the reaction above, we see that when 12 g. of carbon burn

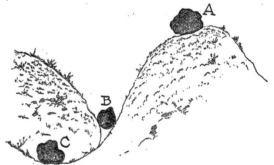

ENERGY CONTENT AND STABILITY. Rock A contains a great deal of energy of position; rock B, which has rolled part way down the hill, contains less; and rock C, at the bottom, contains none at all. In rolling down the hill, rocks B and C imparted their energy to their surroundings. Rock A corresponds to an unstable compound, rock B to a fairly stable one, and rock C to a very stable one.

in 32 g. of oxygen to form 44 g. of carbon dioxide, 96,900 cal. of heat are released and disposed of in some way. Now suppose we wish to decompose the 44 g. of CO_2. We simply have to supply the missing 96,900 cal., or the reaction does not take place. A reaction that is exothermic in one direction is endothermic in the reverse direction; with an equal caloric value.

Heat of formation as a measure of stability. Let us think of several similar rocks on top of a hill. Suppose they all had equal energy of position. If one rock rolls down to the bottom

of the hill, it gives up all of its energy of position. It is now in
its most stable position. If another rock rolls two-thirds the
way down the hill, it loses two-thirds of its energy of position;
but it is not so stable as the other rock, since it can still go
further down hill. Another way of stating these facts is to say
that the more energy released in going down hill, the more
stable the resultant position.

A parallel situation exists in regard to the stability of chemical compounds. In a series of compounds of some element,
the one having the greatest heat of formation is the most
stable, and the one with the smallest heat of formation is the
least stable.

The following series illustrates this point :

HF	HCl	HBr	HI
38,500 cal.	22,000 cal.	8600 cal.	− 6400 cal.

In this series HF is the most stable, then HCl, then HBr, and
finally HI. Hydrogen iodide, as will be noticed, has negative
heat of formation. This means that heat has to be supplied
to make it out of its elements. The uncombined elements
represent a more stable condition than when combined into
the compound. It would naturally be expected that this substance would not keep well, but would decompose spontaneously to the elements. The facts are just as suggested. Hy-
driodic acid is hard to keep as a laboratory reagent. It
almost always looks dark because of the presence of free iodine,
the result of decomposition.

Let us take another illustration of the relationship between
the stability of a substance and its heat of formation :

H_2O 69,000 cal. H_2O_2 46,800 cal.

The figures indicate that hydrogen peroxide is less stable than
water. Such is the fact, as experiment shows. Peroxide
readily loses one of its oxygen atoms to become water.

$$H_2O_2 = H_2O + O$$

Predicting chemical reactions. Heat of formation can be
used to predict whether certain reactions will or will not take
place, on the theory that if the reaction will result in a more

stable situation, it will take place of itself, and if a possible reaction would result in substances of less heat of reaction, it will not take place.   Will the following reaction take place?

$$Cl_2 + 2\ NaBr \longrightarrow 2\ NaCl + Br_2$$
$$86,100 \qquad\qquad 97,900$$

Sodium chloride has the greater heat of formation; hence the right side represents a more stable condition and the reaction will take place spontaneously as follows:

$$Cl_2 + 2\ NaBr \longrightarrow 2\ NaCl + Br_2 + 2(97,900 - 86,100)\ cal.$$

Will the following possible reaction take place?

$$Mg + CaCl_2 \longrightarrow MgCl_2 + Ca$$
$$190,400 \qquad 151,010$$

Since $MgCl_2$ has the lower heat of formation, it represents a lower condition of stability and the reaction will not take place without having energy supplied.

HEATS OF FORMATION

SUBSTANCE	HEAT OF FORMATION	SUBSTANCE	HEAT OF FORMATION
KCl	105,610	HgCl	31,300
KBr	95,310	HgBr	24,500
KI	80,130	HgI	14,300
NaCl	97,900	ZnCl$_2$	97,400
NaBr	86,100	ZnBr$_2$	76,000
NaI	69,080	ZnI$_2$	49,231
CaCl$_2$	190,400	CuSO$_4$	181,700
CaBr$_2$	154,920	ZnSO$_4$	229,600
CaI$_2$	127,400	AgSO$_4$	96,200
CuF$_2$	89,600	FeSO$_4$	234,900
CuCl$_2$	51,400	HgSO$_4$	165,100
CuBr$_2$	32,600		

What we learn from heats of formation.   In the first place, an energy change is used as proof of chemical reaction under conditions where reaction could not be proved otherwise. The liberation of heat on mixing two substances is proof that there has been a chemical reaction.

In the neutralization of equivalent amounts of any acid with any base, it was found that the same heat of neutralization always resulted.   Heat of neutralization means the number of calories liberated when enough acid to contain 1 g. of neutralizable hydrogen is neutralized by enough base to contain 17 g. of hydroxyl, OH.

$$\overset{+}{H}\overset{-}{Cl} + \overset{+}{Na}\overset{-}{OH} \longrightarrow \overset{+}{Na}\overset{-}{Cl} + H_2O + 13,700 \text{ cal.}$$

$$\overset{+}{H}\overset{-}{Br} + \overset{+}{K}\overset{-}{OH} \longrightarrow \overset{+}{K}\overset{-}{Br} + H_2O + 13,700 \text{ cal.}$$

$$\overset{+}{H}\overset{-}{NO_3} + \overset{+}{K}\overset{-}{OH} \longrightarrow \overset{+}{K}\overset{-}{NO_3} + H_2O + 13,700 \text{ cal.}$$

$$\overset{+}{H}\overset{-}{NO^3} + \overset{+}{Na}\overset{-}{OH} \longrightarrow \overset{+}{Na}NO_3 + H_2O + 13,700 \text{ cal.}$$

The conclusion to be drawn from the fact that such different combinations of acids and bases have the same heat of neutralization is that the same reaction takes place in each case.

The reaction is $\overset{+}{H} + \overset{-}{OH}$ yields H_2O in each case.

Electricity instead of heat. In the case of many chemical reactions it is possible to arrange the reaction in such a way as to yield the energy of the reaction as electricity instead of heat. All electrical cells are the practical use of the energy of reactions as electricity. Some cells make use of replacement reactions; and others, oxidation-reduction reactions.

Energy may start chemical reactions. Many chemical reactions need energy to start them. When dynamite explodes,

CHEMICAL REACTION CAUSED BY A BLOW. Many mixtures explode when struck.

it undergoes a chemical change. The powerful shock of the percussion cap is needed to start this reaction. A mixture of potassium chlorate and sulfur can be made to explode by striking it with a hammer. This experiment is rather dangerous, and one should risk striking only a small pinch of the mixture with the hammer. Serious accidents have happened by using larger amounts. There is a record of a chemist being killed by grinding sulfur and potassium chlorate together. Heat is the type of energy most frequently used to start chemical reactions. Every time one starts a fire, he uses heat to begin the reaction. Even light can start a reaction. Flashlight bulbs are often set off by electricity being led through one of them. The rest of them are set off by the light of the one.

Heat energy speeds up a reaction. Most chemical reactions are speeded up with heat. This is the most common method of hurrying such reactions. Many of the modern chemical miracles are carried out at temperatures up to 500°–700° C., temperatures considered impracticable a few years ago.

Producing high temperatures. Temperature is an expression of the intensity of energy. We have three general methods of producing high temperatures : (1) the resistance to heavy electric currents ; (2) the impact of high-speed particles ; and (3) chemical reaction. In the first class, we have different types of electric furnaces and light globes. The second method is illus-

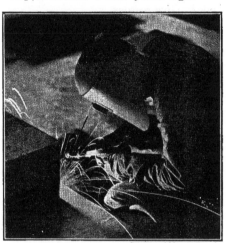

Courtesy of Morgan Eng Co.

ATOMIC HYDROGEN ELECTRIC ARC.

trated by the electric arc in the arc light. Of the third type, we have the oxyhydrogen blowpipe and the Goldschmidt process as described on page 250.

Recently Dr. Langmuir has invented a modification of the oxyhydrogen blowpipe which yields an exceedingly high temperature. Dr. Langmuir's invention uses atomic hydrogen as the fuel instead of molecular or gaseous hydrogen. Gaseous hydrogen in which the atoms have already reacted to form molecules of H—H has already given up some of its heat of reaction. When gaseous hydrogen burns, some of the possible heat of reaction must be consumed in breaking these molecules in two before they can react with the oxygen. In the method of Dr. Langmuir, an electric spark jumps across the oxyhydrogen flame. The energy of the electric current decomposes the hydrogen molecules, so the oxygen does not have to use some of its heat of burning to do this. The result is that this method gives a more intense flame.

REVIEW EXERCISES

1. Would the decomposition of CO_2 be exothermic or endothermic? Give reasons.

2. Define the term "heat of formation."

3. What would it mean if a compound had a negative heat of formation?

4. If a reaction is exothermic, what is the nature of the reverse reaction?

5. In what sense is the loss of energy in rolling down hill a measure of the stability of a rock?

6. What do the heats of formation tell us about the relative stability of similar compounds? Illustrate with the hydrogen halides.

7. What is the rule which determines whether a reaction will take place spontaneously when started or whether it must receive energy during the entire reaction?

8. By studying the table of heats of formation of the different substances involved, decide which of the following reactions (a) take place readily, (b) take place slowly, (c) are impossible:

(1) $Zn + Ag_2SO_4 \longrightarrow ZnSO_4 + 2Ag$

(2) $Br_2 + CaCl_2 \longrightarrow CaBr_2 + Cl_2$

(3) $2 NaI + Br_2 \longrightarrow 2 NaBr + I_2$

(4) $I_2 + 2 KCl \longrightarrow 2 KI + Cl_2$

(5) $Fe + CuSO_4 \longrightarrow FeSO_4 + Cu$

(6) $CaBr_2 + Cl_2 \longrightarrow CaCl_2 + Br_2$

(7) $CaBr_2 + I_2 \longrightarrow CaI_2 + Br_2$

(8) $CuBr_2 + Ca \longrightarrow CaBr_2 + Cu$

(9) $CaCl_2 + Zn \longrightarrow ZnCl_2 + Ca$

(10) $Fe + HgSO_4 \longrightarrow FeSO_4 + Hg$

9. Is the liberation of energy when two substances are mixed proof of a chemical reaction?

10. What proof is there that all neutralization reactions are simply $\overset{+}{H} + \overset{-}{OH} \longrightarrow H_2O$? Explain in detail.

11. Give several illustrations of starting chemical reactions with different kinds of energy.

12. What is the common method of speeding up many chemical processes?

13. Give three ways of producing high temperatures.
14. Describe and explain the atomic hydrogen arc.

REFERENCES FOR SUPPLEMENTARY READING

1. Darrow, Floyd L. *The Story of Chemistry*. Blue Ribbon Books, Inc., New York, 1930. Chap. IV, pp. 152–194, "Chemistry and Power."
   (1) What are our tangible power assets? (p. 155)
   (2) Compare the energy that reaches the United States in one minute with our needs for a year. (p. 155)
   (3) What is our estimated water power? (p. 156)
   (4) How much coal is there supposed to be in the United States? (p. 158)
   (5) How much petroleum have we produced up to 1926 in the United States? (pp. 170–174)
   (6) Discuss oil shale in the United States. (pp. 170–174)
   (7) What is obtained from the hydrogenation of one ton of bituminous coal? (p. 180)
   (8) Discuss low-temperature carbonization of coal. (pp. 189–191)
2. Holmes, Harry N. *Out of the Test Tube*. Ray Long and Richard R. Smith, New York, 1934. Chaps. XXVII and XXVIII, pp. 289–304, "Chaining the Sun.'
   (1) Examine the formula of the green chlorophyll of plants, and state what elements are in it. Is chlorophyll a complicated substance? (p. 290)
   (2) How many stomata does an average leaf have? What do these stomata do? (p. 290)
   (3) Why are chemists studying photosynthesis? (p. 291)
   (4) How much energy falls on the Sahara Desert? (p. 291)
   (5) The sun's energy that falls upon 1 acre of ground in 90 days is equivalent to how much coal? (p. 292)
   (6) Describe the photo cell. (p. 293)

SUMMARY

Energy is that which can do work. Energy exists in various forms, such as electricity, sound, heat, and energy of motion, which can be converted into other forms but cannot be destroyed.

Chemical reactions which give out heat are said to be exothermic, and those which absorb heat are called endothermic. The heat given out when one mole of a substance is formed is called the heat of formation. The heat of formation of a substance is a measure of its stability, the most stable substances being those with largest heats of formation. A substance with a negative heat of formation tends to decompose spontaneously. From the heats of formation, it is possible to predict when reactions will take place without the addition of energy. If a possible reaction will result in liberating large amounts of heat, when once started, the reaction goes readily and usually with a speed in proportion to the heat liberated. On the other hand, a reaction that uses up heat will not take place of itself. The neutralization of equivalent amounts of any acid with any base results in the same quantity of heat, which shows that actually the same reaction has taken place regardless of kind of acid or base. This reaction is $\overset{+}{H} + \overset{-}{OH} \longrightarrow H_2O$.

Small amounts of energy may be needed to start a reaction even if it does liberate much heat. Jarring, heat, and light are used to start reactions. Heat also tends to speed up chemical reactions. One use of chemical action is to produce high temperatures. The Goldschmidt process is a good illustration of this.

←
→

CHAPTER 37

CHEMISTRY IN RELATION TO AGRICULTURE

Introduction. Agriculture has depended upon chemistry for a long time to supply fertilizer for the soil. Only recently, however, has the agriculturist come to realize what the chemist can do for him in the line of making useful and salable products out of agricultural wastes and in improving the quality of nearly all products manufactured from things produced by the farmer.

Fertilizers. Years ago the farmers of the Atlantic States found their farms producing less and less because the soil was becoming exhausted. They usually abandoned the land after it ceased to pay and moved westward to new virgin soil. In Europe there was no convenient place to move to, so the farmers there attempted to restore the fertility of the soil. In this they have been remarkably successful. Soil that has been producing continuously for three hundred years still produces as much as it ever did. This is accomplished by adding fertilizer as rapidly as the necessary elements are removed in the crops. Now the United States also has come to the place where there is practically no new soil available. The old agricultural lands must be restored and kept up with fertilizer.

The requirements of plants. The ancient agriculturists knew little of the relation of fertility to plant growth. They looked for some one substance which alone made plants grow, which they called the *principle of vegetation*. Van Helmont, an early European investigator, performed an experiment which led him to conclude that water was this principle. He put 200 pounds of dry soil in an earthenware vessel and planted a 5-pound willow in it. To this he added only water for 5 years. He now pulled up the willow and dried it and found it to weigh 169 pounds 3 ounces, while the soil had only lost 2 ounces in weight. Van Helmont thought this 2 ounces was too little to consider. He also made the mistake of concluding that the growth was entirely due to water, ignoring the fact that the plant was getting carbon dioxide from the air during all those five years. In view of the limited knowledge of his day, we cannot blame him altogether for overlooking this important fact. Nevertheless his conclusion was wrong.

Discovery of mineral plant foods. Glauber, in 1650, learned that potassium nitrate, which was then called saltpeter, would increase plant growth on certain soils. He concluded that saltpeter was the principle of vegetation. To learn what minerals are in plants, other investigators burned large quantities of plant material and analyzed the ashes. Many elements were found in the residues. By experiment twelve of these elements were proved to be essential to the well-being of plants. It was next proved that plants obtained all their carbon from the carbon dioxide of the air.

Liebig's "law of the minimum." The German chemist Liebig discovered a very important principle, which has since been known as *Liebig's law of the minimum*. He learned that if any one of the essential elements for plant food was missing in a soil, the addition of any amounts of the others would not compensate for the absence of this one element. This idea that the minimum amount of one element may limit the crop is nothing more than common sense. If we lack water and are thirsty, we cannot make up for this lack by eating an excess of dry bread. Liebig's law can be illustrated in this way. Suppose there is only enough phosphorus in a given soil for one-third of a crop of grain; although there may be enough of all the other essential elements for ten crops, as long as no more phosphorus is added, there cannot be over one-third of a crop.

Plant food. The raw materials out of which plant substance is synthesized consist of carbon dioxide, water, oxygen, and suitable compounds of nitrogen, phosphorus, sulfur, potassium, calcium, magnesium, iron, and probably manganese, silicon, and sodium.

Nitrogen in plant foods. Of all the elements needed by the plant which are not supplied by water and the carbon dioxide of the air, nitrogen is needed in the largest amounts and is most frequently deficient in worn-out soils. Most plants require nitrates as the source of their nitrogen. It makes no immediate difference how much nitrogen in other forms may be in the partially decayed plant residues. It must be changed into nitrates before it can be used by most plants. Ammonium compounds can be used by peas, oats, barley, and mustard equally as well as nitrates. Potatoes seem to prefer the ammonium compounds, although nitrates serve very well.

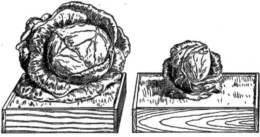

RESPONSE OF PLANTS TO NITROGEN FERTILIZER. The large cabbage grew in soil containing plenty of all the plant foods; the small one, in soil containing plenty of all the plant foods except nitrogen. The soil was deficient in this element.

Nitrogen starvation causes yellowing of the leaf (especially in cold spring weather), absence of growth, and a poor, starved appearance generally. Overabundance of nitrogen, on the other hand, leads to a bright green color, to a copious growth of soft, sappy tissue, liable to insect and fungoid pests, and to retarded ripening. Seed crops, such as barley, which are cut dead ripe, are not intentionally supplied with more than the necessary amount of nitrate; but oats, which are cut before being quite ripe, can receive larger quantities. All cereal crops, however, produce too much straw if the nitrate supply is excessive, and the straw does not commonly stand up well, but is beaten down, or "lodged," by wind and rain. Tomatoes, again, produce too much leaf and too little fruit if they receive an excess of nitrate. On the other hand, crops grown solely for the sake of their leaves are wholly improved by an increased nitrate supply; growers of cabbage have learned that they can not only improve the size of their crops by judicious application of nitrates, but they can also impart the tenderness and bright green color desired by purchasers. Wheat shows a greater increase in straw than in grain as the nitrogen supply is increased.

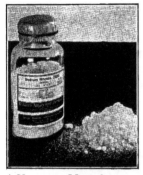

A NITRATE. Most plants require their nitrogen to be in the form of soluble nitrates before they can use it to form proteins.

Before the invention of nitrogen fixation, by which the nitrogen of the air is made into nitrates or other nitrogen compounds useful to the plant, the farmer could restore the nitrogen deficiency in one of three ways. He could allow the soil to stand idle for ten to twenty years, during which time the nitrogen-fixing bacteria would restore the nitrates. He could plant clover, peas, or alfalfa for a year or two. The bacteria that live in the roots of these plants restore nitrates much more rapidly than other nitrogen-fixing bacteria. If he needed immediate restoration of his soil, he could buy the nitrates from Chile or apply barnyard manure.

Today nitrates from the nitrogen-fixation chemical plants are abundantly available for immediate use. (See Chap. 6.) These products can be used directly or made into other nitrogen compounds. If the nitrogen compound applied is not in the form of nitrate, the soil bacteria can make nitrate out of it.

EXPERIMENT

To show the presence of nitrogen in the soil. Materials: Soda lime, mortar, ignition tube, and litmus paper.

Mix 5 g. of soil and an equal bulk of soda lime in a mortar; transfer to a strong test tube. Connect the test tube with a delivery tube which leads into another test tube containing distilled water. Heat cautiously the test tube containing the soil and soda lime with the Bunsen burner, for 5–10 minutes. Test the liquid with litmus paper and note the reaction. Soda lime aided by heat decomposes the organic matter of the soil and forms CO_2, H_2O, and NH_3. The nitrogen in the form of ammonia is distilled and absorbed by the water in the second test tube; the reaction is due to the presence of the ammonia.

What is soda lime? Why does not the ammonia gas escape through the apparatus into the air?

Phosphorus in plant foods. Phosphates are by far the most efficient phosphorus foods known for plants. The effect of a phosphate on the crop is twofold. In the early stages of growth it promotes root formation in a remarkable way. Dressings of phosphates are particularly valuable wherever greater root development is required than the soil conditions normally bring about. They are invaluable on clay soils, where roots do not naturally form well, but, on the other hand, they are less needed on sands, because great root growth takes place on these soils in any case. They are used for all root crops, such as turnips and potatoes, and for shallow-rooted plants with a short period of growth, such as barley. They are beneficial wherever drought conditions are likely to come on, because they induce the young roots to grow rapidly into the more moist layers of soil below the surface.

Later on in the life of the plants, phosphates hasten the ripening process, thus producing the same effect as a deficiency of water, but to a less extent. An application of phosphates may be used to hasten a crop to avoid bad weather. Crops receiving phosphates are golden yellow in color while others are still green. These effects, however, are not nearly so striking as those shown by the nitrogen compounds. There is no obvious change in the appearance of the plant announcing deficiency of phosphate like those changes showing nitrogen starvation or excess.

Deposits of calcium phosphate are extensive and quite numerous. The natural product is so insoluble that it takes a long time for it to become available to the plant. The chemist, however, has learned how to make it soluble and immediately available to the crop by changing it to the superphosphate with sulfuric acid.

EXPERIMENT

To determine the approximate percentage of phosphorus in the home soil. Materials: Conc. nitric acid, ammonium molybdate solution, measuring tube. (The tube should be wide for part of its length and have an internal diameter of 3 mm. for the lower 50 mm. of length.)

Two grams of the fine soil is put into a test tube with 5 cc. of concentrated nitric acid and made to boil for only a couple of seconds. Five cc. of water are quickly added, and the whole mixture thrown on a small filter and washed with a little water, saving all the liquid. Warm this liquid and add 5 cc. of ammonium molybdate solution; stir thoroughly and allow the precipitate to settle. Draw off most of the liquid and transfer the precipitate and the remaining liquid to the measuring tube. Allow it to settle into the small part of the tube and measure the length of the precipitate in millimeters. One millimeter in height of the precipitate obtained indicates .01 per cent of phosphoric acid in the soil.

What is the yellow precipitate? Can you think of any reason for choosing this phosphorus compound in the test?

REVIEW QUESTIONS

1. Contrast the agricultural policy of the early American farmers with that of the European farmers.

2. To what extent was this difference one of circumstances?

3. What was the nature of the earliest service of the chemist to agriculture?

4. How have European farmers kept up the fertility of their soil for centuries?

5. How simple did the ancient farmers consider the problem of fertility?

6. What was discovered by studying the ashes of plants?

7. Relate Van Helmont's experiments and give his erroneous conclusions.

8. What was the first chemical discovered to aid plant growth?

9. What hasty conclusion did Glauber reach as a result of this discovery?

10. State Liebig's "law of the minimum" and illustrate it in detail.

11. Name the elements needed by plants.

12. Where do plants get their food?

13. What element is most frequently deficient in soils?

14. What are the symptoms of nitrogen starvation as shown by plants?

15. How do we obtain nitrogen compounds for fertilizers?

16. What plants can stand being overfed with nitrates and why?

17. What provision does nature make for restoring nitrates to the soil?

18. What part of the plant is stimulated by phosphorus in the early part of its life?

19. What type of crop especially needs an abundance of phosphates?

20. In the latter part of the life of plants what service does phosphorus render to the plant?

21. Where do we get phosphates?

22. How is superphosphate made?

ADDITIONAL EXERCISE FOR SUPERIOR STUDENTS

1. Calculate the weight of calcium superphosphate that can be obtained from 100 lbs. of $Ca_3(PO_4)_2$. (p. 286)

REFERENCES FOR SUPPLEMENTARY READING

1. Darrow, Floyd L. *The Story of Chemistry*. Blue Ribbon Books, Inc., New York, 1930. Chap. VI, pp. 213–252, "Agriculture and War."
   (1) How are agriculture and war related? (pp. 213–214)
   (2) What calamity predicted by Sir Wm. Crookes in 1898 never materialized? (p. 217)
   (3) Outline the history of nitrogen fixation in the United States. (pp. 228–230)
   (4) Discuss the relation of by-product coking to agriculture and war. (pp. 230–233)
   (5) How did the Indians supply phosphorus to the soil? (p. 238)
   (6) Where are deposits of phosphates in the United States? (p. 238)
2. *Proceedings of the Second Dearborn Conference on Agriculture*. The Chemical Foundation, Inc., New York, 1936.

Prepare a report on one of the following topics:
(1) Science in Industry (pp. 8–15)
(2) Research with Terminal Facilities (pp. 18–36)
(3) Tung Oil (pp. 51–55)

Potassium in plant foods. All potassium compounds have practically the same nutritive value. Potassium starvation can be more readily detected than phosphate starvation. The leaf takes on a dull color, and also tends to die at the tips. The most striking effect, however, is the loss of efficiency in making starch. The whole process comes to an end without the potassium salts. All starch-forming crops reduce their sugar production with decreasing potassium supply. A second effect is on the formation of grain. Without potash, as the agriculturist calls the potassium compounds, the individual grains are

A GRANITE MOUNTAIN. Granite, one of the common rocks in mountain ranges, contains the element potassium. The soils on the sides and at the bases of granite mountains are rarely deficient in potassium plant food.

Courtesy of Amer. Potash and Chemical Corp

SEARLES LAKE CHEMICAL PLANT. When the World War cut off the importation of potash, this plant was built for the recovery of potassium compounds from underground brines. It is still in successful operation.

stunted. A third effect of the potash salts is to improve the tone and vigor of the plant. Potash-starved plants are the first to suffer in a bad season, or to die from disease.

Up until the World War the United States imported all of its potassium compounds for both chemicals and fertilizers from Germany. A great deal of potash was used in fertilizer, as potassium is the third element that plants often exhaust from the soil. When the World War cut off the German supply, all potassium compounds jumped to unheard-of prices. This element could no longer be used in fertilizer. Agriculturists became alarmed. Every possible source of potassium compounds was studied to the utmost.

Potassium compounds were discovered in the dust from cement mills. This was caught by electrical precipitation methods, but the supply was small when compared with the country's needs. Sóme potassium compounds could be recovered from the wastes of beet-sugar manufacture. This source also was too limited to amount to much.

Although certain minerals, such as feldspar, which occurs in granite, contain potassium, it is not there in sufficient amount to make separation feasible. In Wyoming was found a mineral deposit that showed some promise of becoming a source of potassium, but no process of getting it out could be found that could compete with peace-time prices.

However, one source of potassium compounds that was developed during this period has been able to survive the after-war competition and grow until it now supplies one-fifth of our total needs. This source is the underground brines in Searle's Lake in the Mohave Desert. At the Trona Plant, Dr. Teeple worked out a very complicated method of separating and recovering the borax and the potassium salts.

Another very promising deposit of potassium has been discovered underground in western Texas. It bids fair to become another extensive source of these salts as soon as successful processes are worked out to raise and purify the desired fertilizer material.

Calcium in agriculture. Some fertilizer substances are added to the soil for their indirect effects. Calcium is one of the elements needed by plants. However, when limestone and gypsum are added to the soil to improve fertility, it is not because there is not enough calcium to serve the needs of the plant for calcium food. The limestone and gypsum improve the living conditions of the plant. Limestone neutralizes any acid condition and releases other plant foods. Gypsum reacts with sodium carbonate, called black alkali, a harmful constituent of some soils, to make less harmful substances.

$$CaSO_4 + Na_2CO_3 \longrightarrow CaCO_3 + Na_2SO_4$$

The Na_2CO_3 is caustic to the living plant and very destructive. The limestone ($CaCO_3$) is entirely harmless and the Na_2SO_4 often included in "white alkali" is much less harmful than the carbonates.

The other elements found in the plant tissue are usually found in sufficient quantities in the soil, the soil water, or the air. They are not ordinarily of concern to the farmer if the plant gets enough water.

Organic fertilizers. Barnyard manure is a complete fertilizer in that it contains all three commonly needed elements in quickly available forms. It is not available in sufficient amounts for all uses. It has one disadvantage in that it contains many grass and weed seeds. It is likely to introduce a lot of undesirable plants. The chemist has helped to prepare other organic fertilizers that are free from grass seeds. One of these fertilizers is the dried blood and waste from the packing houses, and another is fish scraps from sardine and other fisheries.

Chemical use of agricultural by-products. The chemist has succeeded in making salable products out of several agricultural wastes. Among these are cotton seed, cotton linters, bagasse, oat bran, corn cobs, wood waste, and straw.

Cotton seed to cooking fat. After the cotton has been picked, about three-fourths of the weight is in the seeds. The oil is pressed out of the seeds and made into a solid cooking fat by hydrogenation. To hydrogenate the cottonseed oil, the catalyst — finely divided nickel — is mixed with the oil in a covered iron tank, from which all the air has been replaced by hydrogen. While the oil is agitated with a mechanical stirrer, hydrogen gas is forced in under about twenty-five pounds per square inch pressure while the temperature is held at 175°– 190° C. by means of heating coils. When the oil reaches the right consistency, the hydrogenation is stopped. If the oil were completely hydrogenated, it would be as hard as wax, which is harder than the housewife is used to and too hard for convenience.

Insulation board from wood waste. In the pine lumber districts of the South, there is an enormous waste of woody fiber. Even the sawdust represents a lot of wood. Around saw mills everywhere can be seen large unsightly piles of sawdust, from which the rain washes dark resinous liquids to pollute the streams. In some places oxalic acid is made from sawdust. This waste, however, is hardly able to compete with the other sources of this acid. Eventually it was found that if the wood waste were subjected to superheated steam under great pressure and the pressure then suddenly released, the cells of the wood exploded. The disintegrated wood fiber could then be made into an insulating wall board, which is now sold throughout the United States under the name of masonite.

Cotton linters to rayon. When the cotton is ginned — that is, combed away from the seeds — there remains a lot of fine hairs still sticking to the seeds. These are removed by a second ginning and are called cotton linters. These linters are matted together for filling bed mattresses. They make about the poorest kind of mattress, and the demand does not use all of the linters available. The invention of rayon provided a new market. The cotton linters are now made into valuable cloth by the nitrate and copper-ammonium rayon processes.

Useful products from corn waste. Corn is one of our largest agricultural products. The corn cobs, husks, and stalks are not profitable in themselves. Today, several chemists are at work trying to make salable articles from these wastes. Some progress has already been made. Cornstalks have been made into pressed board and paper. Straw has long been used in the manufacture of paper by the different paper-making processes.

When corn cobs are treated with acid and distilled, there is obtained a clear liquid substance, furfural. When first produced, practically no uses of furfural were known. Chemists were put to work studying this substance to find uses for it. Already a half dozen or more uses for furfural have been found, and the number will undoubtedly be increased. Among those already found are: lacquer solvent, a raw material product for manufacturing dyes, embalming fluid for biological specimens, vulcanizing accelerator, fuel for lamps, and a reducing agent for silver mirrors.

Another product made from corn wastes by hydrolysis is the five-carbon sugar, xylose. This sugar has some properties that make it a very promising compound. Before the recent work on corn products, xylose was a chemical curiosity costing $100 per pound. Now it is on the market for about 10 cents per pound. Xylose does not poison people with diabetes as do other sugars. It will be serviceable for mixing with honey as it does not crystallize so readily as other sugars. This property of refusing to crystallize will make xylose useful in ice creams also. Xylose is made by the hydrolysis by acids of the carbohydrate xylan, occurring in various wastes in the following amounts: cornstalks 29–31 per cent, corn cobs 31–37 per cent, peanut shells 23 per cent, and oat hulls 31 per cent.

REVIEW EXERCISES

1. What are the symptoms of starvation for potassium as shown by the plant?

2. How does potassium deficiency affect sugar synthesis by the plant?

3. What potassium industry that grew up in the United States during the World War has survived?

4. Name several sources of potassium which did not survive competition.

5. What supply of potassium compounds is yet to be developed?

6. Why is limestone added to some soils?

7. What does gypsum do to soil containing black alkali, Na_2CO_3?

8. Why is the farmer not concerned about some plant-food elements?

9. Describe the hydrogenation of cottonseed oil.

10. Discuss barnyard manure as a fertilizer.

11. What other organic materials make good fertilizers?

12. Give a list of agricultural by-products.

13. What useful product is made from bagasse?

14. Describe the process for making insulating board from pine waste.

15. Give the copper ammonium process for making rayon. (See Chap. 26.)

16. Name two new products from corn waste and give several uses of each.

17. From what carbohydrate is xylose made?

18. Give the approximate percentage of xylan in several agricultural wastes.

1. Calculate the weight of $CaSO_4$ that can be prepared from 20 lbs. of $CaSO_3$.

2. What weight of sulfur is necessary to prepare 40 lbs. of CaS_2O_3 according to the equation:

$$3 Ca(OH)_2 + 12 S \longrightarrow CaS_2O_3 + 2 CaS_5 + 3 H_2O$$

3. How many pounds of CaS_5 is formed in problem 2?

4. Calculate the volume of oxygen measured under standard conditions that is required in problem 1.

5. What volume would 76 g. of carbon disulfide occupy at 0° C. and 76 cm. mercury?

ADDITIONAL EXERCISE FOR SUPERIOR STUDENTS

1. How many pounds of gypsum, $CaSO_4 \cdot 2 H_2O$ are necessary to react with 100 lbs. of black alkali?

Insecticides and fungicides. One of the chemist's aids to agriculture has been the preparation of various kinds of poisons to kill the insects and the various fungus growths that destroy and injure fruits, berries, and plants of all kinds. The farmer must wage a continuous fight against these pests. His success in most cases will depend upon the diligence and effectiveness with which he suppresses parasitic pests upon plants and animals. These pests may include chicken lice, ticks, potato bugs, plant lice, codling moth, mildew, scabies in sheep, apple scab, and leaf blight.

Insecticides are substances used to kill insects. They may kill insects by being taken internally, by corrosive action on contact, or by being breathed in.

Internal poisons. Internal poisons for insects generally depend upon some compound of arsenic for their poisonous effects, usually arsenious oxide, commonly called *white arsenic*. Soluble arsenic compounds should not be used since they injure the .foliage. The water-insoluble arsenic compound should, while in a finely divided condition, be mixed with water and sprayed on the plant. All insects that feed on the surface, such as the codling moth and the potato bug, will eat some of the arsenic compound and be killed.

Paris green. Paris green is an insoluble green compound of copper, arsenic, and acetic acid, whose chemical name is copper acetoarsenite, $Cu(C_2H_3O_2)_2 \cdot Cu_3(AsO_3)_2$. Paris green has been used for a long time in the United States for the potato bug. It is prepared by adding a hot solution of sodium arsenite to a hot solution of copper acetate.

$$2 Na_3AsO_3 + 4 Cu(C_2H_3O_2)_2 \longrightarrow Cu(C_2H_3O_2)_2 \cdot Cu_3(AsO_3)_2 + 6 NaC_2H_3O_2$$

Paris green separates from the mixture and settles out as a rather fine powder of a clear, green color. The pure compound is practically insoluble in water, but readily dissolves in ammonium hydroxide.

Paris green scorches the foliage of some plants, such as the potato. This can be overcome by mixing limewater with it. The limewater makes any soluble arsenic into insoluble calcium arsenite.

$$2 Na_3AsO_3 + 3 Ca(OH)_2 \longrightarrow Ca_3(AsO_3)_2 + 6 NaOH$$

Contact insecticides. Sucking parasites, like plant lice, must be destroyed by corrosive substances coming in contact with their bodies, or by closing up their breathing tubes. The *lime-sulfur wash* is one of the commonest of the corrosive insecticides. It is used quite extensively as a sheep dip for scabies, for which purpose it is about the best. It is very effective for San Jose scale also.

The lime-sulfur wash for plants is made by boiling quicklime and flowers of sulfur in water. Several compounds are formed, including calcium thiosulfate (similar to ordinary hypo), calcium pentasulfide (CaS_5), calcium sulfite, and calcium sulfate.

$$3 Ca(OH)_2 + 12 S \longrightarrow CaS_2O_3 + 2 CaS_5 + 3 H_2O$$
$$2 CaS_5 + 3 O_2 \longrightarrow 2 CaS_2O_3 + 6 S$$
$$2 CaSO_3 + O_2 \longrightarrow 2 CaSO_4$$

Theory of action of the lime-sulfur wash. The excess of lime loosens up scales and exposes the insects to the action of those substances of insecticidal properties. The finely divided sulfur (from the second reaction), together with the thiosulfate (CaS_2O_3) and sulfite ($CaSO_3$) are effective insecticides.

Kerosene washes. Different hydrocarbons of the paraffin series are good insecticides but are somewhat injurious to the tree. Kerosene on ponds kills the young mosquitoes in the water. Kerosene emulsions with soaps have some use also for spraying to kill oyster-shell scale. Resin soaps kill the orange scale. They are prepared by boiling resin with sodium carbonate and diluting with water.

Gaseous insecticides. Gaseous insecticides are used against pests particularly difficult to attack. Hydrocyanic acid (HCN) is by far the most effective. Hydrocyanic acid gas is liberated as an invisible gas when sodium cyanide is treated with concentrated sulfuric acid.

$$2\,\mathrm{NaCN} + \mathrm{H_2SO_4} \longrightarrow \mathrm{Na_2SO_4} + 2\,\mathrm{HCN}$$

This is an extremely powerful poison, a single breath being fatal. In order that the gas may be retained long enough for complete action, it is applied in tightly closed rooms or in special tents, which may be put over the tree. After an hour or more the inclosure should be opened from the outside and thoroughly aired before being entered, and the strongly acid residue from the reaction should be carefully disposed of. In some places the HCN gas is generated by a machine and pumped into the tents.

Carbon disulfide. Carbon disulfide is an excellent substance with which to fumigate raisins and other fruits. The fumigation should be done in air-tight rooms, which should be closed for 24 hours. Moths are killed in 12 hours, their eggs in 16 hours, but beetles require 24 hours' treatment. Twenty pounds of CS_2 per 1000 cu. ft. are used by exposing in an open pan. Carbon disulfide is also a good poison for the destruction of weevils in grain. The heavy vapors sink through the grain to the bottom of the bin, where they may be released by boring holes in the bottom. Ants, moles, squirrels, and similar pests are exterminated by placing cotton saturated with CS_2 in the runs and covering the openings tightly. Because of its great combustibility carbon disulfide should never be brought near flames.

Spraying grape vines. Various sprays are utilized for the destruction of parasitic plants. Sodium thiosulfate, lime sulfur, and sulfur alone were used in this capacity as early as 1885 against apple scab and pear blight. Flowers of sulfur is blown over grape vines to kill the mildew. The raisin crop in the San Joaquin Valley is largely dependent upon this sulfur. The sulfur probably oxidizes to SO_2, which kills the fungus. Bordeaux mixture has been the chief fungicide since 1883, when a Frenchman named Millardet used it for the downy mildew of the grape. It was accidentally discovered by observing the flourishing condition of grape vines to which lime and copper salts had been applied to prevent the theft of grapes in the province of Bordeaux. The commonest formula for Bordeaux mixture is copper sulfate 6 pounds, quick lime 4 pounds, and water 50 gallons. It is necessary to have lime in excess to keep the copper out of solution. An excess of lime can be detected if the solution is alkaline to litmus. Copper sulfate hydrolyzes to form an acid reaction; hence an acid litmus test shows insufficient lime.

When the slaked lime reacts with the copper sulfate, it forms a basic sulfate, which is partly base and partly sulfate.

$$Ca(OH)_2 + 2\ CuSO_4 \longrightarrow Cu_2(OH)_2SO_4 + CaSO_4$$

When sprayed on the trees, it oxidizes to form copper carbonate and copper sulfate. To the latter is due most of the fungicidal properties. Copper sulfate is not suitable for direct application because its acid reaction injures plants. It is used to kill smut on grains.

REVIEW EXERCISES

1. Name a list of pests farmers have to combat.
2. Name two internal insecticides.
3. What is the trade name for arsenious oxide?
4. Give the formula for Paris green.
5. What must be the feeding habits of the insect in order that arsenic compounds will be effective?
6. How is Paris green kept from scorching the potato?

7. Write the reaction for what takes place in ex. 6.

8. How are sucking insects killed?

9. Name a contact insecticide.

10. Write the reaction between Ca(OH)₂ and sulfur.

11. Write reactions to form two other substances occurring in the lime-sulfur wash.

12. What does the excess Ca(OH)₂ of the lime-sulfur wash do?

13. What three substances kill the insect?

14. Discuss kerosene in connection with mosquito control.

15. What scale is killed by kerosene emulsions with soap?

16. How are resin soaps made, and what scale will they kill?

17. Name one gaseous insecticide.

18. Discuss the use of CS₂ for fumigation of dried fruit and grain products.

19. How does powdered sulfur serve as a fungicide?

20. Relate the history of the discovery of Bordeaux mixture.

21. Name the two active chemicals of Bordeaux mixture.

22. How does Bordeaux mixture attack the insects?

ADDITIONAL EXERCISE FOR SUPERIOR STUDENTS

1. How many pounds of Na₃AsO₃ are needed to make 30 lbs. of Paris green?

REFERENCE FOR SUPPLEMENTARY READING

1. *Proceedings of the Second Dearborn Conference on Agriculture.* The Chemical Foundation, Inc., New York, 1936.

   Prepare a report on one of the following topics:

   (1) Cotton Roads (pp. 65–68) .

   (2) Power Alcohol (pp. 73–111)

   (3) Growing Artichokes in America (pp. 111–120)

Killing peach borers.   To keep borers from the roots of peach trees, paradichlorobenzene is mixed with the soil around the base of the tree.   Being a substance that sublimes, the dichlorobenzene forms a vapor, which penetrates the soil and the scars in the tree where the borers live and kills them.   If the borers are not killed, they soon kill the tree.

Paradichlorobenzene is prepared by the action of chlorine on benzene under the influence of catalysts, such as aluminum chloride or iron salts. In structural formulas the reaction would be:

$$\text{(benzene)} + 4\,Cl \longrightarrow \text{(dichlorobenzene)} + 2\,HCl$$

The term *para* when applied to the replacement of hydrogen atoms of benzene means that the two hydrogen atoms on opposite side of the molecule have been replaced.

Formalin. Formalin, a 40 per cent solution of formaldehyde, is a most efficient fungicide for smut on grain. Formaldehyde is a very common disinfectant or disease germ destroyer. A 5 per cent solution is the common strength used. When the gas is desired to disinfect a room, it is made in one of three ways: The gas is heated under pressure and piped into the tight room; the solution is sprayed on surfaces, such as sheets, and allowed to evaporate; or 6 parts of formalin are poured on 5 parts of potassium permanganate. In the last case, heat is generated by the chemical reaction in the oxidation of the aldehyde to formic acid, and 50 per cent of the aldehyde is liberated as the gas. Ten ounces of formalin are necessary for each 1000 cu. ft. of space in the first two cases, and twice as much in the second.

Paraform. Paraform, $(CH_2O)_3$, is a condensation product of formaldehyde put up as a powder. Two ounces of paraform liberate gas sufficient to disinfect 1000 cu. ft. of space.

Formaldehyde has several advantages: It is a powerful germicide; it is not poisonous for hay and grain; and it is not injurious to fabrics and metals. Its disadvantages are that it condenses in cold weather, and requires a sealed room to hold it long enough to obtain its effects.

Mercuric chloride. Mercuric chloride, or corrosive sublimate, is a poisonous white crystalline salt. Strengths of 1 to

500 or 1 to 1000 are used, the greater strength being necessary to destroy bacterial spores. It is a powerful poison and must be kept away from children. It forms insoluble compounds with proteins, hence raw eggs and milk are given as antidotes. Its action with proteins makes it unsuitable for disinfecting protein substances like blood. Solutions of this salt should be used only in glass or earthenware, as it reacts with tin and other common metals.

Chloride of lime. Chloride of lime or bleaching powder (CaCl · ClO) is both a disinfectant and a deodorizer, acting as an oxidizing agent because of the chlorine which it liberates. Because of its instability it is good only when freshly removed from the air-tight containers in which it must be kept.

$$2\,CaCl \cdot ClO \longrightarrow 2\,CaCl_2 + O_2$$

Phenol. Phenol (C_6H_5OH) or carbolic acid, as its impure water solution is called, is a good disinfectant for everything except bacterial spores. It is only slightly interfered with by albumens, is readily available, and does not destroy metals or fabrics. About 1 part of the strong liquid to 9 parts of water is the proper concentration. A 5 per cent solution will sterilize garments in an hour's time.

There are several substances similar to phenol called cresols. They are even better than carbolic acid. A 2 per cent solution of cresol is equivalent to a 5 per cent solution of phenol. It is cheaper than carbolic acid, is more effective with spores, and is equal to phenol in all other respects.

Creosote. Creosote is the name of an impure mixture of the cresols with a little carbolic acid and other antiseptic substances. It is used to impregnate wood and posts to protect them against bacteria and fungi. Lysol, a household antiseptic, is a good quality mixture of the cresols, the active agents in creosote. Chemically, the cresols belong to the same class of compounds as phenol.

Chemically separated pectin. The chemist has put jelly making on a scientific basis. The housewife often has the disagreeable experience of having her jelly refuse to gel, or get stiff. The chemist found that gelling depends upon three factors. In the first place, the substance that gels is a colloidal substance called pectin. If pectin is absent from the fruit juice, it is absolutely impossible for gelling to take place. Some fruits do not contain appreciable amounts of pectin. Especially when the fruit is overripe, much of its pectin has changed to sugar. The chemist has prepared pure pectin from oranges. This can be added to fruit juices lacking in pectin, in which case they gel readily.

The jelly problem is not quite so simple as merely having the right concentration of pectin. Even when there is sufficient pectin, it will not gel unless there is also sufficient sugar and acid. At a definite acidity, gelling takes place most readily. The commercial producers of jelly take care to control this acidity to the right point, so their product will always be good.

The chemist's relation to canned foods. The farmer depends to a large extent upon selling his products to the canneries. The whole canning industry has depended upon the chemist to produce the metals for containers. Tin cans have held sway for a long time. Recently aluminum cans have been successfully used. This will probably be a good change as far as the United States is concerned. We produce very little tin, but we can produce unlimited quantities of aluminum.

Indirect help to the farmer. Indirectly, the life of the farmer is made richer in a thousand ways. Everything the chemist does for society as a whole includes the farmer in its benefits — better automobiles, better machinery, new foods, and better things of all kinds.

REVIEW EXERCISES

1. What substance is used to combat the peach borer?

2. In paradichlorobenzene, how are the chlorine atoms situated in the benzene ring?

3. How is paradichlorobenzene prepared?

4. Paradichlorobenzene sublimes. Recall what this statement means.

5. How is formaldehyde generated for the fumigation of a room?

6. What is the composition of formalin?

7. What is formaldehyde used to destroy?

8. Recall the formula of formaldehyde and compare it with that of paraform.

9. Give some advantages and some disadvantages of formaldehyde in fumigation.

10. In what concentrations is mercuric chloride effective for killing germs?

11. With what complex organic compound will mercuric chloride react?

12. Is nascent oxygen formed from bleaching powder according to the reaction $CaCl \cdot ClO \longrightarrow CaCl_2 + O$?

13. Discuss phenol as an antiseptic.

14. What two antiseptics are mixtures of cresols?

15. What substance has been prepared to help in jelly making?

16. Name three substances which must be present in satisfactory amounts before the juice will gel.

17. What new can material is being tried out, and why is it more suitable than tin for use in the United States?

18. Write the formulas for (a) normal calcium chloride, (b) basic calcium chloride, (c) acid sodium sulfate.

19. Picture the atomic structure equation for the following:

$$Ca + O \longrightarrow CaO$$

20. Tell of the chemist's indirect help to the farmer.

Reference for Supplementary Reading

1. *Proceedings of the Second Dearborn Conference on Agriculture.* The Chemical Foundation, Inc., New York, 1936.

Write a report on one of the following topics:

(1) Starch and Sugars (pp. 139–165)

(2) Plastics (pp. 166–183)

(3) Cellulose (pp. 184–219)

Summary

Plants require nitrogen, phosphorus, and potassium in fairly large amounts. The chemist has developed a local source of potassium compounds, which now supplies about one-fifth of our agricultural needs. Artificial nitrogen fixation supplies all the nitrogen compounds needed for fertilizer. Furthermore, soluble superphosphate is made from our natural supplies of normal calcium phosphate.

Limestone is often added to the soil, not because the soil contains insufficient calcium but to correct acid soils and release other plant foods. Gypsum can correct black alkali, or sodium carbonate, in soil.

$$CaSO_4 + Na_2CO_3 \longrightarrow CaCO_3 + Na_2SO_4$$

The chemist has aided agriculture by making use of by-products, such as cooking fat from cotton seed, rayon from cotton linters, wall board from bagasse, and paper from straw.

In combating plant pests, the chemist has supplied Paris green for the potato bug and lime-sulfur wash for sucking parasites. This wash contains CaS_2O_3, CaS_5, and $CaSO_3$. Among gaseous insecticides are HCN (made by the action of sulfuric on cyanides) for orange scale, and CS_2, which is effective towards moths.

Flowers of sulfur is used for mildew, and Bordeaux mixture for other fungi. Basic copper sulfate is the active agent in Bordeaux mixture. Formaldehyde as a gas and (especially) in solution is an effective disinfectant. Paradichlorobenzene is used for the peach borer at the roots of the tree.

Other common disinfectants are $HgCl_2$, $CaCl \cdot ClO$, C_6H_5OH, and creosote.

The chemist has put jelly making on a scientific basis and has prepared pectin to be added to fruits which do not contain enough of it. He has given new metal cans to the canned-food industry.

← →

CHAPTER 38

CHEMISTRY AND WARFARE

Introduction. Each decade warfare becomes increasingly dependent on chemistry. In the days of the Greeks and Romans, when wars were fought in hand-to-hand encounters with swords and spears, chemistry had only to furnish the metal for the weapons.

With the invention of gunpowder, warfare underwent a complete change. Men no longer had to engage in hand-to-hand conflict in order to kill one another. Chemistry now took on the added task of supplying the gunpowder. Of every improvement since that time, chemistry has had either direct or indirect charge; whether the improvement was smokeless powder, dynamite, guncotton, T.N.T., stronger armor plate, better guns, or poison gases, the chemist has been at the bottom of it.

Explosives. The Chinese invented gunpowder at some remotely distant date, but they seemed to have been satisfied with using it in firecrackers. Roger Bacon is credited with a later invention of black powder. The practical English soon began to use the newly invented powder for warfare. Thus war came to be dominated by explosives, and it has continued to be dependent on them ever since. Black powder is made from a mixture of sulfur, charcoal, and potassium nitrate. The three substances are ground very fine, mixed into a paste, dried, and finally ground into grains. In exploding, the carbon is burned to carbon dioxide, and the sulfur to sulfur dioxide. The oxygen is furnished by the potassium nitrate. The solid potassium nitrite and some unburned carbon create a great deal of smoke at the time of the explosion. Because the reaction is between adjacent particles, it is necessarily quite slow, so that black powder as compared with the smokeless powder is weak in explosibility.

686

Smokeless powder. With the discovery of cellulose nitrate came the smokeless powders, smokeless because the oxidizing part and the fuel are all inside the same molecule. In other words, one part of the molecule oxidizes the rest of it to products that are all gases. There being no carbon or inorganic residues, as there are from black powder, there is no smoke. The explosion is more violent in the case of the cellulose nitrate than in the case of the black powder because the reaction is faster. The faster the explosion, the higher the temperature produced. The higher the temperature, the more violent the expansion of the gases. The reason the combustion is so much faster in the smokeless powder than in the black powder is that in the latter oxygen has to get out of the KNO_3 molecule and over to the C and S atoms before reaction can take place. The confusion of the billions of oxygen atoms finding carbon and sulfur atoms to attack requires some time. In the cellulose nitrate, since both reactants are already paired off and near each other within the molecule, the reaction progresses many times faster than in the other situation.

Nitroglycerin. The next great advance in explosives came in Nobel's invention of glyceryl trinitrate in 1878. This substance is generally called nitroglycerin, but this name is not correct in the language of the organic chemist. A nitro compound is one that contains NO_2 groups, while one containing NO_3 groups is called a nitrate. This explosive is a trinitrate as it contains three nitrate groups. Glyceryl trinitrate is prepared by treating glycerin with a mixture of concentrated nitric acid and concentrated sulfuric acid according to the following equation:

$$
\begin{array}{l}
CH_2 \,|\, OH \quad H \,|\, NO_3 \qquad CH_2ONO_2 \\
\;| \qquad\qquad\qquad\qquad\;\; | \\
CH \,|\, OH + H \,|\, NO_3 \longrightarrow CHONO_2 + 3\,H_2O \\
\;| \qquad\qquad\qquad\qquad\;\; | \\
CH_2 \,|\, OH \quad H \,|\, NO_3 \qquad CH_2ONO_2
\end{array}
$$

glycerol   nitric acid   glyceryl trinitrate   water

The sulfuric acid dehydrates the other ingredients by removing three molecules of water between the glycerol and the nitric acid. The common name *glycerin* is not chemically correct; it should be glycerol. The substance is an alcohol because it contains OH groups. The chemist gives all the alcohols the ending *-ol*.

The preparation must be made with an elaborate system of cooling and temperature control. The temperature *must* be kept below a certain point, or a serious explosion results. Before the cooling systems were properly designed, plants

manufacturing nitroglycerin often blew up, destroying everything and killing many people. There were so many unexpected explosions in the manufacture and use of glyceryl trinitrate that one European government gave up trying to use it and another government forbade the construction of any plants for its manufacture within the country. Eventually the industry made nitroglycerin manufacture reasonably safe, but it is not safe for any person to try to prepare it without adequate control.

Courtesy of U. S Bureau of Mines
DYNAMITE EXPLODING.

Dynamite. The safe use of glyceryl trinitrate came by the invention of dynamite. Dynamite consists of tubes of sawdust which has absorbed a certain amount of glyceryl trinitrate. In some cases infusorial earth may be used instead of the sawdust. The 25 per cent gelatin is reasonably safe from ordinary shocks. The sharp explosion of the percussion cap is needed to cause it to explode.

T.N.T. The next great improvement in high explosives came in the invention of trinitrotoluene, or T.N.T. as it is popularly known. Glyceryl trinitrate has the disadvantage of exploding unexpectedly unless highly diluted as in the case of dynamite. This of course makes it occupy much more volume and weakens the explosion. Trinitrotoluene is free from this difficulty of danger. A rifle bullet can be fired through a case of T.N.T. without exploding it. The pure substance can be lighted with a match and will burn quietly without explosion. This explosive was the one most used in the World War. The Belgian forts could not withstand the explosive shells of the Germans. The Allies finally drove the enemy from the trenches by a rain of explosive shells each containing T.N.T., in some cases as much as 500 lbs. in one shell.

Picric acid. One other high explosive is picric acid or trinitrophenol. Picric acid is a yellow dye, but it is also an explosive.

Explosives contain nitrogen. It is an interesting fact that all of the explosives used in warfare are nitrogen compounds. Nitrogen, one of the most inactive of the elements, is the constituent of all explosives, from black powder on. The list of explosives includes :

black powder   picric acid   nitroglycerin   trinitrotoluene

All of the high explosives are made with the use of nitric and sulfuric acids.

Chemistry and war machinery. The next war will be a war of machines, high explosives, and chemicals. Everything that improves war machinery promotes success in warfare. Most of the early attempts to fly across the Atlantic resulted in disaster. Now, however, it is not uncommon to fly the Atlantic. The difference is in the engine. Airplane engines are much more dependable than formerly. The improvements in airplanes are so extensive that it is said that one modern airplane adequately armed could whip all the airplanes that took part in the World War. One modern battleship could defeat all the navies of the whole world that existed before the American Civil War. This battleship could whip them all at once. The difference is answered in better armor, better guns, and more speed due to better engines.

In other chapters it has been mentioned how chemistry provided light alloys and helium for the dirigible and many new alloys for the automobile. The new alloys also help the war truck, the tank, and the airplane. The same is true of improved fuels and lubricating oils.

QUESTIONS OF FACT

1. State four fields in warfare in which chemistry plays a large part.

2. What three substances are in black powder?

3. What different use did the English find for gunpowder from that of the Chinese?

4. What chemical reactions take place when black powder explodes?

5. What causes the smoke from black powder?

6. Why is black powder weak as compared with the smokeless powder?

7. What is smokeless powder?

8. Why is it smokeless?

9. Why does smokeless powder react faster than black powder?

10. In what sense is nitroglycerin wrongly named?

11. How is nitroglycerin prepared?

12. Why is it dangerous to try to prepare glyceryl trinitrate in the laboratory?

13. What is dynamite?

14. What is the great advantage of trinitrotoluene over glyceryl trinitrate?

15. Name an explosive that is also a dye.

16. What inactive element is in all explosives?

17. Discuss the importance of chemistry to war machinery.

QUESTIONS OF UNDERSTANDING

1. The following are structural formulas of modern explosives. Name them.

(a)

CH_2—NO_3
|
CH—NO_3
|
CH_2—NO_3

(b)

CH₃ structure:
CH_3
|
C
/ ‖
NO_2—C C—NO_2
‖ |
HC CH
\\ /
C
|
NO_2

(c)

OH
|
C
/ ‖
NO_2—C C—NO_2
‖ |
HC CH
\\ /
C
|
NO_2

REFERENCE FOR SUPPLEMENTARY READING

1. Haynes, Williams. *Men, Money, and Molecules.* Doubleday, Doran, and Co., Inc., Garden City, New York, 1936. Chap. V, pp. 73–92, "Munitions and Molecules."

Chemical warfare.

Although all the great nations had previously agreed not to use poisonous gases in warfare, in the spring of 1915 the Germans surprised the Allies on the Western Front with an attack of chlorine gas discharged from cylinders. The effect upon the unprotected Allied troops was horrible beyond description. Literally thousands of soldiers were lost or rendered helpless. Fortunately for the Allies, the Germans had not anticipated the extent of their success and were in no position to take full advantage of the opportunity. Before their next attack was ready, the Allies' chemists had devised a crude mask as protection against gas.

Later in the year the Allies retaliated by using poison gas themselves, but the element of surprise was now lacking, so no spectacular success was achieved. Throughout the remainder of the War, both sides made active use of poison gases and liquids — particularly in high-explosive shells.

Chemical warfare ha's come to stay. Although treaties may outlaw gas in warfare, history repeats itself, and human nature is ever the same. If nations get into a supreme struggle and either side feels that it can gain an advantage by the use of poisonous gases and liquids, these will be used. At any rate, it is wise policy to assume that they will be and to be prepared for them.

Toxic chemicals. The gases and liquids used in chemical warfare can be divided into two classes of toxic substances: those which produce death; and those which merely incapaci-

Courtesy of Signal Corps, U. S Army
GAS ATTACK ON THE WESTERN FRONT.

tate for the time being. As has been mentioned, the first toxic substance used in the World War was chlorine, which was sprayed from cylinders containing liquid chlorine. The cloud of heavy greenish-yellow gas drifted over the Allied lines, working death and havoc. Gas masks with chlorine absorbents were soon developed, so today a chlorine gas attack would not be very serious to a prepared army.

Phosgene, a lethal gas. A more formidable toxic war chemical was phosgene ($COCl_2$), a poisonous gas made from carbon monoxide and chlorine. Relatively small amounts of this gas, inhaled and followed by exercise, led to sudden collapse — death sometimes following more than twenty-four hours after exposure.

At first phosgene was added to chlorine and floated as a cloud. Later it was enclosed in bombs and thrown into the enemy trenches by means of a Liven projector — a sort of portable mortar. Eventually it was put with a small opening charge of high explosive into regular artillery shells and fired long distances.

Lachrymators. Another type of poisonous gas introduced by the Germans was the so-called "tear gas," which so affected the eyes as to cause temporarily blindness and incapacity for effective action. Xylyl bromide $(C_6H_3(CH_3)_2CH_2Br)$ was the first tear gas, or lachrymator, used. Ethyl iodoacetate $(CH_2ICOOC_2H_5)$ and others followed. The lachrymators are not lethal gases. Since the war they have been used to protect banks, to disperse mobs, and to rout criminals from buildings.

Sneezing gas. Temporary disability was also produced by the sneezing gases These substances produced violent sneezing, accompanied by intense pain and irritation of the nose and throat. They were mostly arsenic compounds, which usually produced arsenic poisoning as an aftereffect. The principal substance in this class is diphenylchlorarsine, $(C_6H_5)_2AsCl$. This compound was used in one type of gas shell.

Mustard gas The most effective of the war gases was known as mustard gas. Incidentally, this chemical is not a gas at all, but a liquid, dichloroethyl sulfide, $(CH_2ClCH_2)_2S$. It was responsible for more casualties than all the other war gases put together. There are a number of reasons why this was so. Mustard gas was particularly insidious in that it had little odor and produced no discomfort to give warning of danger. It not only affected the lungs and eyes but also attacked the skin and the mucous membrane of all exposed parts of the body. Frequently it caused sores in the armpits. It was the most persistent of all the war gases. It penetrated the soldiers' clothes and their bedding. It even entered the very soil, from which it was later evaporated by the heat of the sun. Since gas masks cannot be worn indefinitely, a mustard-gas attack literally rendered a locality uninhabitable for a considerable time. Soldiers were overcome by the poison upon removing their masks as much as twelve hours after an attack. Fairly adequate protection from mustard gas could be obtained by conscientiously wearing a suitable mask and by treating the clothing and bedding with chemicals to counteract the poison.

The following account of a gas attack with shells of mustard gas is taken from an officer's report. (V. Lefebure, *The Riddle of the Rhine*, page 67. Chemical Foundation, New York.)

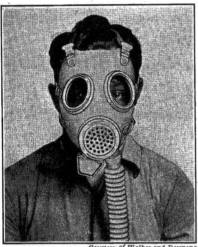

Courtesy of Walker and Downing

GAS MASK. Chemical warfare has made the gas mask an absolute necessity.

"I was gassed by dichlordiethyl sulfide, commonly known as mustard stuff, on July 22nd. I was digging in (Livens Projectors), to fire on Lambartzyde. Going up we met a terrible strafe of H. E. and gas shells in Nieuport. When things quieted a little I went up with the three G. S. wagons, all that were left, and the carrying parties. I must say that the gas was clearly visible and has exactly the same smell as horseradish. It had no immediate effect on the eyes or throat. I suspected a delayed action and my party all put their masks on.

"On arriving at the emplacement we met a very thick cloud of the same stuff drifting from the front line system. As it seemed to have no effect on the eyes I gave orders for all to put on their mouthpieces and noseclips so as to breathe none of the stuff, and we carried on.

"Coming back we met another terrific gas shell attack on Nieuport. Next morning, myself and all the eighty men we had up there were absolutely blind. The horrid stuff has a delayed action on the eyes, causing temporary blindness about seven hours afterwards. About 3000 were affected. One or two of our party never recovered their sight and died. The casualty clearing stations were crowded. On August 3rd, with my eyes still very bloodshot and weak, and wearing blue glasses, I came home, and went into Millbank Hospital on August 15th."

Smoke screens. Chemistry contributed another feature that was characteristic of the World War — that is, the smoke screen. The smoke screen was used to hide troop movements and to protect battleships. Sometimes the smoke was accompanied by poisonous gases, and sometimes it was not. At the sight of smoke, which often preceded an infantry charge, the

Courtesy of Signal Corps, U. S. Army

SMOKE SCREEN.

Germans would withdraw into their dugouts to get ready for the expected charge. On one occasion the Allies surprised them by attacking in the cloud of harmless smoke, in which instance they were upon the unsuspecting Germans before they had a chance to fire upon the charging Allied soldiers.

Phosphorus was the main substance used in smoke screens. It can be dissolved in carbon disulfide. When sprayed into the air, the phosphorus rapidly oxidizes to a white smoke.

Helium. The chemists of the United States have succeeded in recovering helium in amounts large enough to inflate the dirigible war balloons. This gas, being inert, will not explode. The future only can tell how much of an advantage this fact will give our airships. It should be a large factor if dirigibles work out to be an effective type of military aircraft.

The future of chemical warfare. The World War has demonstrated several things. (1) Nations in supreme conflict resort to chemical warfare even though there may be treaties to the contrary. (2) Chemical warfare is extremely effective. Whole companies and even whole battalions were disabled in the World War. (3) The nation with the best organized peacetime chemical industries will be the strongest in a protracted war.

War is supported by chemistry in all its phases — explosives, war machinery, and war gases; and all such aids to armies depend on chemistry. An army of chemists is just as essential as abundant resources and a trained army of soldiers.

REVIEW QUESTIONS

1. Classify war gases into two divisions.

2. What was the first war gas used in the World War? How was it applied?

3. What is the effect of chlorine (a) where the soldiers have no gas masks? (b) where they have gas masks?

4. Describe the toxic effects of phosgene.

5. Name one war gas that was a tear gas.

6. What advantage has a tear gas (a) in war? (b) in quelling mobs?

7. Name a substance that was a sneeze-producing chemical.

8. What was the chemical that produced the greatest number of casualties in the World War?

9. Describe the effect of mustard gas; its persistence.

10. What substance was used in smoke screens?

11. What part did helium play in warfare?

12. Discuss the place of chemistry in future wars.

13. Identify the following structural formulas for war gases:

(a)   (b)   (c)

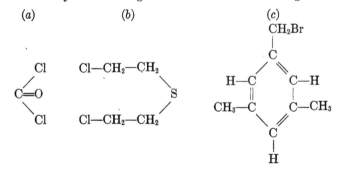

REFERENCES FOR SUPPLEMENTARY READING

Prepare a report on some phase of chemical warfare.

1. Fries, Amos A., and West, Clarence J. *Chemical Warfare.* McGraw-Hill Book Co., Inc., New York, 1921.
   (A thorough discussion of all phases of chemical warfare.)
   Poisonous Gases (pp. 1–195)
   Gas Masks and Protective Measures (pp. 195–285)
   Smokes (pp. 285–336)
2. Lefebure, Victor. *The Riddle of the Rhine.* The Chemical Foundation, New York, 1929.
   (A complete discussion of all phases of chemical warfare.)
   Gaseous Warfare in World War. (Chaps. I–IX)
   Lines of Future Development. (Chap. X)

SUMMARY

Black powder, the first explosive, is a mixture of sulfur, carbon, and potassium nitrate. The potassium nitrate furnishes the oxygen to burn the sulfur and carbon to the gaseous oxides. Smoke results because of imperfect combustion. Glyceryl trinitrate, commonly called nitroglycerin, being too explosive for use pure is soaked into sawdust or earth to make the safer dynamite. Trinitrotoluene, being a safer explosive than nitroglycerin, can be used pure.

Besides explosives, the chemist has supplied good motors, good battleships, good armor, and war gases. The most lethal or deadly of the war gases is phosgene ($COCl_2$). Of the lachrymotors xylyl bromide and ethyl iodoacetate are among the most effective. The principal sneezing gas is diphenylchlorarsine, $(C_6H_5)_2AsCl$. Mustard gas, the most effective war chemical, is $(ClC_2H_4)_2S$, which is a corrosive blistering liquid.

For smoke screens, phosphorus is used, which forms white phosphorus pentoxide. Helium, also, has been supplied for inflating balloons. In the future, the nation with the best chemical industries and largest corps of research chemists will probably have the upper hand.

←
→

CHAPTER 39

THE CHEMISTRY OF COMMON THINGS

Introduction. There are many things about our homes that enrich our daily lives that are directly dependent on chemistry. Most of these have been mentioned in the earlier chapters of this book. However, there are a few that have been missed. The topics in this chapter, being simply those things that were overlooked, will not be considered in any particular order.

Fire in ancient times. In the early days of the race, and even up to the beginning of the nineteenth century, fire was difficult to maintain at all times. Matches, as such, were not invented until 1826 and were not readily obtained in out-of-the-way places for a long time after that. Some primitive races tried to keep a fire supply always on hand at a centrally located place. This was done by the Romans by making the vestal virgins custodians of the sacred fire, which was always kept burning.

Other peoples kept the fire going in their homes in large fireplaces, where there were always live coals, at least. In the evening the backlog and other large sticks would get well started burning; and at bedtime the partially burned logs, together with a large quantity of live coals, would be buried in the ashes. In the morning there would remain the partially charred logs and live coals. These could readily be raked together and fanned into a flame.

Pioneer methods of starting fire. If the early pioneer had the misfortune of having his fire go out, he would borrow fire from a neighbor, if there was one close. Before the day of matches a fire could be started by striking a piece of flint with a hard object and allowing the spark to ignite a piece of dry tinder, a cottonlike fungus found on some woods. The Indians could start fire by friction — that is, by rotating a

698

stick against a piece of dry wood. Sometimes Boy Scouts learn this process, but it is not at all easy.

Phosphorus matches. The first matches used the element phosphorus as the active constituent. If yellow phosphorus is exposed to the air, it soon takes fire spontaneously, because it oxidizes so readily. If just a little phosphorus is mixed with an oxidizing agent, such as potassium chlorate, some easily combustible substance, such as sulfur, and some binding material, which is usually glue, the mixture will not ignite until it is s_t_ar_t_e_d. Starting will take place if the match is "struck" or is held against a hot object. These matches were often called friction matches or sulfur matches. Friction

Photo by Cowling

EARLY METHOD OF MAKING FIRE.

generated enough heat to start the phosphorus burning. The burning phosphorus ignited the other combustible materials in the head of the match, which soon became hot enough to ignite the stick.

Phosphorus. Phosphorus is one of the nonmetallic elements. It occurs in the nucleus of every living cell and as calcium phosphate constitutes a large part of the bones. Fortunately for wornout soils, large deposits of calcium phosphate occur in several places. Phosphorus was first obtained in quantity from bones, by heating with sand and coke in a furnace.

$$Ca_3(PO_4)_2 + 3\ SiO_2 \longrightarrow 3\ CaSiO_3 + P_2O_5$$
$$2\ P_2O_5 + 10\ C \longrightarrow 4\ P + 10\ CO$$

Yellow phosphorus. Phosphorus exists in two allotropic forms, yellow phosphorus and red phosphorus. Yellow phosphorus is a very active substance. It reacts with the oxygen of the air so rapidly that it has to be kept under water. It is soluble in carbon disulfide. This variety of phosphorus is very

poisonous. Workers in match factories using yellow phosphorus often got a disease of the bones. Yellow phosphorus is used in war for making smoke screens to hide battleships and moving columns of soldiers from the enemy. Because of injury to the workmen in match factories, the Government passed a law forbidding the use of yellow phosphorus.

YELLOW AND RED PHOSPHORUS. Yellow phosphorus has to be kept under water as it ignites spontaneously when exposed to air. Red phosphorus does not take fire spontaneously but is easily ignited by friction.

EXPERIMENT

Burning phosphorus under water. (Do not handle phosphorus.) Put a small piece of yellow phosphorus into a large test tube and half fill the tube with water. Warm this test tube of water and phosphorus and set it into a beaker of water which is heated to 80° C. Insert a tube connected to an oxygen generator and slowly bubble oxygen against the heated phosphorus, which will now burn brilliantly under water.

EXPERIMENT

Smoke rings from phosphorus. Put 200 cc. of concentrated sodium hydroxide solution into a small flask. Into this solution put a piece of yellow phosphorus about the size of a pea. Fit the flask with a two-holed rubber stopper with an inlet tube reaching nearly to the bottom of the solution and an outlet tube which just reaches through the cork but not to the solution. The other end of the outlet tube should reach into the water of a pneumatic trough.

Pass illuminating gas through the solution until all air is driven out of the flask and tubes. Bring the solution to a boil by a burner under it. Now turn off the gas. Soon the bubbles rising from the water in the trough will ignite spontaneously and produce smoke rings. The inflammable gas is hydrogen phosphide.

Red phosphorus. On standing, yellow phosphorus gradually changes into a red powder. This red phosphorus no longer takes fire easily, and is not soluble in carbon disulfide. Red phosphorus is not poisonous. For this reason, it has replaced the other variety in making matches.

The strike-anywhere match. The modern match that can be ignited by rubbing it against any rough surface uses a sulfide of phosphorus instead of the free element. Potassium chlorate is still the oxidizing agent. Paraffin is the easily oxidizable substance. Phosphorus sesquisulfide (P_4S_3) is the starter of combustion and glue the binder. Other fillers are often added to the match head, such as ground glass to increase the friction, and hence the heat of friction, and clay or plaster of Paris to protect from moisture.

Reducing fire hazards. Matches have been responsible for many destructive fires, so numerous attempts have been made to reduce the possibility of fire. One device was to soak the match stick in sodium sulfate solution to prevent an afterglow when the flame is extinguished. When the wood is not thus treated, the small after-spark often enlarges and starts a fire.

Another attempt to increase the safety of matches is illustrated by the bird's-eye type of match. The major portion of the head of the match has the phosphorus sesquisulfide omitted from it. On the end of the head is placed a much smaller drop of some more material containing the phosphorus compound. To ignite the match, it must be rubbed on the end. Stepping on the match does not ignite it, as the friction does not touch the part containing the phosphorus. Mice gnawing at these matches are not apt to ignite them. The greatest step in the direction of safety — although not always of convenience — has been the safety match.

Safety match. In the safety match, the red phosphorus, the ground glass, and some of the oxidizing agents are put on the box. The combustible material is placed on the match. This match will not ordinarily ignite when rubbed against any

SAFETY MATCHES. As no phosphorus is put in the head of the match, it can be lighted only by striking against the side of the box.

surface other than the prepared surface on the box. When the head of the match is rubbed on the box, the friction between the surfaces is enough to ignite the phosphorus, which in turn ignites the easily combustible head of the match. As the match leaves the box, it continues burning; but the box does not, as the layer of phosphorus is not thick enough to maintain the combustion. Safety matches have the disadvantage of being useless if the box gets wet or lost.

QUESTIONS OF FACT

1. Relate how the Romans maintained a supply of fire.
2. Explain how the early pioneers kept fire overnight.
3. How was fire started with tinder?
4. How did the American Indians start a fire?
5. What element is most easily oxidized?
6. When were matches invented?
7. Give the constituents of the first matches.
8. From what is phosphorus obtained?

9. Why was a law passed prohibiting the use of yellow phosphorus in matches?

10. Describe red phosphorus and contrast its properties with those of the yellow variety.

QUESTIONS OF UNDERSTANDING

1. Discuss the strike-anywhere match.
2. What has been done to reduce fire hazards from matches?
3. Explain the safety match.
4. What is one disadvantage of the safety match?

PROBLEMS

1. Calculate the weight of phosphorus obtainable from one ton of $Ca_3(PO_4)_2$.

2. Calculate the weight of SiO_2 necessary to react with 200 pounds of calcium phosphate, according to the equation:

$$Ca_3(PO_4)_2 + 3 SiO_2 \longrightarrow 3 CaSiO_3 + P_2O_5$$

REFERENCE FOR SUPPLEMENTARY READING

1. Martin, Geoffrey. *Triumphs and Wonders of Modern Chemistry.* D. Van Nostrand Co., New York, 1922. Chap. XIV, "The Phosphorus Group of Elements."

   (1) Relate the occasion of the discovery of phosphorus. (p. 314)
   (2) What remarkable property of phosphorus made it exhibited over Europe as a wonder of nature? (pp. 314–315)
   (3) Describe the effects of phosphorus poisoning. (pp. 317–318)
   (4) State some of the objections to matches made with yellow phosphorus. (p. 318)
   (5) What difficulty was experienced in trying to replace yellow phosphorus with the red variety? (pp. 318–320)
   (6) Describe what a modern match-making machine does. (pp. 320–321)
   (7) How many persons could be poisoned by the phosphorus in one man? (p. 321)
   (8) Where do fishes get their phosphorus? (p. 322)
   (9) Describe a great discovery of a source of phosphorus. (pp. 323–324)
   (10) Relate the story of an atom of phosphorus. (pp. 325–329)

Paper making. The first people to make paper were the Chinese. Hoping to keep the process secret and always within the empire, the Chinese guarded it zealously. However, in the eighth century some captured Chinese were compelled to reveal the secret, which slowly spread until it reached Europe in 1150 and America in 1690.

All paper was made from rags until the demand became so great that rags could not be obtained fast enough to supply it; then wood became the raw product for all but the best papers, such as photographic papers, which must retain their tensile strength when wet and which are still made from rags.

The principle of paper making is quite simple, although the machines for making it rapidly in large quantities are not so simple. Whether wood or rags is the starting point, the material is treated chemically to remove everything except the cellulose fibers. It is then shredded and beaten in water until the floating fibers give it the appearance of milk. These are now matted by catching them on a vibrating screen, and then dried into a sheet. To make the fibers cling together more firmly and to make the surface smooth, the better papers are sized with glue, rosin, and white pigments.

The different processes of making paper from wood differ in the chemicals used to remove the lignin, the binding substance in the wood. The sulfite process uses sulfurous acid and calcium acid sulfite, the soda process uses $NaOH$, and the sulfate process uses a mixture of $NaOH$ and Na_2S to remove the lignin.

Burlap bags and gunny sacks. The jute plant is somewhat similar to the flax, although greatly exceeding the latter in height. Whereas three feet is a fair height for flax, the jute of India grows to an average height of twelve feet. Jute fibers, also being the bast fibers of the plant, are consequently much coarser than linen. We find them in burlap bags, gunny sacks, and other cheap wrappings. Jute is the cheapest as well as the weakest and least durable of the cellulose fibers of commerce. However, it is the most easily spun and ranks next to cotton in the quantity used. Enough gunny sack cloth is made in India to wrap a strip one yard wide around the earth three hundred times. Jute must be rotted in much the same way as is flax to remove the pulp between the fibers. The rest of the process of making the cloth is also not greatly different from the method used with flax.

Tent cloth, rope, and Panama hats.   Among the other vegetable fibers are hemp — similar to jute in appearance — from
which tent cloth, sail cloth, fire hose, and fishing nets are made.
Another fiber, sisal, used in making binder twine, comes from
the tough fibers which run the length of the sisal plant of Yuca-
tan.   The sisal is a variety of desert century plant.   Fine quality rope comes from the manila hemp, a plant resembling the
banana tree.   Panama hats, from Central America as the name
suggests, are made from fibers in the stems of the toquilla plant.

Oilcloth.   Oilcloth has been a useful substance for a long
time.   Oilcloth is made by spreading layer after layer of linseed oil on cloth and printing some pigment design on it.   The
linseed oil oxidizes into a tough film.

Linoleum.   Linoleum is similar to oilcloth in that it is made
from linseed oil.   Cork dust is put into the linseed to give body
to the linoleum.   In the cheaper form, the design is printed
on the surface, but in the case of inlaid linoleum, blocks of the
colored linoleum are worked together with fresh oxidized oil.
The advantage of inlaid linoleum over printed linoleum is that
the design of the former is the same down to the burlap base;
whereas the design of the latter is only on the surface and soon
wears away.

Leather.   Leather is made from the skins of animals by a
process called tanning.   In tanning, the proteins of the skin
are put through a toughening process that makes the material
more flexible and less affected by water.   The raw hide is stiff
and somewhat brittle when dry, but limp as a dishrag when wet.
The leather is not so absorbent of water as the raw hide and can
withstand a great deal of water.

Curing hides.   After the hides are removed from animals,
they are attacked by the bacteria of decay.   To keep the hides
from spoiling, the butcher coats the raw side of the skin with
solid salt, which forms a concentrated salt solution in the water
of the hide.   As long as this solution is six per cent or more of
salt, decay will not take place.

Soaking and fleshing.   At the tannery the salt is washed
off, and the hide is allowed to soak up water until it contains

as much as it originally did before any was lost by drying or removed by the salt. Now all fat and pieces of flesh are scraped from the under side of the skin.

Liming. The skin is now soaked in a solution of calcium hydroxide containing a little sodium sulfide, which loosens the thin epidermis, or outer skin, and also the hair, which is now scraped off.

THE MANUFACTURE OF LEATHER.   *Courtesy of Ohio Leather Co.*

Bating. The lime treatment causes the skin to swell up and become rubbery. In order to correct this condition, the skin is next put into a solution of pancreatic enzyme and salt.

Pancreatic enzyme is a complex organic substance which occurs in the pancreas of animals. Before chemists put tanning on a scientific basis, pigeon manure was used in the bating solution. By discovering exactly what produced the effect of bating and finding a cleaner way to get the substance, the chemist has eliminated many of the foul odors and nasty solutions which were formerly found at the tanneries.

Vegetable tanning. After bating, the hide is soaked in a tank containing tan bark and leaves rich in tannin. The tannin unites with the protein of the under skin (collagen) in the ratio of 60 parts tannin to 100 parts collagen, to form leather.

Chrome tanning. In chrome tanning, the bated hide is soaked in sulfuric acid and salt for a while and then put into a solution of basic chromium sulfate, $Cr(OH)SO_4$. The chromium compound unites with the collagen to form a superior grade of leather, which is not affected by boiling water.

Final treatment of leather. Leather prepared by either process of tanning will get hard unless given a final treatment of oil. When a suitable oil is worked into the leather, it remains pliable.

REVIEW QUESTIONS

1. Who invented paper making?

2. How did the secret of paper making get out of China?

3. When did the secret of paper making first reach Europe and America?

4. What was the raw product from which all European papers were made for years?

5. From what are photographic papers made?

6. Explain the purpose of each of the following in paper making:
    (a) chemical treatment of raw materials
    (b) shredding and beating
    (c) matting

7. In making paper explain: (a) sulfite process; (b) soda process; (c) sulfate process; (d) sizing; (e) beating; (f) what paper is.

8. What plant is the source of each of the following: (a) burlap; (b) rope; (c) Panama hats; (d) fish nets; (e) gunny sacks?

9. Describe the making of oilcloth.

10. What is inlaid linoleum?

11. What part does chemistry play in leather making?

12. In connection with tanning leather, tell what each of the following are: (a) liming; (b) bating; (c) vegetable tanning; (d) chrome tanning.

13. What is the final treatment of leather, and what is its purpose?

REFERENCES FOR SUPPLEMENTARY READING

1. Allen, Nellie B. *Cotton and Other Useful Fibers.* Ginn and Company, Boston, 1929. Chap. VI, pp. 211–217, "Jute and Sisal Fibers" [burlap and gunny sacks], and pp. 218–257, "Hemp Fibers" [tent cloth, fishing nets, rope].
   (1) Describe the production of burlap.
   (2) Tell about hemp and its products.
2. Howe, H. E. *Chemistry in Industry,* Vol. I. The Chemical Foundation, Inc., New York, 1925. Chap. XII, pp. 157–175, "The Making of Leather."
   (1) What substances occur in a cowhide?   (p. 157)
   (2) Describe the epidermis.   (p. 158)
   (3) Describe the derma.   (p. 160)
   (4) What does the lime used in removing hair from the hides do to them?   How is this effect counteracted?   (pp. 166–167)
   (5) What is meant by vegetable tanning?   (p. 167)
3. Howe. *Chemistry in Industry,* Vol. II. Chap. XIII, pp. 221–231, "Matches."
   (1) Write an account of the manufacture of matches.
4. Howe. *Chemistry in Industry,* Vol. II. Chap. X, pp. 171–189, "Glues and Gelatins."
   (1) In what sense are glue and gelatin alike?   (p. 171)
   (2) Describe the manufacture of gelatin.   (pp. 172–176)
   (3) Discuss the uses of gelatin and glue.   (pp. 177–178)
   (4) How can glue be injured?   (p. 179)
   (5) Discuss the relation of gelatin to photography.   (p. 183)
   (6) What are some other substances used as glue?   (pp. 183–188)
5. Howe. *Chemistry in Industry,* Vol. II. Chap. III, pp. 37–57, "Casein."
   (1) How was casein first obtained from milk?   (pp. 37–38)
   (2) Considering fat as the principal product of milk, compare the amounts of casein and milk sugar.   (pp. 38–39)
   (3) What elements occur in casein?   (p. 39)
   (4) What are some of the properties of casein?   (p. 40)
   (5) How is self-soured casein produced?   (pp. 40–42)
   (6) Describe the making of acid casein.   (pp. 42–44)
   (7) Discuss the preparation of rennet casein.   (pp. 44–45)
   (8) Compare the caseins prepared by different methods.   (pp. 45–48)
   (9) Tell about the different articles made from casein.   (pp. 48–56)

Soaps.   Soap is one of the common articles around the home. It is the king among cleansers.   Nothing can take its place. Soap is made by melting fat with steam coils and running in a sodium hydroxide solution.   After the fat has reacted with the water and alkali, there are two useful products in the mixture, glycerol and soap. An excess of salt is added to the solution to reduce the solubility of the soap and cause it to separate from the liquid. The glycerol remains dissolved in the salt-alkali solution, from which it can be separated with more or less difficulty.

The Twitchell process. To make the recovery of glycerol simpler, an American chemist named Twitchell invented a process of decomposing the fat into glycerol and fatty acids by the use of acid. The glycerol can be separated, and the fatty acids subsequently neutralized to soap.

Special soaps. Toilet soaps are made to lather more freely by including coconut oil. The floating soaps have air blown into them. Soaps for kitchen use often contain alkaline substances, such as sodium carbonate or sodium silicate. Shaving soaps, on the other hand, must be free from all alkali, or they will sting the face. Transparent soaps are made by dissolving the dried soap in alcohol. The excess alcohol is evaporated off, leaving the soap transparent.

Perfumes and flavors. Perfumes and flavors are closely related for the reason that the senses of taste and smell are closely related. Closing the nose restricts one's ability to taste to such an extent that nothing tastes very good when the nose is stopped up with a cold. By perfumes are meant those substances which appeal only to the sense of smell, while the flavors are put into food and drink to be taken internally. Some substances, however, may serve both purposes.

The chemist improves perfumes. Originally all perfumes were obtained from plants or (in a few instances) from animals. The processes of extraction were wasteful and inefficient. For instance, lard was used to extract the perfume *attar of roses* from the rose petals. Because of the greasy nature of lard it had to be separated from the final product. With the methods available before the chemist took charge, some of the lard remained and afterwards became rancid. Moreover, during one step of the process the wash water took out one of the valuable constituents. To keep from wasting this substance, the water was sold under the name of *rose water*.   When the chemist took charge, he learned that the substance absorbed in the rose water is phenylethyl alcohol.   This substance he was able to synthesize.   This made it possible to replace it in the original perfume, and make the odor of the latter equal to that of the original rose.

The perfume industry, which depends upon the flowers, requires large acreages of these flowers.   For instance, it takes from 250,000 to 750,000 roses to produce one pound of attar of roses which sells for $160.   Although knowing that the chemist could help the perfume industry, its leaders were reluctant to hire them, for they remembered what happened to the dyeing industry when the chemists took charge.   At one time thousands of acres of the indigo plant were grown in India for dye production.   However, when the chemist learned to synthesize indigo, the indigo plant was no longer needed.   Likewise, at one time thousands of acres of the madder plant were grown in France for the dye, alizarin.   In this case, also, the chemist synthesized the dye more cheaply than it could be obtained from the plant, so that growing the plant came to a stop.   It is little wonder the perfume makers feared for the thousands of dollars invested in growing roses and other fragrant plants.

The chemist, however, did not abolish the industry of growing plants for perfumes, but he did improve it in many ways.   He found better solvents to extract the fragrant oils.   He invented steam distillation, a process of distilling high-boiling oils, which would not distill by themselves.   In addition to this, the chemist studied the active ingredients of most of the perfumes and flavors.   Many of these have been synthesized.   Among them are : citral from lemon oil ; menthol from peppermint oil ; geraniol, the alcohol in rose oil ; cinnamic aldehyde from cinnamon ; and camphor.   Synthetic camphor from turpentine today competes successfully with the product of the Formosa camphor tree.

New perfumes.   Perhaps the greatest contribution of the chemist to the perfume industry has been the discovery of new substances with delightful odors.   One of these, ionone, is very similar to violets and sold at one time for $1280 per pound, but the price now has been reduced to $10 per pound. Another similar achievement is artificial musk, trinitrobutyltoluene:

$$
\begin{array}{c}
CH_3 \\
| \\
C \\
NO_2C \diagup \quad \diagdown\!\!\!= CNO_2 \\
HC \quad\quad C\!-\!C_4H_9 \\
\diagdown C \diagup \\
| \\
NO_2
\end{array}
$$

This substance at one time sold for $160 per pound but is now only $4 per pound. Another synthetic perfume, benzyl acetate, has the odor of jasmine. Its formula is $CH_3CO_2CH_2C_6H_5$.

The flavors. In the study of esters, page 487, several of the common fruit flavors were mentioned — amyl acetate of the pear, ethyl butrate of the pineapple, etc. Some of the flavors are esters, some alcohols, and some aldehydes. Dozens of the natural flavors have been synthesized, yet natural flavors have not been completely duplicated, due to the fact that nature uses mixtures which are hard to imitate. However, in time the chemist will probably be completely successful.

REVIEW QUESTIONS

1. How is soap made?
2. Why is the mixture in a soap vat saturated with salt?
3. In what soaps is alkali desirable, and in which kinds must it be absent?
4. How does the Twitchell process differ from the old method of making soap?
5. What is the advantage of the Twitchell process?
6. Why is coconut oil put into toilet soaps?
7. How are some soaps made to float in water?
8. What other useful substance is obtained in soap making?
9. How are transparent soaps made?
10. In what sense are perfumes and flavors related?
11. How was attar of roses originally made?

12. What was one defect of the method?

13. What substance is in rose water?

14. Discuss three ways the chemist has helped the perfume industry.

15. What constituent in attar of roses is now synthesized?

16. Why were perfume producers reluctant to hire research chemists?

17. Did their fears materialize?

18. Name a few natural odors which have been synthesized.

19. To what perfume does synthetic ionone have a similar odor?

20. What was the original price of this perfume? What is its present price?

21. In artificial musk, name the radicals attached to the benzene ring.

22. Give the formula for the perfume that has the odor of jasmine.

23. Why are the synthetic flavors inferior to the natural?

24. To what classes of substances do most of the flavors belong?

REFERENCE FOR SUPPLEMENTARY READING

1. Howe, H. E. *Chemistry in Industry*, Vol. I. The Chemical Foundation, Inc., New York, 1925. Chap. XVI, pp. 253–293, "Perfumes and Flavors."

   (1) To what extent do perfumes have antiseptic properties? (p. 255)

   (2) How are odorous substances obtained from the plants? (pp. 260–264)

   (3) What parts of the plants may be used? (pp. 260–261)

   (4) Name some natural perfumes of animal origin. (pp. 264–267)

   (5) Name several fragrant substances obtained from the plants. (pp. 268–272)

   (6) Relate how chemists have changed cheap substances to valuable substances. (pp. 272–274)

SUMMARY

In the head of a modern match are certain easily oxidized materials, potassium chlorate (which supplies oxygen for burning), a compound of phosphorus and sulfur, and glue (to hold these materials together). The heat of friction starts the phosphorus sulfide to burning, which immediately ignites the rest of the materials. The phosphorus used in making the match is obtained by heating silica and carbon with calcium phosphate. Phosphorus exists in two forms, the poisonous spontaneously combustible yellow phosphorus and the non-poisonous red phosphorus. In the safety match the phosphorus compound is not put into the head of the match but is placed on the box instead. In this way the match will not light unless it is struck on the box.

Paper, one of our most important common substances, is made by matting together many small cellulose fibers by straining them out of water with a vibrating sieve. The different paper processes differ in the chemicals used in removing the lignin of the wood from the small cellulose fibers.

Burlap and gunny sacks are made from the fibers of the jute plant. Rope comes from a similar plant called hemp.

Oilcloth and linoleum depend upon the fact that boiled linseed oil hardens by oxidation into a tough, flexible material. In oilcloth, the waterproof surface is produced by dipping the cloth several times in the oil and allowing time for oxidation after each dipping. In linoleum the oxidized linseed oil is used as a binder for cork dust.

The chemist has put leather making upon a completely scientific basis. In place of ashes, which are of variable alkalinity, used by the ancients for removing hair, he uses calcium hydroxide, which is uniform in strength. In place of pigeon manure for bating the hide, he has substituted a sanitary enzyme. He also is able to prepare oak-bark solutions with just the right concentration of tannin for making the best leather out of the collagen in the hide. He has also discovered chrome tanning, in which basic chromium sulfate replaces tannin and produces a better leather for some purposes.

Soap, the king of cleansers, is made by boiling fats in strong alkali solution. When finished, the solution is saturated with salt to make the soap separate from solution as a cake, leaving the glycerol dissolved in the water.

In the field of perfumes and flavors the chemist has evolved better methods of extracting the fragrant substances from the flowers; has synthesized many of the flavors, and many new substances with delightful odors.

CHAPTER 40

←→

THE FUTURE OF CHEMISTRY

Introduction. In evaluating a science or an invention or an occupation, one must take into account its future outlook. Thirty years ago the village blacksmith was often the most prosperous man in the town. Each little community had at least two blacksmith shops. Blacksmithing was considered a most promising trade. Today, blacksmiths are few in number, and most of them are hard pressed to make ends meet. A generation ago a well-known science magazine listed a number of desirable inventions, which would bring a fortune to the successful inventor. One of these was a self-locking buggy-whip holder. Today, such an invention would not be worth patenting, as buggy whips are no longer used.

The outlook for chemistry was never brighter than it is now. Such a foundation of chemical knowledge has been laid that nothing seems impossible any more. The accomplishments of chemistry in the last hundred years are as marvelous as fairy tales. Moreover, all this is probably only the beginning. The chemical achievements of the next hundred years will undoubtedly eclipse those of the past.

Future of inorganic chemistry. The future will search out the elements that are little known now and put their production upon a commercial basis. Many new uses will be found for the new metals, which will provide better materials for special purposes. As a result, there will be better cooking utensils, better cutlery, better farm machinery, better automobiles, better airplanes, and better alloys for innumerable uses.

The chemicals in large deposits, such as the Dead Sea, the Great Salt Lake, and the ocean, will be recovered and made use of. The future chemist will produce better and cheaper chemicals, drugs, pigments, cements, acids, and alkalies.

A RESEARCH CHEMIST AT WORK. In a large measure the scientific and industrial future of the world rests on the shoulders of the research chemist.

Future of photography. The outstanding need of photography today is a simplified color process, by which prints in natural color can be produced by the amateur without too great expense. The chemist is hard at work now in this field. He is already far enough along to feel certain that success is just around the corner. Success will probably come by the discovery of dyes that are sensitive enough to certain kinds of light so that a triple-dyed emulsion can be developed and the color made permanent. The successful process may be a selective bleaching method or some sort of toning method, in which the colors are developed in the emulsions.

Future of analytical chemistry. Easier and more definite tests for the common ions are being invented from year to year. Comparable tests for all the ions and for useful natural substances will be discovered. A microtechnique has been commenced that will be perfected to the point that an analysis can be carried on with only a drop of the substance. This procedure will be a great help to medical research, especially in dealing with the powerful and mysterious regulators in the human body known as hormones, and equally the important substances in the foods called vitamins.

Future of dyes and textiles. Synthetic dyes have far surpassed the natural dyes in all respects. New synthetic dyes will continue to be invented. Every month there will continue to be a score or more patented. Synthetic fibers will continue to be improved. No doubt nitrogen will eventually be introduced into the molecule of the fiber, thus making rayon more like silk and wool. Synthetic furs will probably replace natural furs, thus enabling the ladies to have the beauty and comfort of furs without necessitating the death of some animal. Even synthetic leather is not beyond the realm of possibility.

Future contribution of chemistry to health. In the near future chemistry will not only put nutrition on a sounder basis, but it will no doubt solve the structure of some of the vitamins and perhaps synthesize them in the laboratory. Vitamin C has already been synthesized and some wonderful cures are reported. The synthesis of vitamins will insure a cheap supply, so they can be supplied in correct amounts to all foods. Great progress has been made in the knowledge of children's diet. This has saved the lives of thousands of babies. Similar information regarding adult diet will no doubt become available. Many diseases will be avoided by avoiding the cause, and many others by keeping the body resistance up to par.

Future aid of chemistry to medicine. The chemist has accomplished some noteworthy successes in salvarsan (the specific for syphilis), Baeyer 205 (a specific for sleeping sickness), the antiseptics — mercurochrome, metaphen, and hexylresorcinol — and the anesthetics — ethylene and procaine. These indicate what he can do if he is given a chance. It is a pity that there is so little financial support of medical research. Imagine what a cure for cancer would mean to thousands of people who now face the future with little hope. Enough has already been done to make us confident that almost any disease can be conquered.

Future of chemistry and agriculture. Chemistry will continue to improve methods of analysis of agricultural products. It will insure an adequate supply of fertilizer. It will find out how to make new salable products from the agricultural wastes, such as fruit seeds, corn cobs, corn husks, peanut shells, cottonseed bran, and oat bran, and (above all) from sawdust and wood waste.

Future of petroleum chemistry. Petroleum products will be used to a better advantage. Gasolines will have increased antiknock properties. Many unsaturated and ring hydrocarbons that are now burned up will be used for synthesizing new substances. Lubricating oils will be selected with properties to fit the exact needs of the situation. If the natural oil supply ever becomes exhausted, the chemist will step in and fill the gap with synthetic oils by the hydrogenation of soft coal by the Bergius process, or by some better method.

Future of rubber. The future of rubber chemistry will probably be along two lines. First, there will be an increase in the abrasion resistance and lasting qualities of natural rubber through the use of new accelerators and antioxidants. The

most radical changes, however, will come through the introduction of synthetic rubber. Duprene, the synthetic chlorine
rubber from acetylene, is here as a commercial product now.
Synthetic rubber is also being used in Germany and Russia.
If the price of natural rubber increases materially, synthetic
rubber will come increasingly into competition with it. The
president of one the leading tire and rubber companies recently
made the prediction that all transportation will soon be on
rubber-tired wheels. Even tractors used in plowing are beginning to use rubber tires.

Chemistry and war in the future. Chemistry will continue
to take a greater part in supporting war in all its phases,
poisonous gases included. The newspapers have carried
accounts of poisonous gases that would kill every living thing
in a large city. The death-dealing properties of these gases
are fortunately not so bad as suggested, but they are undoubtedly bad enough.

In other lines, chemistry will produce better armor and
better war materials of all kinds. Perhaps the greatest influence of chemistry on war will be in connection with chemical
engineering — that is, the production of explosives and other
war materials in large quantities. The colleges and several of
the chemical magazines are placing more and more emphasis
on chemical engineering. It is one thing to produce a successful laboratory synthesis and quite another thing to produce it
in carload lots. Many of the simple processes of the laboratory become serious problems in the plant. Filtering, stirring,
heating, and transferring materials, all have become difficult
problems requiring specialized equipment and special knowledge.

Investors and chemical research. In large industries today
the chemist is in control of nearly every step of the manufacturing process. He is responsible for the quality of materials
from the time the supplies are purchased until the finished
product is sold. The company that fails to make this provision
for maintaining the quality of its materials is likely to be
crowded out of business by its more progressive competitors,

who have made use of chemical control. Moreover, each manufacturing organization should have a corps of research scientists to see if better products can be made. Part of these people must be chemists. Not only must the directors of companies be able to appreciate the value of chemical research, but investors will do well to invest their money in those companies that are known to make use of the possibilities of chemical research. . In the recent world-wide depression, those companies with the most elaborate chemical research organizations were able to pay dividends while the others were suffering losses.

Chemistry in the United States. America being a pioneer country where there was plenty of room for settlement, where there were boundless natural resources for development, interest in scientific knowledge lagged behind Europe for a long time. Up until 1900 a person wishing to get a doctor's degree usually went to Germany as there was a feeling that as scientific institutions the German universities were superior to our own. This belief was more or less true at that time. Now, however, the situation is entirely changed. The American universities are second to none. They are doing outstanding research in nearly every field. A doctor's degree in our better universities gives the holder as much or more prestige than one from the best of the European universities. Notable discoveries in the pure sciences develop in the United States as frequently as anywhere, and industrial research is developing here, as fast or faster than anywhere else in the world. The United States can feel proud of her scientists and especially her chemists.

REVIEW QUESTIONS

1. What may we expect as the future of inorganic chemistry?
2. What desirable chemical achievements may be solved in photography?
3. What great improvements are taking place in analytical chemistry?
4. How may chemistry help us avoid disease?
5. Name several achievements of chemistry as an aid to medicine.
6. How is chemistry aiding agriculture?

7. What improvements may chemistry make in petroleum products?

8. What can the chemist do for the automobilist if the crude oil supply gives out?

9. What is the outlook for synthetic rubber?

10. What are some of the problems of the chemical engineer?

11. Are there many things for the future chemists to do?

12. Will financiers be compelled to consider the importance of chemical research?

REFERENCES FOR SUPPLEMENTARY READING

1. Darrow, Floyd L. *The Story of Chemistry.* Blue Ribbon Books, Inc., New York, 1930. Chap. XII, pp. 391–450, "American Progress in Chemistry."
   (1) How does the chemical industry of the United States compare with that of any other country? (p. 391)
   (2) Name three substances manufactured in Virginia. (p. 393)
   (3) When were laboratory chemicals and apparatus first imported into the American colonies? (p. 393)
   (4) What substances were encouraged by the provincial congress? (p. 393)
   (5) When was the first dyestuff plant established? (p. 394)
   (6) Who invented the oxyhydrogen blowpipe? (p. 394)
   (7) When was the first protective tariff on chemicals? (p. 394)
   (8) When was natural gas first used? (p. 394)
   (9) When was the sulfite paper process invented? (p. 396)
   (10) What four schools taught advanced chemistry in 1853? (pp. 396–397)
   (11) Outline the history of E. I. du Pont de Nemours. (p. 403)
   (12) Discuss the history of one of the following chemical products: (a) metals; (b) electrochemistry; (c) dyestuffs; (d) glass; (e) industrial alcohol. (p. 403–450)

2. Holmes, Harry N. *Out of the Test Tube.* Ray Long and Richard R. Smith, New York, 1934. Chap. XXXII, pp. 362–373, "Have You a Chemist on Your Board?"
   (1) What is the present status of chemical research in America? (p. 362)
   (2) How many research laboratories are mentioned on page 363?
   (3) How is research related to profits? (p. 363)
   (4) What German practice is being adopted by American bankers? (p. 363)
   (5) Illustrate the value of research. (p. 365)
   (6) Discuss the value of pure research. (pp. 366–368)

3. Howe, Harrison E. *Chemistry in the World's Work*. D. Van Nostrand Co., New York, 1926. Chap. XVI, pp. 223–240, "The Trend and Purpose of Modern Research."

    (1) Are some industrial chemists working on pure science which offers no immediate rewards? (p. 229)

    (2) What is the trend of chemical research? (pp. 229–230)

    (3) Why is the corrosion problem becoming increasingly important? (pp. 230–231)

    (4) What promises to be one of the most important peacetime results of the wartime gas service? (p. 231)

    (5) What new principle is suggested for causing insecticides to stick to the leaf? (p. 232)

    (6) Why will synthetic resins be more in demand in the future? (p. 234)

    (7) With what is research confronted in relation to paper? (pp. 235–236)

SUMMARY

To sum up the possible improvements that chemistry should make to our present living conditions, we may say that there is not a thing of any kind that we use that cannot be improved either directly or indirectly by the chemist. There is not a waste product of any industry, including agriculture, that may not be made into useful substances by the chemist. The beautiful and useful things of nature may be imitated and even improved. Who would dare to say that the chemist will not be able to synthesize the diamond, the ruby, and the emerald? Who would dare to state that elastic panes of glass are not a possibility? A prophet in any line is risking the chance that his prophecy will not come true, but perhaps he is taking less risk in chemistry than in any other line.

CHAPTER 41

CHEMISTRY 'AS A LIFE WORK

Introduction. In contemplating a life work, many a modern youth thinks he would like to be a ship's radio operator, an aviator, or a news-reel photographer. This is a perfectly natural attitude, for these occupations suggest the romance of far-away places, and promise the thrills associated with personal risk. Ambitious young people should not, however, fail to consider the possibilities and satisfactions inherent in scientific work. While it is true that adventures in science are usually intellectual rather than physical, they are nevertheless just as real. Moreover, the opportunities for serving others are far greater in this field than in most others. Finally, there is promise of adequate financial recognition for one's efforts.

Nature of the chemist's work. In choosing an occupation, one should consider other things besides salary — important as that is. Success has been defined as "doing what you want to do and getting paid for it." To have an occupation that is enjoyable is a highly desirable thing. Chemistry has many attractive features and some drawbacks. Of course, chemical work is mostly indoor work. No doubt there are days when the chemist longs for the sunshine and fresh air of the out-of-doors. On the other hand, when it is dark and cold and wet outside, the chemist in his warm laboratory can turn on the light and comfortably work with his test tubes and solutions. He can rest assured that, rain or shine, he will get his pay check at the end of the week. Another good thing about the chemist's work is that in most cases it has variety — new tasks to perform, new problems to solve.

The chemist, like everyone else, has his ups and downs. Some phases of his work are highly interesting, while others are not. Some chemical jobs are better than others. Compared

A MODERN INDUSTRIAL LABORATORY.

with other professions, however, chemistry is as good as any and better than most.

Salaries of chemists. In chemistry more than in most vocations, the most highly trained men receive the highest pay. To get to the top salary, a person should have a Ph.D. degree, which is granted only at the end of three years of specialized training after graduation from college. Although the young chemist may have to accept a beginning salary of $1500–$2000 a year, if he is properly trained, he may expect rapid advancement for the next ten years.

A study was made of a thousand chemical-engineering graduates of the Massachusetts Institute of Technology. After they had been out working for fourteen years, the highest fifth were receiving an average of $13,000 per year, while the lowest fifth were receiving $3000 per year.

There is practically no upper limit to the salary of chemists who may become general managers of large plants. The strictly chemical jobs usually stop at $5000 per year, but they are usually the stepping stones to positions of higher pay.

Occasionally one hears of a chemist's getting only $60 or $70 per month. Usually one will find that this person is only a high-school graduate making routine tests that require very little knowledge.

Chances of employment. In times of extreme depression some men in every profession or occupation are unable to find work. The chemists, however, usually fare better than civil engineers or electrical engineers. In fact, the chemical industries, as a whole, fared better than other industries during the world-wide depression after the World War.

Opportunities for chemists. Aside from regular employment at a fixed salary, the chemist has the chance of developing new products and new processes. In the earlier chapters of this book we have noted many instances where chemical discoveries have brought resourceful men honor and renown, wealth and leisure.

Nor must the young student imagine that all the important discoveries have already been made. In a recent magazine article, Dr. Charles F. Kettering, Director of Research of General Motors, stated that there are now more interesting problems to be solved than ever before in the history of the world. One of these problems is that of studying the green chlorophyll of the grass to discover how it uses the sun's energy to synthesize organic substances. All the heat and power now stored in coal, oil, and natural gas has been stored by the chlorophyll of plant life. If we knew this secret, we could get enough energy directly from the sun to run all our machinery.

Parting message to the student. You have now finished the first course in chemistry. If you think chemistry is hard and uninteresting, you will probably not study it further. If, on the other hand, you like chemistry and your high school does not give a second year of the subject, you will probably study it in college. The first course in college chemistry is somewhat like this course except that it goes deeper into the subject. The last part of the laboratory course is usually qualitative analysis — a scheme for detecting the different elements and ions in an unknown mixture. In some colleges qualitative analysis is made an independent course from general chemistry.

After the first course, it is customary to divide chemistry into different courses, such as quantitative analysis, the measuring of the percentages of elements in compounds, mixtures, and minerals; organic chemistry, the chemistry of the carbon compounds — those substances that are so closely related to life, health, and happiness; and physical chemistry, which deals with the laws of chemistry. Knowledge is said to be power. This is especially true of chemistry. The chemist who knows the most will be the most successful. The chemical engineer will need other courses besides those that have been mentioned.

If you take up further work in chemistry, we wish you every success. This will only come by a great deal of hard work.

THE SIMPLIFIED UNITS OF SCIENCE

Need for the new scientific units. In chemistry it is often necessary to measure different quantities of length, weight, volume, and temperature. The student no doubt remembers having wrestled with the difficult and irregular units of linear measure, cubic measure, liquid measure, avoirdupois weight, etc. He is therefore in a position to appreciate the impatience of the chemist with such a cumbersome system of weights and measures. The scientist has discarded the old units in favor of a convenient and unified decimal system of weights and measures, in which reduction from one unit to another becomes a mere matter of shifting the decimal point. Although the new units may appear strange and forbidding at the start, after a little practice the student will find them much easier to use than the old ones.

There is another excellent reason for scientists adopting the metric units — although this reason may not appear so obvious to the student. These units are in international use, and a familiarity with them makes it easy for a chemist to understand the work of foreign chemists, who publish their findings in metric units.

POUND AND KILO-GRAM.

The metric system of measures and weight. The fundamental unit of measure is the *meter*, which is 39.37 inches (a little over a yard) long. Of the many secondary units that have been derived from the meter, only a few are of importance to the chemist. These important ones are shown in the table on the following page.

The conversion factors of the last column are approximations, except for the measures of length, which are exact by definition. Some of the statements of equivalence in the second column are not absolutely true either. For example, one milliliter actually equals 1.000027 cubic centimeters. These slight inaccuracies need not, however, disturb us at this stage of our work.

METRIC UNITS

UNIT	METRIC EQUIVALENT	U. S. EQUIVALENT
Length		
meter (m.)	fundamental unit	39.37 inches
centimeter (cm.)	.01 m.	.3937 inches
millimeter (mm.)	.001 m.	.03937 inches
Volume		
liter (l.)	1000 cubic centimeters (cc.)	61.025 cu. in.
milliliter (ml.)	1 cc.	.061 cu. in.
Weight		
kilogram or kilo (kg.) . .	1000 grams or weight of 1 l. of water	2.246 lbs.
gram (g.)	weight of 1 cc. of water	15.432 grains (gr.)
milligram (mg.)	.001 g.	.015 gr.

How the metric system works. Suppose you need to change 138 inches to feet. You must divide 138 by 12, getting 11½ as a result. However, to change 138 centimeters to meters, it is merely

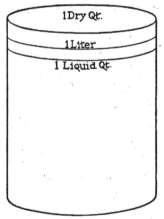

necessary to move the decimal point two places to the left, giving 1.38 as the answer. Now let us try a more difficult problem. What is the weight of 25 cubic inches of water? Since there are 1728 cubic inches in one cubic foot, and a cubic foot of water weighs 62.4 pounds, then 25 times 62.4 divided by 1728 equals the desired answer, 0.9 pounds, approximately. Compare the foregoing with the ease of finding the weight of 25 cubic centimeters of water. Since a cubic centimeter weighs a gram, the 25 cubic centimeters weigh 25 grams.

LIQUID QUART, LITER, AND DRY QUART.

A scientific thermometer. Along with the metric system of weights and measures, the scientist has adopted an improved thermometer, a thermometer whose significant points have some meaning. You have no doubt heard the expression, "forty miles from nowhere." That just about describes our common household thermometer, the Fahrenheit thermometer. It measures temperature from nowhere; its zero has no significance and cannot be determined directly. Its hundred-degree mark also has no significance, and the size of its degree was arbitrarily fixed.

COMPARATIVE TEMPERATURE READINGS

SIGNIFICANT POINT	FAHRENHEIT READING	CENTIGRADE READING
Water boils	212°	100°
Body temperature	98.6°	37°
Recommended room temperature . .	68°	20°
Water freezes	32°	0°
Oranges freeze on trees	23°	− 5°
"Zero weather"	0°	− 17$\frac{7}{9}$°
Propane boils	− 40°	− 40°

The centigrade scale, on the other hand, has its zero point at the temperature at which pure water freezes or ice melts. This temperature is fixed and easily determined. Likewise, its hundred-degree mark is at the boiling point of water, another unchanging and easily determined temperature. Each degree is just one-hundredth of the distance between these points. Since we are more familiar with the Fahrenheit thermometer, it may be just as well to compare readings for a few important temperatures.

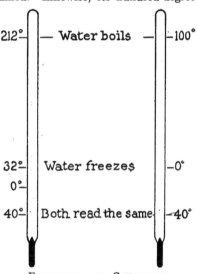

FAHRENHEIT AND CENTIGRADE THERMOMETERS.

An examination of the two scales shows that there are 180 degrees (212 − 32) between the freezing and boiling points of water on the Fahrenheit thermometer , as compared with 100 degrees on the centigrade thermometer. This indicates that 1° C. = $\frac{9}{5}$° F. or that 1° F. = $\frac{5}{9}$° C. To change a given number of Fahrenheit degrees to centigrade degrees, multiply by $\frac{5}{9}$; to change a given number of centigrade degrees to Fahrenheit degrees, multiply by $\frac{9}{5}$.

Here is a simple method of changing from one reading to the other :

(1) Add (algebraically) 40 to the reading.

(2) Multiply by $\frac{9}{5}$ (to change from C. to F.) or by $\frac{5}{9}$ (to change from F. to C.).

(3) Subtract 40 from the product of (2).

If the student wonders about the reason for this method, he should note that the two scales have the same reading at $-40°$. By adding 40 degrees as suggested in step (1), we simply count the degrees from the "coincident zero" instead of from the usual zeros. Step (2) changes to degrees of the other scale, and step (3) changes back to the customary zeros.

Why not adopt the metric system? The student may wonder why the metric units are not in common use here as they are in Continental Europe. The reasons are mainly historical. At the time the metric units were devised, there were many conflicting local and national units in use throughout Western Europe, so that there was a very real need for a common system to facilitate trade. America, on the other hand, had a single system in use all over the country — the English weights and measures. In the United States the land has been surveyed, and a whole industrial system built up in terms of the old units. To introduce the metric system as a whole would involve the junking of billions of dollars' worth of machinery and equipment, and the entire rewriting of our records.

GLOSSARY

absolute zero. The temperature at which a body would have no heat; a temperature 273° below centigrade zero.

acid anhydride. The oxide formed by the loss of H_2O from the molecule of an acid.

acid salt. A compound, part acid and part salt.

adrenaline. The hormone secreted by the adrenal glands, which acts as a heart stimulant.

agate. A form of silicon dioxide colored with small traces of other substances.

alabaster. An exceptionally pure form of gypsum.

alcohols. Organic hydroxyl compounds.

aldehyde. An organic compound containing the group CHO.

alkali. A base or a substance that hydrolyzes to form a base.

alkaline-earth metals. A family of elements intermediate in properties between the alkali metals and the earth metals; it includes beryllium, magnesium, calcium, strontium, barium, and radium.

alkanes. The saturated hydrocarbons.

alkenes. The unsaturated hydrocarbons related to ethylene.

alkynes. The unsaturated hydrocarbons related to acetylene, C_2H_2.

alloy. A metallic substance formed by melting two or more metals together.

alpha iron. A soft magnetic form of iron.

alpha rays. A stream of positively charged helium atoms expelled from radioactive substances.

alum. A crystallized double sulfate of a monovalent and a trivalent metal, such as $K_2SO_4 \cdot Al_2(SO_4)_3 \cdot 24\ H_2O$.

amalgam. An alloy of mercury with some other metal.

amethyst. A form of silicon dioxide with a trace of other substance as coloring.

amines. Ammonia with the hydrogens replaced by alkyl groups.

amphoteric. A substance which may act as either an acid or a base.

anesthetic. A drug which causes unconsciousness or deadens pain.

anhydrous. Entirely free from water.

annealing. Slowly cooling glass to free it from internal strains.

aqua regia. A mixture of three parts hydrochloric acid and one part nitric acid.

argyrol. A colloidal silver-protein compound.

atomic numbers. The number of an element in a row determined by its X-ray spectrum.

atomic weight. The weight of an atom in atomic units; the unit is one-sixteenth the weight of the oxygen atom.

atoms. The smallest units of an element that are chemically united to form the molecules.

atropine. A plant product which dilates the pupil of the eye.

Avogadro's hypothesis. Equal volumes of all gases under equal temperature and pressure contain the same number of molecules.

babbitt. An alloy of tin, antimony, and copper used for bearings in machinery.

baking powder. A mixture of baking soda and some acidic powder containing a little starch and often a little egg albumen.

baking soda. Sodium bicarbonate, $NaHCO_3$.

balancing a chemical equation. Seeing that there are the same numbers of every atom involved, on both sides of the arrow.

base. A substance which in solution tastes bitter, turns pink litmus blue, and produces hydroxyl ions.

basic salt. A compound part base and part salt.

bauxite. Aluminum ore, crude Al_2O_3.

bell metal. An alloy of copper and tin.

benzene. A light liquid, C_6H_6.

beriberi. A deficiency disease caused by a lack of vitamin B in the food.

beryl. Beryllium aluminum silicate, $Al_2Be_3(SiO_3)_6$.

Bessemer steel. Low-grade steel made by the rapid Bessemer process.

beta rays. A stream of negative electrons expelled from radioactive substances.

black alkali. Sodium carbonate in the soil.

bleaching powder. Calcium hypochlorite and chloride.

Boyle's law. The statement that at constant temperature the volume of a gas varies inversely as the pressure.

brass. An alloy of copper and zinc.

bronze. An alloy of copper and tin.

burning. The combining of oxygen with other substances accompanied with light and heat.

burnt alum. Powdered alum which has had all of its water of crystallization expelled by heat.

calcite. Calcium carbonate in large crystals.

calomel. Mercurous chloride.

calorie. The gram calorie is the quantity of heat necessary to raise the temperature of one gram of water one degree centigrade. The kilogram calorie is one thousand times the gram calorie.

caramel. A substance produced by heating cane sugar until it turns red.

carat. Unit of twenty-fourths used in expressing the composition of gold.

carbohydrates. The energy foods including the sugars and all substances which hydrolyze into sugars only.

carborundum. Silicon carbide, next to diamond in hardness.

Carrel-Dakin solution. An antiseptic solution of sodium hypochlorite.

cast iron. Impure iron coming from the blast furnace.

catalyst. A substance which speeds up a chemical reaction without itself being used up.

caustic potash. Potassium hydroxide.

caustic soda. Sodium hydroxide.

cellophane. Cellulose made into sheets.

celluloid. Cellulose nitrate dissolved in camphor.

cellulose. The woody fiber of plants, a carbohydrate which is difficultly hydrolyzed to glucose.

centigrade. The scientific thermometer scale.

chalcedony. A form of silicon dioxide, SiO_2, with a trace of other substances for coloring.

chalk. A form of calcium carbonate.

Charles's law. At constant volume, the pressure of a gas varies directly as the absolute temperature; or at constant pressure, the volume of a gas varies directly as the absolute temperature.

Chile nitrate. Sodium nitrate.

Chile saltpeter. A name for sodium nitrate from Chile.

chloride of lime. A mixed calcium chloride and calcium hypochlorite.

chloroform. $CHCl_3$, used as an anesthetic.

chlorophyll. The green coloring matter in plants, which uses the energy of the sun to synthesize sugars and starch out of water and the carbon dioxide of the air.

chlorox. Trade name for sodium hypochlorite.

chrysoprase. A form of silicon dioxide colored with traces of other substances.

clay. Hydrated aluminum silicates, $H_2Al_2(SiO_4)_2 \cdot H_2O$.

coal gas. Gas made by heating bituminous coal.

cocaine. A substance occurring in coca leaves, which is a local anesthetic.

collodion. Pyroxylin dissolved in a mixture of ether and alcohol.

colloidal state. Condition in which suspended particles are between .0001 mm. and .000001 mm. in diameter.

combination. A reaction in which two substances unite to form a single substance.

combustion. A synonym of burning.

compound. A substance formed by the chemical union of two or more elements.

conductor. A substance which permits the flow of electricity through it.

conservation. In chemistry often suggests indestructibility.

contact process. The process of making sulfuric acid by the use of a catalyst to oxidize SO_2 to SO_3.

copal. A fossil gum resin from Zanzibar.

corrosive sublimate. Mercuric chloride.

cracking of oils. Changing hydrocarbons into hydrocarbons of different boiling points by intense heat.

cream of tartar. Potassium acid tartrate, $KHC_4H_4O_6$.

creosote. A mixture of cresols, or methyl hydroxy benzenes.

crude oil. An impure mixture of hydrocarbons occurring naturally.

cryolite. The mineral Na_3AlF_6, used as a solvent in the Hall process in the metallurgy of aluminum.

crystallization. A process of forming geometrically shaped crystals, usually from solutions or the melted substance.

crystals. Geometrically shaped solids characteristic of the substance.

cut glass. A brilliant, heavy glass containing lead.

cyclo propane. A general anesthetic with a formula C_3H_6.

dammar. A resin from Malaysia.

Daniell cell. A cell using copper-sulfate solution around the positive copper pole and zinc sulfate around the negative zinc pole, with the two solutions separated by a porous earthenware jar.

decomposition. The process of breaking a compound into simpler substances, often into its constituent elements.

dehydrating agent. A substance that removes water from another substance.

dextrin. A substance or substances formed by toasting starch.

dioxide, trioxide, tetroxide, pentoxide. Oxides containing 2, 3, 4, and 5 oxygen atoms respectively per molecule.

disaccharide. A sugar which can be hydrolyzed to two simpler sugar units.

distillation. The process of vaporizing a substance and condensing the vapor back to the liquid state.

double decomposition. A reaction in which two chemical substances exchange parts.

ductile. The property of a metal which enables it to be drawn out into a wire.

dyes. Complex colored organic compounds containing acid and basic groups.

dynamite. Sawdust or infusorial earth with glyceryl nitrate absorbed in it.

effloresce. The process of losing water of crystallization spontaneously.

electric current. Negative charges going one way, positive charges going the other way, or both kinds going opposite ways simultaneously.

electrolysis. The production of chemical reactions by means of an electric current.

electrolyte. A substance which makes water a conductor when dissolved in it.

electron. The fundamental negative charge.

electropotential series. An arrangement of the metals in a series in which each one will be the negative pole when used in a cell with any metal following it in the series.

element. One of the ninety-two simple substances that cannot be further simplified by ordinary chemical reactions.

emerald. A high quality beryl tinted green, a beryllium aluminum silicate.

emulsoid colloid. A substance whose particles in the colloidal state absorb more or less of the dispersing medium.

enamel. A varnish that has been baked on.

enamel ware. Iron vessels coated with glass.

endothermic reaction. A reaction which requires heat to bring it about.

energy. That which can do work.

enzyme. A complex organic catalyst occurring in plants and animals and their products.

equation. A chemical equation is a balanced symbolic representation of a chemical reaction involving molecules, and sometimes atoms and radicals.

equilibrium. A condition in which two reactions in opposite directions are taking place simultaneously without the relative proportions of the constituents changing.

esters. Compounds in which the hydrogen of the organic acids are replaced by groups, such as CH_3.

eutectic mixture. The lowest melting alloy of any two metals.

evaporation. The escape of a liquid from its surface to form vapor.

exothermic reaction. A reaction which gives out heat.

Fahrenheit. The thermometer scale used on the common household thermometer.

fats. The glycerin esters of the higher fatty acids.

feldspar. Potassium aluminum silicate, $KAlSi_3O_8$.

fixation of nitrogen. The process of causing the atmospheric nitrogen to unite with other elements to form compounds.

flint. A form of silicon dioxide colored with traces of other substances.

flotation. A process of concentrating metallic ores by means of an oily foam on water.

flowers of sulfur. Sulfur powder obtained by condensing sulfur vapor.

flux. Substances added to metallic ores to react with the gangue to form slag.

foamite. A mixture used in fighting oil fires.

formula. A chemical formula is a symbolic representation of the numbers and kinds of atoms in the molecule of a substance.

fractional distillation. The process of separating two liquids of different boiling points by distillation and catching fractions of short boiling ranges.

Fraunhofer lines. Dark lines across the sun's spectrum, caused by the absorption of light by the outer atmosphere of the sun.

fungicide. A chemical used to kill fungi.

gamma rays. Electromagnetic X rays given off by a radioactive substance.

gangue. Rock not part of metallic ore.

garnet. Calcium aluminum silicate, $Ca_3Al_2(SiO_4)_3$.

Gay-Lussac's law. Gaseous volumes react in ratios of small integers.

German silver. An alloy of copper, nickel, and zinc.

ginning. Combing the cotton loose from the seeds.

glass. A complex mixture of silicates of calcium, sodium, aluminum, and often boron and lead.

gluten. The protein in flour.

glycogen. A carbohydrate formed in the livers of animals.

grain alcohol. Ethyl alcohol, C_2H_5OH.

gram-molecular volume. 22.4 liters, which will hold under standard conditions as many grams as there are units in the molecular weight of the substance.

graphite. A form of carbon.

gypsum. Crystallized calcium sulfate.

Haber process. A process of nitrogen fixation in which the nitrogen combines with hydrogen to form ammonia.

halogen. Salt-former — chlorine, bromine, iodine, or fluorine.

hard water. Water containing dissolved calcium and magnesium compounds.

heat of formation. The number of calories liberated when one mol of the substance is formed from its elements.

heat of fusion. The number of calories necessary to melt one gram of the substance.

hematite. Iron oxide ore, Fe_2O_3.

hexavalent. An element having a valence of six.

hexylresorcinol. A synthetic antiseptic of high potency.

hormones. Substances secreted by the ductless glands in the animal body, which serve as regulators of the body processes.

humid. Rich in water vapor.

humidity. Moisture in the air.

hydraulic mining. Washing gold-bearing gravel with a swift stream of water.

hygroscopic moisture. Invisible moisture on the surface of apparently dry substances; the finer the particles, the greater the amount of hygroscopic moisture.

hygroscopic substance. One which absorbs moisture from the air, often enough to form a solution.

hypnotic. A sleep-producing drug.

hypo. A term in photography applied to the fixing agent, sodium thiosulfate.

hypothesis. Something thought to be true but which is not susceptible to direct proof.

Iceland spar. Calcite of exceptionally clear crystals.

impurities. Any substance mixed with the substance under consideration.

incandescence. The state of emitting light when heated to a high temperature.

insecticide. A chemical used to kill insects.

insoluble. When a solid will not distribute itself into a liquid or when one liquid will not mix with another, they are said to be insoluble.

insulator. A substance which will not permit electricity to flow through it.

insulin. The hormone, secreted by the pancreas, which prevents diabetes.

ionization. Process of an electrolyte separating into charged ions in solution.

ions. Atoms or radicals in solution carrying positive or negative charges.

isooctane. The hydrocarbon used as standard in antiknock rating.

isotopes. Atoms of an element which have slightly different atomic weights, but the same chemical properties.

jasper. A form of silicon dioxide colored with traces of other substances.

Javelle water. A solution of sodium hypochlorite.

kaolin. Hydrated aluminum silicate, $H_2Al_2(SiO_4)_2 \cdot H_2O$.

kindling temperature. The temperature at which a substance bursts into flame, and below which burning ceases.

lac. A resin secreted by a plant louse.

lachrymotor. Tear gas.

lactic acid. The acid in sour milk.

lactose. The sugar of milk.

lake. The precipitate of a dye by a mordant in the absence of the cloth, used as a pigment in painting.

latex. The milky juice of the rubber tree.

law. A single statement that summarizes a large number of similar scientific facts.

law of the minimum. A deficiency of any element required by a plant will limit the plant growth even if there is an excess of other plant foods.

leather. A substance made by the chemical reaction between the protein collagen and tannin.

lime. Calcium oxide.

linoleum. A substance made from cork dust held in an oxidized linseed oil binder.

litmus. An organic dye obtained from a lichen plant. It is pink in acid and blue in basic solutions.

lye. A mixture of sodium and potassium hydroxides.

malleable. The property of metals enabling them to be rolled into thin plates.

maltose. The sugar produced in malting grains.

marble. Calcium carbonate, which has been crystallized into small crystals.

mass. A measure of the quantity of a substance more universally usable then weight, which is the earth's pull on matter.

mass action. The promoting of a chemical reaction by adding an excess of one of the reactants.

matter. Any kind of substance.

mayonnaise. Colloidal olive oil in lemon juice or vinegar with white of an egg as protective colloid.

mercurochrome. An antiseptic dye containing mercury.

metal. Any element whose oxide combines with water to form a base. Metals are opaque and lustrous and are good conductors.

metallurgy. The process of getting a metal out of its ores.

methane. The hydrocarbon, CH_4.

mica. Hydrogen potassium aluminum silicate, $H_2KAl_3(SiO_4)_3$.

milk of lime. An emulsion of calcium hydroxide and water.

milk of magnesia. An emulsion of magnesium hydroxide and water.

miscible in all proportions. Is applied to two liquids which are soluble in each other in any proportions.

molal solution. One mole of a substance dissolved in 1000 grams of water.

mole. The molecular weight in grams of a substance.

molecular weight. The number of times a molecule is as heavy as the atomic unit, one-sixteenth the oxygen atom.

molecules. Very small particles of any compound; the smallest that can be said to represent the compound, and the ones which compose the gas when the substance is vaporized.

monel metal. A noncorrosive alloy of two parts nickel and one part copper.

monoclinic sulfur. Needlelike prisms of sulfur stable between $96°$ and $117°$ C.

monovalent. Having a valence of one.

monoxide. An oxide containing only one atom of oxygen per molecule.

mordant. A substance which reacts with the dye to form an insoluble salt in the cloth.

mortar. A mixture of calcium hydroxide, sand, and water used in laying brick.

mosaic gold. Stannic sulfide.

mustard gas. A war liquid, dichlorodiethyl sulfide, $(ClC_2H_4)_2S$.

naphthalene. The hydrocarbon used in moth balls.

nascent. The condition of an element which has just been released from chemical union.

natural gas. A mixture of the three lower hydrocarbons.

naval stores. Tar, pitch, rosin, and turpentine.

negative electricity. Produced on a rubber rod by rubbing it with a cat skin.

neutralization. The process of a base reacting with an acid to destroy the characteristic properties of both.

neutrino. An uncharged unit of atomic mass formed by the loss of a positron from a proton.

neutron. The fundamental unit of mass.

nitrifying bacteria. The bacteria which change ammonium compounds into nitrates in the soil.

nitroglycerin. Glyceryl trinitrate, $C_3H_5(NO_3)_3$.

nodules. Sores or enlargements on the roots of leguminous plants in which colonies of nitrogen-fixing bacteria live.

nonelectrolyte. A substance which will dissolve in water without making it a conductor of electricity.

oil cloth. Cloth covered by a waterproof layer of oxidized linseed oil.

onyx. A form of silicon dioxide colored with small traces of other substances.

opal. A hydrated form of silicon dioxide often colored by small traces of other substances.

open-hearth furnace. A furnace for the slow process of making high-grade steel.

orthorhombic sulfur. Diamond-shaped crystals of sulfur, the stable form under 96° C.

oxidation. The uniting with oxygen.

ozone. A very active form of oxygen with a molecule O_3.

paint. Pigment held suspended in drying oil, which oxidizes to a hard film in the air.

paint drier. Catalyst to speed up the oxidation of drying oils.

paper. A substance made of matted cellulose fibers.

paraform. A polymer of formaldehyde, $(CH_2O)_3$.

Paris green. Copper arsenite with copper acetate.

partially miscible. Is a term applied to liquids that partially dissolve each other.

passive state of a metal. The surface made resistant by chemical treatment.

pectin. A carbohydrate of fruits which causes jelly to set.

pellagra. A deficiency disease caused by an insufficiency of vitamin G in the diet.

pentavalent. An element having a valence of five.

periodic law. The properties of the elements are periodic functions of their atomic weights.

permalloy. Iron containing 30–80 per cent nickel used in the cores of transformers.

permanent hardness. The hardness in water caused by all soluble compounds of magnesium and calcium except the bicarbonates.

peroxide. Often applied to hydrogen peroxide, but in general a compound with two oxygen atoms per molecule in which one of them is loosely held and easily lost.

phenolphthalein (fenol taleen). An organic dye which is colorless in acid solution and pink in basic solution.

phlogiston. The name given to an imaginary substance formerly supposed to escape from substances as they burn.

phosgene. Carbonyl chloride, $COCl_2$.

photographic film. Cellulose nitrate or cellulose acetate.

physical change. Changes such as changes in temperature and shape, but in which no new chemical substance is formed.

picric acid. An explosive and yellow dye, trinitrophenol.

placer mining. Mining where the gold is washed out with a rocker or shoveled into a sluice.

plaster of Paris. A white powder formed by heating gypsum enough to drive out part of its water of crystallization and making it capable of setting into a solid when mixed with water.

Portland cement. A pulverized mixture of tricalcium silicate, dicalcium silicate, tricalcium aluminate, calcium ferro-aluminate, with a little calcium sulfate and magnesium oxide.

positive electricity. Produced on a glass rod by rubbing it with silk.

positron. The fundamental positive charge.

precipitate. An insoluble solid which separates from a solution as a result of cooling or of adding a reagent.

primary cell. A cell in which zinc is the negative pole; it cannot be recharged.

producer gas. Carbon monoxide and nitrogen obtained by partially burning coal. ,

protective colloid. A colloidal substance which makes another colloid more stable.

proteins. Complex nitrogenous substances in animals and plants; the cell-building materials in our foods.

proton. The fundamental unit of mass combined with the positron.

purification. The process of removing other substances from the substance being purified.

pyroxylin. The lower nitrates of cellulose.

quartz. Silicon dioxide, SiO_2.

radical. A group of elements which form either the negative or positive part of a compound and which change as a unit in double decompositions.

radioactivity. The property of certain heavy metals of emitting rays which penetrate matter, destroy cancer cells, make minerals fluoresce, and discharge an electroscope.

radon. A radioactive gas formed by the degradation of radium.

rayon. Cloth made by chemical transformations of cellulose.

reaction, chemical. A change in which new chemical substances appear as the original substances disappear.

reduction. The removal of oxygen from an element or elements to which it was chemically joined.

replacement. A reaction in which one element takes the place of another in a compound.

replacement series. An arrangement of the metals in the order in which they replace other metals from their compounds.

rickets. A deficiency disease of the bones due to an insufficiency of vitamin D in the diet.

roll sulfur. Rolls of sulfur $1\frac{1}{2}$ in. in diameter made by casting molten sulfur.

saccharin. A substance 550 times as sweet as sugar but of little heat value.

safety match. A match without a phosphorus compound, and which ignites by rubbing against phosphorus on the box.

sal ammoniac. The trade name for ammonium chloride.

salt. Sodium chloride, or table salt; more generally, the product of the neutralization of an acid with a base, or a substance which in solution forms ions, which include neither hydrogen nor hydroxyl.

saltpeter. Potassium nitrate.

sardonyx. A form of silicon dioxide, SiO_2, colored with traces of other substances.

saturated. When a liquid has absorbed all of another substance that it is capable of holding at a given temperature in the presence of some of the solid solute, it is said to be saturated with that substance.

saturated hydrocarbons. Those that take part in only a few chemical reactions and then with difficulty.

scurvy. A deficiency disease caused by an insufficiency of vitamin C in the diet.

sedative. A substance which lessens the sensitiveness of the nerves to pain.

serpentine. Hydrated magnesium silicate, $Mg_3Si_2O_7 \cdot 2 H_2O$.

shatterproof glass. Two pieces of plate glass cemented together by a transparent synthetic substance.

shellac. Lac dried in thin sheets.

slag. Light rock scum floating on the metallic iron in a blast furnace.

slaked lime. Calcium hydroxide.

smokeless powder. Cellulose nitrate.

soaps. Sodium and potassium salts of the high-carbon organic acids, called the fatty acids.

solder. A low-melting alloy of lead and tin used to join pieces of copper or tin.

solute. The solid where a solid is dissolved in a liquid, and the liquid in smaller amount when two liquids dissolve each other.

solution. A mixture in which one substance is dissolved in another, usually a liquid.

solvent. The liquid if a solid is dissolved in a liquid, and the liquid in larger amount if two liquids dissolve each other.

specific heat. The number of calories necessary to raise the temperature of one gram of the substance one degree centigrade, usually less than one.

spectra. The colors which make up any light.

spectroscope. An instrument which separates light into its colors.

spontaneous combustion. Combustion that has not been started by applying a flame or burning materials.

stalactite. An iciclelike structure of calcium carbonate in limestone caves.

stalagmite. A slender cone-shaped structure of calcium carbonate, sticking up from the floor of a limestone cave.

standard conditions. Conditions under which gaseous volumes are considered, usually 0° C. and 760 mm. mercury pressure.

starch. A carbohydrate in most plants, which hydrolyzes into glucose.

steel. Iron to which has been added small amounts of carbon and often other metals.

strong acid or base. An acid or base which largely separates into ions in dilute solution.

sublimation. Condition of a solid changing directly to a vapor without going through the liquid state.

sublime. To vaporize directly from the solid state.

substantive dye. A direct dye for silk and wool.

sucrose. Cane sugar.

sugars. Aldehyde or ketone alcohols, many of which have a sweet taste.

superphosphate. Calcium acid phosphate.

supersaturated solution. An unstable situation where by careful cooling a liquid is induced to hold more of a solute than it normally requires for the prevailing temperature.

suspensoid colloid. A substance in the colloidal state whose charged particles do not absorb the dispersing medium.

symbiotic nitrogen-fixing bacteria. Bacteria which live in the roots of the leguminous plants and which fix the nitrogen of the air into nitrates.

synthesis. The process of uniting elements chemically to form compounds.

talc. A hydrated magnesium silicate, $H_2Mg_3(SiO_3)_4$.

tempering. Fixing the hardness of steel by heating to a definite temperature and chilling.

temporary hardness. Hardness in water caused by dissolved calcium and magnesium bicarbonates.

tetravalent. An element having a valence of four.

thyroxine. The hormone, secreted by the thyroid gland, which regulates the combustion processes in the body.

tincture of iodine. Iodine dissolved in alcohol.

T.N.T. The explosive, trinitrotoluene.

toning. Changing the tone of a photographic print.

topaz. Aluminum fluorosilicate, $Al_2F_2SiO_4$.

tourmaline. Hydrogen borosilicate, $H_2B_2Si_4O_{21}$.

trivalent. An element having a valence of three.

type metal. An alloy of lead, antimony, and tin.

ultramicroscope. A microscope for looking through beams of light inside a colloidal suspension to view the small particles as splashes of light.

unsaturated hydrocarbons. Those that react readily by addition.

valence. The holding power of an atom for other atoms in terms of hydrogen as 1.

vitamin. Complex organic substances which occur in food in small quantities but which are absolutely necessary for good health.

volatile. Easily vaporized.

washing soda. Sodium carbonate, Na_2CO_3.

water gas. A mixture of hydrogen and carbon monoxide formed by spraying steam on very hot coke.

water glass. Soluble sodium silicate, Na_2SiO_3.

water of crystallization. Water driven out of crystals by heat accompanied by the crystals crumbling up into powder.

water of hydration. Water that is chemically united to substances, including water of crystallization but usually applied to non-crystalline substances, such as clay.

weak acids or bases. Those that form only a few ions in dilute solution.

weight. The force with which a body is pulled toward the earth.

white arsenic. Arsenious oxide.

white lead. The white pigment, basic lead carbonate.

wrought iron. The purest form of iron produced in a furnace.

xylyl bromide. A tear gas $C_6H_3(CH_3)_2CH_2Br$.

zinc white. The white pigment, zinc oxide.

zwieback. Heavily toasted bread.

INDEX

747

Date Due

THE ELEMENTS

[The names of the more important elements are printed in bold-face type.]

Name	Symbol	Atomic Number	Approx. Atom. Wt.	Name	Symbol	Atomic Number	Approx Atom. Wt.
Actinium	Ac	89	227?	**Mercury**	Hg	80	200.6
Alabamine	Ab	85	221?	Molybdenum	Mo	42	96.0
Aluminum	Al	13	27.0	Neodymium	Nd	60	144.3
Antimony	Sb	51	121.8	Neon	Ne	10	20.2
Argon	A	18	39.9	**Nickel**	Ni	28	58.7
Arsenic	As	33	74.9	**Nitrogen**	N	7	14.0
Barium	Ba	56	137.4	Osmium	Os	76	191.5
Beryllium	Be	4	9.0	**Oxygen**	O	8	16.0
Bismuth	Bi	83	209.0	Palladium	Pd	46	106.7
Boron	B	5	10.8	**Phosphorus**	P	15	31.0
Bromine	Br	35	79.9	**Platinum**	Pt	78	195.2
Cadmium	Cd	48	112.4	Polonium	Po	84	210?
Calcium	Ca	20	40.1	**Potassium**	K	19	39.1
Carbon	C	6	12.0	Praseodymium	Pr	59	140.9
Cerium	Ce	58	140.1	Protoactinium	Pa	91	231.0
Cesium	Cs	55	132.9	**Radium**	Ra	88	226.0
Chlorine	Cl	17	35.5	Radon	Rn	86	222.0
Chromium	Cr	24	52.0	Rhenium	Re	75	186.3
Cobalt	Co	27	58.9	Rhodium	Rh	45	102.9
Columbium	Cb	41	92.9	Rubidium	Rb	37	85.5
Copper	Cu	29	63.6	Ruthenium	Ru	44	101.7
Dysprosium	Dy	66	162.5	Samarium	Sm, Sa	62	150.4
Erbium	Er	68	167.6	Scandium	Sc	21	45.1
Europium	Eu	63	152.0	Selenium	Se	34	79.0
Fluorine	F	9	19.0	**Silicon**	Si	14	28.1
Gadolinium	Gd	64	156.9	**Silver**	Ag	47	107.9
Gallium	Ga	31	69.7	**Sodium**	Na	11	23.0
Germanium	Ge	32	72.6	**Strontium**	Sr	38	87.6
Gold	Au	79	197.2	**Sulfur**	S	16	32.1
Hafnium	Hf	72	178.6	Tantalum	Ta	73	180.9
Helium	He	2	4.0	Tellurium	Te	52	127.6
Holmium	Ho	67	163.5	Terbium	Tb	65	159.2
Hydrogen	H	1	1.0	Thallium	Tl	81	204.4
Illinium	Il	61	146?	Thorium	Th	90	232.1
Indium	In	49	114.8	Thulium	Tm	69	169.4
Iodine	I	53	126.9	**Tin**	Sn	50	118.7
Iridium	Ir	77	193.1	Titanium	Ti	22	47.9
Iron	Fe	26	55.8	**Tungsten**	W	74	184.0
Krypton	Kr	36	83.7	Uranium	U	92	238.1
Lanthanum	La	57	138.9	Vanadium	V	23	50.9
Lead	Pb	82	207.2	Virginium	Vi	87	224?
Lithium	Li	3	6.9	Xenon	Xe	54	131.3
Lutecium	Lu	71	175.0	Ytterbium	Yb	70	173.0
Magnesium	Mg	12	24.3	Yttrium	Y	39	88.9
Manganese	Mn	25	54.9	Zinc	Zn	30	65.4
Masurium	Ma	43	———	Zirconium	Zr	40	91.2